Sulfur Dioxide and Vegetation

Sulfur Dioxide and Vegetation

Physiology, Ecology, and Policy Issues

Edited by
**William E. Winner, Harold A. Mooney, and
Robert A. Goldstein**

Stanford University Press, Stanford, California
1985

Stanford University Press, Stanford, California

© 1985 by the Board of Trustees of the Leland Stanford Junior University. Copyright in this work does not extend to those parts of it written as works of the United States Government, namely papers 1, 2, 11, 13, 21, 25, 28, and 29.

Printed in the United States of America

CIP data appear at the end of the book

To the memory of Philip C. Miller,
a pioneer in the integrative and quantitative approach
to the study of plants and plant communities

Preface

This volume was developed from a symposium held at Asilomar, California, from November 28 to December 2, 1982. The purpose of this meeting was (1) to summarize our current knowledge of the effects of SO_2 on the productivity of plants, (2) to bring about an integration of the various research approaches—from biochemical to ecological—used to determine the impact of SO_2 on plant productivity, (3) to assess the accuracy of the technology currently used to study SO_2 effects on plant productivity, and (4) to consider present pollution management issues and to establish research priorities for the future.

This volume represents a summary of the presentations and discussions that took place at the meeting. Issues related to managing air quality are addressed first, and set the stage for discussion of the effects of SO_2 on plant productivity. The analysis of plant responses to SO_2 begins with biochemical considerations and ends with an analysis of the long-term responses of plant communities. How the effects of SO_2 on the biochemistry and physiology of plants translate into changes in plant growth is discussed next, and how these changes in turn affect the structure, function, and productivity of plant communities is then explored. Finally, a summary provides reviews of methods and techniques used in air-pollution studies, highlights important concepts from the foregoing chapters, and identifies conceptual cross-links between topics that are related but pertain to different levels of biological organization.

The editors at Stanford University Press made a substantial contribution to the success of this project. We also thank the Electric Power Research Institute, the Air Quality Division of the National Park Service, and the California Energy Commission for providing the support for this project.

W. E.W.

H.A.M.

R.A.G.

Contents

PART 4. SO₂ EFFECTS ON PLANT COMMUNITIES

PART 5. SUMMARY

Contributors

Richard M. Adams is Associate Professor, Department of Agricultural and Resource Economics, Oregon State University. He received his Ph.D in Agricultural Economics from the University of California, Davis, in 1975. His research interests focus on the economics of resource and environmental use. He is currently working with plant scientists, meteorologists, and other economists in assessing the economic consequences of air pollution to U.S. agriculture for the U.S. Environmental Protection Agency.

F. A. Bazzaz received his Ph.D. in Plant Biology at the University of Illinois in 1963 for his work on plant succession. His research interests include the physiological and life history characters of plants in disturbed habitats and the response of plants to environmental contaminants such as Pb, Cd, SO_2, and to increased levels of global CO_2. He is currently professor of Biology at Harvard University.

J. N. B. Bell received his Ph.D. in Botany in 1969 from the University of Manchester. He has worked for the last 14 years on the impact of air pollution on plants, and is currently a Senior Lecturer in Plant Ecology in the Department of Pure and Applied Biology, and Assistant Director of the Centre for Environmental Technology, Imperial College, London.

James P. Bennett received his Ph.D. in plant ecology and air pollution from the University of British Columbia in 1975. He continued researching the effects of O_3, SO_2, and H_2S on vegetable crops while teaching at the University of California, Davis. Currently he is a research ecologist for the Air and Water Quality Division of the National Park Service, Denver, where he is responsible for the Service's Biological Effects Research Program.

Valerie J. Black received her Ph.D. from the University of Aberdeen in 1974 for a study of the sensitivity of Bryophytes to pollutants. While at the University of Nottingham, she investigated the effects of gaseous air pollutants on physiological processes in crop plants. She is currently a lecturer in ecology at the University of Technology, Loughborough, Leicestershire, where her research interests include the effects of acidic precipitation on ecosystems.

Roger W. Carlson received his Ph.D. in Forestry from Yale University in 1976. His research interests include the physiological response of plants to environmental stress, including moisture, light, toxic metals, and air pollution. He is currently Assistant Professor of Plant Ecology at the Institute for Environmental Studies, University of Illinois, Urbana.

Carl W. Chen is with Systech Engineering, Inc., Lafayette, California. He received his Ph.D. in Environmental Engineering from the University of California, Berkeley, in 1968. He has developed a kinetic model of fish toxicity threshold and numerous aquatic ecosystems models. From 1978 to 1982 he led the development of the acid rain model for the Integrated Lake-Watershed Acidification Study (ILWAS), a model that followed atmospheric depositions from tree canopy, through soils, to the lake.

D. V. Crawford is Senior Lecturer in Soil Science at the University of Nottingham School of Agriculture. He graduated in chemistry from the University of Oxford and specializes in soil chemistry research, including herbicide retention, heavy metal availability to plants, and the cycle of sulfur in agricultural systems.

Thomas D. Crocker is Professor of Economics at the University of Wyoming. He received his Ph.D. in agricultural economics from the University of Missouri. He has served on the U.S. Environmental Protection Agency's Science Advisory Board and several National Academy of Sciences committees and panels. His major research interests are the development of methods to value non-marketed, particularly environmental, goods, and the description of the properties of alternative allocation systems.

David Fowler received his Ph.D. in Environmental Physics at the University of Nottingham in 1976 for studies of SO_2 deposition on field crops. As a senior scientific officer in the air pollution research group of the Institute of Terrestrial Ecology since 1976, he has studied the effects of gaseous pollutants on cereal crop growth, dry deposition of SO_2 onto forests and other natural surfaces, and rainfall chemistry and acidic deposition.

S. G. Garsed received his Ph.D. in Botany from Sheffield University for work on the uptake and metabolism of SO_2 by plants. Since 1975 he has been Research Assistant in the Department of Pure and Applied Biology at Imperial College, London, working on various aspects of plant response to environmental stresses.

Steven Gherini is Chief Engineer and Director at Tetra Tech, Inc., Lafayette, California. After an undergraduate training in civil engineering at Stanford University, he received his graduate education in aquatic chemistry at Harvard University. His interests include mathematical modeling of the transport and reaction of pollutants in surface and ground waters, acid-base processes, experimental design, and the analysis of field and laboratory data.

Robert A. Goldstein is a Senior Project Manager in the Environmental Assessment Department, Electric Power Research Institute, Palo Alto, California. He received his doctorate in 1969 from Columbia University. From 1969 to 1975 he was a research staff member of the Ecological Sciences Division at Oak Ridge National Laboratory. Since 1977 he has directed the Integrated Lake-Watershed Acidification Study (ILWAS). His major professional interests are ecosystem analysis and mathematical ecological modeling.

S. K. Gregson gained her doctorate at London University for research on the mobility of lead in polluted soils. In 1980 she joined the University of Nottingham School of Agriculture as a research assistant to develop techniques for studying the availability of atmospherically derived soil sulfur to agricultural crops. In particular, she has used soil lysimeters and radioactively labeled solutions to quantify the sulfur cycle in the outdoor environment.

Allen S. Heagle received his Ph.D. in Plant Pathology in 1967 from the University of Minnesota. He started his research on air-pollutant effects on crops in 1968 with the U.S. Department of Agriculture. He is currently a research plant pathologist in the Air Quality Research Program with the U.S.D.A. Agricultural Research Service, and Professor of Plant Pathology and Crop Science at North Carolina State University at Raleigh.

Walter W. Heck received his Ph.D. in Plant Physiology in 1954 from the University of Illinois. He started research on the effects of various air pollutants on plants in 1959 at Texas A&M University. He is currently a research plant physiologist and research leader of the Cooperative Air Quality Research Program with the U.S.D.A. Agricultural Research Service, and Professor of Botany at North Carolina State University at Raleigh.

Herbert C. Jones III is Supervisor, Research Section, Air Quality Branch, Office of Natural Resources, Tennessee Valley Authority. He received his Ph.D. in Botany, specializing in plant physiology, from the University of Florida. He is responsible for the management and direction of research on the dispersion, long-range transport, and chemistry of pollutants emitted from TVA's coal-fired power plants and the effects of power plant pollutants on vegetation.

Teresa Jones (formerly Davies) graduated in 1977 from the University of Wales, Swansea, and received her Ph.D. from the University of Lancaster in 1980. She now works in the Plant Growth Regulation Section of ICI Plant Protection Division, Jealott's Hill, United Kingdom.

David F. Karnosky is Professor of Forestry at Michigan Technological University, and Director of the Center for Intensive Forestry in Northern Regions. Trained in forestry and forest genetics at the University of Wisconsin, where he received his Ph.D. in 1975, he has studied extensively the nature of variation in air pollution responses in trees. He is co-chairman of the International Union of Forestry Research Organizations Working Group on "Genetic Aspects of Air Pollution."

Richard W. Katz received his Ph.D. in Statistics from Pennsylvania State University in 1974. While at Oregon State University, he served as a visiting statistician for the Corvallis Environmental Research Laboratory of the U.S. Environmental Protection Agency. He is currently a scientist with the Environmental and Societal Impacts Group of the National Center for Atmospheric Research in Boulder, Colorado.

Theodor Keller studied forestry at the Swiss Federal Institute of Technology, Zurich, and soils at the University of Wisconsin, where he received his Ph.D. in 1957. He joined the Swiss Federal Institute of Forestry Research in 1959, building up and leading a research division on air pollution. He currently heads a research group on bioindications.

J. R. Kercher is an Ecological Systems Analyst in the Environmental Sciences Division, Lawrence Livermore National Laboratory, Livermore, California. He received his Ph.D. in 1973 from Cornell University. Since then he has worked on the simulation and mathematical analysis of a wide range of ecological problems, including the effects of air pollutants on vegetation.

D. A. King received his Ph.D. in botany from the University of Wisconsin at Madison in 1979 and was employed as a postdoctoral research assistant at Stanford University and at the University of California, Davis. His research interests include modeling the factors influencing plant performance. He is currently Associate Research Professor at the University of Portland and works at the U.S. Environmental Protection Agency's Corvallis Environmental Research Laboratory.

Robert Kohut is a Plant Pathologist with the Boyce Thompson Institute for Plant Research at Cornell University. He received his Ph.D. at the Pennsylvania State University in 1975 and subsequently studied as a postdoctoral research fellow at the University of Minnesota. His current research includes evaluating the effects of air pollutants on plant growth and productivity and on plant community dynamics.

T. T. Kozlowski is Director of the Biotron, a controlled-environment research laboratory, and also is the Wisconsin Alumni Foundation Senior Distinguished Research Professor of Forestry at the University of Wisconsin in Madison. He received his Ph.D. degree from Duke University in 1947, and an honorary D.Sc. degree from Université Catholique de Louvain in Belgium in 1978. His research emphasizes effects of environmental stresses on the physiological ecology of woody plants.

W. K. Lauenroth is an Associate Professor of Range Science and Senior Researcher in the Natural Resource Ecology Laboratory at Colorado State University. He received his Ph.D. in Systems Ecology from Colorado State University in 1973. His research interests in ecological systems include responses to air pollution exposure.

T. A. Mansfield received his Ph.D. in 1961 from the University of Reading, Berkshire, and is now Professor of Plant Physiology at Lancaster University. He has studied the physiological responses of plants to air pollutants since 1968, concentrating on SO_2 and NO_x, whose occurrence in the United Kingdom is widespread.

S. B. McLaughlin, Jr., is a senior research staff member in the Environmental Sciences Division at Oak Ridge National Laboratory, Oak Ridge, Tennessee. He received his Ph.D. in Forest Tree Physiology in 1970 from Duke University, where he worked on the effects of fluoride on the physiology of six plant species of the Southeastern United States. His principal research has since focused on the physiological responses of crops and trees to natural and anthropogenic stress.

D. G. Milchunas is Research Associate, Natural Resource Ecology Laboratory and Range Science Department, Colorado State University. He is a systems ecologist working in areas of disturbance-stability theory and rangeland succession, resource partitioning in plants, nutrient cycling, plant-animal interactions, and ruminant nutrition. His research in air pollution includes organism, population, and system-level responses.

Harold A. Mooney is the Paul Achilles Professor of Environmental Biology at Stanford University. His interests are in the environmental physiology of plants, particularly photosynthetic responses. He has conducted studies on plants from a wide variety of ecosystem types, including tundra, deserts, and the tropics.

Richard L. Olson, Jr., received his Ph.D. in Bioengineering from Texas A&M University in 1984. His research interests include the modeling of the ecology and physiology of plants. He is currently a Postdoctoral Research Associate at the Laboratory for Air Pollution Impact to Agriculture and Forestry at Virginia Polytechnic Institute and State University.

David M. Olszyk received his Ph.D. in horticulture and botany from the University of Wisconsin in 1979 and studied as a National Research Council Postdoctoral Research Associate at the U.S. Environmental Protection Agency's Corvallis Environmental Research Laboratory in Oregon. He is currently an assistant research plant physiologist with the University of California, Riverside, Statewide Air Pollution Center.

Ronald J. Oshima is the Program Manager of the Environmental Hazards Assessment Program in the California Department of Food and Agriculture. He is responsible for characterizing the movement and fate of pesticides in California to more effectively regulate use and to assess the impact of air pollution on California agriculture. His air pollution research has focused on quantification of plant responses from air pollutants with emphasis on photosynthate partitioning and yield.

Galen Peiser received his Ph.D. in Plant Physiology from the University of California, Davis, in 1977. As a postdoctoral research associate he studied the biochemical effects of SO_2 on plants and the biosynthetic pathway of ethylene. In 1984 he joined Native Plants, Inc., in Salt Lake City, Utah, as a senior research scientist studying the physiology and biochemistry of fruit ripening.

K. P. Preston is Assistant Professor in the Department of Geography at the University of New Orleans. His doctoral research in the Department of Geography at the University of California, Los Angeles, concerned the effects of SO_2 and O_3 on aboveground plant growth in a range of species of coastal sage scrub.

S. M. Rowlatt completed his doctorate at Liverpool University on the geochemistry of sulfur in lake and stream sediments. Until 1981 he was research assistant at the University of Nottingham School of Agriculture, where he studied the pathways of atmospheric sulfur to agricultural crops. His work included the analysis and interpretation of throughfall chemistry. Since 1980 he has been a research scientist at the Government Fisheries Laboratory, Burham-on-Crouch, Essex.

P. J. W. Saunders is a Senior Principal Scientist in the Natural Environment Research Council, Swindon, Wiltshire. After training at Exeter University and Cambridge, he worked in the Pollution Research Unit of the University of Manchester. At the N.E.R.C., he has headed the Terrestrial and Freshwater Life Sciences Group and is currently Head of the Policy and Planning Group. He is also Chairman of the C.E.C. Coordinating Group for air pollution research.

Peter J. H. Sharpe received his Ph.D. in Biological Sciences in 1970 from the University of New South Wales. He is presently a Professor of Bioengineering, Department of Industrial Engineering, at Texas A&M University, and is consultant to several research-oriented companies. His research interests include ecosystems analysis and modeling of population dynamics, physiological ecology, pest management and environmental impact, and space industrialization.

H. H. Shugart was Senior Scientist, Environmental Sciences Division, Oak Ridge National Laboratory at the time he wrote his contribution to this volume. He is now Professor, Department of Environmental Sciences, University of Virginia. His research interests have included several areas of plant and animal ecology but have usually involved dynamic models. A recent area of particular interest has been computer simulation of forest succession.

G. E. Taylor, Jr., received his Ph.D. in Biology from Emory University in 1976 and was a National Academy of Sciences–National Research Council Postdoctoral Associate in plant physiology from 1977 to 1979. His research interest is plant stress physiology and ecology. He is currently on staff in the Environmental Sciences Division of Oak Ridge National Laboratory.

David T. Tingey received his Ph.D. in plant physiology and cell biology from North Carolina State University in 1972. He has conducted physiological and greenhouse studies on the effects of O_3, SO_2, and NO_2 singly and in combination on physiological processes and plant growth. Currently he is a plant physiologist with the U.S. Environmental Protection Agency, Corvallis Environmental Research Laboratory, Oregon.

M. H. Unsworth graduated in physics from the University of Edinburgh. Until 1982 he was in the Environmental Physics group at the University of Nottingham School of Agriculture, where his research included radiation climatology, atmosphere-surface exchange of pollutant gases and droplets, and physical analysis of physiological responses of plants. He now heads the Institute of Terrestrial Ecology's Edinburgh Research Station, which specializes in air quality research, tree biology, and freshwater ecology.

Susanne von Caemmerer received her Ph.D. from the Australian National University. She has worked as a post-doctoral fellow at the Carnegie Institution. At present she is with the Research School of Biological Sciences of the Australian National University.

L. B. Weeks is a Master's candidate in the Department of Geography at the University of California, Los Angeles. His thesis research concerns the effects of SO_2 and O_3 on root growth in a range of species of coastal sage scrub.

Alan R. Wellburn received his Ph.D. in biochemistry from Liverpool University in 1966 and is currently Senior Lecturer in plant biochemistry at Lancaster University. In addition to studying the biochemical effects of atmospheric pollution on vegetation, he also conducts research on chloroplast development and metabolic changes associated with water stress.

W. E. Westman is a Senior Research Scientist at the NASA Ames Research Center in Moffett Field, California. He received his Ph.D. in Ecology and Evolutionary Biology from Cornell University in 1971. His current research involves application of remote sensing to ecosystem studies. He is the author of a recent text on ecology, impact assessment, and environmental planning.

Mary E. Whitmore is a Research Fellow in the Biology Department at the University of Lancaster, studying the physiological effects of SO_2 in water-stressed plants. She received her Ph.D. from the University of Lancaster in 1982 for research into the effects of SO_2 and NO_2 on the growth of grasses.

Kimberlyn Williams is a Ph.D. candidate in the Department of Biological Sciences at Stanford University. Her research interests lie in the area of physiological plant ecology. Past research has included studies of plant carbon balance and SO_2 exchange of leaves.

William E. Winner received his Ph.D. in Biology from the University of Calgary in 1978 and studied as a postdoctoral scholar at Stanford University. His research interests include the ecological and physiological responses of plants to air pollutants. He is currently Assistant Professor of Plant Pathology and Director of the Laboratory for Air Pollution Impact to Agriculture and Forestry, at Virginia Polytechnic Institute and State University.

Shang Fa Yang is Professor and Biochemist in the Department of Vegetable Crops at the University of California, Davis. His research interests include ethylene physiology and biochemistry, post-harvest physiology, biochemistry of sulfite oxidation, and the effects of SO_2 on plant metabolism.

A Note on Units

In scientific usage, SO_2 concentrations are most commonly expressed in μg m^{-3} or as a volume-to-volume ratio such as parts per million (ppm), parts per hundred million (pphm), or parts per billion (ppb). The use of these units dates back to the early 1900's and is recognized by most scientists who study air pollutants. Units of μg m^{-3} or ppm are usually inscribed on the meters of SO_2 analyzers and are used to describe SO_2 concentrations by the U.S. Environmental Protection Agency and other regulatory agencies.

Despite the widespread use of μg m^{-3} and ppm, the unit mol m^{-3} is the proper one for expressing gas concentrations within the International System of Units (Système International d'Unités, or SI). The International System has been adopted by most industrialized nations in a series of agreements about the use of weights and measures. SI units have been used by the U.S. National Bureau of Standards since 1964 and are the unit system prescribed by almost all scientific journals. The contributors to this volume have been encouraged to use SI units to define gas concentrations. When units such as μg m^{-3} or ppm are used, we have parenthetically inserted the SI equivalent. We have done this to provide a standardized unit for air-pollution concentrations throughout the text, and we offer the reader a tabulation of equivalents for conveniently converting units from one form to another. The following conversions assume a temperature of 20°C and a pressure of 760 mm Hg.

Gas	Volume/volume, ppm	Density (SI), μg m^{-3}	Concentration (SI), μmol m^{-3}
SO_2	1	2,660	41.6
O_3	1	2,000	41.6
NO	1	1,250	41.6
NO_2	1	1,910	41.6
H_2O	1	749	41.6

Sulfur Dioxide and Vegetation

Introduction

William E. Winner, Harold A. Mooney, and Robert A. Goldstein

SO_2 is an industrial air pollutant often associated with coal combustion, fossil fuel refining, and ore smelting, and consequently is an effluent of concern in most industrialized nations. The effects of SO_2 on terrestrial vegetation range from the spectacular, such as those found for areas that are denuded of vegetation because of emissions from a nearby SO_2 source, to the subtle and poorly understood, such as those associated with estimating the consequences of low SO_2 doses on plants. Although regulations throughout the United States and other developed countries prevent severe environmental degradation in the vicinity of SO_2 point sources, those interested in environmental affairs are still concerned about the complex ways in which SO_2 may affect plants.

SO_2 is only one of several pollutants released from industrial smokestacks, chimneys, and automobile exhaust pipes. Others include nitrogen oxides, hydrocarbon compounds, carbon monoxide, and aerosol particles. In addition, these pollutants and their derivatives can react to form a class of pollutants known as photochemical oxidants. The photochemical oxidants most threatening to plants are ozone and PAN (peroxyacetyl nitrate). Thus SO_2 is only one of several air pollutants that can affect plants.

SO_2 is distinguished from other air pollutants in several ways. The acute effects of SO_2 emissions on plants were recognized long before the effects of other pollutants, and as a result SO_2 impacts have received more attention than the impacts of other pollutants. In addition, SO_2 is among the pollutants produced in the greatest amounts. About 30 million tons of SO_2 were emitted in the United States during 1977, whereas emissions of hydrocarbon compounds and nitrogen oxides were estimated to be about 25 million tons each (U.S. Environmental Protection Agency, 1980b). Annual production of photochemical oxidants depends upon the abundance of chemical precursors and a number of meteorological factors, and is difficult to calculate.

Economic estimates of air-pollution damage offer an additional perspective for viewing the environmental effects of air pollutants. Annual air-pollution damage to plants was estimated at $86 million in the late 1960's (Benedict, Miller, and Olson, 1971) and at $1.8 billion in more recent years (Jacobson, 1982). This increase in economic loss does not reflect an increase in air-pollution emissions; it reflects, rather, a greater understanding of the environmental effects of air pollutants and improvements in the economic tools necessary for making such estimates. Although SO_2 and SO_2 plus O_3 account for much of these losses, it is difficult to estimate the costs of all impacts from only SO_2 and its derived compounds. The difficulty in calculating SO_2 damage to vegetation is that subtle and indirect yield reductions are hard to detect, and ecological effects are hard to evaluate in economic terms.

This volume is intended as a reference for those interested in SO_2-caused changes in plant physiology and ecology. The chapters that follow provide descriptions of our current understanding of how vegetation responds to SO_2 and projections about research areas and technical advancements that represent future challenges. In the remainder of the Introduction we (1) briefly describe SO_2 chemistry and emissions trends, (2) give, from the American point of view, a short history of research on SO_2 effects on vegetation to provide a background for the chapters that follow, and (3) outline some of the current research challenges.

SO_2 Chemistry and Emissions

One of the reasons that SO_2 effects are difficult to determine is that this pollutant molecule may undergo a number of chemical reactions before contacting vegetation. Since SO_2 and its derivatives differ in phytotoxicity, it is important to characterize the environmental chemistry and emissions trends of SO_2.

Chemical properties. SO_2 is a colorless, nonflammable gas that, at normal barometric pressures, condenses to a liquid at $-10°C$ and solidifies at $-72°C$. The solubility of SO_2 in water is 17.7% (w/w) at $0°C$ and 8.5% (w/w) at $25°C$. When SO_2 dissolves in water it forms $SO_2 \cdot H_2O$, which exists as

$$SO_2 \cdot H_2O \rightleftharpoons H^+ + HSO_3^- \rightleftharpoons 2H^+ + SO_3^=$$

The state of SO_2 in this series of reactions is driven to the left as solution pH decreases (Lyons and Nickless, 1968). HSO_3^- and $SO_3^=$ are favored at pH values ranging from 5 to 9, increasing acidity favoring HSO_3^-, which is thought to be the more phytotoxic chemical species (Puckett *et al.*, 1973).

SO_2 released into the air is normally oxidized by a complex series of chemical reactions; these reactions have recently been reviewed by Calvert *et al.* (1978) and Moller (1980). If SO_2 is absorbed into leaves, the oxidation reactions take place in mesophyll tissues and may be detrimental to cells. If SO_2 is adsorbed to leaf surfaces or other objects, the oxidation of SO_2 may be influenced by the presence of other chemicals and by environmental conditions such as temperature. These factors may also influence the rate of SO_2 oxidation in the atmosphere. These oxidation reactions may proceed with SO_2 in the gas phase. In this case SO_2 becomes SO_3 gas, which is short-lived and subsequently goes into solution with atmospheric water to form sulfuric acid. The conversion of SO_2 to SO_3

may take from 8 to 48 hours and is influenced by a number of chemical and environmental factors. SO_3 is hydrophylic and exists for only a fraction of a second under normal environmental conditions. SO_2 in air may also go directly into solution with liquid-phase water and form sulfite ions, which are then oxidized to sulfate.

SO_2 and its diverse group of reaction products can be deposited in a number of physical forms. Aside from gaseous SO_2 deposition, sulfite and sulfate ions can be deposited with particulate aerosols and various forms of precipitation. Of these forms of deposition, SO_2 gas is the most toxic to plants.

SO_2 emissions. Emissions of SO_2 and sulfate particulates play a major role in the global sulfur cycle. Current estimates suggest that the annual flux of sulfur between the earth and the air is between 144 (Granat, Rodhe, and Hallberg, 1976) and 217 Tg per year (1 Tg $= 10^6$ tonnes) (Friend, 1973). Natural sources of atmospheric sulfur include organic decomposition in terrestrial and oceanic ecosystems that produce H_2S, volcanic activity, and sea spray. On a global basis, about 65 Tg per year of sulfur are annually emitted as SO_2, which accounts for between 11% (Friend, 1973) and 45% (Granat, Rodhe, and Hallberg, 1976) of the total burden of atmospheric sulfur.

SO_2 emissions totaled about 31 million tons (2.8×10^7 tonnes) in the United States during 1978 (U.S. Environmental Protection Agency, 1980a). About 60% of these emissions came from coal-burning electric power plants, about 20% came from smelting and industrial combustion of coal and oil, and the remaining emissions came from a number of industrial sources, including fossil-fuel refining and commercial sulfuric acid production. Ninety-five percent of the emissions were from stationary sources located in the eastern half of the United States. More than 40% of SO_2 emissions were from states in the upper Midwest, including Minnesota, Ohio, Illinois, Indiana, Michigan, Pennsylvania, and West Virginia (U.S. Environmental Protection Agency, 1980b).

The level of SO_2 production in the United States during 1978 was similar to production levels estimated during the 1970's (U.S. Environmental Protection Agency, 1978b). However, though production held constant at about 22 million tons (2.0×10^7 tonnes) per year between 1940 and 1960, it increased by about 30% during the 1960's. Projections for SO_2 emissions through the year 2000 vary with a number of scenarios for energy use and emissions control (U.S. Environmental Protection Agency, 1978a). However, estimates that assume a "business as usual" approach to development of new energy sources and pollution control suggest that annual emissions in 2000 will range from 35 to 40 million tons (3.2×10^7 to 3.6×10^7 tonnes). Thus, SO_2 production trends show historical increases, and additional increases are forecast. It seems that SO_2 will continue to be a pollutant of concern to those interested in our environment.

Historical Aspects of Air-Pollution Research

An emerging global issue in the early 1900's. The first large-scale attempts to manage air quality date back to at least 1306, when Edward I of England restricted the use of low-grade coal (peat) as a fuel source in order to reduce smoke production. Even today, the combustive oxidation of sulfides in coal throughout Europe and North America results in SO_2 production.

Prior to 1910, ore smelters were the largest sources of SO_2 in Germany and the United States. Sulfides in ores were oxidized during smelting and resulted in the release of SO_2. In Germany it was recognized early that smoke contained both solid particulates and gases, and that gases were more damaging to plants (Wislicenus, 1914). Stoklasa (1923) suggested that SO_2 might reduce plant growth without causing visible foliar injury.

The ore smelters in the United States and Canada were much larger than those in Europe. The consequences of the emission of SO_2 and particulates (including metals) from smelters at Copper Hill, Tennessee, Anaconda, Montana, Salt Lake City, Utah, Trail, British Columbia, and Sudbury, Ontario, were severe and are reviewed (Swain, 1949; Thomas, 1951). Large areas around these smelters are still denuded of vegetation today, long after SO_2 emissions have ceased or been restricted by stringent emissions controls. Although SO_2 is only one of several components of smelter emissions that can lead to soils sterilization, these sites are important reminders of the impact air pollutants can have on landscapes.

Implications of early research. The devastation of landscapes around ore smelters throughout Europe and North America during the late 1880's and early 1900's created an interest in controlling the ecological and agricultural impacts of air-pollution emissions. This interest resulted in the development of emissions controls (Swain, 1949), which included the construction of lead chambers for the large-scale conversion of SO_2 gas to sulfuric acid, the design of huge electrostatic precipitators to capture small air-pollution particles, and the installation of huge stacks to disperse pollutants in low concentrations over large areas. In addition, strategies for managing air-pollution emissions were instituted. In some cases air pollutants were not emitted when meteorological factors could lead to high pollution concentrations at ground level.

Concern about the impacts of air pollution also spawned a great deal of scientific research on SO_2 impacts on vegetation. Sophisticated experimental apparatus had been developed prior to 1950 and included continuously recording analyzers for SO_2 (Thomas and Cross, 1928; Thomas and Abersold, 1929; Thomas, 1932) and CO_2 (Thomas, 1933) concentrations. These instruments led to field studies in which photosynthesis, respiration, and transpiration were monitored for alfalfa and wheat (Thomas and Hill, 1937a,b). Growth cabinets were built for laboratory SO_2 fumigation studies in which plants were raised in controlled environments (Swain and Johnson, 1936). In addition, flowthrough greenhouses were developed for field fumigation studies. These greenhouses were bottom-less, could be positioned over field-grown crops, and were designed to straddle prepared beds of artificial planting media in which irrigation and nutrient con-centrations were controlled (Thomas et al., 1943b). Finally, fumigation experi-ments were done with radiolabeled SO_2 (Thomas et al., 1944; Harrison, Thomas, and Hill, 1944). Up-to-date versions of this type of research equipment are cur-rently in use in modern air-pollution research laboratories.

The substantial literature published prior to 1950 on the effects of SO_2 on crops supported some general conclusions. The idea that SO_2-caused crop-yield reductions were directly related to the extent of visible leaf injury was widely held (Holmes, Franklin, and Gould, 1915; Hill and Thomas, 1933; Thomas, 1951, 1956). In addition, scientists concluded that SO_2 absorption into leaves was regulated by stomata (Wells, 1917), that absorbed SO_2 was oxidized to non-toxic sulfate, which could then be reduced via assimilatory sulfate-reduction pathways (Thomas, Hendricks, and Hill, 1914), that the overall nutritional status of plants determined their potential to convert SO_2 to less toxic chemical forms (Thomas et al., 1943a), and that when SO_2 absorption exceeded detoxification, SO_2 absorption was related to photosynthetic changes, changes in transpiration (Thomas and Hill, 1935), and visible foliar injury (Thomas and Hill, 1937b).

The concept that foliar injury and yield losses from SO_2 were related was widely accepted during the 1950's. Research and technological effort was di-rected toward (1) establishing the link between visible foliar injury and crop yield losses and (2) minimizing the geographic extent of acute forms of injury to vegetation. Once the consensus was reached that SO_2-caused visible foliar injury could be equated to crop yield losses, scientific interest in SO_2 fumigation stud-ies declined. This consensus was reached in the early 1950's and led to the prac-tice of industrial representatives' appraising SO_2-caused visible foliar injury to crops and paying farmers for projected yield losses.

Rekindled interest. Interest in air-pollution research, at low levels during the 1950's, gained momentum in the 1960's. The research efforts initiated at this time have continuously expanded and led to our current understanding of plant re-sponses to gaseous pollutants. The new flurry of research focused not on SO_2 but rather on the impacts of ozone and other photochemical oxidants. Of the various oxidant pollutants, O_3 was found to be the most ubiquitous, covering large regions throughout eastern North America and California. Early studies during this period were aimed at characterizing the different patterns of visible foliar injury caused by O_3 on many crops, including grapes (Richards, Middleton, and Hewitt, 1958), tobacco (Heggestad and Middleton, 1959), and beans (Taylor et al., 1960). Al-though emphasis was placed on crops, white pine (Berry and Ripperton, 1963) and ponderosa pine (Miller et al., 1963) were identified as native trees for which ambient air pollution caused visible foliar injury and growth reductions. Foliar injury caused by O_3 was found to differ from injury caused by SO_2; O_3 resulted in stipple or bronzing on upper leaf surfaces, whereas SO_2 resulted in interveinal necrosis (Thomas, 1961).

Studies of air-pollution-caused foliar injury led to research advances in the late 1960's. Air-pollution exposure doses were found not to be linearly related to foliar injury (Heck, Dunning, and Hindawi, 1966), environmental conditions and leaf age were found to affect the development of foliar injury following fumigations (Heck, 1968), and some approaches were developed for protecting leaves from air-pollution injury (Ordin *et al.*, 1962; Taylor and Rich, 1962). Also during this period, the question of whether plants can sustain yield losses in the absence of visible injury was reopened. A few researchers, including Bleasdale (1952a), Koritz and Went (1953), and Todd (1958), had kept the idea alive even though the consensus was that air-pollution-caused leaf loss was proportional to yield loss. The development of open-top chambers (Heagle, Body, and Heck, 1973) eventually led to field experiments that conclusively showed that visible foliar injury may or may not reflect changes in vegetative growth or fruit production (Heagle, Body, and Neely, 1974).

Research during the 1970's led to studies on the environmental impacts of many pollutants, including SO_2, photochemical oxidants, hydrocarbons, nitrogen oxides, carbon monoxide, and particulate matter. All of these pollutants are known to cause visible leaf injury following severe fumigations and to reduce plant growth rates when applied in low doses. By the late 1970's researchers recognized that many different pollutants existed, that species and cultivars varied widely in their sensitivity to air pollution, and that plant responses to an air pollutant were related to the presence of other air pollutants, environmental conditions, and plant age. Many scientists, including contributors to this volume, are currently working to clarify the relationships among air pollution, environment, and plant ecology, physiology, and productivity.

The Challenges Today

The impressive early research efforts in assessing the impacts of SO_2 on plant productivity ended abruptly in 1950. Interest in this topic was renewed in the 1960's, and has been sustained to the present time. The contributions to this volume emphasize research progress made during the past 10 years. Although it is apparent that substantial progress has been made, research is continuing at a fast pace in all areas. This continued effort reflects the fact that scientists still cannot satisfactorily answer important questions associated with the impact of SO_2 on vegetation. Complicating the problem is the fact that plants stressed by SO_2 are generally also stressed by other air pollutants (such as ozone) and environmental factors such as light, temperature, and the availability of water and nutrients. Some of the chapters in this volume address research on interactions between SO_2 and other stress agents, and should stimulate further work in this important area.

Part 1
Issues of Pollution Management

1

Assessing SO$_2$ Effects on Vegetation: Viewpoints from an Industrial Scientist

Herbert C. Jones III

The major demonstrable effect that emissions of sulfur dioxide (SO$_2$) by utilities have had on vegetation has been visible foliar injury. Such injury, which may cause reductions in growth, yield, or quality, is rapidly being eliminated as most power plants achieve compliance with emission limits defined by their local State Implementation Plan (SIP). For existing plants with allowable emission rates of 1.8 to 2.3 kg of SO$_2$ per 10^6 BTU (4 to 5 lbs of SO$_2$ per 10^6 BTU) or less, and reasonably tall stacks (good engineering practice, GEP), or new plants meeting the New Source Performance Standard (NSPS), the probability of causing visible effects is nil (Jones *et al.*, 1979). The remaining questions about the direct effects of SO$_2$ on vegetation are: Does intermittent exposure to SO$_2$ alone, or to mixtures of SO$_2$ and other pollutants, at *ambient* concentrations less than those that cause visible injury, cause permanent economic losses? If so, how much more emission control is needed to mitigate these effects? Billions of dollars in costs for additional controls, legislation, standards development, enforcement, and waste management rest on the resolution of this issue. Controversy over this issue is already causing costly delays in the construction of at least one major coal-fired power plant in the west, even though it has been designed to limit emissions to the very low rate of 0.09 kg of SO$_2$ per 10^6 BTU (0.2 lb of SO$_2$ per 10^6 BTU) or about 7,257 kg (7.3 metric tons) per day.

It is important that this issue be resolved in a way that instills confidence in the farmer or forester that his crops are being adequately protected, and in the private investor or rate payer that any additional costs for controls are justified; the broad issue is no longer whether emission controls are needed, but how much control is needed. Resolution of the issue will require well-designed experiments to document the occurrences of effects under ambient conditions, or under conditions that adequately simulate ambient conditions. However, the diversity of opinion among researchers about what constitutes well-designed experiments, ambient conditions, or simulated ambient conditions seriously complicates resolution of the effects issue. If we are ever to reach technical agreement on the primary issue, we must reach agreement on the following secondary issues.

1. What constitutes, in terms of SO_2, O_3, and NO_2, exposure regimes representative of ambient conditions? Is it meaningful to measure responses based on continuous (or almost continuous) exposure to SO_2 concentrations in excess of the background of 0.02 ppm (0.8 μmol m^{-3}), where the background is defined as the average hourly SO_2 concentration in areas far from large sources of SO_2; or on simultaneous exposure to mixtures of SO_2 and O_3, or to mixtures that contain more than 0.08 to 0.10 ppm (3.2 to 4.1 μmol m^{-3}) of NO_2?

2. What is the minimum acceptable experimental design, particularly in terms of replication, number of years of data, and meaningful parameters of growth or yield for field experiments? For greenhouse or laboratory experiments? Are unreplicated experiments, or 1 or 2 years of data, acceptable for field experiments? Is a significant difference in weight per seed of practical value when differences in number of seed and total yield are not significant?

3. Are changes in physiology and biochemistry ultimately manifested in permanent losses in growth, yield, or quality? Under what conditions relative to the rural ambient environment?

4. How can dosage for intermittent exposures during a growing season or decades be described in terms having practical application? Is dose cumulative? For all exposures or for exposures in excess of some threshold level?

These issues are not new. They have been around for more than 40 years, and still remain largely unresolved. Visible effects are being eliminated. If the more subtle effects of air pollution, if they exist, are to be eliminated, we researchers must provide data that can be used by policymakers, and we can do so only by making a conscientious effort to answer these questions. I would ask you to consider these issues in your deliberations at this conference, and to ask how your research might be designed to answer such questions. I will offer some suggestions and experimental approaches that may be useful. If we could reach consensus on some of the subissues, agreement on the primary issue would be hastened.

Experimental Design, Exposure, and Measurements

Regardless of whether you view your research as basic or as applied science, we all have one thing in common: most, if not all, of our work is being funded by organizations or agencies who want to obtain information for making decisions about the management of our air resource. They want information that is as accurate and applicable to real-world conditions as possible. For our research this means that:

1. The pollutant exposure regimes employed in our experiments should be representative of the range of exposure conditions experienced in the ambient environment.

2. The design of experiments, and the collection, analyses, and interpretation

TABLE 1.1
Five-Year Average Growing Season Cumulative Frequency Distribution for Hourly Average
SO$_2$ Concentration, Cumberland Steam Plant, All Monitors, 1977–81

SO$_2$ concentra- tion class,[a] ppm (μmol m^{-3})	Hours SO$_2$ concentration equaled or exceeded class value					
	Morning, 0700–1300		Afternoon, 1400–1700		Daylight	
	No.	Cumulative	No.	Cumulative	No.	Cumulative
0.1 (4.1)	10	13	14	19	24	32
0.2 (8.2)	2	3	3	5	5	8
0.3 (12.3)	0	1	1	2	1	3
0.4 (16.4)	1	1	0.5[b]	1	1.5	2
0.5 (20.4)	0	0	0.5[b]	0.5	0.5	0.5

[a]0.1 ppm = 0.06 to 0.15 ppm (2.4 to 6.1 μmol m^{-3}); 0.2 ppm = 0.16 to 0.25 ppm (6.5 to 10.2 μmol m^{-3}); etc.
[b]The same average frequency for each class.

of data, should follow procedures that maximize the integrity of results and minimize the probability of arriving at erroneous conclusions.

3. Dosages that elicit responses should be quantifiable in terms that will be
useful in developing control strategies.

4. Responses should be measured in ways that can be quantified in economic
terms.

We would like to suggest some minimum acceptable criteria for SO$_2$ research
that we think would greatly increase both the usefulness and the acceptability of
results.

Representative ambient exposure regimes. The combination of tall stacks and
reduced emission rates from most power plants has reduced the frequency, duration, and concentrations of ground-level SO$_2$ exposures. As more of the base-
generating load is taken over by modern plants meeting NSPS, ground-level exposures will be reduced even further. If effects occur in the future, they will
therefore result from intermittent exposure to concentrations of SO$_2$ that are
much less than those that cause visible injury.

The data in Table 1.1 illustrate a five-year growing season, daylight hours, exposure regime for a 2,600-MW power plant before it came into compliance with
an allowable emission rate of 2.3 kg of SO$_2$ per 10^6 BTU (5 lbs of SO$_2$ per 10^6
BTU) and average daily emissions of about 1,179 (metric) tons. The exposure
regime for this power plant, which has not violated ambient air-quality standards, can be considered "worst case" compared to what might be expected for
power plants sited in nonmountainous terrain and meeting more restrictive emission limits. Extrapolating from the data, we find that the upper limits for exposure regimes that are typical for ambient conditions in the vicinity of a rural-
sited power plant would consist of about 21 intermittent exposures, averaging
about 1.5 hours apiece, spread over a 6-month growing season. The maximum 1-
hour and 3-hour average SO$_2$ concentration would not exceed 0.5 ppm (20.4

μmol m^{-3}). The total hours of exposure to concentrations equal to or greater than 0.1 ppm (4.1 μmol m^{-3}) would not exceed 32. In areas with more unstable or more windy conditions during the growing season, exposure would be less.

Vegetation growing within the area exposed to greater SO$_2$ concentrations near point sources is also exposed to other pollutants, such as ozone and NO$_2$; hence there is considerable concern about the interactive and synergistic effects of exposure to mixtures of SO$_2$ and other pollutants. It is important to the researcher to understand the conditions under which exposure to mixtures of pollutants can occur in the rural ambient environment, as well as the conditions that limit the dosage. In addition to SO$_2$ and other pollutants, one constituent of a power-plant plume is nitrous oxide (NO). The proportion of NO to SO$_2$ varies (depending on the boiler type) between about 1:4 and 1:8; as SO$_2$ emissions are reduced, this ratio will increase. The NO in the plume reacts very rapidly with background ozone (O$_3$) to produce NO$_2$ + O$_2$, depleting the O$_3$ (Meagher *et al.*, 1981). The upper limit for production of NO$_2$ is determined by the background O$_3$ concentration and dispersion. For point sources sited in rural areas, this reaction has two practical consequences.

1. SO$_2$ cannot coexist with O$_3$ until the reaction NO + O$_3 \rightarrow$ NO$_2$ + O$_2$ is complete. By the time this occurs, SO$_2$ concentrations have been reduced, by dispersion, deposition, and transformation, to not much more than background values.

2. NO$_2$ levels cannot exceed background O$_3$ levels; so NO$_2$ will rarely exceed 0.06 to 0.08 ppm (2.4 to 3.2 μmol m^{-3}).

Design of experiments. The experimental approach consists of four basic steps.

1. Finding out what information is already available on dose/response for the particular crop about which you wish information.

2. Identification of the additional information you wish to obtain, and development of hypotheses.

3. Design of the field and laboratory experiments to obtain the data you need to test these hypotheses, including identification of the statistical models you will use to test the hypotheses.

4. Collection, reduction, analysis, and interpretation of data and reporting results.

I am sure all of you are familiar with these steps, but I wish to draw your attention to them, because they often seem to be ignored today. They are important if research is to be conducted efficiently and if the results are to be of high quality.

In my opinion, all research on pollution effects, whether done in the field or done in the greenhouse, funded by either the taxpayer or the electric-rate payer, should meet certain minimum criteria, as follows.

1. The species selected for study should be of economic or ecological importance, except where special research, e.g., on genetics, is warranted.

2. All research should include measurements of productivity or quality that are related to the intended use of the crop, e.g., total weight of seed for seed crops, total fresh and dry weight for root or leaf crops, cubic volume for timber.

3. For field experiments, because of the large differences in edaphic, climatic, and biotic factors within a site and between years, treatments should be randomized, replicated at least three times, and repeated for at least three years, before firm conclusions are drawn.

4. Care and thought should be exercised in the selection of controls for comparing the productivity of crops exposed to pollutants with that of crops not exposed to the pollutants. In most field experiments conducted today, either some kind of pollutant-exclusion (e.g., open-top chambers) or pollution-gradient (e.g., ZAPS) methodology is used to make these comparisons. With the pollution-exclusion method, either all pollutants or individual pollutants may be excluded. Where all ambient pollutants are excluded, the individual pollutants may be added back alone, or in mixtures, to compare the effects on productivity of the individual pollutants (or their interactions) with the effects of other pollutants. With pollution-gradient methods, the individual pollutants (or mixtures of them) are released into the ambient air and transported by wind over the crop. The crop is exposed to any ambient concentrations of pollutants that are present plus the added pollutant(s). In measuring the effects that emissions from power plants, or from any other point source sited in rural areas, have on crop productivity, what must actually be measured is how much productivity has been reduced by the power plant's emissions, beyond any reduction caused by exposure to ambient levels of non-power-plant pollutants. The correct control is, therefore, productivity of plants exposed to ambient levels of non-power-plant pollutants, not productivity of plants grown in air filtered to remove all pollutants. The dosages of pollutants employed in mixtures must be in the same relative proportions that would occur in ambient air when the power-plant plume is on the ground; i.e., ozone should be depleted to less than 0.025 ppm (less than 1.0 μmol m^{-3}) for SO$_2$ dosages of more than 0.05 ppm (2.1 μmol m^{-3}), NO dosages of from 0.25 to 1.0 times the SO$_2$ concentration, and NO$_2$ dosages of less than or equal to ambient ozone concentrations. When one is comparing the productivity of crops grown under conditions in which all pollutants are excluded with productivity of crops grown in ambient air near a point source, a similar comparison must be made in a remote area where pollutants of the type emitted from the point source do not exceed background concentrations. The data in Table 1.2 illustrate the erroneous conclusions that might be drawn from such comparisons. From 1971 through 1975, soybeans were grown in filtered and unfiltered air in small, portable, compartmentalized greenhouses near the Widows Creek Steam Plant. Yields in filtered air averaged 21% higher than those in unfiltered air during the period of study. One might conclude that the power plant was causing the reduced yields. The greenhouses were moved in 1976 to a remote area where SO$_2$ concentration did not exceed 0.01 ppm (0.4 μmol m^{-3}). During the period 1976

TABLE 1.2

Yields of Soybeans Grown in Compartmentalized Filtered Air / Ambient Air Greenhouses near a 1,900-M Coal-Fired Power Plant and at a Site Remote from Stationary Sources of Pollution

(grams dry weight)

Year	Cultivar	Filtered air		Ambient air		Percent reduction
		Per plant	Total	Per plant	Total	
POWER PLANT SITE						
1971	Bragg	12.2	—	8.2	—	32%
1972	Hood	18.4	—	15.9	—	16[b]
1973	Hood	18.1	—	13.8	—	24[b]
1974	Hood	14.7	1,159	12.1	932	17[b]
1975	Essex	12.4	987	9.8	733	21[b]
Average 1971–75		15.2		12.0		21
REMOTE SITE[c]						
1977	Essex	—	1,080	—	856	21%
1978	Essex	—	842	—	607	28[b]
1979	Essex	—	864	—	617	29[b]
Average 1977–79		928		698		25

[a]Yield reduction significant at $p = 0.10$.
[b]Yield reduction significant at $p = 0.05$.
[c]SO_2 hourly average concentration less than 0.02 ppm (0.8 μmol m^{-3}).

to 1979, yield in filtered air averaged 25% more than that in unfiltered air, which suggests that the power plant was not responsible for the yield reduction.

5. Instrumentation should be calibrated frequently (at least biweekly) over the range of potential exposures for ambient studies or the actual range of exposures employed in controlled studies.

6. Pollutant sampling lines should be heated, from the sampling point to the instrument intake, to prevent condensation of moisture in the lines and resulting erroneous measurements.

7. The efficiency of pollutant-exclusion systems for removing or lowering the concentration of pollutants in study plots should be documented for meteorological conditions that will cause maximum exposure.

8. For field experiments, cultural practices and cultivars recommended for the particular area should be followed.

9. Use of subsamples or aliquots of a treatment as replicates should not be considered legitimate.

Even though other criteria could be recommended, if these were applied in both field and greenhouse experiments, the effects data generated would be significantly more useful and more acceptable to all interested parties.

Expression of dose. Classically, dose has been expressed as concentration of a pollutant times duration of exposure in hours. For acute exposures, dose is normally expressed as a time-average concentration, e.g., 0.5 ppm (20.4 μmol m^{-3}) of SO$_2$ for 3 hours. This expression of dose is a useful predictor for occurrences of visible injury caused by exposure to short-term, high-concentration episodes. Because ground-level exposure near a point source results from many interacting factors—meteorology, stack height, terrain, various plant-operating variables—long-term growing season or annual averages in themselves usually do not adequately reflect shorter-term acute or chronic exposures that have been demonstrated to cause permanent injury. For example, annual average SO$_2$ concentration might not exceed 0.02 ppm (0.8 μmol m^{-3}) in the vicinities of two power plants; but, because of the factors listed above, the emissions from one plant might cause significant effects and the emissions from the other none.

For multiple or chronic exposures, some investigators (McLaughlin, 1981) have suggested using cumulative dose, which is calculated by multiplying the pollutant concentration by the duration of the exposure in hours and summing the individual dosages. Others (Oshima, 1975; Lefohn and Benedict, 1982) have suggested, for ozone, accumulating dosages that exceed a threshold level. Cumulative dose appears to have merit for sites that are often exposed to concentrations of ozone high enough to cause visible injury (Heck *et al.*, 1982a); in these cases the vegetation was exposed nearly every day to O$_3$ for five or more hours. Cumulative dose has not been shown to have merit for intermittent ambient exposures to SO$_2$ or for exposures that simulate the ambient environment.

Response measurements. So far, the only measurements of plant response that are useful for decision makers in managing air quality are those that have economic meaning for the crop in question; these include measurements of growth, yield, and quality. The secondary standards, which have been promulgated to protect human welfare, have been based largely on concentrations of a pollutant that can be demonstrated to cause permanent losses in productivity or quality. Many investigators have demonstrated physiological and biochemical responses that appear to be transient. However, the cumulative effects of these transient responses on growth, yield, or quality are not known.

Future Research Directions

Does intermittent exposure to ambient concentrations of SO$_2$ and other air pollutants cause reductions in yield? If we are to answer this question, by providing data that will be of use in managing air quality, our experimental approach to the problem must be modified. What we need to know is: how many incremental exposures to given pollutant dosages can a crop tolerate before its productivity is reduced? Research designed to answer this question should provide data that could be used to develop a statistical-type secondary standard for the number of given short-term dosages that cumulatively would reduce the productivity of crops and forests. Furthermore, the results should be useful in evaluating alternative control strategies for existing plants. Since a reduction in emissions should proportionally reduce the frequency of ground-level exposures of given concentrations, one could use frequency distributions based on historical data to calculate what reduction in emissions would be needed to avoid exceeding various dosages of a pollutant. Similarly for new sources, models could be used to generate frequency distributions to assess the probability that damage may occur. Adequate information is available for designing a set of experiments that would answer the question posed here and resolve the critical issue set forth in the introduction. The objective of these experiments would be to find out what cumulative dose would have to be exceeded for a reduction in yield to occur. We would suggest the following experimental approach.

Ambient monitoring data from a worst-case power plant (defined as a plant that meets all NAAQS, does not cause visible effects, and has an emission rate representative of most SIP's) of 1.8 to 2.3 kg of SO_2 per 10^6 BTU would be used to design a first set of dose/response experiments. The average duration of SO_2 exposures of 0.1 ppm (4.1 μmol^{-3}) or more and the average frequency distribution of 1-hour average SO_2 concentrations in increments of 0.10 ppm (4.1 $\mu mol\ m^{-3}$) during a five-year period for daylight hours during the growing season would be measured. Data for the 1,400-MW Colbert Power Plant, which has an emission rate of 1.8 kg of SO_2 per 10^6 BTU (4.0 lbs of SO_2 per 10^6 BTU), is summarized in Table 1.3. The five-year frequency distribution for Colbert for hourly average SO_2 concentrations that equaled or exceeded 4.1, 8.2, 12.3, 16.4, and 20.4 $\mu mol\ m^{-3}$ was 52, 19, 11, 7, and 4 hours respectively. On the average, there were about 35 exposures to 4.1 $\mu mol\ m^{-3}$ of SO_2 or more during the five growing seasons. The average duration of ground-level exposure for Colbert was about 1.5 hours, which agrees with that for other power plants in the TVA system. However, since exposures can occur back-to-back, and because we wish the first set of experiments to be worst case, we would select a duration of three hours for each incremental weekly exposure. The experimental approach is illustrated in Fig. 1.1. The total area within the figure represents the total cumulative exposure to SO_2 concentrations of 4.1 $\mu mol\ m^{-3}$ or more for the six-month growing season. The total hours of exposure (52 hours) to 4.1 $\mu mol\ m^{-3}$ or more

TABLE 1.3
Five-Year Average Cumulative Frequency Distribution for Ground-Level SO_2 Exposure, Colbert Steam Plant, All Monitors, May Through October, 1977–81

SO_2 concentration class, ppm ($\mu mol\ m^{-3}$)	Total hours SO_2 concentration equaled or exceeded class value				Daylight (cumulative)
	Morning, 0700–1300		Afternoon, 1400–1900		
	No.	Cumulative	No.	Cumulative	
0.10 (4.1)	17	28	16	24	52
0.20 (8.2)	5	11	3	8	19
0.30 (12.3)	2	6	2	5	11
0.40 (16.4)	2	4	1	3	7
0.50 (20.4)	2	2	2	2	4

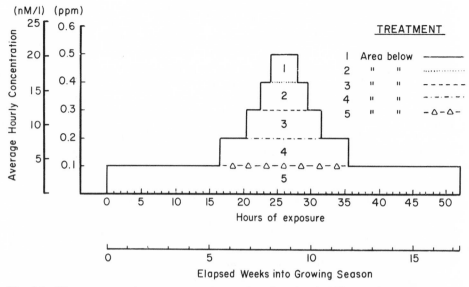

Fig. 1.1. Worst-case growing-season exposure regime, based on total hours of exposure to SO₂ concentrations equal to or exceeding a value for daylight hours during a 6-month growing season. Each treatment consists of the area below the designated line, applied in sequential weekly 3-h exposure for the total number of hours within each hourly concentration increment.

defines the number of weekly 3-hour exposures to be applied during that growing season. The experimental approach is to successively eliminate the higher dosages until a dosage is reached that does not result in a reduction in yield, assuming that one of the treatments will cause a reduction in yield. The experiments would therefore consist of five treatments. Treatment 1 would be the total dosage for the season. Treatment 2 would eliminate exposures that averaged 6.4 μmol m⁻³ h⁻¹ or more; treatment 3, those that averaged 12.3 μmol m⁻³ h⁻¹ or more; treatment 4, those that averaged 8.2 μmol m⁻³ h⁻¹ or more; and finally, treatment 5, those that averaged 4.1 μmol m⁻³ h⁻¹ or more. The total cumulative dosages for treatments 1 through 5 would be 701, 619, 504, 369, and 213 μmol m⁻³ h, respectively.

The individual exposures would be given weekly in three triangular-shaped 1-hour exposures with a peak-to-mean ratio of 2, as illustrated in Fig. 1.2. Concentrations of hourly exposures within some 3-hour increments may differ, because the total hours within a concentration category may not be divisible by 3 or because of the way we have chosen to group the data. The triangular hourly exposure configuration is a reasonable approximation of ambient exposure regimes, which commonly consist of fluctuating peaks of varying durations. The sequence of the individual exposures presented here is conservative or worst case, because the successively higher dosages are grouped toward the center of

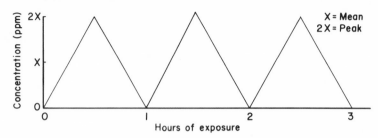

Fig. 1.2. Example of weekly 3-h exposure regime given in 1-h increments with a peak (2x) to mean (x) ratio of 2. The value of x may vary during hourly increments within some 3-h exposures.

the growing season; under ambient conditions the dosages would be more or less randomly dispersed over the growing season. Treatments in excess of 4.1 μmol m^{-3} could be shifted backward or forward in the growing season to correspond to maximum sensitivity of a species, if desired.

The plants would also be exposed concurrently to ozone, and simultaneously to NO and SO_2. The ozone exposure would be representative of the growing-season regime based on five years of monitoring data, and would be applied daily. NO would be introduced with the SO_2 at a rate of 1 part NO to 3 parts SO_2. The NO_2 concentration would be that resulting from the depletion of ozone by the reaction $O_3 + NO \rightarrow O_2 + NO_2$. Thus, on days when the plants are exposed to SO_2, ozone concentrations would be depleted, by the preceding reaction, to zero.

The results of these experiments would answer the following questions.

1. Does intermittent exposure to SO_2, in mixtures with other major pollutants that exist in the ambient environment or in power-plant plumes, under conditions representative of the ambient environment, cause reductions in the productivity of major crops?

2. Does intermittent exposure to SO_2 cause reductions in productivity even where there is no visible foliar injury?

3. Is cumulative dose a useful predictor of SO_2 effects on productivity?

When the minimum-dosage treatment that can cause a reduction in yield has been established, the next step would be to find the minimum number of incremental exposures within the next higher concentration category that would be required to cause injury. For example, suppose treatment 2 (Fig. 1.1) caused no reduction in yield, but treatment 1 did. The next set of experiments would then consist of three treatments: treatment 2, plus a 1-hour exposure to 0.50 ppm (20.4 μmol m^{-3}), 1-hour average; treatment 2, plus 2 hours of exposure to the same concentration; treatment 2, plus 3 hours of exposure to the same concentration. This procedure would show how many hours the concentration would have to exceed 0.4 ppm (16.4 μmol m^{-3}) to reduce productivity.

Assessment of Technique

In my opinion, if we are to answer the questions about the permanent effects of intermittent exposure to low concentrations of SO$_2$, we are going to have to change our experimental approach to the problem. Our results may be accurate for the experimental conditions under which they are obtained, but for the most part they are not applicable to the rural environment where crops and forests are grown. Consider the following dialogue, which, although hypothetical, is based on fact. Place yourself in the position of a person who is responsible for issuing permits to construct and operate a large, coal-fired power plant. He is concerned about whether the emission controls designed to limit ground-level maximum hourly average concentrations to 0.1 ppm (4.1 μmol m^{-3}) or less of SO$_2$ are adequate to protect the major crop of the area, okra. He has therefore convened a hearing at which several effects experts are to testify on dosages affecting the yield of okra. The testimony most probably would go as follows.

Expert No. 1 testifies, "I found that exposure of okra to 0.15 ppm (6.1 μmol m^{-3}) for six hours per day for 30 days caused a significant, 40% reduction in yield."

Expert No. 2 says, "I exposed okra to 1.5 ppm (61.4 μmol m^{-3}) and 2.0 ppm (81.8 μmol m^{-3}) for three hours, and obtained reductions in yields of 6 and 18%, respectively."

A third expert testifies, "I have assessed the impact of the power plant on okra yields. There is consensus among scientists that dose is cumulative. Experts 1 and 2 have found that cumulative doses of 27, 6.0, and 4.5 ppm h (1,098, 245, and 184 μmol m^{-3} h) caused reductions in yields of 40, 18, and 6%, respectively. I plotted these data and fitted a curve to them. The curve shows that we would expect reduction in yields for cumulative exposures between 2 and 3 ppm h (82 and 123 μmol m^{-3} h). Exposure to concentrations averaging only 0.03 ppm (1.2 μmol m^{-3}) for 100 days will result in a total dose of 3 ppm h (123 μmol m^{-3} h); there is a high probability that reductions in yields will occur."

A fourth expert says, "You can't use the data that way. It's like saying that drinking a fifth of whiskey in an hour will have the same effect as drinking it at a rate of an ounce every two weeks."

Another expert testifies that, "Neither exposure to 0.15 ppm (6.2 μmol m^{-3}) for 6 hours a day for 30 days nor exposure to 1.5 or 2.0 ppm (61.4 or 81.8 μmol m^{-3}) for 3 hours is representative of exposures in the rural ambient environment."

Another expert adds, "The data are for field experiments that were conducted only one year," and someone else adds, "The treatments were not randomized and replicated."

Such results must be very confusing to the people who are trying to apply them in setting standards, evaluating alternative emission-control strategies, or estimating the effects of a new power plant on agriculture and forests. Their usefulness could be improved greatly if we did a better job of (1) identifying what the problems are in the ambient environment, (2) designing the experiments to provide the solution to the problem, and (3) returning to the practice of "good science," i.e., randomization, replication, and repetition. There are ample amounts of monitoring data for rural ambient levels of all the pollutants for most regions of the country, and we know enough about the emission rates, dispersion, and chemistry of the pollutants to design experiments that closely simulate ambient exposure conditions, and that would provide dose/response data applicable to the real world and meaningful in economic terms.

2

Regulatory Uses
of SO$_2$ Effects Data

James P. Bennett

Studies of the effects of SO$_2$ on vegetation generate many types of data, ranging from responses at the biochemical level all the way up to the ecological level
of biological organization. Data on effects can take many forms, including
changes in rates, resources, and appearances of plants. Many of these types of
data are detailed elsewhere in this volume, and do not need extensive explanation
in this chapter. An often unacknowledged consequence of generating data from
air pollution experiments is that these data may be misused by industrial groups,
environmentalists, and regulatory agencies, who may be selective in their use of
the literature. Such use of the data is often deceptive, and interferes with the
process of discovering the real nature of pollution effects.

In our society today, industrialists and environmentalists are typically viewed
as being on opposing sides in discussions of SO$_2$ effects, whereas the regulatory
agencies fall in the middle by attempting to weigh and balance the multitude of
data on effects. Regulatory agencies, therefore, are in the position of trying to
discover the real nature of pollution effects on plants, in an attempt to get at the
truth of the problem and resolve the differences between the opposing sides. By
definition, regulatory agencies do not take adversarial positions on the issues,
and endeavor to examine all the facts with careful scrutiny and scientific methods.

This chapter will present a short overview of the uses of data on biological
effects by regulatory agencies, the techniques used to analyze the data, and future research directions that interest regulatory agencies. Its purpose is to inform
basic researchers about what happens to their data after it has been published.
The opinions and conclusions expressed here are those of the author and do not
necessarily represent official policy.

Regulatory agencies involved in air pollution issues include municipal, county,
state, and federal governments. Air pollution control agencies, energy and utilities commissions, natural resource divisions, and conservation agencies are the
typical agencies that will use data on SO$_2$ effects. For several reasons this chapter
will focus on agencies at the federal level of government. First, air pollution does
not recognize political boundaries, and travels across many cities, counties, and
states; so the broadest form of control and regulation of air pollution will reside

in the largest government body we have, i.e., the federal government. Second, many states are adopting laws and implementing rules and procedures that follow the outline and content of the federal laws. In fact, the Clean Air Act requires states to adopt air pollution control plans (called "state implementation plans") to achieve air quality equal to or more stringent than the federal air quality standards required by the Act. Third, the federal laws and regulations are of interest to all people because of their nationwide application. At the federal level, a distinction is made between the agency responsible for controlling air pollution and setting ambient air quality standards, and another group of agencies responsible for protecting natural resources from damage. The first agency, the Environmental Protection Agency, has a mandate under the Clean Air Act to set air quality standards that protect human health and welfare from adverse effects, and to set procedures for controlling air pollutant emissions. The second type of agency, generically called the Federal Land Manager, has a mandate under agency-implementing legislation to protect the natural and cultural resources under its jurisdiction from adverse effects of air pollutants. Federal Land Managers include the Secretaries of the Departments of Agriculture and Interior, and the agencies involved include the U.S. Forest Service, U.S. Fish and Wildlife Service, and the U.S. National Park Service. The Clean Air Act acknowledges the responsibility of the Federal Land Manager, and provides a mechanism for involvement in a process for permitting new pollution sources.

Three Uses of SO_2 Effects Data

Three uses of data on SO_2 effects will be detailed in this section: standard setting, project permit review, and research program guidance.

Standard setting. Damage to vegetation by SO_2 is theoretically prevented by setting a secondary ambient air quality standard. The secondary standard is intended to be set at a concentration and averaging time below which no unacceptably adverse effects on plants will occur. How is the standard set? Usually, the literature on the effects is collected and reviewed, and a concentration and averaging time is selected following a screening process (Padgett and Richmond, 1983). The latter process should use a set of criteria for the acceptance or rejection of effects studies, including the following:
1. Types of effects, in ranked order.
2. Realistic SO_2 exposure concentrations.
3. Realistic exposure durations and frequencies.
4. Ecologic/economic importance of taxa affected.
5. Type of experimental fumigation and/or field study.
6. Use of a control treatment.
7. Clear explanation of growth and exposure conditions.
Criterion 1 requires decisions about the importance of effects, including, e.g.,

changes in enzyme levels in a test organism in a laboratory fumigation apparatus, the decline of a major forest species in a mixed coniferous forest, or the appearance of visible injury symptoms on a common species along a roadside. Should the standard be set to protect species from all types of effects? Or just biochemical effects? Should all species be protected? Or a certain proportion of species?

Criterion 2 involves a judgment about whether or not to use studies that expose plants to unrealistically high SO$_2$ concentrations, e.g., greater than about 2.0 ppm (83 μmol m^{-3}) for 1 hour. Concentrations higher than this are almost never observed in the real world at ground level, and if they were would most likely kill vegetation very quickly. Setting the standard at this concentration would be meaningless if the concentrations never occur in the ambient air. The important question is to decide how far below this concentration to go when selecting studies for review. E.P.A. has used the following concentrations for screening studies for consideration: for short averaging times (one minute to one hour), concentrations must be 1 ppm (41.6 μmol m^{-3}) or less, and for longer averaging times (one week to one year), concentrations must be 0.1 ppm (4.2 μmol m^{-3}) or less (U.S. Environmental Protection Agency, 1982c).

Criterion 3 allows a screening of experimental studies that use unrealistic exposure durations and frequencies. In the real world, high concentrations occur for short periods of time, but low concentrations are of longer duration. The typical exposure regime of a fixed concentration (e.g., 0.40 ppm, or 16.6 μmol m^{-3}) for 8 hours per day for 5 days a week for any number of weeks usually cannot be related to real world fumigations. The more recent attempts to fumigate stochastically, with concentrations fitting a lognormal distribution, are much closer to real world fumigation (e.g., Male, Van Sickle, and Wilhour, 1978). A decision has to be made during the standard-setting process whether or not to include fumigation regimes that are never observed in the ambient environment. This is important because it guides the decision maker in setting the averaging time and the level of the standard.

Criterion 4 requires a judgment on the importance of affected species in two separate realms: the natural world and the man-made world. The decision maker must decide if the species is important ecologically in its particular habitat, if the species is important economically in its place in human culture, or both.

Criterion 5 enables the agency to decide if studies performed in highly controlled fumigation chambers and laboratories are relevant to real world situations. The response of plants in the natural environment is moderated by many uncontrolled factors. Studies in controlled environment chambers remove these factors, and generate results that are the outcome of the pollution treatment only. It is generally thought that the use of controlled environment chambers leads to overestimates of effects, although there are reports to the contrary. It is now possible to theorize that chambers may underestimate air pollution effects under certain conditions. Studies in the field with open-top chambers, wall-less pollution delivery systems, and/or pollution exclusion systems are closer to real world situations, because they can incorporate natural sources of variation in the plant environment. This criterion is important because it allows the reviewers to narrow the literature they are using to set the standard.

Criterion 6 guarantees that the experimental work being considered includes the use of a control treatment in the design. This allows the SO_2 treatment and a treatment with no SO_2 to be compared. Without this, no estimate can be made of the SO_2 effects on the plants.

Criterion 7 allows the decision maker to include only those studies containing clear explanations of the materials and methods used in the experimental work. Both growth and pollutant exposure conditions must be defined in a way that allows comparisons of plant conditions before, during, and after pollutant treatment to be made. This information also helps determine whether or not the work can be repeated if necessary.

The current federal secondary SO_2 standard is 0.50 ppm (20.8 μmol m^{-3}) for 3 hours, not to be exceeded more than once per year. No long-term secondary standard exists, but a primary (health-related) annual standard has been set at 0.03 ppm (1.25 μmol m^{-3}). According to E.P.A., the original secondary standard, promulgated in 1970, was set on the basis of a single study reporting effects on a species of *Fraxinus* (U.S. Environmental Protection Agency, 1982c). This study was conducted according to the following criteria (in the same order as those above):

1. Visible injury symptoms on leaves were the measured effects.
2. An SO_2 concentration of 0.54 ppm (22.46 μmol m^{-3}), which can occur in ambient air.
3. The effects occurred following an exposure duration of 3 hours.
4. A plant of ecological value was the test organism.
5. An experimental fumigation treatment was used.

In more recent documents on the newer SO_2 standard, it is evident that E.P.A. is changing these criteria. Consideration is now being given to growth and yield effects, longer exposure durations, and field studies.

A logical approach for establishing the boundaries for setting a secondary SO_2 standard would be to use the highest observed ambient concentrations for the upper limit, and the lowest concentrations known to affect plants for the lower limit. There is, clearly, no need to set a concentration standard above what is observed in the atmosphere, which is about 0.05 ppm (2.08 μmol m^{-3}) for an annual average and 2.0 ppm (83.2 μmol m^{-3}) for one hour. However, basing the lower limit on the lowest concentration that affects plants could be debated. Some might argue that a margin of error is needed to protect the most sensitive genotypes or species. On the contrary, others may argue that the lowest reported effects are worst-case examples, and would rarely, if ever, occur in the real world. Nevertheless, using the lowest known effects would set the lower limit near 0.015 ppm (0.624 μmol m^{-3}) for an annual average and between 0.50 and

1.00 ppm (20.8 and 41.6 μmol m^{-3}) for a 1-hour average. To conclude, a "window" for establishing a secondary standard can have the upper limit based on air quality and a lower limit based on effects on plants. The actual standard is set somewhere between these values, at a value that depends on political and economic factors.

Project permit review. The Federal Land Manager has a responsibility under the Clean Air Act to review Prevention of Significant Deterioration (PSD) permit applications from new sources of air pollution that wish to locate near lands under the jurisdiction of the federal government. Certain federal lands were given two levels of protection from the effects of SO₂ and total suspended particulates. The most stringent level, given class I area designation, was applied to international parks, national wilderness areas and memorial parks that exceed 5,000 acres, and national parks that exceed 6,000 acres. All other federal lands were designated class II areas. The level of SO₂ protection ("maximum allowable increase") provided in the Act is found in Table 2.1. These are not the same as standards. These are increments of pollutants above existing concentrations that are allowed in the area in question from the new source applying for the permit. Their purpose is to allow industrial growth to occur while maintaining ambient air quality. The Federal Land Manager is responsible for evaluating the effects on plants of the increments plus the existing background concentrations by means of the permit review process shown in Fig. 2.1.

The first step in this process is to check the air quality modeling that the applicant submits. If the modeling predicts increases greater than those allowable, the applicant must demonstrate to the Federal Land Manager that no adverse effects will occur in the class I area near the source. If the demonstration is satisfactory, the source may be granted a variance from the increments and be allowed to proceed. If the demonstration is not satisfactory, i.e., the Federal Land Manager is convinced that adverse effects will occur, no variance may be granted. Even if the air quality models predict allowable increases within the increments inside the class I area, the applicant must determine if resources sensitive to air pollution (called "air quality related values" [AQRVs] in the Act) occur in the class I area, and if they will be damaged by the allowable increase plus the background concentrations. This section of the Act, called the "AQRV test," is the most important part of the permitting process. It allows the Federal Land Manager to

TABLE 2.1
Levels of Protection from Effects of SO₂ Provided in the Clean Air Act

Average time	Maximum allowable SO₂ increase in μg m^{-3} (μmol m^{-3})	
	Class I	Class II
Annual arithmetic mean	2 (0.03)	20 (0.31)
24-hour maximum	5 (0.08)	91 (1.42)
3-hour maximum	10 (0.16)	512 (8.00)

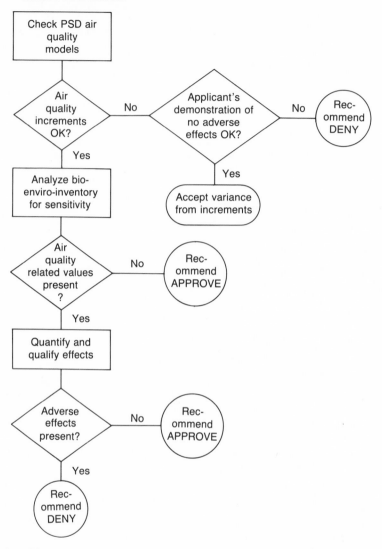

Fig. 2.1. Flowchart of decisions and analyses needed by Federal Land Managers to process Prevention of Significant Deterioration (PSD) permits.

decide if sensitive flora and fauna are present, using any literature on effects that is available, and to decide what kinds of effects are adverse. The Federal Land Manager need not compare the predicted background plus increment concentrations with only the National Ambient Air Quality Standards (NAAQS). Instead, he or she may review any literature that shows effects resulting from concentrations below the NAAQS on resources in the federal lands in order to decide if adverse effects will occur. Projects may then not get approval if the Federal Land

Manager convinces the permitting agency (E.P.A. or a state with PSD authority) that adverse effects on AQRVs will occur whether or not the increment is exceeded.

The current NAAQS concentrations and averaging times for SO_2 are 0.03 ppm (1.25 μmol m^{-3}) annual average, 0.14 ppm (5.82 μmol m^{-3}) for 24 hours, and 0.50 ppm (20.8 μmol m^{-3}) for 3 hours. The first two sets are the health-related primary standards, whereas the last set is the secondary standard. Some studies of effects on native plants below these concentrations and averaging times are known in the literature, but more studies like them are needed by the Federal Land Manager.

The Department of the Interior has recently taken two actions relevant to these permitting and standard-setting processes. In one, the Department defined adverse effects for all PSD permit reviews, and then evaluated waiver applications from some new sources wishing to locate near Theodore Roosevelt National Park in North Dakota. Although no adverse effects were found, ecological effects of SO_2 on lichens and bryophytes were found to be potentially adverse in the park if SO_2 concentrations there continued to increase. The second action was the National Park Service's review of the proposed E.P.A. staff document on the SO_2 ambient air quality standard. The N.P.S. review concluded that a secondary annual SO_2 standard is definitely needed, based on reports in the literature that document effects on lichens and mosses at annual concentrations below the current annual primary standard of 0.03 ppm (1.25 μmol m^{-3}).

Research program guidance. Several federal agencies have established research and development programs in the area of air pollution effects on vegetation. The E.P.A. is the leader in this area, and has traditionally funded basic research on everything from physiological to ecological effects. This research has guided E.P.A. in establishing standards. The U.S. Department of Agriculture has also funded a great deal of effects research in the past several decades, on both agricultural crops and forests. This research has been primarily focused on estimating economic effects of air pollutants on plants.

Recently the Department of the Interior, through the National Park Service, has expanded its research program on vegetation effects. Its focus is entirely on effects on native plants in the national parks from low concentrations of SO_2 and O_3. In this program, the emphasis is mostly on field studies in the parks; these include biomonitoring plots, baseline studies of sulfur concentrations, surveys of bioindicator species, growth studies, population studies, and ecological studies. Several laboratory studies are being performed, with fumigation regimes using concentrations and averaging times less than the NAAQS.

The NAAQS for SO_2, given earlier, are set for three averaging times. The example fumigation cited earlier of 0.4 ppm (16.6 μmol m^{-3}), 8 hours a day, 5 days a week, for, e.g., 2 weeks, would expose plants to concentrations of 0.4 ppm for 3 hours, 0.13 ppm (5.4 μmol m^{-3}) for 24 hours, and 0.004 ppm (0.166

μmol m^{-3}) for an annual average. All of these concentration averages are less than the three standards, yet most plants exposed to this fumigation will probably die. Studies need to be designed that use lower concentrations and longer averaging times than the standards, but produce less drastic effects on plants.

Effects studies primarily help the research program manager in selecting fumigation concentrations and averaging times, test species, experimental designs, fumigation hardware, and response parameters. The way these are selected depends on the type of contractual arrangements the funding agency is allowed to enter into with the principal investigator. The types of contractual arrangements include grants, contracts, cooperative agreements, interagency agreements, and purchase orders. Differences between these contracts depend on the involvement of the funding agency, the transfer of funds, and prior performance of the principal investigators. Most grants and contracts are awarded following a competitive procurement process. An advertisement is announced for the contract or grant, investigators submit proposals, and the proposals are reviewed by the funding agency. Under certain unique circumstances, contracts or grants may be awarded to investigators on a sole-source basis. The justification for such arrangements typically include unique services, short performance periods, special equipment, or substantial prior investment.

Techniques for Using Effects Data

Regulatory users of data on effects manage and analyze the data by using several techniques: data base building and analysis, risk analysis, and field recognition techniques.

Data base building and analysis. Several types of data bases are of use to agencies that need data on SO$_2$ effects; these include literature, effects, geographic, and floristic.

Literature data bases are the simplest type of data base. They consist of compilations of literature references on any subject, and include author citations, year of publication, titles, journals, page references, publishers, key words, abstracts, and other bibliographic information. Searching through the data base can be by key words or by recognition fields. The most well-known literature data base on air pollution effects on vegetation is the APIBE Bibliography entitled "Biological Effects of Air Pollutants on Plants, Animals and Microorganisms." The data base has been funded since the early 1970's by E.P.A.'s Environmental Research Laboratory at Corvallis, Oregon, and is currently maintained by the Oregon State University Library and Computer Center. It contains approximately 10,000 literature references.

An effects data base is focused on the results reported in the literature. It contains actual data on pollutant concentrations, averaging times, effects on plants, and species names. Since it contains quantitative data, it can be accessed for data manipulation and summarization. Plotting of data and statistical analyses are possible. The National Park Service maintains data bases of this type in order to review effects for the permit review process and the standard-setting process.

A geographic data base contains qualitative and quantitative resource information organized according to the location of the area where the data were taken from. The scale of the data can range from arbitrary physical distance parameters to political or physiographic units. The Geoecology Data Base at Oak Ridge National Laboratory is scaled to county level information, and contains almost 100 files of data on agriculture, climate, vegetation, forestry, air quality, natural areas, human population, water, terrain, and wildlife (Olson, Emerson, and Nungesser, 1980). The most useful feature of this data base is that it enables one to algebraically manipulate and factor quantitative data in two dimensions. The Geoecology Data Base contains vegetation and air pollution data, and has been used to map areas of the country containing SO_2-sensitive vegetation that is at risk. The National Park Service is currently using the data base to construct air pollution sensitivity indices for the class I national parks, so that they can be ranked for overall sensitivity.

A floristic data base contains taxonomic information on plant species organized hierarchically. The National Park Service has developed a national park flora data base to use in the permitting process (Bennett, 1982). The data base contains floras of the class I parks, and is structured for access by the System 2000 data-base-management software system. Queries can be made to the data base about the presence of SO_2-sensitive species in any park or in all the parks, the number of SO_2-sensitive species in any park, and the locations of SO_2-sensitive species.

Risk analysis. The effects of SO_2 on vegetation occur as probabilistic events. It is never known for certain where or when a certain SO_2 concentration may occur. There is always a probability associated with the occurrence because physical models do not exist that can totally predict the concentrations. Most air quality models today are combinations of statistical models based on empirical data and physical submodels based on meteorological principles. Consequently, there is always an error associated with the air quality predictions.

The same thing is true of plant effects data. Most studies of SO_2 effects are experimental. Consequently they consist of a design that includes combinations of air pollution and environmental factors, and replicate samples of plants to be exposed. The results are analyzed statistically, and a probability of occurrence by chance is estimated by means of statistical distribution tables. Thus, our knowledge of effects is based on estimates, which come from experiments whose results are judged to be chance or nonchance events. Some physical models of plant response to SO_2 have been developed, but they are highly customized and limited, and are not amenable to general use.

Nevertheless, it is possible to estimate the chances of effects of a particular pollutant in a particular place by factoring the probabilities of component events to arrive at an overall probability of the occurrence. Let us consider an example.

1. Suppose that Study *A* resulted in a 13% decrease in the growth of species *I* after exposure to 0.30 ppm (12.48 μmol m^{-3}) of SO$_2$ for 7 days, and that this decrease occurred only five times out of a hundred (0.05 probability level) because of chance, i.e., occurred 95 times out of 100 because of SO$_2$.

2. Suppose also that Study *B* predicted that the SO$_2$ concentration in location *X* where species *I* occurred would be 0.30 ppm for 7 days, but that it occurred as a non-overlapping event ten times out of a hundred (0.10 probability level) because of chance, i.e., occurred 90 times out of 100 because of real SO$_2$ emissions and meteorological events.

3. Suppose further that Study *C* found that sensitive individuals of the plant species *I* occur only 5% of the time in a population.

4. If we suppose that the preceding three events are independent, then the probability of the effect occurring is Prob.(1) $\times$ Prob.(2) $\times$ Prob.(3). Substituting values in the equation gives $(0.95) \times (0.90) \times (0.05) = 0.043$. Therefore the probability that 5% of the population of species *I* will grow 13% less following exposure to 0.30 ppm (12.48 μmol m^{-3}) of SO$_2$ from a pollution source for 7 days in location *X* is four times in 100. Since there are 52 non-overlapping 7-day periods in a year, this would probably occur about once every two years.

The preceeding is called a formal risk analysis. It consisted of four steps, as follows:

1. Identification of the adverse effect.
2. Identification of the air pollutant exposure conditions.
3. Relating the exposure to the effect.
4. Estimating the overall risk.

It is possible to improve on risk analysis by incorporating more factors and by encouraging the use of better experimental designs. Risk analysis is becoming more acceptable to Federal Land Managers as a decision tool, because it is comprehensive, logical, practical, open to evaluation, politically acceptable, compatible with many agencies, and easy to learn (Fischhoff *et al.*, 1981). The Park Service is currently using this tool in the permit review process.

Field techniques. There is a long history of studies of visible injury symptoms on leaves from SO$_2$. These have provided an adequate compendium of diagnostic aids for identifying SO$_2$ injury to many species in the field. However, such symptoms indicate only that acute fumigation episodes have occurred, and are not very useful in estimating economic or ecologic damage from the injury or the subtle effects of fumigations that do not produce visible injury. It is now common knowledge that fumigations at low concentrations can cause decreased agricultural crop yields, increased sulfur levels in foliage, altered dry matter distributions, decreased rates of physiological processes, decreases in plant species diversity, biomass, and cover, and even mortality. Techniques to detect these changes are increasing and becoming more popular. They include such standard ecological methods as quadrat analysis, transect analysis, nutrient pool analysis, limiting factor analysis, elemental analysis, and population analysis. These techniques have been around for many years, but were not used previously by plant pathologists, although they were used in such kindred disciplines as plant ecology, soil geochemistry, and agronomy.

Regulatory agencies are increasing their use of these techniques. Field inspections are now incorporated in the permit review process to see if air pollution effects are occurring in federal lands before new sources are permitted near the area. Field studies are also useful to the standard-setting process, because effects are related to real-world ambient air pollution regimes. And finally, field research is being funded more by Federal Land Managers to discover how air pollution affects native plants in their natural habitats.

Directions for Future Research

Regulatory agencies are in need of five types of data in the future in order to improve their performance: effects at low, ambient concentrations; effects of combinations of pollutants; effects on native plants; effects on nonvascular plants; and effects on rare and endangered species.

Effects of low concentrations. The distribution of hourly concentrations of SO$_2$ typically approximates a lognormal distribution in most parts of the country. This type of distribution occurs as a function of the sources of SO$_2$, which are mostly point sources. An example of this distribution is shown in Fig. 2.2, which is based on three years of data from Theodore Roosevelt National Park. The solid line shows the distribution of ambient air concentrations. To the right of this line are shown points representing studies in the literature that reported effects on vegetation either known or suspected to occur in the park. The dotted line in the figure connects those points closest to the observed SO$_2$ concentrations for the given averaging times. It is immediately evident that the lines for the ambient air concentrations and the effects concentrations are not parallel, and appear to converge near the longer averaging time intervals, i.e., annual averages, and the one-day averaging times. If future sources were to be located far away from the park, the ambient air concentration line would likely move to the right (i.e., a change in intercept and not slope), closer to the effects line. If future sources were to be located nearer to the park, the ambient-concentration line would likely shift in slope as well, showing more higher concentrations for the shorter averaging times and higher annual averages. In both cases, effects at the lower concentrations and longer averaging times are more likely to occur in the future. It is self-evident that we are greatly in need of more studies at these concentrations.

Combination effects. It is common knowledge that air pollution is a mixture of many toxic gases, not a single, pure gas. In the real world, plants are exposed to more than SO_2 alone. In most areas of the country, plants are exposed to mixtures of SO_2 plus O_3, NO_x, hydrocarbons, particulates, and other pollutants. If the regulator tried to estimate air pollution effects by simply multiplying the individual effects of single pollutants, he or she would inevitably overestimate the effects. Complete vegetation mortality has occurred in the vicinities of smelters and other large sources of SO_2 in the past, but such events are unlikely to occur again in the future given today's pollution controls. In all areas of the country, most vegetation is still alive, and constitutes obvious evidence that mixtures of pollutants do not cause simple multiplicative effects. Interactive effects are obviously subtle and complex. The regulator, however, is often faced with reviewing a permit for a new source that emits only one pollutant, knowing full well that other pollutants are present, and that the plants in the impact area in question will respond to both the predicted pollutants plus the existing pollutants in the area. The same problem occurs in the setting of standards. The regulator is forced to decide if standards should be for single gases, or whether binary standards are needed for pollutants in the presence of other pollutants. Again it is self-evident that we are greatly in need of more mixture effects studies.

Native plants. The majority of past research on SO_2 effects on plants has concentrated on plants of economic value. This was a function of our concern for costing out the externalities of industrial and urban growth on other sectors of our economy. Natural ecosystems, however, have not typically been regarded as a sector of the economy and consequently have been ignored. Recent concern about the degrading quality of the natural environment, however, has focused more interest on the ecological value of natural systems. With ambient concentrations increasing in rural areas because of pollution source development in urban *and* rural areas, native plants in remote areas, such as national parks, are being exposed more and more to air pollution. Federal Land Managers, therefore, are increasingly interested in data about SO_2 effects on native plants in the areas under their jurisdiction.

Nonvascular plants. Fig. 2.2 and Table 2.2 document a group of studies on mosses and lichens where effects were observed at concentrations averaging half of the current annual SO_2 standard. It is generally accepted that the nonvascular plants are more sensitive to SO_2 than vascular plants, because of their lack of a cuticle, their low photosynthetic rates, and their low amount of stored photosynthates to buffer stress, among other reasons. Nonvascular plants are undeniably important in ecosystems for stabilizing soil, nitrogen fixation, and colonizing new substrates, and as a food source for lower orders of animals. Very little is known about the effects of SO_2 on the processes in ecosystems that are per-

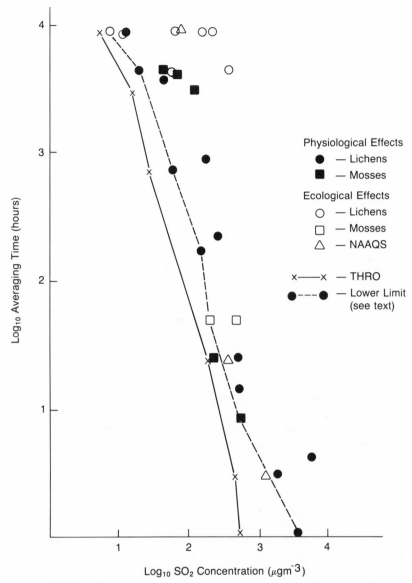

Fig. 2.2. Effects of SO_2 on plants at various concentrations and averaging times.

TABLE 2.2
SO$_2$ Annual Averages Known to Produce Effects on Nonvascular Plants

Species	Annual average in μg m^{-3} (μmol m^{-3})	Effect	Source
Mosses			
Orthotrichum affine	13 (0.203)	Absence threshold	Gilbert (1970)
Bryum argenteum	13–26 (0.203–0.406)	Absence threshold	Taoda (1972)
Pleurozium schreberi	17 (0.266)	Absence threshold	Winner and
Hylocomium splendens	17 (0.266)	Absence threshold	Bewley (1978)
Hypnum cupressiforme	45 (0.703)	Absence threshold	Gilbert (1970)
Barbula acuta	60–65 (0.94–1.02)	Absence threshold	*Ibid.*
Lichens			
Parmelia caperata	5–9 (0.08–0.14)	Injury threshold[a]	Will-Wolf (1980)
Cladonia rangiferina	14 (0.22)	Photosynthesis decrease threshold	Tomassini *et al.* (1977)
Parmelia caperata	26 (0.406)	Absence threshold	LeBlanc (1969)
Parmelia caperata	39 (0.609)	Absence threshold	Showman (1975)
Parmelia reducta	52 (0.810)	Absence threshold	*Ibid.*
Parmelia caperata	60 (0.94)	Absence threshold	Hawksworth and Rose (1970)
Parmelia subreducta	65 (1.02)	Absence threshold	Showman (1975)
Parmelia sulcata	65 (1.02)	Absence threshold	*Ibid.*
Physcia millegrana	65 (1.02)	Absence threshold	*Ibid.*
Physcia stellaris	65 (1.02)	Absence threshold	*Ibid.*

[a] Discoloration, algal plasmolysis, and bacterial infection.

formed by nonvascular plants. An increasingly common problem confronting the Federal Land Manager in permitting actions is that he or she does not know what might happen to the ecosystem if the lichens and mosses disappear. It is self-evident that more research is needed in this area.

Rare and/or endangered species. The national parks are becoming the last sites for many rare and/or endangered plant species as more and more sites on private land are destroyed. With their populations shrinking, any additional stress, including air pollution, may threaten the existence of these species. Although their role in the ecosystem may be unknown, their loss by any cause is unacceptable. The Park Service has begun a program of screening rare and endangered species for air pollution sensitivity (Bennett, 1981). A hierarchical protocol has been developed, and the first studies have begun on species from Great Smoky Mountains National Park. Species of interest include those on federal and state lists, species identified by local native plant societies, and species identified by the Park Service.

3

Regulations and Research on SO$_2$ and Its Effects on Plants in the European Communities

P. J. W. Saunders

The sovereign states of Europe have diverse attitudes toward SO$_2$ pollution and its control. Nevertheless, the environmental problems created by uncontrolled SO$_2$ emissions are widely recognized and discussed openly in various European forums. Certain European institutions have a critical role to play in the formulation and imposition of international regulations to control SO$_2$ pollution, and in the coordination of related monitoring and research programs. This paper reviews progress made with such activities as of 1983, and indicates possible future developments, especially in research, that will influence policies for SO$_2$ pollution control in the longer term. For developments since 1983, readers are referred to the *Official Journal of the European Communities*.

The views expressed throughout this paper are strictly my own, and have no official standing unless otherwise stated.

Institutions, Policies, and Regulations

European institutions aim for a consensus among their members about how particular environmental problems are to be dealt with. Few, however, possess the financial, political, and legal powers to enforce these agreements. The European Communities (or "Common Market") is a notable exception, and is playing an increasingly important role in the development and enforcement of policies for SO$_2$ pollution control and related research. It is, therefore, essential to understand the structure and functions of the "Common Market" and its relations with other European institutions.

The European Communities. The original "Common Market" of six nations was created under the Treaty of Rome to remove trade barriers and to create a wide and uniform economic community. The Treaty embraced three separate communities: for coal and steel (E.C.S.C.), for atomic energy (Euratom), and for economic unity (E.E.C.). The Treaty was amended in 1965 to amalgamate these organizations into the European Communities (E.C.) under a single Council and a single Commission within the oversight of the European Parliament and the legal framework of the European Court of Justice. The E.C. now consists of ten member states (Italy, West Germany, France, the Netherlands, Luxembourg, Eire, the United Kingdom, Belgium, Denmark, and Greece) and two applicants (Spain and Portugal).

The European Parliament, whose members are elected directly from E.C. countries, has few practical powers other than the right to question senior E.C. officials, but expresses opinions that have increasing political significance. Effective political power rests with the Council of Ministers, which operates at several levels (e.g., prime minister or equivalent, agriculture ministers, research ministers, permanent representatives) and decides on all budgetary and policy measures within the E.C. This arrangement ensures that elected national governments control the degree to which their sovereignties are surrendered to the common cause.

The Commission of the European Communities (C.E.C) is the civil administration of the E.C. and is composed of a series of Directorates-General, each being equivalent to a department of state, and each headed by a Commissioner. The Commissioners and the President of the Commission are nominated on rotation and for set periods of time by the national governments. The most senior permanent officers of the C.E.C. are the Directors-General, each reporting to his respective Commissioner. It is important to note that only the C.E.C. can make formal proposals to the Council for E.C. action, and that the C.E.C. is responsible for implementing and enforcing all Council decisions within the framework of the Treaty of Rome.

With the C.E.C., Directorate-General XI (Protection and Improvement of the Environment) is responsible for the formulation and implementation of environmental regulations. Directorate-General XII (Science, Education, and Research) is responsible for the formulation, organization, funding, and coordination of environmental research, much of which is developed specifically to meet the policy requirements of Directorate-General XI and of the Directorates-General responsible for energy, agriculture, health, and other matters. Directorate-General XII also supervises the E.C. Joint Research Centers in Italy, Germany, and the Netherlands, and is often represented at scientific conferences by qualified C.E.C. scientists. Both Directorates-General may employ consultants, and consult with independent experts when required.

The research programs of the C.E.C. take three main forms:

1. Direct action: research in one of the E.C. Joint Research Centers.

2. Indirect action: research commissioned in a laboratory of a member state.

3. Concerted action: coordination of nationally funded projects in member states.

In addition, there are provisions for concerted actions involving nonmember or "third" countries. A program of research may embrace all these forms of action.

Other European institutions. The financial, legal, and political powers of the E.C. far exceed those of all other European institutions and yet, because it embraces only part of Europe, the E.C. must also work closely at the political level with sister institutions. For instance, the Eastern European equivalent of the E.C. is the Council for Mutual Economic Assistance (COMECON), embracing the U.S.S.R. and its allies. There have been discussions between the E.C. and COMECON on environmental issues of common interest, but little progress has been achieved because of fundamental differences in philosophy and because COMECON lacks significant political and economic powers (Füllenbach, 1981).

The E.C. has much more common ground with other international organizations, such as the U.N. Food and Agriculture Organization (F.A.O.), the U.N. Economic Commission for Europe (E.C.E.), and the Organization for Economic Cooperation and Development (O.E.C.D.), whose meetings may be attended by C.E.C. officials as observers or as official E.C. delegates alongside those of member states. In certain circumstances (e.g., at Law of the Sea Conferences), the C.E.C. may be empowered to negotiate a common policy for its members on a major issue of wider international importance. Normally, however, the C.E.C. uses its representatives to gather information and to keep in touch with international developments that affect E.C. policies.

It is instructive to note the great interest displayed by the C.E.C. in the O.E.C.D.'s study (1979) of long-range transport of air pollution (L.R.T.A.P.), which involved active collaboration in research by Norway and Sweden with the United Kingdom and other E.C. countries. Similar interest was shown in the E.C.E./F.A.O.-sponsored symposium in Warsaw on air-pollution effects on vegetation, which contributed to the subsequent O.E.C.D. (1981) report on the costs and benefits of sulfur-pollution control. The C.E.C. is now keeping a close watch on the European Monitoring and Assessment Program (E.M.E.P.) sponsored by the E.C.E. and the World Meteorological Organization (W.M.O.). E.M.E.P. is a classic example of the ability of U.N. organizations to bring member states of the E.C., COMECON, and the Scandinavian bloc together, here in a study of the deposition of sulfur pollutants, and of their fluxes and trajectories across national boundaries. Such information is an essential prerequisite for the formulation and negotiation of pan-European measures to regulate emissions of SO₂.

It must be remembered, however, that all these European institutions are essentially political and economic organizations, with science very much in a supporting role. The C.E.C. has, in fact, been subject to considerable criticism on the grounds that E.C. research programs do not receive adequate scientific scrutiny, and that E.C. environmental regulations often lack credibility in the eyes of the scientific community. The C.E.C. now proposes to introduce peer-review systems for its research activities, and is also seeking independent scientific advice on environmental issues from national academies in the E.C. and from the

European Science Research Councils (E.S.R.C.) to promote fundamental and strategic research on an international basis throughout Europe. For instance, the foundation is sponsoring a study of the dynamics of European forest ecosystems, with major contributions from scientists in Sweden, the United Kingdom, Denmark, West Germany, the C.E.C., Spain, Finland, Norway, and France.

E.C. policies. E.C. environment policies are enunciated in a series of Environment Action Programs. The Second Program (E.C., 1973), in force until the end of 1982, had among its objectives the following:

1. The setting of biological standards for lead pollution.

2. The setting of criteria for atmospheric pollutants (lead, SO_2, NO_2, particulates, photochemical oxidants, asbestos, hydrocarbons, vanadium).

3. The setting of standards, the exchange of information on and common approaches to monitoring systems, specific actions to control certain emissions in the industrial and energy sectors, and measures to protect frontier zones.

4. Measures to control the siting of power stations.

This program was designed primarily to reduce and control pollution and nuisances. Resolutions of the Council indicated specific objectives (e.g., reduction of sulfur content of fuel oil), but mandatory directives were confined to suspended particulates, SO_2, and lead in air. These directives are discussed later in terms of current regulations.

The Council also indicated that it planned further action on the environment in 1981 (E.C., 1981a), but political, technical, and financial arguments delayed the formal promulgation of the Third Environment Action Program (1982–86) until 1983 (E.C., 1983a,b). This program, although still aiming to reduce and control pollution and nuisances, has assumed a more preventive character. For example, much emphasis is placed on the evaluation of new chemicals and on the introduction of environmental-impact assessment systems. The Council's common policy aims not only to protect the environment, but also to ensure that natural resources are well-managed, and that environmental considerations are brought into the planning and organization of economic and social development in the E.C. It is significant that the Council sees the introduction of "clean technologies" as a way both to protect the environment and to create new industries and jobs.

On air pollution, the C.E.C. is instructed to continue its efforts to establish air-quality standards at community and regional levels, to devise emission standards, to develop and apply mathematical models of dispersion, and to harmonize and standardize monitoring systems. Specific references are made (E.C., 1983a,b) to NO_x, hydrocarbons, fluorine, cadmium, mercury, CO, acid rain, SO_2, suspended particulates, and chlorofluorocarbons. There is no mention of effects of these pollutants on plants and plant communities, but such effects are primary considerations in the formulation of standards and control measures.

The E.C. Action Programs are underpinned by a series of research and development programs in the field of environment protection and climatology designed to provide a scientific basis for current and future E.C. policies and regulations for protecting the environment. The Third Research and Development Program (1981–85) covers five main areas, as follows: (1) sources, pathways, and effects of pollutants, (2) reduction and prevention of pollution and nuisances, (3) protection, conservation, and management of natural environments, (4) environment information management, and (5) complex interactive systems—human-environment interactions. This program (E.C., 1981b) covers a wide range of topics (e.g., effects of cadmium on humans, CO_2 cycles, organic pollutants in water supplies) studied within the E.C. Joint Research Centers and in national laboratories under contract to, or coordinated by, the C.E.C.

In practice, only the first area refers specifically to air pollutants and their effects on plant health, and relatively little work involves direct research on the effects of SO_2. Indeed, the C.E.C. reduced its efforts in this field at the end of the Second Research and Development Program (1976–80), on the assumption that SO_2 pollution was not a priority problem. The opportunity to rectify this error has presented itself with the formal review of the Third Program for the phase 1984–85, during which approximately 50% of the projects commenced in 1981 will be continued, and the residual funds, augmented, will be deployed toward new research on the following problems:

1. Effects of pollutants, including the direct and indirect effects of air pollutants (SO_2, NO_x, acid precipitation, O_3), on plants and terrestrial and aquatic ecosystems, especially by acid deposition.

2. Reduction and prevention of pollution, by development of new technologies to reduce waste production and to treat wastes.

3. Protection, conservation, and management of the environment, including studies of the sulfur, nitrogen, and phosphorus cycles.

The revised program (E.C., 1983b) places great emphasis on new research on the effects of SO_2 and other pollutants on plants and ecosystems. The C.E.C. proposes to increase funding for direct and indirect actions, and also to mount a new concerted action to coordinate the already-extensive national efforts in West Germany, the U.K, France, and the Netherlands, and to link up with work in Norway, Sweden, and Finland, and in North America.

These factual descriptions of E.C. environment programs do not reveal their intensely political origins, in which politicians, lobbyists, civil servants, and scientists fight protracted battles to protect their own interests, most of which are national in character. Priorities tend to be determined to some extent by the nationalities of the current Chairman of the Council of Environment Ministers, and of the President of the Commission.

For example, the West Germans tend to become very conscious of environmental issues when their "green" or "ecological" party holds the balance of power at state and national levels. France, with a strong system of centralized government, uses environmental priorities to further national policies for economic development. Denmark has much common ground with Norway and Sweden, and, along with West Germany, is one of the most active of the E.C. states campaigning for action on acid deposition. The Belgians are legalistic in the extreme, and the British generally resent any measure that threatens their pragmatic approach to pollution control. This dichotomy is best illustrated by reactions to C.E.C. proposals for uniform SO_2-emission standards; the Belgians back the C.E.C. proposals without question, but the British, supported by Eire, argue strongly in favor of their policy of "best practicable means," by which allowable emissions from individual factories are calculated in terms of the age of the plant, available technology, economic circumstances, and local environmental circumstances.

The Netherlands have particular problems with photooxidant pollution, which threatens their agricultural and horticultural industries, and are usually staunch supporters of any E.C. monitoring and control programs. Italy and Greece have little scientific and technical expertise in air-pollution problems and, therefore, rely heavily on the E.C. and the Joint Research Centers for advice. The geographical and climatological characteristics of Italy are also peculiar, since the north of the country has pollution problems (e.g., SO_2, lead) similar to those of France, Britain, and Germany, but those in the south are more akin to those of Greece. Greece, in its turn, does not recognize any problems of indoor pollution, on the grounds that their outdoors and indoors are contiguous, and argues strongly for measures to protect ancient monuments from the ravages of air pollution.

Against this background, the C.E.C. has the difficult task of formulating policies and regulations acceptable and applicable to all member states. Successful E.C. actions are, inevitably, the lowest common denominators that can be achieved by protracted bargaining, including political tradeoffs not only between member states, but also between policies for different sectors of E.C. activities. For example, a reduction in contributions to E.C. agricultural subsidies might be traded for a new E.C. air-quality standard.

E.C. regulations. The regulations that emerge from the E.C. Environment Action Programs result from decisions of the Council to implement particular control measures. The latter may be nonmandatory guidelines or resolutions, which the Council makes as declarations of intent, or may be directives, which are mandatory on member states and enforceable by the C.E.C.

The First and Second Environment Programs (E.C., 1973) included proposals for resolutions and directives to control the sulfur content of fuel oils and motor-vehicle fuels by rationalizing the supply of low sulfur fuels in polluted areas, by promoting the development of desulfurization and other measures for the selective reduction of sulfur emissions, and by promoting the more efficient use of fossil fuels. These proposed directives were not implemented, because member states have very different mixes of fuels (e.g., coal, lignite, nuclear, hydro, oil) and different fuel-consumption rates. Also, those states dependent on oil (e.g.,

TABLE 3.1
E.C. Standards for SO_2 and Suspended Particulates

Period	Calculated as	SO_2 range in $\mu g\ m^{-3}$ ($\mu mol\ m^{-3}$)	Total suspended particulates	Black smoke particles (alternative to TSP values)
Limit values				
Year	median daily	80–120 (1.25–1.87)	40	80
Winter (1 Oct to 31 Mar)	median daily	130–180 (2.03–2.81)	60	130
Year	98 percentile of peak daily value	250–350 (3.90–5.46)	150	250
Guide values				
Year	arithmetic mean daily	40–60 (0.62–0.94)	—	40–60
Day	peak mean daily	100–150 (1.56–2.34)	—	100–150

SOURCE: E.C. (1980).

NOTE: The standards are given in the form and units as published by the C.E.C. The C.E.C. conducted calibration exercises which resulted in (1) the option to measure suspended particulates as either black smoke or TSP, and (2) guide values for nonstandard monitoring systems currently measuring SO_2 and black smoke particles as arithmetic and peak mean daily values.

France) would pay a much heavier premium for low-sulfur oils than, say, Britain and Germany, which rely on coal and lignite to generate much of their electricity.

The specific directive requiring the installation of desulfurization systems (e.g., fluidized bed combustion) proved unworkable for somewhat similar reasons. However, the U.K. Chief Alkali Inspector has since stated that such systems will probably have to be introduced in all major new coal-burning plants in the U.K. because of international concern about acid rain (H.S.E., 1982).

The C.E.C. has been successful in formulating directives, mandatory from April 1, 1983, for associated ground-level concentrations of SO_2 and for suspended particulates (Table 3.1). These cover median daily values for the year, for the winter months, and for peak daily (98 percentile) concentrations. It is important to note that these E.C. standards are derived from existing standards of the World Health Organization and of the U.S. Environmental Protection Agency designed to protect human life. The limit values for SO_2 are given as ranges rather than as absolute standards in order to accommodate variations and inaccuracies in monitoring techniques. Particulate matter may be measured either as black smoke particles or as total suspended particulates. In either case, the values in Table 3.1 are absolute standards, being based on the assumption that exposure to particulates is most injurious to human health. The limit values are accompanied by guide or target values for those monitoring systems currently measuring SO_2 and black smoke particles as arithmetic and peak mean daily values.

The directives are supported by regulations for the monitoring and reporting of pollutant concentrations. Large areas of the E.C. are apparently exposed to

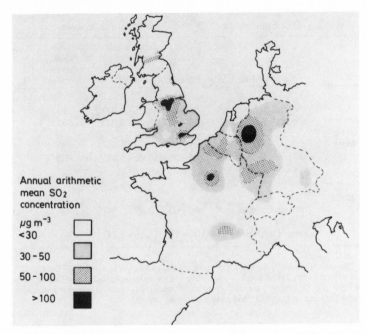

Fig. 3.1. Distribution of SO_2 over western Europe, based on 1974 emission data. Adapted from Fowler and Cape (1982).

concentrations in excess of the limit values; to the areas shown in Fig. 3.1 and Table 3.2 must be added parts of Italy and Greece. In the U.K., numerous urban areas are at risk, including parts of Greater London, Derby, Newcastle-upon-Tyne, Manchester, and Leeds. There is much uncertainty in these estimates, however, because of the differences in distribution of monitoring sites in each area. This problem is compounded by the fact that twenty different techniques are employed to measure SO_2, eight to measure total suspended particulates, and seven to measure black smoke, within the E.C. The C.E.C. is, therefore, using the regulations on monitoring to allocate a coordination and rationalization task to the E.C. Joint Research Center in Italy, with the long-term aims of improving comparability of monitoring data and of developing a common E.C. monitoring system.

The Commission is empowered under the Third Environment Action Program to continue its long-standing efforts to control various forms of air pollution. A directive on lead in air (proposed in E.C., 1973) is likely to be approved in a more stringent form by the Council in the near future. The Commission is considering, in addition, possible directives on NO_x and on oxidant pollutants.

Current and proposed E.C. directives have little overt relevance to problems of SO_2 pollution and plant health, but they are used with monitoring data as indicators of areas of potential SO$_2$ damage to plants within the E.C. (Unsworth and Ormrod, 1982). O.E.C.D. studies (1979, 1981) indicated that SO$_2$ pollution is a major environmental problem within and outside the E.C. (C.E.C., 1983). This situation is highlighted by reports of acid-rain problems in Scandinavia, Switzerland, and E.C. countries. The need for directives that protect both the environment and human health becomes obvious when it is realized that SO$_2$ and NO$_x$ interact to cause significant damage to plants (Mansfield, 1976) and that NO$_x$ emissions contribute about 30% of the acid precipitation in western Europe. Furthermore, there are indications that damage to some forests in Germany is the result of exposure to acid precipitation in the presence of elevated concentrations of NO$_x$ and SO$_2$ that coincide with episodes of O$_3$ and oxidant pollution.

Politicians and administrators in most E.C. countries continue to believe that Europe does not suffer from "Los Angeles smog" or its equivalent. However, work in the Netherlands and the U.K. has shown conclusively that, under certain climatic conditions, elevated concentrations of O$_3$ and photooxidant pollutants can occur over wide areas in summer months. These episodes cause damage to sensitive plants (Ashmore, Bell, and Reily, 1978; Posthumus, 1982).

These comments seem critical of E.C. policies and of the C.E.C., but the latter is well-aware of the need for tough action to protect the environment. A draft directive on environmental impact assessment, for instance, makes specific reference to air-pollution damage to plants (E.C., 1983b). The Commission's monitoring and research programs are also directed increasingly toward better coordination and harmonization of E.C. and national efforts, so that a strong scientific case can be developed for each new directive and regulation. The political and technical difficulties facing the C.E.C., however, are formidable, because national programs differ greatly in character and objectives.

TABLE 3.2
Rural Areas of Western Europe and North America Exposed to SO$_2$ and NO$_2$
(Estimated areas in hectares)

Category	Western Europe[a]	Great Britain	Canada	U.S.A.
Total area	1.1×10^8	2.3×10^7	6.8×10^7	4.4×10^8
Annual arithmetic mean SO$_2$, μg m^{-3} (μmol m^{-3})				
30 (0.47)	2.6×10^7	1.0×10^7	[b]	3.2×10^7
50 (0.78)	5.7×10^6	2.3×10^6	[b]	[c]
100 (1.56)	1.3×10^6	2.4×10^5	[b]	[c]
Annual arithmetic mean NO$_2$, μg m^{-3} (μmol m^{-3})				
25 (0.54)	2.6×10^7	1.0×10^7	[c]	3.2×10^7
40 (0.87)	5.7×10^6	2.3×10^6	[c]	[c]
80 (1.74)	1.3×10^6	2.4×10^5	[c]	[c]

[a] Spain and Italy are omitted.   [b] Data not significant.   [c] Insufficient data.

Monitoring and Research in E.C. Countries

Monitoring. Most monitoring systems in Europe are designed to sample and measure pollutants that can or may affect human health. The diversity of instruments and techniques in use has already been mentioned. Sampling devices are rarely distributed on a grid system (e.g., in Germany), but are usually concentrated in the more heavily polluted industrial and urban areas, with few, if any, sites in truly rural areas. The sensitivity and frequency of measurements, and the restricted range of pollutants sampled, also limit greatly the extent to which the results may be used in plant studies. For example, the U.K. volumetric for monitoring SO_2 is cheap and produces time-averaged data suitable for the E.C. directives (Table 3.1), but becomes increasingly inaccurate below 100 μg of SO_2 m^{-3} of air per 24 h (1.56 μmol m^{-3}) and gives no information on short-term concentrations (E.C., 1980). Nevertheless, such data can be manipulated to estimate the distribution of pollutants such as SO_2 and acid precipitation (Fowler and Cape, 1982; Barrett *et al.*, 1982).

Some E.C. countries operate more sophisticated systems, with full automation and much greater sensitivity and accuracy, but such systems are generally restricted to (1) areas under special investigation because of local pollution problems and (2) specific sites operated for research purposes. Fowler and Cape (1982) describe work of the second kind at Devilla Forest in Scotland, where SO_2, NO_x, O_3, and acid precipitation are monitored in detail, and where direct and continuous measurements of SO_2 fluxes between the forest canopy and the atmosphere are made. The analysis and interpretation of data derived from such systems are complex tasks, but the results contribute enormously to our understanding of the behavior of atmospheric pollutants and their derivatives in terrestrial ecosystems.

There remains a need for carefully designed networks to monitor acid precipitation (Nicholson *et al.*, 1980a; Barrett *et al.*, 1982), SO_2, NO_x, and O_3 in rural areas. At present, information on NO_x and O_3 concentrations is especially sketchy, ambient concentrations usually being estimated from very limited emission data. Despite these problems, Fowler and Cape (1982) estimate that approximately 2.3 M hectares of Britain are exposed to annual means of 50 μg of SO_2 and 40 μg of NO_2 m^{-3} per 24 h (0.78 and 0.87 μmol m^{-3}, respectively), and that about 5% of the land area of Europe is exposed to 50 μg of SO_2 m^{-3} (0.78 μmol m^{-3}) (Fig. 3.1). Table 3.2 makes some interesting comparisons of conditions in Europe, Canada, and the U.S.A.

Biological monitoring is used quite extensively in the E.C. For instance, a wide range of indicator plants is used in a network of stations in the Ruhr area of Germany as part of a state (rather than federal) effort to detect industrial pollution by metals, fluorides, SO_2, NO_x, hydrochloric acid fumes, and other phytotoxicants as well as by O_3. The Germans have also developed an elaborate scheme for the protection of plant life by ranking plant species according to their

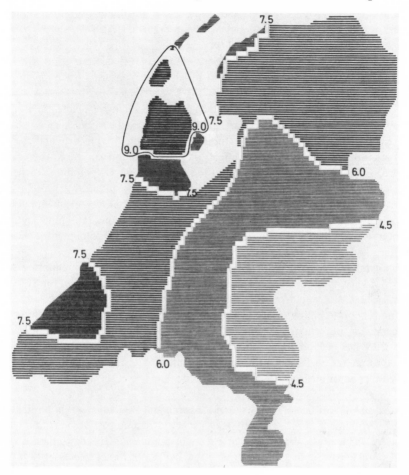

Fig. 3.2. Distribution of the mean ozone-effect intensity on tobacco variety Bel W$_3$ (as percentage of leaf area damaged, indicated by the numbers), measured weekly during the vegetation period from June 6 until October 28, 1977, in the Netherlands. Adapted from Posthumus (1982).

relative sensitivities to acute damage by individual pollutants. The scheme does not allow, however, for exposures to mixtures of gases or for variations in climatic and edaphic conditions, and has not found favor with the C.E.C.

A somewhat similar approach to routine monitoring was developed in the Netherlands, where sensitive indicator and accumulator plants are used at many sites, to detect phytotoxic concentrations of fluorides, NO$_x$, SO$_2$, and O$_3$. Fig. 3.2 illustrates the results of O$_3$ monitoring by Posthumus (1980).

Monitoring data for O$_3$ are generally scarce in Europe. The use of indicator plants, including the ubiquitous Bel-W3 tobacco, has distinct advantages in such circumstances (Ashmore, Bell, and Reily, 1980; Posthumus, 1980). The frequency of episodes of O_3 pollution is of the order of 100 h y^{-1} during which concentrations exceed 0.05 ppm (2.08 μmol m^{-3}) over the U.K., but with great differences between years and locations. Many of the central, southern, and western areas of Europe are potentially at risk from exposure to such phytotoxic concentrations.

It should be noted that sensitive indicator and accumulator plants are used in the U.K. less for routine monitoring and more for limited surveys to establish the extent and nature of pollution, as a prelude to more precise physical and chemical monitoring, and to studies of effects on plants in the field. The work of Ashmore, Bell, and Reily (1980) typifies this approach for O_3 pollution. Lichens, bryophytes, and fungi have also been used as indicators of various air pollutants and of damage (Heagle, 1982). Such organisms are often useful integrators of relatively long-term exposures, and can be used to study long-term trends in pollution in particular regions or around industrial sources of a pollutant (Perkins, Millar, and Neep, 1980).

In summary, the C.E.C. is faced with a mixture of different monitoring systems of varying sophistication and intensity. A start has been made, for SO_2 and particulates, to bring some harmony within the E.C. The C.E.C. is also giving some support to biological monitoring of O_3. However, the nettle of developing a scheme for uniform and sophisticated monitoring of all major air pollutants in the E.C. has yet to be grasped. In developing such a scheme, the C.E.C. will have to recognize the need for detailed monitoring in rural areas in support of research and of measures to protect plant life.

Research. The E.C.'s Third Sectoral Research and Development in the field of the Environment (E.C., 1981b) made specific mention of research into the sources, pathways, and ecological effects of pollutants. The first phase of the program, however, included relatively few studies of direct effects of SO_2 and other pollutants on plants and terrestrial ecosystems. Work at the Joint Research Center in Italy ceased in 1981, and few projects in national laboratories were carried through from the Second Program. It is known, however, that some contracts were placed by the C.E.C. for research by experienced groups in Germany, France, the Netherlands, and the U.K. For instance, Bell (U.K.) and Posthumus (Netherlands) continue to study O_3 effects, Mansfield (U.K.) is commissioned to study the effects of SO_2/NO_x interactions, and Bonte (France) and Perkins (U.K.) are working on fluoride pollution problems. These contracts represent, in fact, the token stakes of the C.E.C. in much bigger, national programs.

The C.E.C. is fortunate, however, in being able to review the Third Program as it enters its second phase (1984–85) and thus respond positively to current public and political concern about air pollution, especially acid deposition and its effects on plants and on terrestrial and aquatic ecosystems. The C.E.C. proposes (1983) to make this a priority area, into which it will divert much of its existing funds; additional funds are being sought to double this effort. The C.E.C. is also seeking to strengthen its influence on national programs, by expanding an existing concerted action on the physicochemical behavior of atmospheric pollutants, and by creating a new concerted action on the pathways and effects of acid deposition and other pollutants. The latter will involve collaboration with countries outside the E.C.

The extent to which the C.E.C. can fulfil its ambitions will depend on the willingness of the Council of Ministers (both Environment and Research) to approve the necessary funds. Also, the concerted actions may be effective only in part if the C.E.C. cannot persuade some of the more prestigious national research institutions not only to submit to central coordination without earning thereby some financial gain, but also to divert resources from research on pressing national problems, particularly when the C.E.C.'s priorities conflict with those of national funding agencies. It is important in this context to understand the concerns of the moment and how they are tackled in various European countries. Nonetheless, the C.E.C. is now moving in the direction of NCLAN with the active support of member states.

Emission controls are rudimentary in COMECON states, and extensive areas of agriculture and forestry suffer acute injury and damage from exposures to high concentrations of SO$_2$, fluorides, acid aerosols, and many other pollutants. Research in these countries tends, therefore, to parallel work carried out in the U.S.A. and in western Europe in the period 1910–50. In western Europe and Scandinavia, on the other hand, acute injury is uncommon, except close to major industrial complexes, and attention is focused on the subtle effects of long-term exposures to mixtures of pollutants at low concentrations. There is thus a fundamental difference between East and West on priorities for research.

Stocklasa's "Hidden Injury Hypothesis" has never been a contentious issue in western Europe, especially since the early 1950's, when the first reports of reduced photosynthetic activity and productivity in plants exposed to relatively low ambient concentrations of SO$_2$ and smoke began to appear (e.g., Bleasdale, 1952b). Symptoms of visible injury on plants, other than on selected indicators, are thus not regarded as acute indicators of yield or quality losses in crops. Also, it is doubtful that the E.P.A.'s National Crop Loss Assessment Network could be replicated in Europe, because of the extreme differences in agricultural systems, crop varieties, soils, and climates between individual European countries. These differences have forced scientists in western Europe to take the long-term approach of studying the fundamental processes involved in the deposition, movement, and effects of air pollutants in terrestrial and aquatic ecosystems.

The "fundamentalist" approach is perhaps best exemplified in the U.K., where the Research Councils, including the Natural Environment Research Council (N.E.R.C., 1976), exist primarily to support fundamental research of intellectual interest and scientific merit. Awards are made on the basis of peer judgment of the scientific worth of individual projects. The Councils do, of course, recognize the strategic and applied implications of their research, and their own laboratories often undertake applied research on contract to industry and government departments. However, they are independent bodies, and they use their shares of the science funds from the Department of Education and Science as they wish. A classic example of divergence in policy occurred in the mid-1970's, when the U.K. Departments of Environment and Agriculture decided that O_3 pollution was not a problem in the U.K., but the N.E.R.C. continued to support university research on the distribution and effects of ozone on plants because of the intellectual and scientific interest of the proposed research (Ashmore, Bell, and Reily, 1980). Analogous approaches may be found in some other European countries that permit a degree of freedom in research independent of the political issues of the moment. The framework of the intellectual independence varies from country to country, but it is often based on the need to achieve greater understanding of the dynamics of ecosystems and of the geochemical processes of the natural environment, and it is a major source of continuity and rigor in European research.

For example, the physical processes of deposition have been subjected to intense investigation in the U.K. following the early work of Chamberlain (1960) on deposition velocities. Eddy-correlation techniques have been used subsequently, but the construction of a fast-response sulfur analyzer has recently permitted the direct and instantaneous measurement of fluxes to and from a forest canopy, observations that are to be extended to measurements of NO_x, O_3, and other gases (Fowler and Cape, 1983). Unsworth and Mansfield (1980) have also demonstrated the influence of wind speed on the deposition and uptake of SO_2 and NO_2, and the resultant effects on plants, as has Davison (1982) in working with fluorides. These and other workers have measured accurately the uptake of gaseous and particulate pollutants by plants in the U.K.

A wide variety of systems are used in Europe for the controlled fumigation of plants in the laboratory and the glasshouse. There is intense research and speculation, especially in the U.K., on the advantages and disadvantages of various air-flow, pressure, temperature, light, and humidity regimes (Mansfield, 1976; Unsworth and Ormrod, 1982). The results achieved by these various systems are not always consistent, and attempts are being made now to calibrate certain systems in order to compare calculated with actual deposition velocities and doses of pollutants. A similar problem has emerged with systems used, mainly in France and the U.K., to conduct controlled fumigation of whole plants or crops in the field, either *in situ* or in enclosures such as open-top chambers. Again, there are plans in the U.K. to compare open-top and *in situ* systems with support from the N.E.R.C. and the Departments of Agriculture and Environment.

All fumigation and exclusion systems present the investigator with the basic problem of finding out precisely how the experimental regimes differ from the real environment, and the extent to which such differences influence plant responses to pollutants. For instance, differences in growth rate during the long-term exposure of grasses to low concentrations of SO$_2$ can have profound effects on the response of the plant, as has been shown by several studies in the U.K., the Netherlands, and Germany. A coincidence of low growth rate and low temperature appears to enhance the sensitivity of grasses to SO$_2$, NO$_2$, and O$_3$. Likewise, wind speed influences the size of the boundary layer, deposition and dose rate, pollutant uptake, and transpiration. Unsworth and Mansfield (1980) describe wind-tunnel systems developed in the U.K. to study this complex phenomenon, which is important in the study of air-pollution effects in the windy and exposed conditions of much of western Europe.

Much effort is also devoted to investigating the changes that occur on the leaf surface (e.g., erosion of cuticular areas); stomatal behavior in relation to gas exchange; transpiration and photosynthesis; the nature and relative importance of stomatal and cuticular resistance; and the mechanisms by which pollutants are taken up by leaf tissues. The best European work of these kinds has been conducted as part and parcel of experimental fumigations and field studies in which plants are grown and exposed to pollutants under conditions that are as realistic as possible. However, we still know little about the chemical, physical, and biological processes that occur on the leaf surface itself.

Pioneering work by Mudd (Mudd and Kozlowski, 1975) in the U.S.A., and by Zeigler (1973) and Wellburn (Wellburn, Majernik, and Wellburn, 1972; Wellburn, 1982) in Europe, has given some fascinating indications of the biochemical mechanisms by which plants may be damaged by pollutants and, in some cases, by which plants may avoid injury. However, work on fluorides, SO$_2$, NO$_x$, O$_3$, and peroxyacetyl nitrate (PAN) has revealed few common denominators, in the form of particular enzymes or reaction products in exposed plants, that might be used as indicators of incipient damage or yield losses. The administrator's dream of a rapid and nondestructive means of diagnosing such effects will thus probably not be realized. On the other hand, Wellburn is now attempting to study directly the pathways and fluxes of sulfur within plant cells and their organelles, and his results may elucidate some of the biochemical mechanisms that influence the sensitivity of plants to SO$_2$.

Acid rain is undoubtedly the major issue of the moment in Europe, with reports of damage to forests in Germany and France, and of acidification of lakes in Sweden, Norway, Germany, and the U.K., notably in Scotland. The official U.K. attitude toward acid rain has been essentially one of either indifference or protest at the economic cost of better controls on SO$_2$ emissions, and had been at odds with the views of many E.C. and other states in western Europe. However, the Research Council system has permitted U.K. scientists to conduct independent research on aspects of acid rain, sometimes in conjunction with, but not always in agreement with, their colleagues in the power industry. Many of these scientists do not believe the official U.K. line to be tenable, and share the concern of scientists elsewhere about the potential scale of acidification of the environment.

Recent work (Nicholson *et al.*, 1980a; Barrett *et al.*, 1982) has shown that the distribution of acid rain in Britain is distinctly skewed, with total deposits of acid being highest in the high rainfall areas of the northwest, whereas the rain in the northeast is less in quantity but greater in acidity. The latter area also suffers exposure to significant concentrations of SO_2 and NO_2, the ratio of both wet to dry deposition and sulfate to nitrate ions being approximately $2:1$. These ratios differ throughout western Europe, with wet deposition tending to predominate in Norway and Sweden, and dry deposition predominating in Germany and other industrial countries to the south. The extent to which these differences increase the acidity of rain as it passes through the tree canopies deserves further research, as do the marked differences between tree species in how much they increase acidity.

Although some experiments indicate increased erosion of cuticular waxes, there is little firm evidence of direct injury to vegetation by acid rain in Europe. The reported deaths of trees in several countries would appear to involve other phenomena, including the coincidence of flushes of acid rain (following periods of drought) with high concentrations of O_3 and other gaseous pollutants. We are not at present geared to study such extreme conditions of exposure to pollution (Fowler and Cape, 1982). Attention must also be paid to the complex physical, chemical, and microbiological changes that occur in soils in response to acid deposition. Are such changes large enough to cause long-term effects on tree roots and, therefore, tree growth? How do they interact with the natural processes of acidification that occur beneath coniferous trees on some soils? These changes are important, because they influence directly the nature of surface runoff and drainage waters from such soils in the headwaters of upland catchments.

On a catchment scale, more complex factors, such as changes in land use, forestry and agriculture management, and land drainage, come into play; these factors may be even more important than acid deposition in the long-term acidification of streams, rivers, and lakes in many parts of Europe. For instance, changes in the E.C. and national policies for the support of agriculture may induce a major reduction in grazing by animals in favor of afforestation in upland areas. In the U.K., the demands placed on the uplands by grazing, wildlife conservation, forestry, water catchment, and leisure usage already conflict enough to require later resolution by firm government action. The most appropriate policies for government will have to take into account the interactions between acid deposition and land-use practices. Research to provide the necessary data base is being initiated in Scotland, and there are proposals to extend the work to other areas of the E.C.

Future Patterns of Research

The Council and national governments of the E.C. tend to focus on what they see as applied problems relating to immediate issues of pollution control. Most central funding bodies relentlessly seek immediate answers, both to long-term environmental problems and to long-term research problems. Matters are not helped by shifts in research priorities according to changing political pressures. In such circumstances, fundamental and strategic research on air-pollution problems suffers, and scientists are forced to speculate from an inadequate data base on the likely impact of, say, acid deposition. Most administrators and politicians have yet to accept the uncertainties and high margins of error associated with such speculations. Inherent in that acceptance is the recognition of the need to reduce uncertainties by supporting appropriate long-term fundamental and strategic research.

The C.E.C. is aware of these problems, but lacks the machinery and the political power to do more than maintain a small core of strategic research in its environmental programs. In the longer term, however, it must tackle a wide range of issues, which are reviewed briefly in the following.

Dispersion and distribution. The mass balances for SO$_2$ in western Europe deserve more study, especially to determine the trajectories of sulfur pollutants and their interactions with climatic and meteorological variables. Similar, intensive work is needed on other pollutants, notably NO$_x$ and O$_3$, if we are to predict with any accuracy the behavior of the "body pollutant" and the fate of its components. The lack of data on sources and emissions of NO$_x$ in Europe must be remedied. We also need improvements generally in the monitoring of pollutants in rural areas; some of these would be relatively easy to achieve (e.g., measurements of pollutants in precipitation), but others, such as continuous measurement of gaseous pollutants, would be more expensive. The C.E.C. must also press hard for a modest European network of standardized monitoring devices.

Atmospheric loads, fluxes, and deposition. Direct measurements of the fluxes of pollutants between the atmosphere and plant surfaces, and of net dry deposition of SO$_2$, NO$_2$, O$_3$, and other gases to a variety of plant species and assemblages, must be carried out if we are to define and quantify conditions of exposure. It is essential also to quantify stomatal and cuticular resistances under realistic conditions, and to discover the mechanisms involved.

Effects of pollutants on plants and ecosystems. Our aim must be ultimately to quantify the effects of ambient mixtures of gaseous pollutants on plant productivity under rigidly controlled conditions by controlled fumigation and exclusion experiments in the laboratory, the glasshouse, and the field, with the different experimental systems operated in parallel. Wind, temperature, and moisture regimes typical of those in Europe will have to be employed. The costs of such work will be large, but the expenditure is essential now to understand the effects of the systems themselves as well as the effects of pollutants on plant productivity. Physiologists must tackle the problem of interpreting how rate of growth influences plant response in such systems. There is also a strong case for incorporating studies of more subtle effects (e.g., on pollen germination) and of biochemical mechanisms within plant cells (Bonté, 1982; Wellburn, 1982).

I have not mentioned the differing effects of SO_2 and other pollutants on mixed communities of plants, because most agricultural and forestry systems in Europe can be regarded as monocrop cultures. However, the composition of mixed woodlands and of permanent grasslands is undoubtedly influenced and changed by exposure to pollution (Last, 1982). Such changes offer exciting opportunities for novel research in population dynamics and in the physiological, biochemical, and genetic factors that determine the sensitivity of plants to pollution.

Transfer of pollutants to soils and surface waters. The nature and spatial distribution of precipitation containing pollutants derived from washout, rainout, and dry deposition, together with those in leachates from foliage, deserve the closest attention, because of the suspected variations that occur in the mass balances of ions moving from plant canopies into the soil. Such work must investigate the modifying influences of variation in species and morphology on the total ionic inputs to soils.

The behavior in soils of sulfur, nitrogen, fluorine, and other elements derived from the atmosphere also deserve investigation in order to understand the ways in which they affect normal nutrient-cycling processes. We need, in fact, to know more about the natural evolution of soils, especially in forests, and the roles played by hydrogen ions in the leaching of calcium and aluminium into drainage waters. It is interesting to note also that the exposure of foliage to gaseous and other pollutants can have large, adverse effects on root mass and form. We know little about such effects and their consequences for soil formation and nutrient cycling in various soils.

Freshwater systems. It was once often assumed that drainage waters pass into streams, rivers, and lakes without influence from atmospheric pollution. The acid-rain phenomenon has destroyed that illusion! The scientist is faced now with the very taxing problem of providing the scientific basis for rational decisions that maximize plant productivity in catchments that are exposed to air pollution and yet do not cause substantial changes in the volume and quality of water catchments, or inhibit the optimal use of land for agricultural and forestry purposes. We need to learn now precisely how different soil systems react to different management regimes and air-pollution conditions, and how the latter influence aquatic systems.

Methods and techniques. I have mentioned some of the more important methods and techniques in use in Europe already in this paper. Certainly the diversity of air-pollution problems faced by each country would alone spawn a variety of technical approaches. Uniformity of technique is thus not to be expected, and, from a scientific standpoint, would be deplorable. Nevertheless, calibration is necessary, as I mentioned in discussing field and laboratory fumigation systems. The development by the C.E.C. of contact groups has also fostered collaboration and intercalibration between scientists in member states. These contact groups have tapped the existing systems of informal coordination that exist, for instance, in the U.K. and in Scandinavia. In consequence, there is a ready exchange of data and information between active researchers.

Undoubtedly, the experimental fumigation of plants in the laboratory, glasshouse, and field is a topic of great discussion in Europe. Field fumigation systems in France and the U.K. owe much to U.S. developments. This applies to open-top and certain closed-circuit chamber designs, although the variety in design in the U.K. is quite startling! Solar domes appear to be unique to the U.K., as do wind-tunnel systems. Future designs must seek to recreate (1) a realistic airflow, and (2) plant-growth conditions representative of the external environment.

We will probably not soon be able to estimate accurately the current and future effects of air pollutants on plant productivity in terms of yield loss. Early estimates (O.E.C.D., 1981) were valiant attempts, but suffered badly from inadequate data bases. We need more information on concentrations of ambient pollutants in rural areas under typical conditions of climatology, meteorology, and crop management. We must check the crude estimate of scientists that overall yield losses lie in the range 1 to 5% in most years, allowing for beneficial effects (Cowling, Jones, and Lockyer, 1973; Cowling and Koziol, 1982), but excluding effects on forest systems. Scientists must not lose sight of the ultimate goal: to quantify, with a reasonable degree of accuracy, the magnitude of such losses.

Acknowledgments

I am most grateful for the support of the U.K. Natural Environment Research Council in the preparation of this paper, and for the permissions of Dr. D. Fowler, Dr. N. Cape, and Dr. A. C. Posthumus to reproduce certain figures and tables from their publications.

4

Yield Response Data in Benefit-Cost Analyses of Pollution-Induced Vegetation Damage

Richard M. Adams, Thomas D. Crocker, and Richard W. Katz

Many parts of our society regularly confront the problem of "optimal" ecosystem management. A farmer's choice of crop to grow, a householder's choice of yard plantings, and multiple-use management of the national forests can all be cast as problems of ecosystem management. From such a viewpoint, the ecosystem is seen as a process that produces desired outputs and services, and that may be facilitated by some human inputs (e.g., fertilization) and hindered by others (e.g., atmospheric pollution).

Just as a manufacturing firm requires a description of its production process in order to allocate its resources efficiently, the ecosystem manager must have a description of the process by which the desired outputs are produced, and of what effects any factors that he may control have on these outputs. A model of the internal workings of the ecosystem must be fundamental to any economic version of its optimal management. Questions about the optimal combination of inputs with which to produce one or more types of ecosystem output have therefore traditionally been attractive research topics for agronomists, foresters, and economists.

This paper is one of a series (see Adams and Crocker, 1980 and 1981; Adams, Crocker, and Thanvibulchai, 1982; Crocker, 1982b) in which we consider how plant scientists and economists might better coordinate their approaches to assessing pollution damages. Our general procedure has been to explain and present examples of the economics approach, but emphasizing the role played by plant-science information. We hope that greater understanding of their role in an economics approach will enable plant scientists to help design economics studies that better exploit existing plant-science research and provide guidance for future research as well. To this end, we present here a heuristic explanation of a method for estimating the part that precision in plant-science information plays in predicting the economic consequences of the effects of pollution on vegetation. Katz, Murphy, and Winkler (1982) have applied this method to meteorological information.

In the first section, we discuss differences between the agronomic and the economic perspectives on the problem of designing ecosystem production processes. We then set forth a general economic framework in which natural-science studies of plant damage are a necessary component. In the third section, we provide estimates of supply and demand parameters and of yield responses to ozone air pollution for four major crops in the United States, and estimate the shifts in these parameters that would be caused by changes in ozone concentrations. Point estimates of these shifts allow one to make point estimates of the resulting changes in economic surpluses. In the fourth section, we discuss the differences that the degree of precision in plant-science information makes to these estimates of changes in economic surpluses. We have decided on an expository treatment, avoiding detailed arguments, formulations, and proofs. Nevertheless, we provide enough references and information to allow the curious reader to recreate our empirical results.

The Agronomic and the Economic Perspectives

Although drawing upon known biological principles, the agronomist is a type of engineer who focuses on the design and operation of processes that can produce desired outputs with a minimum expenditure of valuable material inputs. The input prices that determine costs are taken as given. The agronomist thus seeks to solve a constrained optimization problem involving detailed knowledge of both the biological possibilities and the expected costs of material inputs. If all nonmaterial inputs, such as labor, were free, these agronomic solutions would alone, in economic terms, decide the optimal process design. However, nonmaterial inputs are also costly, and both material and nonmaterial costs can change. Agronomic results thus identify the technically efficient set of materials combinations, within the larger set of ultimate biological and technological limits, that will produce a given amount and type of desired output.

Economists who study agriculture also focus on cost minimization, but they assume that the agronomist has already solved his own constrained optimization problem. The economic theory of cost and production describes the effects of variable input prices on cost-minimizing combinations of material and nonmaterial inputs. From the set of cost-minimizing combinations, the theory provides rules for identifying that combination having the lowest material and nonmaterial costs, and it specifies how this "minimum of the minima" will be altered as relative input prices change. It identifies economically efficient input combinations. In this fundamental sense, then, economics portrays the results of agronomic reoptimization in terms of the effects of changes in biological and technological possibilities, and of changes in the relative prices of inputs, on the cost-minimizing combination of inputs. In dealing with pollution, it describes how *differences* in pollution effects alter the costs of the alternative physical and biological ways an economic agent has to meet his or her objectives.

The agronomist viewed as engineer considers all input combinations consistent with known biological and physical laws, since his or her objective is to develop and ultimately implement a detailed plan for a particular production process. Subsequent efforts to improve on this plan will center on any input or subset of inputs that might allow substantial cost reduction. In contrast, most basic scientific studies of plant damages from air pollution have concentrated on how pollution affects a single input combination in some production process. Many have also studied such effects independently of *any* cost-minimizing production process. Hence, at best, the focus has been on a small set of the feasible input combinations. These highly detailed studies typically assume that other input types and quantities remain constant. They therefore frequently fail to provide the design information the economist needs in order to estimate the changes in cost-minimizing input combinations that pollution induces.

A piece-by-piece approach to the study of how pollution affects vegetation makes truly awesome the task of covering and synthesizing the feasible or even the technically efficient input combinations. In agriculture alone, thousands of different types of inputs exist. Each input type is, in turn, embodied in one or more production processes or crops, which may appear in a variety of cultivars and which can be put to many distinct uses. Moreover, there may be environmental cofactors, such as moisture and temperature, that act in concert with pollution to aggravate or to soften its impact. Basic-science studies of how pollution affects plants have neither received nor provided much guidance about which of these embodiments, varieties, or uses have any economic significance.

Upon encountering such overwhelming detail, the economist's traditional response is to aggregate these innumerable forms into a much smaller number of composites. Often the degree of aggregation (e.g., capital, labor, land) is no less bewildering than the detail with which the plant scientist struggles. In effect, both the plant scientist and the economist try to make their tasks manageable by imposing separability assumptions on the processes being studied. The plant scientist goes to extremes by setting all inputs but one at a biologically optimal level. Because this practice removes confounding sources of stress, the plant scientist can then view the effect of pollution independently of its surroundings. Similarly, the most intemperate of economists who practice aggregation treat choices within a particular input (e.g., land) or output (e.g., corn) category as separable from the quantities of inputs or crops outside the category. The plant scientist often fails to ask whether all details are economically important, and the economist rarely considers whether his or her aggregation of data masks technically feasible or economically worthy ways to deal with the effects of pollution on ecosystems. One key problem, then, is to reconcile the plant-science and the economics approaches by developing criteria for deciding how much plant-science detail must be retained in any particular circumstance. When an objective of plant-science studies of pollution is to provide information useful for estimating economic consequences, the basic question is whether more or less plant-science detail will alter the economic estimates in a nontrivial way. Both the plant scientist and the economist must refine their knowledge in such a way that the relations between pollution and damages to vegetation and ecosystem processes are defined in dimensions that correspond to those in which real-world decision agents choose to define them. Because the agronomist and the forester as engineer are responsive to economic phenomena, their insights can be highly instructive in attempts to capture these dimensions.

However, no degree of care in depicting the biological and technological detail and the relative input prices of production processes will capture all the economic consequences of pollution-induced plant damages. The effects on the consumer who ultimately uses these plant materials must also be weighed. Like the producer, the consumer may also substitute other materials, alter his or her time allocations, or simply do nothing, either because there is no economically worthy alternative or because his or her use of the plant or ecosystem is insensitive to pollution damages.

The Economic Assessment Problem

The objective of efforts to assess the economic consequences of pollution-induced plant damages is to estimate differences in the sums of consumer surpluses and producer quasi-rents caused by two or more policy-relevant pollution levels. Consumer surplus is the difference between the maximum a representative consumer would be willing to commit himself to pay for a given quantity of a commodity and what the consumer in fact has to pay. Similarly, a producer quasi-rent is the difference between what the owner of the inputs to a particular activity receives for supplying a particular output quantity and the minimum he must receive in order to be willing to commit to that supply. The sum of consumer surplus and producer quasi-rent is thus a measure of the net benefits associated with the consumption and the production of a commodity. The observable unit prices of other commodities that provide equal satisfaction set an upper bound to the consumer's maximum willingness to pay; the observable earnings that the inputs could obtain in other activities set a lower bound on the minimum reward the producer must receive. Maximum willingness to pay represents demand; the minimum necessary reward defines supply.

Fig. 4.1 traces the set of factors that must be accounted for if a complete assessment of the economic consequences of pollution-induced plant damages is to be performed. Assume that a farmer is trying to decide what kind and how much of a crop to grow. One or more of these prospective plantings is susceptible to air-pollution damages. For a given level of expected air pollution exposures, the portion of the figure lying to the left of the leftmost dotted line represents the problem of finding the least costly way of producing a given crop. It answers this question: Given the alternative ways I have to produce a certain amount of this crop, which ways allow me to employ the minimum combinations of those costly

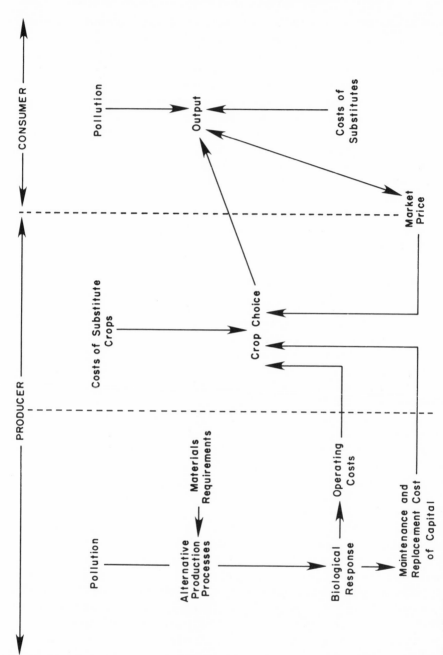

Fig. 4.1. The sources of economic consequences.

inputs biologically and physically necessary for production? Variations in pollution levels will alter these minimum necessary combinations. Identification of these minimum necessary combinations for each pollution level enables the farmer to maximize the output of the crop obtained from a given expenditure on inputs. However, solution of this problem cannot decide which of the alternative crops to grow: it can only show the least costly plan for growing a particular crop, and the manner in which this plan will vary as air pollution varies.

The portion of Fig. 4.1 lying between the dotted lines depicts the problem of choosing among the alternative crops, given prior identification of the least costly means of producing each crop. In this portion, the producer is allowed to substitute crops that are more or less prone to yield and/or quality reductions from expected air-pollution levels. The problem of this middle portion can be contrasted with the problem of the first portion, where the producer could only manipulate the process used to produce the crop. Basically, the producer has a set of alternative crops dictated by biological and physical science knowledge of the laws of nature, as well as by institutional constraints (the laws of man) limiting allowable processes for producing these crops. For each known and allowable process for each crop, the producer estimates the least-cost process, and for given output prices and pollution levels, then selects the combination of crops expected to generate the maximum quasi-rent.

In the rightmost portion of Fig. 4.1 resides the consumer. There are two routes whereby pollution-induced vegetation damages can influence consumer behavior, thus altering the surplus that the consumer obtains from the output in question. First, an increase in vegetation damage (leading to a decrease in yield) will increase costs and therefore increase the minimum price the producer must receive in order to be willing to commit himself to supply any given quantity of the output (and vice versa for a decrease in damage). In addition, altered levels of air pollution may affect the attributes of the output, thus changing the consumer's willingness to pay and the consumer surplus acquired from any quantity of the output. The change in cost implies a shift in the producer's supply function, whereas the change in willingness to pay is represented by a shift in the consumer's demand function. Both result in a change in the market price of the output.

Given that producers and consumers have already adapted in order to minimize their prospective losses, or maximize their prospective gains, Fig. 4.2 depicts one example of the changes an air-pollution increase can have on consumer surplus and producer quasi-rent. The air-pollution increase reduces the desirable properties of the output, making smaller the consumer's willingness to pay and causing his or her demand function to shift from D_0 to D_1. It simultaneously increases the least cost of producing any particular output quantity, thereby causing an upward shift in the supply function from S_0 to S_1. Market price for the output drops from P_0 to P_1, partly because of the greater relative magnitude of the shift in the demand function, and partly because of its lesser relative slope. Consumer

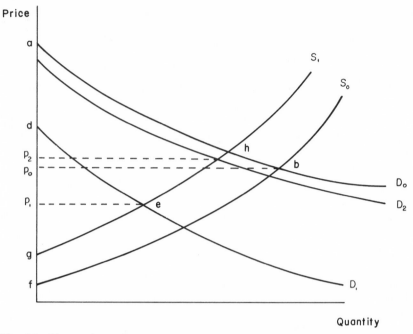

Fig. 4.2. Changes in producer's quasi-rent and consumer's surplus.

surplus was the triangle aP_0b; it is now the triangle dP_1e. Producer quasi-rent was the triangle fP_0b; it is now the triangle gP_1e. Total economic surplus from the production and use of the output in question is thus reduced by the area enclosed by $fghb$ plus the area enclosed by $adeh$.

Of course, other relative shifts in demand and supply relations, reflecting different effects in the production and consumption sectors, will yield other results. For example, if the demand curve shifts to D_2 rather than D_1, market price will rise to P_2. If the supply shift from S_0 to S_1 is small enough, producers could thus actually see an increase in their quasi-rents. Qualitative results are unchanged, however: alterations in producer quasi-rents and in consumer surplus result from the pollution-caused changes in the two sectors. These two examples illustrate the issues of concern here: consumers and producers can bear very different economic gains or losses, depending on the relative shifts of the demand and supply functions. Moreover, the distribution of these economic consequences can differ drastically, depending on how the slope of the demand function is related to the slope of the supply function.

The preceding observations describe, in the broadest of brush strokes, the problem of calculating what social benefits may follow from controlling pollution-related vegetation damages. In the following sections, we illustrate how plant-science information aids in calculating these costs and benefits.

Estimating Economic Surpluses: An Illustration

The combined production of corn, soybeans, cotton, and wheat made up 64 percent of 1980 cropped acreage, 70 percent of 1980 total crop value, and 65 percent of the value of 1980 agricultural exports for the United States (U.S. Department of Agriculture, 1981). The obvious economic importance of these crops has probably motivated the attention that the Environmental Protection Agency's National Crop Loss Assessment Network (NCLAN) research program on the yield-response effects of air pollution has devoted to them. On the basis of the results to date of the NCLAN and other yield-response research, Heck *et al.* (1982) conclude that ozone has the greatest effect on crop production of the various air pollutants known to harm vegetation. Furthermore, ozone in potentially damaging concentrations is thought to be the most widespread of these pollutants.

Estimates of specific parameters are required to implement the general economic framework we described earlier. In particular, values for the demand and supply parameters must be available. Moreover, the parameters of the functions that relate crop yields to ozone (the yield-response functions) are necessary if the shifts in demand and supply functions, and the changes in consumer surpluses and producer quasi-rents that they lead to, are to be provided. Clearly, assumed and actual (current ambient) ozone concentrations by regions are needed. The demand and supply relations are estimated by standard econometric procedures. The yield-response information is taken from NCLAN Research Management Committee (1980, 1981); the E.P.A.'s SAROAD data base is the source of the ambient ozone information. Each of these required bits of information is discussed in more detail in the following.

Supply and demand relations. For corn, wheat, and soybeans, empirical versions of the following equilibrium supply and demand relations were estimated:

$$P_C = f(Q_D, R, T, WP, P_O, N_{-1}), \tag{1}$$

$$Q_S = f(B_C, S, P_j, T, QL), \tag{2}$$

$$Q_S = Q_D, \tag{3}$$

where P_C is crop price; Q_D and Q_S are, respectively, supply and consumption levels in equilibrium; R is per-capita disposal income; T is time; WP is a weighted price of other feed grains; P_O is oilseed-meal price; N_{-1} is cattle on feed in the previous period; S is stocks; P_j is an index of production costs; and QL is lagged production. The posited structure for cotton was similar to (1), (2), and (3), except that time, weighted price, and cattle numbers were excluded. Prices and income are always expressed in actual dollars. For estimation, each expression was assumed to be linear in the original variables.

The system (1)–(3) was estimated for corn, soybeans, and wheat by the Zellner (1962) Seemingly Unrelated Regression (SUR) procedure. Two-stage least squares (2SLS), with the Cochrane-Orcutt (1949) iterative procedure for serial

TABLE 4.1
Supply and Demand Parameters for Corn, Cotton, Soybeans, and Wheat

Crops	Supply coefficients		Demand coefficients	
	Own price (actual dollars)	Elasticity [a]	Own price (actual dollars)	Price flexibility [a]
Corn [b]	0.0008 ($t = 3.211$)	0.31	−0.000089 ($t = 1.850$)	0.24
Cotton [c]	0.9980 ($t = 1.821$)	0.47	−0.1306 ($t = 1.602$)	0.50
Soybeans [b]	0.0037 ($t = 4.662$)	0.48	−0.0015 ($t = 2.296$)	0.38
Wheat [b]	0.0044 ($t = 3.743$)	0.59	−0.00081 ($t = 1.810$)	0.52

[a] Evaluated at mean quantity and price for 1960–80 period.
[b] Estimates from Seemingly Unrelated Regression (SUR) supply and demand blocks for four feed grains (corn, barley, wheat, and grain sorghum) and soybeans.
[c] Quantities in 500-pound bales. Estimates from two-stage least squares and the Cochrane-Orcutt iterative procedure applied to expressions which were linear in the original variables.

correlation, was used for the cotton system. The supply and demand parameters for corn, wheat, and soybeans were taken from SUR supply and demand blocks for major livestock feeds and feed grains. Simultaneities involving cotton stocks required 2SLS estimation for that crop. Data cover the period 1960 to 1980, and are taken from the U.S.D.A. *Agricultural Statistics* annuals.

The estimated supply and demand parameters are presented in Table 4.1. They are consistent with theoretical expectations. For the supply estimates, the own-price coefficients show the average increase in the quantity of the crop that a one-dollar increase in crop price causes producers to be willing to supply. The elasticity coefficient is simply ($\partial \log Q_S / \partial \log P_C$), the percentage change in quantity supplied relative to the percentage change in own price. Similarly, the demand coefficients for own price show the reduction of willingness to pay in dollars caused by a one-unit increase in the availability of a crop. The price-flexibility coefficient is ($\partial \log P_D / \partial \log Q$), the percentage change in own demand price relative to the percentage change in crop availability.

Yield-response information. As noted earlier, changes in air-pollution concentrations will cause the supply function of Fig. 4.1 to shift. In this study, we presume that ozone causes no changes for the consumer in the desirable properties of the crop; that is, the demand function of Fig. 4.1 does not shift. The magnitudes of the yield-response parameters that drive our estimates of shifts in supply functions are reported in Table 4.2. (The variables used in Table 4.2 are derived from Eq. 5, discussed at the beginning of the next section.) When yield-response information for more than one cultivar was available in the NCLAN data sets, regional differences in responses were accounted for by testing for the homogeneity of slopes across cultivars. Where regionalization was statistically justified, regional yield response was weighted by the same region's market share when summing to arrive at the national yield response. Note, however, that we shall not address the potentially large uncertainties inherent in the extrapolation to regional and national levels of data generated under the specific edaphic and climatic conditions of a few NCLAN sites.

Table 4.3 reports, in the form of yield-ratio statistics, the changes in yields predicted for alterations in ambient ozone under the current SNAAQS standard of 0.12 ppm (4.99 μmol m^{-3}) not to be exceeded more than once a year. These relative yields thus represent estimates of the biological consequences of alternative regulatory options. They trigger the economic consequences. In terms of biological yield, they match or exceed the losses that Boyer (1982) attributes to insects, diseases, and weeds.

The pollution exposure (dose) in each of the yield-response expressions was

TABLE 4.2
Estimated Values of Yield-Response Parameters for Corn, Cotton, Soybeans, and Wheat

Crop:[a] j	Sample size: n	Intercept: $\hat{\alpha}$	Slope: $\hat{\beta}$	Standard error: $s(\hat{\beta})$
Corn	20	174.0	−685.0	128.0
Cotton	12	1098.4	−3708.0	228.52
Soybeans	30	21.4	−93.1	7.6
Wheat	16	5.0	−12.0	2.6

NOTE: Yield is in grams per harvested plant. Ozone is in parts per million by volume.
[a] Estimated from data reported in NCLAN Research Management Committee (1981, 1982). Estimates were obtained by ordinary least squares for an expression linear in the original variables.

TABLE 4.3
Yield-Ratio Statistics for Corn, Cotton, Soybeans, and Wheat

Crop: j	Action (ozone standard), in ppm by volume:[a] a_i	Estimated yield ratio:[b] $\hat{Y}_{ij}$	Standard error of ratio: $s(\hat{Y}_{ij})$
Corn	0.10	1.025	0.00963
	0.08	1.053	0.01926
	0.14	0.948	0.00963
Cotton	0.10	1.042	0.00329
	0.08	1.085	0.00659
	0.14	0.958	0.00329
Soybeans	0.10	1.059	0.00481
	0.08	1.118	0.00961
	0.14	0.941	0.00481
Wheat	0.10	1.028	0.00329
	0.08	1.056	0.00659
	0.14	0.972	0.00329

NOTE: Units are identical with those in Table 4.2.
[a] 1 ppm O_3 = 41.6 μmol m^{-3}.
[b] The yield ratio is the estimated yield for the action i relative to the current standard of 0.12 ppm (4.99 μmol m^{-3}). These ratios were calculated from the information in Table 4.2.

measured as a seven-hour seasonal mean concentration of ozone. The seven-hour period is from 9:00 A.M. to 4:00 P.M., the period in which stomatal activity and hence plant sensitivity to pollution is greatest. In order to transform the mean seven-hour dose to the same basis as the SNAAQS, ambient ozone is assumed to be log-normally distributed. Thus, for example, a seasonal seven-hour concentration of .07 ppm (2.91 μmol m^{-3}) is treated as being a SNAAQS concentration of .14 ppm (5.82 μmol m^{-3}).

By adjusting the supply condition for each crop, one can give economic meaning to the biological consequences of the alternative hypothetical ambient ozone standards of 0.08, 0.10, and 0.14 ppm (3.33, 4.16, and 5.82 μmol m^{-3}). As noted in the discussion of Fig. 4.1, the supply shifts that the biological consequences induce are registered in movements of the economic surplus measures. Thus, by comparing changes in economic surpluses across successful attainments of the alternative ambient standards, we are able to assess the differences in societal benefit across standards, including the current standard of 0.12 ppm (4.99 μmol m^{-3}).

Calculation of expected economic surpluses. Integration of the areas under the crop supply and demand functions at relevant price and quantity equilibria provides measures of the economic surpluses associated with each ambient standard. Given the assumed linear forms of the market relations, the integration can be accomplished by using the expression

$$w(Q) = \tfrac{1}{2}(\hat{b}_j + \hat{e}_j)Q^2 \qquad (4)$$

where Q is the quantity of the crop in question, $w(Q)$ is the sum of consumer surplus and producer quasi-rent, $\hat{b}_j$ is the absolute value of the own-price demand coefficient for the *j*th crop in Table 4.1, and $\hat{e}_j$ is the own-price supply coefficient. The estimated surpluses are reported in Table 4.4, where the consumption level for each crop is taken to be the 1978–80 arithmetic mean. The table states that the total 1980 economic surplus for the national production and consumption of the four crops is $44 billion at the ambient ozone concentrations associated with the standard of 0.12 ppm (4.99 μmol m^{-3}); it does *not* say that this standard is responsible for the entire $44 billion. The last column, labeled "Change in expected surplus," is the feature of interest. Its entries represent the change in economic surplus predicted to result from altering the concentrations associated with the current ambient standard of 0.12 ppm (4.99 μmol m^{-3}).

Our economic model in expressions (1)–(3) does not register the different cropping patterns that growers might select as ambient ozone levels change. Some of the ways in which the grower might maximize gains from an ozone decrease and minimize losses from an ozone increase have been omitted. This causes the estimated gains of economic surplus in Table 4.4 from an ozone decrease to be understated, and the estimated losses from an ozone increase to be overstated. We do not know the extent of under- or overstatement. Its repair requires a more complete model of grower-decision processes.

TABLE 4.4
Sum of Economic Surpluses for Corn, Cotton, Soybeans, and
Wheat with Alternative Secondary Ozone Standards
(Billions of 1980 dollars)

Ambient standard (ppm)[a]	Expected surplus	Change in expected surplus
0.12	43.726	—
0.10	46.125	2.399
0.08	49.271	5.545
0.14	39.918	−3.808

NOTE: Calculated from information in Tables 4.1 and 4.3.
[a] 1 ppm O_3 = 41.6 μmol m^{-3}.

Economic Surplus and the Precision of Yield-Response Information

The values of the yield-response parameters in Table 4.2 were estimated by using the following "simple normal linear" regression model (Zellner, 1971):

$$Y_k = \alpha + \beta X_k + \varepsilon_k, \tag{5}$$

where $k = 1, 2, \ldots, n$, with the error term, ε_k, being independently and normally distributed with zero mean and constant variance. In (5), X_k denotes the pollutant quantity applied to the kth experimental plot, and Y_k is the observed crop yield for that plot. The expression is one representation of the randomized NCLAN data-generating process. It allows one to test the null hypothesis that ozone does not affect crop yields (i.e., $\beta = 0$). In classical statistics, the object of the inferential exercise, β, is considered to be an unknown constant. Application of standard hypothesis tests allows one to reject, or to fail to reject, this hypothesis, i.e., to treat it as "true" or "untrue" at various levels of confidence in the data.

However, treating β as if it were equal to $\hat{\beta}$ has caused a good deal of potentially useful information to be laid aside. Though the NCLAN data are being allowed to speak for themselves, a deaf ear has been turned to much of what they are trying to say. In particular, the stochastic features of the NCLAN experiments are of interest. Unlike classical techniques, Bayesian techniques (Bayes, 1764) of statistical inference enable one to exploit the aforementioned uncertainty. (Zellner, 1971, provides a formal treatment and defense of Bayesian techniques; Winkler, 1972, is an excellent introductory presentation.) Treating β as a random variable rather than as a constant allows the calculation of the entire probability distribution of the economic surpluses, not just the expected (mean) surpluses. Rather than assuming the estimated yield-response parameters to be the "true" ones, we acknowledge statistical uncertainty (imprecision) by explicitly including it in the analysis. The inclusion provides insight about the reductions in the variability of economic surplus that additional observations on the yield responses of corn, soybeans, wheat, and cotton can provide. These variability reductions have policy worth only if they might alter a decision about ambient ozone concentrations.

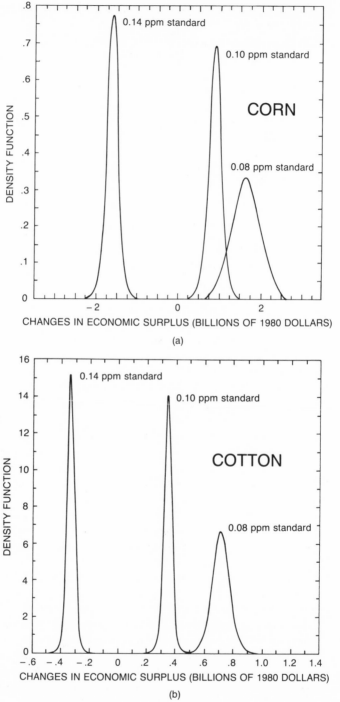

Fig. 4.3. Distribution of changes in economic surplus, corn, cotton, soybeans, and wheat.

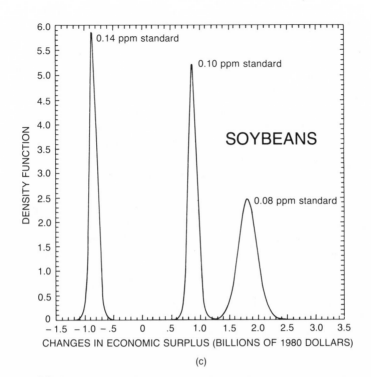

(c)

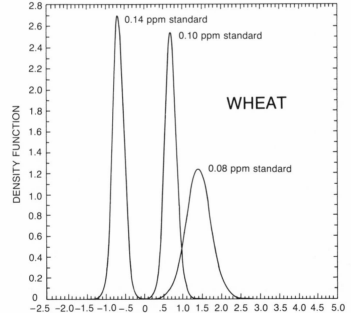

(d)

Fig. 4.3 displays the uncertainties in economic surpluses caused by imprecisions in the estimated parameters of the yield responses of Tables 4.2 and 4.3. For each of the four crops, the figure shows the density functions of the differences in economic surpluses between the current ambient standard of 0.12 ppm (4.99 μmol m^{-3}) and alternative standards (reading from left to right in the figures) of 0.14, 0.10, and 0.08 ppm (5.82, 4.16, and 3.33 μmol m^{-3}). The scales on the vertical axes are probabilities per unit of output. Two features of these functions are noteworthy, particularly when one remembers (recall Table 4.2) that the linear yield-response estimates on which these economic-surplus distributions are founded involve from 10 to 28 degrees of freedom.

First, the mass of the surpluses at and near the means is very great for each crop and ambient-ozone combination. Only the distributions for the 0.08 ppm (3.33 μmol m^{-3}) (on the far right for each crop) exhibit much variability. Under the assumptions of this analysis, additional yield-response observations for the four crops would increase the mass at the mean of each distribution, but this increase would probably contribute very little to any standard-setting policy decisions based on these distributions. This judgment is reinforced by the fact that except for the 0.10 and 0.08 ppm (4.16 and 3.33 μmol m^{-3}) surplus distributions for corn and wheat, there is no overlap between distributions; so the probability

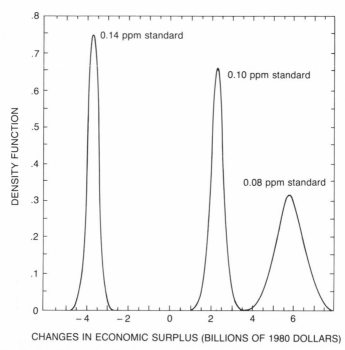

Fig. 4.4. Overall distribution of changes in economic surplus, corn, cotton, soybeans, and wheat.

of taking a "wrong action" is quite small, given that the surpluses are to be used to discriminate among the economic implications of the named alternative ambient-ozone standards. Even the two overlaps are each about one standard deviation from the mean of at least one of the overlapping distributions. If, as is in fact done in Fig. 4.4, one aggregates the surplus distributions for each ambient standard across the four crops, the force of any policy concerns about these two instances of overlapping is tempered, simply because the extent of the overlapping becomes nearly nonexistent.

Conclusions

We have sketched and empirically illustrated an analytical framework within which the economic worth of yield-response information may be reckoned. Worth depends on reductions in the variability of economic surplus for which uncertainty about biological yield-response parameters can be assigned responsibility. The framework has been applied to four major field crops that have dissimilar growing requirements (Heck *et al.*, 1982). Moreover, as Table 4.1 illustrates, the market relations for the four crops are diverse. Finally, the framework was used to evaluate discrete ambient-ozone standards that differed by between 17 and 33% from the current standard of 0.12 ppm (4.99 μmol m^{-3}). Relatively small variations in ozone levels should, in principle, make precision more important in yield-response estimates. Better precision improves one's ability to discriminate among the economic consequences of the alternative standards. We have found for all four crops, ignoring the grower alterations of cropping patterns, that about 30 or fewer linear yield-response observations similar to those found in Heck *et al.* (1982) are adequate to discriminate among the differences in economic surpluses that are generated by the ambient-ozone standards of 0.14, 0.12, 0.10, and 0.08 ppm (5.82, 4.99, 4.16, and 3.33 μmol m^{-3}) used in this analysis.

In evaluating our conclusions, the reader must remember that our treatment employs linear yield-response functions like those in Heck *et al.* (1982). The issue of model uncertainty has therefore been disregarded. When one is dealing with small perturbations in yield, linear forms can serve as reasonable and highly tractable approximations to actual nonlinear responses. However, some of the yield changes presented in Table 4.3 approach 10 percent or more of the base yields. One might doubt whether changes of this magnitude are properly viewed as "small." Nonlinear forms, such as the quadratic, could provide better statistical fits to the yield-response data. When censoring is present, as it probably is when one is dealing with perennial crops such as citrus and alfalfa, the various time-to-failure models set forth in Kalbfleish and Prentice (1980) are deserving of attention. Our framework is applicable to many nonlinear yield-response functions. Nonetheless, if a given precision is to be attained, most nonlinear models use more degrees of freedom than the linear form.

Finally, there are no grounds at this time to extrapolate our results in Figs. 4.3 and 4.4 on the worth of more yield-response information to other response-functional forms, crops, or ecosystems. In order to judge when such an extrapolation would be proper, one would need a formal analysis of the properties of yield response and of demand and supply relations that are influential for economic surplus measures. This has not yet been undertaken for the effects of pollution on vegetation. For classes of vegetation effects involving many decades, depletion effects, and episodic effects, the supply-and-demand portion of the judgment is unlikely to be available any time soon. In these circumstances, the worth of more precise yield-response information will be greater than in our illustration, because it will be the *only* technical information that the policymaker will have about the potential societal effects of changes in concentrations of pollutants.

Acknowledgments

This research was partially funded by the U.S. Environmental Protection Agency. Scott Atkinson and William Winner have provided helpful comments. Only the authors are responsible for any remaining errors or ambiguities.

Part 2
SO$_2$ Effects on Plant Metabolism

5

SO$_2$ Uptake and Transport

S. G. Garsed

The effects of SO$_2$ on plant productivity occur on time-scales of months or years. However, most of the experimental work on the uptake and transport of SO$_2$ has been of short duration. In some areas of study one can make acceptable estimates of long-term effects from short experiments, providing that a reasonable range of environmental conditions are considered. For instance, a series of short-term measurements of SO$_2$ fluxes to standing crops will allow approximate values of the annual deposition rate to be calculated, providing that the measurements cover a whole year. In other spheres, for instance, studies of sulfur biochemistry, there are almost no long-term data; so it remains difficult to predict long-term effects.

Relationship Between the Uptake of SO$_2$ and Plant Injury

Plant damage results from the accumulation of sulfur dioxide or its derivatives during some period of time. However, for a constant product of concentration $\times$ time, acute injury increases if the concentration component is increased, and accumulation increases if the time is extended (Guderian, 1970). Thus before the effects of SO$_2$ concentration and time of exposure on plant productivity can be compared, we must look briefly at the effects of these two factors on flux and on the type of injury that may occur.

Until recently most workers were content to relate the effects of SO$_2$ on plants solely to the measured gas concentrations in the air, but as Unsworth, Biscoe, and Black (1976) point out, the concentration of SO$_2$ supplied is only one of several factors that determine the actual flux to the leaf. These authors provide a concise analysis of the use of resistance analogs in the estimation of pollutant flux. Briefly, the aerodynamic resistance (r_a) and the boundary-layer resistance (r_b) determine how much SO$_2$ arrives at the leaf, whereas the stomatal (r_s) and cuticular (r_c) resistances control its entry into the plant, and "mesophyll resistance" (r_m) is the residual resistance to absorption inside the plant. (The subject is discussed more comprehensively by Fowler in this book.)

Taylor, McLaughlin, and Shriner (1982) use the term "effective pollutant dose" to describe the quantity of SO$_2$ available to affect the physiology of the plant. In the short term, a measure of the total internal flux probably does give a reasonable estimate of the effective pollutant dose. However, the "dose" experienced by the plant in the long term will be different, because the SO_2 and its derivatives can accumulate and become involved in the sulfur metabolism of the plant. Thus not only is the chemical nature of the active species likely to be different, but also the total sulfur budget of the plant must be considered. The differences in the chemical nature of the sulfur species in the short and long term also mean that the physiological and injury responses will differ.

Several workers have shown that there is no relationship between sensitivity to acute injury and the likely sensitivity of the growth response under long-term exposure to low concentrations of SO_2 (Horsman et al., 1979; Ayazloo and Bell, 1981; Garsed and Rutter, 1982a). Their work also indicates that there must be more than one mechanism of injury by SO_2. Acute injury can be explained by the plant being unable to cope with a high flux of SO_2, with consequent damage to cellular structures that leads to the characteristic symptoms. For the longer-term effects of SO_2, I propose that "chronic injury" be considered in terms of two different processes. The first process involves direct injury by SO_2, but at concentrations too low to cause acute injury. The effects of such invisible damage may accumulate enough after a long period to reduce growth. The other process is caused not by SO_2 itself, but by the accumulation of sulfate, which can reduce buffering capacity and has been implicated in the premature senescence of leaves in polluted plants (Grill and Härtel, 1972).

We cannot yet distinguish the two types of injury. They may well occur together and even interact. However, the theoretical distinction shows that the long-term effects of SO_2 on crops involve not only the immediate metabolic fate of SO_2, but also its subsequent distribution in the plant and mechanisms of loss. Since a high capacity to prevent the accumulation of sulfate may be an important mechanism in the avoidance of long-term damage (just as high stomatal resistance may lead to the avoidance of acute injury), mechanisms by which sulfur may be lost from plants are also considered in this paper.

Techniques for Measuring SO_2 Uptake by Plants

Measurement of deposition to vegetation by meteorological techniques. In this approach two values are normally calculated: the deposition velocity (Vg), measured in cm s^{-1}; and its reciprocal, deposition resistance (r), measured in s cm^{-1}. Two techniques are used: the profile (gradient) method, which is suitable for short crops, and the eddy-correlation method, which gives better resolution for forests, where the vertical concentration gradients are small (Galbally, Garland, and Wilson, 1979). The methods are discussed by Hosker and Lindberg (1982). The practical applications of these techniques include the measurement of the influence of environmental factors on flux, assessment of the filtering effects of different vegetation types, and calculations of sulfur inputs to ecosystems.

Measurement of sulfur contents of vegetation. In theory, analysis of the sulfur content of vegetation should give an indication of the sulfur load experienced by the plant. This method has been used frequently under laboratory conditions to measure the short-term uptake of SO$_2$. In these circumstances, it probably gives a fairly reasonable estimate of the net flux to the plant. Besides carrying out analyses of total sulfur, many authors have separated the sulfur into various fractions, in order to follow the changes undergone by the absorbed SO$_2$ (e.g., Thomas and Hendricks, 1944; Jäger and Steubing, 1970).

Interpretation of the results from longer-term laboratory studies is more difficult. As the plants grow, the accumulated sulfur is translocated and diluted, and over a longer period leaves senesce and are shed. Garsed, Farrar, and Rutter (1979) have found a relationship between leaf sulfur content and time of abscission. There are also suggestions that exposure to SO$_2$ can reduce the uptake of sulfate by the roots (e.g., Faller, 1972), but the evidence is still contradictory. The subject is discussed in more detail by Cowling and Koziol (1982). Finally, sulfur may be lost from the plant by leaching, emission of H$_2$S, or leakage from the roots (discussed later). Despite these reservations, sulfur analysis can provide useful data to complement laboratory growth experiments, particularly if a time course is followed.

In view of the problems in interpreting data on plant sulfur content in laboratory experiments, where the SO$_2$ concentrations and environmental conditions are known, it is hardly surprising that interpretation of field data is normally very limited. The most common application is the definition of "zones of accumulation" around point sources. Modifications to the basic technique have been used by Malcolm and Garforth (1977), who determined the sulfur/nitrogen ratio of conifers from nonpolluted sites, and then compared the mean value obtained with the ratios of trees growing in the vicinity of a brickworks. Sulfur accumulation was then expressed as the quantity of sulfur in excess of that expected by reference to the nitrogen content. Another technique in which a ratio method is used is the comparison of the isotopic abundance ratio of ^{34}S to ^{32}S, or δ ^{34}S (Krouse, 1977). Gaseous sulfur compounds emitted from the stacks of a natural-gas processing plant in Alberta have a higher ratio of ^{34}S to ^{32}S than soils or natural waters. Comparisons of the δ ^{34}S values of vegetation near the gas plant have shown that lichens obtain their sulfur primarily from the atmosphere, but *Picea glauca* obtains its sulfur from both soil and atmosphere (Case and Krouse, 1980).

Schwela (1979) exposed standardized cultures of *Lolium multiflorum* for two-week intervals at various sites in the Rhein-Ruhr area between 1973 and 1975. He then calculated deposition velocities for several pollutants from the foliar concentrations of various elements. These data agreed reasonably well with other deposition data. Such a technique allows sequential estimates of deposition velocity to be made over the whole year, and several sites to be compared simultaneously.

Depletion experiments in chambers. This technique involves the estimation of the flux of SO_2 to a plant by comparing the concentrations of SO_2 at the inlet and outlet of a chamber containing a known amount of plant material. Although simple in concept, the apparatus requires careful design as well as close attention to details of environmental and physiological factors affecting flux. In particular, it is important to ensure that there are no unquantifiable losses of SO_2 by adsorption to fittings. Various designs are described by Rogers *et al.* (1977), Black and Unsworth (1979c), Winner and Mooney (1980a), and Taylor and Tingey (1981). An analysis of some of the factors affecting the mass balance of such systems is given by Unsworth (1982). The best use is made of the technique when many variables are measured together, for instance, SO_2, CO_2 and water-vapor fluxes, to allow comparisons of SO_2 flux with those of other gases (e.g., Black and Unsworth, 1979b; McLaughlin and Taylor, 1981). Most such experiments have been of short duration, and performed on material with little or no previous exposure to SO_2. There is a need for comparisons with plant material that has previously been exposed to SO_2 for a substantial period.

Studies with $^{35}SO_2$. There are some instances of the use of $^{35}SO_2$ in the open, either for the measurement of deposition velocity (e.g., Garland, Clough, and Fowler, 1973) or to label crops for studies of sulfur movement (e.g., Raybould, Unsworth, and Gregory, 1977). Most of the studies with $^{35}SO_2$ have been done within some form of enclosure, but unfortunately have been performed with little regard for the environmental variables influencing flux (see previous section). For instance, not all authors include fans in their enclosures to circulate the air, and authors have often been concerned not so much with flux *per se*, but with introducing sufficient $^{35}SO_2$ into the plant for subsequent chemical analysis. $^{35}SO_2$ is a difficult gas to handle because of its reactivity and most workers are reluctant to contaminate expensive pieces of equipment, so data on such parameters as stomatal resistance are rarely supplied. Garland and Branson (1977) overcame this problem by enclosing pine shoots first in a clip-on cuvette connected to a porometer, and, after taking a reading, quickly replacing the cuvette by another, connected to a source of $^{35}SO_2$, and fumigating the shoot for a brief period.

Many workers have used closed systems for fumigation, particularly when the $^{35}SO_2$ has been obtained in a break-seal ampule. The disadvantage of releasing $^{35}SO_2$ from ampules is that a high concentration occurs on release of the gas, although subsequent depletion is normally rapid (Garsed and Read, 1977a). One of the factors contributing to the high depletion rate in closed chambers is high humidity, which causes condensation in the chamber and can affect deposition to the plant. There is a pronounced "speckling" of leaves of *Vicia faba* exposed to $^{35}SO_2$ at 95% R.H., whereas leaves exposed at 55% R.H. showed more even labeling (Fig. 5.1). Similar observations have been made with *Medicago sativa* and *Lolium perenne*. The most likely explanation for the "speckling" is that discrete water droplets form on the lamina under conditions of high humidity and

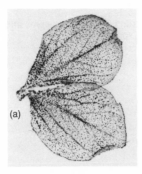

Fig. 5.1. Autoradiographs showing the influence of relative humidity on the deposition of ^{35}SO$_2$ on leaves of *Vicia faba*: (a) plant exposed to ^{35}SO$_2$ at 95% R.H.; (b) plant exposed to ^{35}SO$_2$ at 55% R.H. Source: J. N. B. Bell (unpublished).

provide a preferential sink for the ^{35}SO$_2$. If so, measures of flux under such conditions need to be interpreted very carefully.

Open-circuit systems allow better control of humidity and CO$_2$ concentration, and a lower and more constant concentration of ^{35}SO$_2$ can be used. The ^{35}SO$_2$ may be generated by addition of Na$_2$^{35}SO$_3$ to acid, or it may be released from a cylinder, the latter method giving the most accurate control of ^{35}SO$_2$ concentration.

Factors Affecting Flux to the Plant

Pathways to the plant. Under conditions favorable to rapid CO$_2$ and water-vapor exchange, the pathway of SO$_2$ to unwetted foliage is largely through open stomata, and the cuticle forms a relatively minor sink for SO$_2$. In depletion experiments lasting several hours, Black and Unsworth (1979b) found that less than 10% of the total flux was accounted for by adsorption on the cuticle. There are similar reports for conifer foliage (Garland and Branson, 1977; Caput *et al.*, 1978). However, in the short term a much greater proportion of SO$_2$ may be deposited on the cuticle, since Garland and Branson (1977) found that pine shoots washed 0.05 min after exposure to ^{35}SO$_2$ lost 70% of their radioactivity, but 20 minutes later less than 10% could be recovered. In experiments with *Lolium perenne*, the proportion of ^{35}S washed from leaves was much greater 10 to 35 min after exposure (Table 5.1, Experiment 2) than after 25 h (Table 5.1, Experiment 1). Although experimental conditions and the time intervals were different, the parallel to Garland and Branson's results is striking. Another feature of the results reported in Table 5.1 is that the only hirsute species studied (*Holcus lanatus*) had both the lowest total flux and the smallest proportion of removable radioactivity. Although this may indicate an ability to "avoid" gaseous SO$_2$, it does not necessarily follow that the species has the same capacity to "avoid" aqueous SO$_2$.

TABLE 5.1

Comparisons of Percentage Radioactivity Washed from Grass Leaves after Exposure to $^{35}SO_2$ in Two Experiments

Details of experimental material	Total ^{35}S flux to plant (cpm cm^{-2} × 10^3)	Percentage of total ^{35}S washed
Experiment 1[a]		
Lolium perenne cv S23 High sulfur, high nitrogen	98	10.9%
Lolium perenne cv S23 High sulfur, low nitrogen	43	13.1
Lolium perenne cv S23 Low sulfur, high nitrogen	58	22.5
Lolium perenne cv S23 Low sulfur, low nitrogen	83	15.8
Experiment 2[b]		
Holcus lanatus "tolerant"	44	24%
Holcus lanatus "sensitive"	48	28
Festuca rubra "tolerant"	455	73
Festuca rubra "sensitive"	302	63
Dactylis glomerata "tolerant"	73	62
Dactylis glomerata "sensitive"	81	63
Lolium perenne "tolerant"	271	65
Lolium perenne "sensitive"	283	67

SOURCES: Experiment 1, Ayazloo, Bell, and Garsed (1980). Experiment 2, Ayazloo, Garsed, and Bell (1982).

[a] Plants exposed to 320 μg m^{-3} (5.0 μmol m^{-3}) of $^{35}SO_2$ for 6 h. Leaves were washed 25 h after end of fumigation. Plants were grown in high and low sulfur and nitrogen treatments. Values are means of nine replicates.

[b] Plants exposed to 275 μg m^{-3} (4.3 μmol m^{-3}) of $^{35}SO_2$ for 45 min. Leaves were washed 10 to 35 min after end of fumigation. Plant material was taken from clones established from field samples. "Tolerant" and "sensitive" indicate responses in acute injury screening. Values are means of five replicates.

Attempts to distinguish between internal and external sinks for SO_2 have also been made by comparing uptake in light and darkness (Elkiey and Ormrod, 1981b). These authors exposed their plants in an unstirred chamber and concluded from silicon-paste impressions that the stomata were "95% closed." The uptake in the dark was 20 to 50% of that in the light (much greater than the values for cuticular uptake given by Black and Unsworth, 1979b), and it is probable that much of the SO_2 deposited in darkness was actually inside the leaf. Garsed and Read (1977a) found that only 34 to 55% of the ^{35}S taken up by *Glycine max* in darkness could subsequently be washed from the leaves, and Jarvis (1971) indicates that complete stomatal closure at night can never be assumed.

Although it is possible to estimate the deposition of SO_2 to the external surfaces of the plant, the identification of the subsequent pathways is more difficult. Studies with isolated cuticles (Lendzian, 1983) show that permeability to SO_2 is much greater than for CO_2 and greatly exceeds that for water vapor. Thus the cuticle appears to be less of a barrier to SO$_2$ than may have been thought previously. Autoradiographic studies indicate that SO$_2$ accumulates preferentially in the guard cells in spinach (Weigl and Ziegler, 1962). The pattern of distribution of the ^{35}S resembles that of ectodesmata, which occur with greatest frequency in the guard cells, and are also common above the veins and around the bases of trichomes (Franke, 1971). Since ectodesmata are believed to be polar routes through the cuticle (Franke, 1971), they would be expected to have an affinity for inorganic ions. It appears to be more than coincidence that damage to *Phaseolus vulgaris* and *Helianthus annuus* by simulated acid rain appeared first near the stomata, along the veins, and around the trichomes (Evans, Gmur, and Da Costa, 1977). There is clearly a need for much more information on the uptake into the leaf of SO$_2$ that has been deposited onto dry cuticular surfaces, whether they remain dry or are subsequently rewetted.

Influence of humidity. The influence of humidity on the sensitivity of plants to acute injury has been recognized for a long time. Katz and Ledingham (1939) considered humidity to influence sensitivity to SO$_2$ more than temperature or sunlight. The main influence of humidity on SO$_2$ flux is exerted through its effects on stomatal resistance (see Black, this volume), but factors independent of stomatal resistance are also involved. Studies on the influence of humidity on SO$_2$ deposition to barley leaves (Spedding, 1969) showed an approximate five-fold increase in Vg as the stomata opened (80 to 85% R.H.), but there were also increases in Vg independent of stomatal opening (Table 5.2). It is probable that the change in Vg between 40 and 50% R.H. was caused by a change in the physical properties of the cuticle, since filter papers also showed an increase in Vg with increasing humidity. The increase in Vg above 90% R.H. may be a conse-

TABLE 5.2
Average Deposition Velocities (Vg) for SO$_2$ to the First Leaf of
Barley Seedlings in a Variety of Conditions and Comparisons
with Filter Paper

Relative humidity (percent)	Stomatal condition	Average Vg (cm s^{-1})
10–20%	closed	0.0028
30–40	closed	0.0019
50–60	closed	0.010
70–75	closed	0.012
75–80	closed	0.015
80–85	closed	0.011
80–85	open	0.066
85–90	open	0.065
90–95	open	0.18
11	filter paper	0.010
81	filter paper	0.031

SOURCE: Spedding (1969).

quence of condensation on the leaf surface, as already shown in Fig. 5.1. In addition to the effects of humidity on stomatal and surface resistances, there may also be effects on the mesophyll resistance, since McLaughlin and Taylor (1981) found that the ratio of SO_2 flux to CO_2 flux was nearly three times greater at 78% R.H. than at 35%.

Influence of surface water. Most work on the uptake or effects of SO_2 has been done with dry foliage. Such studies neglect an important pathway for SO_2 in temperate oceanic climates: deposition into surface moisture. Fowler and Unsworth (1974) followed the pattern of deposition of SO_2 to a wheat crop for several nights by means of micrometeorological techniques. If the night was windy and no dew formed, the surface resistance remained virtually constant, but on still nights the formation of dew caused the surface resistance to fall almost to zero and Vg to increase severalfold. After midnight the surface resistance increased again, which the authors suggested was caused by the decreasing solubility of SO_2 as the pH of the dew fell.

The flux of SO_2 to surface moisture during long periods can be maintained in two ways, either by preventing the fall in pH and maintaining the solubility of SO_2 or by removal of the SO_2 from the moisture by absorption into the plant or oxidation to sulfate. Leaves possess a substantial capacity to buffer droplets on the surface (Lepp and Dickinson, 1976), particularly if they are senescent or damaged (Rowlatt, Crawford, and Unsworth, 1978). There may also be external influences on the pH of the leaf surface; for instance, deposition of SO_2 to forests may be greater near stock-raising areas, where ammonia concentrations are high (Van Breemen *et al.*, 1982). Thus any factor that contributes to the maintenance of the pH of the leaf surface may increase the flux of SO_2 to a crop. It appears that foliage has a substantial capacity for the permanent capture of SO_2 from surface water. A qualitative example is shown in Fig. 5.2, in which the increased depletion of atmospheric SO_2 that occurred when foliage of *Pinus sylvestris* was sprayed with water was not balanced subsequently by an increased emission of gaseous sulfur as the foliage dried. We do not yet have quantitative measures of the fluxes of SO_2 from surface water into the plant. The chemical reactions of SO_2 in the surface water and the chemical form in which the sulfur is absorbed will depend largely on the pH of the water. Uptake of sulfur anions with a valence state of 4 (hereafter called S IV; cf. sulfate, which is S VI) is favored by low pH, since the less-charged species SO_2 and HSO_3^- enter more readily than SO_3^{2-} (Puckett *et al.*, 1973). Under normal conditions it is probable that S IV enters the cell largely as HSO_3^-, with a small but important quantity of undissolved SO_2 (Hocking and Hocking, 1977). Whatever the form in which the sulfur species are absorbed from surface water, there is evidence that the flux can be sufficient to affect the health of the plant. Harcourt and Farrar (1980) found that the growth of *Hordeum distichum* was reduced when sprayed with simulated acid rain containing sulfite, but the acidity itself had no effect. Elkiey and Ormrod

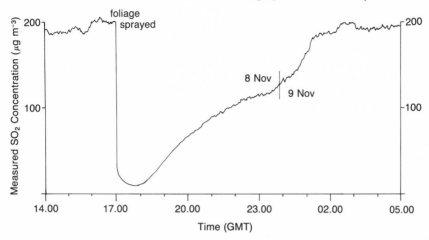

Fig. 5.2. Effect of spraying pine foliage with water on the measured SO$_2$ concentration in the fumigation chamber. Three-year-old *Pinus sylvestris* plants in a chamber were sprayed with water (*p*H approx. 7) for 1 min. The trace records the SO$_2$ concentration measured amid the foliage. There were two air changes per minute in the chamber. Source: S. G. Garsed (unpublished).

(1981d) demonstrated increased visible injury in *Poa pratensis* if the plants were misted twice daily during fumigations with SO$_2$, NO$_2$, and O$_3$. However, interpretation of their results in the present context must remain equivocal, since they also reported increased stomatal opening.

SO$_2$ concentration. The gas concentration can affect flux both by changing stomatal resistance (Black, 1982) and mesophyll resistance (Taylor, McLaughlin, and Shriner, 1982). The latter showed that for *Phaseolus vulgaris* the ratio of flux to SO$_2$ concentration decreased severalfold as the SO$_2$ concentration increased from 0.2 to 0.8 ppm (8.3 to 33.3 μmol m^{-3}). Since the flux of ozone, which was supplied simultaneously, was not affected by SO$_2$ concentration, they concluded that the change in SO$_2$ flux occurred at the mesophyll level. There are three possible explanations for the effect, as follows.

1. A high flux of SO$_2$ saturates the water in the intercellular spaces and limits the absorption of further quantities of SO$_2$. If so, a strong relationship between the flux of a gas and its solubility would be expected. Comparisons of data of solubility of a range of gases (Bennett, Hill, and Gates, 1973) and their fluxes to a plant canopy (Bennett and Hill, 1973a), together with a study of the relationship between solubility and flux of various sulfur-containing gases (Taylor, McLaughlin, and Shriner, 1982), suggest that within wide limits there is a general relationship between $\log_{10}$ solubility and flux to the plant. However, the relationship is not absolutely consistent in the data of Bennett and colleagues or in that of Taylor *et al.* (shown in Fig. 5.3). The line drawn by Taylor *et al.* suggests that

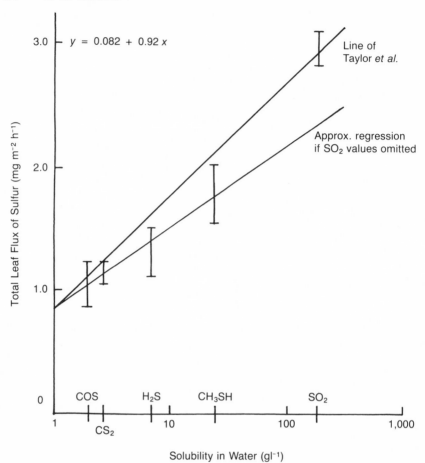

Fig. 5.3. Total flux of sulfur to foliage of *Phaseolus vulgaris* for five coal-conversion gases at an ambient concentration of 500 μg m^{-3} (7.81 μmol m^{-3}): carbonyl sulfide, COS; carbon disulfide, CS$_2$; hydrogen sulfide, H$_2$S; methyl mercaptan, CH$_3$SH; and sulfur dioxide, SO$_2$. Redrawn from Taylor, McLaughlin, and Shriner (1982).

the relationship between log$_{10}$ solubility and flux is strongly linear. However, if the line is drawn through the mean values of the gases other than SO$_2$, it is apparent that the flux of SO$_2$ is greater than extrapolation of the data for the other gases would suggest. Consequently factors in addition to solubility must contribute to the mesophyll resistance.

2. The permeability of the plasmalemma is affected by SO$_2$. High concentrations of SO$_2$ cause serious damage to membranes, inducing permeability changes and loss of cell turgor, but how much do direct effects of air pollutants on the plasmalemma affect the internal resistance? Srivastava, Joliffe, and Runeckles

TABLE 5.3
Leakage of Potassium from *Lolium perenne* Exposed to SO$_2$
(percent of total potassium)

Clone	"Tolerant"[a]	"Sensitive"[a]	"Helmshore"[b]	"S23"[b]
SO$_2$	3.58%	3.22%	1.38%	1.30%
Control	1.73	1.26	1.94	1.37
p	<0.001	<0.001	n.s.	n.s.

[a] Data of Ayazloo, Garsed, and Bell (1982); SO$_2$ exposure of 39 μmol m^{-3} for 2 h.
[b] Data of Bell and Mudd (1976); SO$_2$ exposure of 2.2 μmol m^{-3} for 81 d.

(1975b) investigated the influence of high concentrations of NO$_2$ (1 ppm and above) on mesophyll resistance. Resistance increased with the time of exposure or as the concentration increased. They argued that, if saturation with NO$_2$ was the factor limiting uptake, then pollution-free intervals between the exposures would allow the absorbed NO$_2$ to be metabolized and the flux should return to its previous rate. However, if the decrease in flux was the result of physiological changes in the membranes, the flux should remain depressed after the pollution-free interval. It remained depressed. If a similar finding were reported for SO$_2$, it would suggest that part of the decrease in flux observed by Taylor, McLaughlin, and Shriner (1982) may have been a consequence of membrane changes. Studies with *Lolium perenne* (Table 5.3) indicate that permeability changes occurred (in the absence of visible injury) when plants were fumigated with 2,500 μg m^{-3} (39 μmol m^{-3}) of SO$_2$ for 2 hours (Ayazloo, Garsed, and Bell, 1982). However, it is questionable whether SO$_2$ causes major changes in membrane permeability or mesophyll resistance at low concentrations, since exposure of *L. perenne* to 140 μg m^{-3} (2.2 μmol m^{-3}) for 81 days produced no increase in K$^+$ leakage (Bell and Mudd, 1976). Furthermore, Black and Unsworth (1979b) concluded that between 53 and 530 μg m^{-3} (0.83 and 8.3 μmol m^{-3}) of SO$_2$, the internal (mesophyll) resistance of *Vicia faba* was very low.

3. The continued flux of SO$_2$ derivatives from an aqueous solution into the cell requires the maintenance of a concentration gradient. Such a gradient may be maintained by chemical transformations near the site of entry, or by transport to other parts of the cell. Although the work discussed in the preceding paragraph suggests that membrane damage may be a factor in r_m at high concentrations of SO$_2$, the possible contribution of concentration gradients to the degree of mesophyll resistance at various SO$_2$ concentrations has not yet been investigated.

Difference between annual and perennial crops. Most of the studies of SO$_2$ flux to plants have been performed on actively growing material either in the field or under artificial conditions approximating summer days as closely as possible. It has already been seen that, under such conditions, the SO$_2$ flux is closely correlated with stomatal resistance and that the cuticular component of uptake is fairly small. For annual crops growing under dry conditions, extrapolations from such measurements to provide an estimate of sulfur "load" during the growing season will probably give a reasonably accurate value, providing that losses from the roots, gaseous emission, or leaf death are not of major importance. For annual plants growing in wet conditions, estimates of uptake via the surface water and of losses by leaching also need to be taken into account. However, the plants still fulfill their life cycle under conditions generally favorable to rapid growth. Crops with perennial aboveground parts, on the other hand, spend approximately 50% of their lives under conditions unfavorable to fast growth at a time when SO_2 concentrations are at their highest. Therefore estimates of annual flux and sulfur "load" for perennial crops calculated by extrapolating from results obtained under summer conditions can underestimate seriously the level of stress imposed in winter.

Although the stomata are the major pathway of uptake in summer, their activity in winter is low. Fowler (1980) makes the point that uptake of SO_2 through the stomata is a discontinuous process, unlike cuticular uptake, and that during the whole year the two pathways may make similar contributions to flux. The greater emphasis on uptake across the cuticle in winter means that surface water assumes increasing importance. Not only is the concentration of gaseous SO_2 usually higher in winter, but there are also appreciable concentrations of dissolved S IV in winter rainwater (Pena *et al.*, 1982), so that wet deposition may also make a contribution to flux. The high stomatal resistance in winter is probably a result of the restriction in the growth and physiological activity of plants by low temperature, low irradiance, and short days. Although the low stomatal conductance under such conditions might make the plants relatively resistant to acute injury, because of the high degree of avoidance, grasses, at least, become much more sensitive to effects on growth (Davies, 1980a; Bell, Rutter, and Relton, 1979). The most likely explanation for the effects on growth is that physiological activity is too low during the winter to allow dilution of the sulfur absorbed or its incorporation into protein. In the experiments of Davies and Bell *et al.*, the plants were watered from below and the foliage would have remained dry throughout the exposures. However, in the field during winter (in Britain, anyway), the foliage is likely to be wet much of the time. If the stomata are largely inactive, then deposition to a crop in winter via surface water will exceed that via the stomatal pathway severalfold. It is therefore possible that the effects measured by Davies (1980a) and by Bell, Rutter, and Relton (1979) may be serious underestimations of effects of winter pollution under conditions of dew and light rain. On the other hand, continuous heavy rain might have a protective effect by leaching sulfur compounds from the plants.

Evergreen trees continue to absorb SO_2 throughout the winter, even in subzero temperatures (Materna and Kohout, 1967). Keller (1981a) has shown that SO_2 causes an increase in the sulfur content of *Picea abies* throughout the winter, although the rate was much less than in the summer. The uptake of SO_2 was paralleled by a decrease in net assimilation rate, together with increases in peroxi-

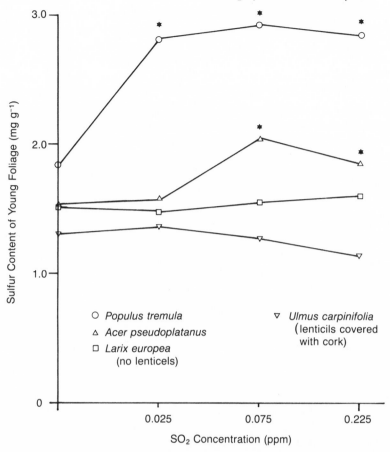

Fig. 5.4. Effect of winter-time SO$_2$ concentrations on the sulfur content of young foliage in the following spring. The trees were exposed to SO$_2$ for between 77 and 102 days. Asterisks indicate a difference from clean-air control significant at $p < 0.05$. Note logarithmic scale on x-axis. Plotted data from Keller (1981b).

dase activity and sensitivity to frost. These effects appeared even at SO$_2$ concentrations as low as 65 μg m^{-3} (1.02 μmol m^{-3}), which many authors still regard as a "safe" level (e.g., I.E.R.E., 1981).

Even deciduous trees can absorb SO$_2$ in winter. Those with active lenticels may absorb enough SO$_2$ to cause elevated sulfur contents in the following year's foliage, but those without lenticels or with obstructed lenticels showed no such increase in sulfur content (Fig. 5.4). Heavy accumulation of ^{35}S has been found in the lenticels of deciduous trees (Godzik, 1968; Materna and Kohout, 1969). The latter authors also found heavy labeling in bud scales, indicating that the apical meristems are not invariably protected from the influence of SO$_2$. The buds will also receive atmospheric sulfur indirectly by translocation. Keller (1978a) has observed increased bud mortality in *Fagus* sp. exposed to SO_2 overwinter.

Distribution Inside the Plant

There are several routes available to the SO_2 products that cross the plasmalemma (Fig. 5.5). The major factor determining the metabolism of SO_2 is the relative proportion of S IV and S VI anions in the cell. This proportion depends first of all on the concentration of SO_2 entering the plant. Table 5.4 shows the results of some labeling experiments with $^{35}SO_2$. All the studies show radioactivity in sulfate, "soluble" and "insoluble" fractions (although the precise terminology varies from author to author). With very high concentrations of SO_2, large quantities of S IV were recovered, but at lower concentrations S IV appeared only in the dark. In general the proportion of radioactivity represented by organic material increases as the concentration of SO_2 supplied becomes less. If sulfur-deficient plants are exposed to $^{35}SO_2$, there is a rapid incorporation of radioactivity into organic compounds even at relatively high SO_2 concentrations (Thomas, 1948; Fried, 1948).

The persistence of S IV anions at high SO_2 concentrations will lead to acute injury. Katz and Ledingham (1939) and Thomas, Hendricks, and Hill (1950) described how "free sulfurous acid" could be recovered from leaves exposed to enough SO_2 to cause acute injury, but not from leaves in which acute injury did not appear. They concluded that below a certain threshold, SO_2 would be oxidized soon after entering the cell, and consequently the plant was protected from injury. The idea that oxidation of S IV may protect the cell from acute injury has received support from more recent work in which correlations have been shown between resistance to acute injury and activities of both sulfite-oxidizing enzymes (Kondo *et al.*, 1980) and rates of sulfite oxidation in leaves (Miller and Xerikos, 1979). However, the concept of an *absolute* threshold for SO_2 injury is now being questioned. There is a growing body of evidence of direct effects of SO_2 on both biochemical and physiological processes in the absence of visible injury. This is in contrast to Thomas' and Katz' interpretation of the relationship between sulfurous acid and injury as absolute; i.e., there is no injury to the plant if there is no free "sulfurous acid." The difference between the viewpoints lies in the fact that visible injury is primarily a consequence of extensive SO_2 damage to the plasmalemma; yet current evidence suggests that the incoming SO_2 can damage sensitive physiological processes at concentrations too low to be detected chemically or to cause irreparable harm to the plasmalemma. Furthermore, the possibility that the oxidation process itself may be harmful to the plant cannot be ignored.

Besides the effects of SO_2 concentration on the proportion of S IV, Table 5.4 also shows a large effect on the distribution of radioactivity between inorganic

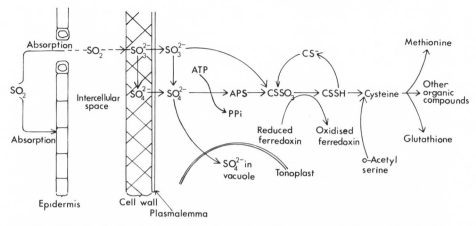

Fig. 5.5. Routes for the formation of organic sulfur compounds from sulfur dioxide. APS = adenosine-5'-phosphosulfate. CS, CSSO₃, and CSSH represent a free-carrier protein and its adducts with SO₃ and SH groups, respectively. Redrawn from Garsed and Read (1977b).

and organic fractions. The difference indicates that rate-limiting processes are involved in the formation of organic materials. The basic steps for the formation of organic sulfur compounds occur in the chloroplasts (which are also the major sites for oxidation of S IV). In brief, the process involves the "activation" of sulfate by reaction with ATP to form adenosine-5'-phosphosulfate (APS), which is then sequentially reduced, via a carrier-bound SO_3 form, to a carrier-bound sulfhydryl. Under normal conditions the bound sulfide is transferred to a carbon skeleton (o-acetyl serine) to form cysteine, from which the other organic sulfur compounds are synthesized (Schiff and Hodson, 1973). S IV may enter the pathway directly, without prior oxidation to S VI, and there is evidence for preferential use of S IV rather than S VI (Winner *et al.*, 1981).

Two major rate-limiting steps can be identified. The first is the reduction process itself, which is dependent on reduced ferredoxin (FdH_2) produced by Photosystem 1. The pathway can proceed at a slow rate in darkness by the formation of FdH_2 by ferredoxin NADP oxidoreductase (Hennies, 1975). This accounts for the small proportion of organic materials labeled in darkness (Table 5.4). The second is the transfer of the sulfhydryl moiety to the carbon skeleton. If the rate of reduction exceeds the availability of carbon skeletons, H_2S may be emitted (Wilson, Bressan, and Filner, 1978).

The extreme concentration-dependence of the pathways followed by SO_2 makes interpretation of labeling experiments difficult. Moreover, one cannot be sure how much the results of such experiments actually represent the effects of the SO_2 in the context of the sulfur budget of the cell. The first problem is that incoming sulfate is metabolized in preference to the endogenous sulfate, which is

TABLE 5.4

Selected Data on the Distribution of Radioactivity in Leaf Fractions after Exposure of Plants to $^{35}SO_2$

SO₂ concentration[a]		Duration	Time after exposure	Irradiance during exposure	Percent "sulfite" (S IV)	Percent sulfate	Percent organic (sol. + insol.)	Percent others
μmol m^{-3}	μg m^{-3}							
1. Spinacea oleracea								
2,109 (max)	135 × 10³	30 min	2 h	20 klux	50%	18%	9%	23%
			7 h	20 klux	20	43	13	24
2. Glycine max								
100 (max)	6,380	60 min	24 h	10 klux	Not detected	54	42	4
107 (max)	6,830	60 min	24 h	Dark	1.7	81	15	2
3. Lolium perenne								
67.0 (max)	4,290	2 h	3 h	Not recorded	3	70	12	15
4. Lolium perenne								
4.3	275	45 min	10–35 min	191 μE m^{-2} sec^{-1}	0	15	85	—
5. Helianthus annuus Leaf 3[b]								
62.4	3,993	5 d	Immediate?	Not recorded	Trace	76	24	—
41.6	2,662	5 d	Immediate?	Not recorded	Not determined	76	24	—
20.8	1,331	5 d	Immediate?	Not recorded	Not determined	53	47	—
8.9	568	5 d	Immediate?	Not recorded	Not determined	31	69	—
0	0	5 d	Immediate?	Not recorded	Not determined	10	90	—
6. Medicago sativa (S-deficient)								
4.2	266	56 h	Immediate?	Not recorded	Not determined	8	92	—

SOURCES: 1, Weigl and Ziegler (1962). 2, Garsed and Read (1977b). 3, Garsed and Read (1977c). 4, Ayazloo, Garsed, and Bell (1982). 5, Faller, Herwig, and Kühn (1970b). 6, Fried (1948).

[a] SO₂ concentrations in ppm have been converted to μg m^{-3} using the conversion 1 ppm = 2,662 μg m^{-3}. In experiments in sealed chambers, (max) refers to the initial concentration of SO₂.

[b] In these experiments the SO₂ concentrations refer to the quantity of carrier SO₂ that was mixed with the $^{35}SO_2$. There is also a suspicion that the plants had a low sulfur supply in the nutrient solution.

located largely in the vacuole (Kylin, 1960). Although incoming SO$_2$ probably behaves similarly, no quantitative comparisons of SO$_2$ fluxes and sulfate pool sizes have yet been reported. The second problem is that most of the experiments involve pulse feeding; and the distribution of radioactivity following a pulse of ^{35}SO$_2$ to a plant that has had no previous exposure to SO$_2$ may not reflect the true situation of plants in the field that are receiving SO$_2$ almost continuously. The plants that received low concentrations of ^{35}SO$_2$ as shown in Table 5.4 had most of their radioactivity in organic forms; yet plants exposed to low SO$_2$ concentrations in the field accumulate largely sulfate (e.g., Malcolm and Garforth, 1977), although small increases in both protein and nonprotein thiols have also been reported (Grill et al., 1980; Grill and Esterbauer, 1973). Rennenberg et al. (1982) describe a "futile" sulfur cycle in plants. They argue that the cell needs the intracellular cysteine concentration to be kept relatively stable. The cysteine produced by the sulfate-reduction pathway may be used for the synthesis of other sulfur-containing compounds, stored in the vacuole or transported out of the cell. Excess cysteine, however, is degraded to release H$_2$S, some of which may then be oxidized back to sulfate, and the cycle repeated.

Since the accumulation of sulfur in a short-term pulse experiment is trivial compared with that during long-term exposure in the field, there is a great need for well-designed tracer experiments to investigate the role of continuous SO$_2$ exposure in the long-term biochemical sulfur balances of plants.

Translocation

The main soluble derivatives of sulfur dioxide that are found in the leaves also appear in the phloem. Studies with both ^{35}SO$_2$ (Garsed and Read, 1977b) and ^{35}SO$_4^{2-}$ (Rennenberg, Schmitz, and Bergmann, 1979; Bonas et al., 1982) show that sulfate is the main form of sulfur translocated, followed by glutathione, cysteine, and methionine. Sulfite may also be translocated if high concentrations of Na$_2$SO$_3$ are applied to the leaves (Garsed and Mochrie, 1980), but we don't know whether significant quantities are mobile after exposure to low concentrations of SO$_2$. Sulfur-dioxide products can be translocated very rapidly. Thus more than 20% of the total ^{35}SO$_2$ derivatives were exported from soybean leaves within 1.5 hours, and 50% after 24 hours (Garsed and Read, 1977b). Translocation occurs in both light and darkness; the absolute quantities are greater in the light, but the proportion may be greater in darkness, because of the smaller concentration in the source leaves (Garsed and Read, 1977a). The pattern of translocation relative to leaf age and physiological activity generally resembles that for ^{14}CO$_2$ products, with preferential accumulation in meristematic regions of vegetative plants (Garsed and Read, 1974) and in the seeds of fruiting plants (Thomas, 1948). Mature leaves may also import some radioactivity, but it is uncertain whether this is direct leaf-to-leaf transfer or whether ^{35}S is translocated back from the roots. A substantial proportion of the ^{35}S translocated to the roots may be lost into the nutrient solution (De Cormis, Cantuel, and Bonté, 1968; Garsed and Read,

1977b), even if the plants are poorly supplied with sulfur (Fried, 1948), which suggests that loss from the roots may be of major importance in sulfur cycling. Translocation of $^{35}SO_2$ has also been studied in pasture grasses and trees. The movement of sulfur compounds in the grasses appears to be comparable with that in the legumes (Garsed and Read, 1977c), but translocation of ^{35}S products by trees appears to occur more slowly. Jensen and Kozlowski (1975) found that four days after 1-year-old seedlings of broad-leaved trees were exposed to $^{35}SO_2$, the leaves still retained more than 89% of the total radioactivity. Moreover, there was little activity in the nutrient solutions. Godzik (1976) suggested that the poor mobility of $^{35}SO_2$ products in trees was the result of physical barriers. Thus in *Quercus* sp. radioactivity accumulated in the epidermal layer and the bundle sheaths; in pine needles, the endodermis and vascular tissue were particularly heavily labeled.

Loss of SO_2 Derivatives from Plants

For a crop growing for a long period in the field, the overall sulfur budget is likely to have more effect on its performance than any individual exposure to SO_2 (Garsed, Mueller, and Rutter, 1982). The sulfur may be lost in various ways, as follows.

Leaching. Rain removes sulfur from both inside and outside the leaf. Raybould, Unsworth, and Gregory (1977) concluded that nearly 90% of the sulfur gained by rain passing through a wheat canopy could be attributed to leaching. Senescent leaves are particularly susceptible to sulfur loss into both rain and dew (Rowlatt, Crawford, and Unsworth, 1978). The manner in which the leaves are leached can determine the quantities of sulfur removed; so intermittent rain may have effects different from those of long-term immersion (which may cause waterlogging). For instance, Rice *et al.* (1979) found that up to 85% of the sulfur could be leached from leaves immersed for four days in tapwater. The physiological condition of the leaves after removal was not described. Although Raybould, Unsworth, and Gregory (1977) could attribute most of the sulfur enrichment of rainfall to leaching, they found the total sulfur in the throughfall could account for only one-third of the overall S lost by their wheat crop between anthesis and harvest, and suggested that the remainder was lost by emission of H_2S or by leakage from the roots.

Gaseous emissions. Emission of H_2S has been shown to occur in plants after they have been exposed to SO_2 or have been allowed to absorb unusually high concentrations of sulfate through damaged roots or cut stalks (De Cormis, 1968; Bressan, Wilson, and Filner, 1978). This has been interpreted as a mechanism by which the plant "rids" itself of excess sulfur. Rennenberg *et al.* (1982) identify two sources of H_2S: a light-dependent process via the S-assimilation pathway; and a light-independent process involving the desulfhydrylation of cysteine. The quantity emitted as a consequence of exposure to SO_2 depends on concentration. Bell and Clough (unpublished data) found emission of both $H_2^{35}S$ and $^{35}SO_2$ from *Vicia faba* exposed to 53 μg m^{-3} (0.83 μmol m^{-3}) of $^{35}SO_2$. Similarly, Hällgren and Fredricksson (1982) found H_2S emissions to be linearly related to SO_2 concentration between 53 and 210 μg m^{-3} (0.83 and 3.28 μmol m^{-3}) in *Pinus sylvestris*. No H_2S was detected in the absence of SO_2. Emission was light-dependent, and in periods of high photon flux could exceed SO_2 absorption. The light-dependence suggests that substantial variation may occur with season. Wilson, Bressan, and Filner (1978) found that plants raised in a greenhouse during winter were much less active at emitting volatile sulfur than those grown in summer. In view of the increased sensitivity of grasses under low irradiance (Davies, 1980a), the possible physiological link between slow growth and rates of H_2S emission needs to be investigated. Besides the effects of irradiance on H_2S emission, leaf age is also important. Young leaves are not only much more resistant to acute injury from SO_2 than mature leaves (discussed by Guderian, 1977), but the greater resistance is caused by a physiological mechanism rather than by better avoidance of SO_2 (Wilson, Bressan, and Filner, 1978). Furthermore, comparisons of fast-growing and slow-growing plants indicate that the sensitivity difference is a consequence of age *per se* rather than of developmental stage. Recent work by Sekiya, Wilson, and Filner (1982) demonstrates that young leaves produce H_2S far more efficiently than mature ones, and Schmutz and Brunold (1982) show an interesting parallel in that the sulfate-reduction pathway is at its most active in young leaves, but declines rapidly when the leaves are fully expanded. These recent findings indicate that to understand the metabolism of sulfur dioxide in whole plants, we must consider the influence of environmental factors on the age structure of the plants and of ontogenetic changes in the physiological activities of the constituent parts.

Leakage from the roots. In experiments with $^{35}SO_2$, radioactivity can almost invariably be found in the nutrient solution a short time after exposure (Garsed and Read, 1977b). Such leakage can also occur in sulfur-deficient plants (Fried, 1948), but the mechanisms of the process and its magnitude and role in sulfur cycling remain unstudied.

Death and loss of plant parts. Increased senescence and leaf death attributable to SO_2 are commonly reported for both herbaceous and woody crops, but there have apparently been no quantitative estimates of the importance of organ death in the overall loss of sulfur from the plant.

Assessment of Techniques

I will deal only with tracer studies. Almost all the tracer studies discussed here (including several of my own) are of limited value, because little attention was paid to the details of the environmental variables that affect flux. Sealed systems in particular allow no control of relative humidity, temperature, or CO_2 concentration. The most dubious experiments are probably those performed in the field in which enclosures have been left on plants in bright sunlight for several hours with no air movement. Furthermore, the SO_2 concentrations used in such experiments normally have been high ($>2,000 \ \mu g \ m^{-3} = 31 \ \mu mol \ m^{-3}$) and frequently have not been measured. If $^{35}SO_2$ is to be applied in the field, it is best if the plants either remain unenclosed or be enclosed for the shortest possible time in a stirred, open-circuit vessel, such as a cuvette. Because the stomatal and cuticular pathways are so dependent on environmental factors, data on humidity, temperature, and irradiance inside the vessel, together with measurements of stomatal resistances to water-vapor loss, should be provided, as an absolute minimum. To avoid contamination of equipment, the measurements could be made, if necessary, immediately before release of $^{35}SO_2$ or in a parallel "control" chamber receiving the same concentration of SO_2 under identical conditions.

In the laboratory, the care given to the control and measurement of environmental variables in chambers containing $^{35}SO_2$ should be comparable to that in the most carefully planned depletion experiments. The only reproducible way to release low concentrations of $^{35}SO_2$ at a steady rate is by means of a gas cylinder. With such a technique it should be possible to use an open-circuit system with a high degree of accuracy. Above all, the concentrations of $^{35}SO_2$ should be of the order of those expected in the field.

In the last 10 years our knowledge of the factors affecting both gas uptake and sulfur biochemistry has increased greatly. It is now time to perform a comprehensive analysis of the uptake and metabolism of $^{35}SO_2$ with the attention to detail now being used in depletion experiments.

Directions for Future Research

The emphasis of this paper has been on the long-term sulfur "load" experienced by plants, but most of the work on the uptake and transport of SO_2 has been done as short-term experiments on plants that have had no previous exposure to the pollutant. It is particularly unfortunate that studies on gas flux and metabolism are normally completely divorced from growth experiments, since it is then difficult to relate individual processes to changes in the whole plant. Therefore future experiments should compare the fluxes and metabolism of SO_2 in plants that have had previous exposure to SO_2 with those in unfumigated controls, and where possible the effects should be related to changes in plant growth. No studies of the metabolism of $^{35}SO_2$ have yet taken into account the size of the various pools of endogenous sulfur that may accumulate under different conditions. A comprehensive study of the uptake and metabolism of low concentrations of $^{35}SO_2$ in relation to normal and abnormal sulfur metabolism is therefore necessary. There is virtually no information on the physiological responses to SO_2 of plants growing in winter conditions. An integrated study of the relative importance of the gaseous and liquid-phase pathways to plants growing under such conditions, and the effects on metabolism and growth, should be a high priority if the overall effects of SO$_2$ on perennial crops are to be reliably estimated. It is remarkable that in a careful study of sulfur losses from a wheat crop, two-thirds of the sulfur was unaccounted for (Raybould, Unsworth, and Gregory, 1977; Unsworth *et al.*, this volume). The need for intensive studies of sulfur loss and cycling is all too evident. In addition, the need for studies of sulfur metabolism and sulfur balance in relation to leaf age and plant ontogenetic changes is clearly indicated.

6

SO$_2$ Effects on Stomatal Behavior

Valerie J. Black

It has been recognized that stomata respond to many toxic chemicals, but our understanding of their reactions to air pollutants is incomplete. Studies of these stomatal responses to pollutants are important for several reasons. First, stomata control the major proportion of water lost from plants, and thus any agent that influences this function may alter the water relations and energy balance of leaves, and thus affect cellular turgor, growth, and the operation of important physiological and biochemical processes. Second, pollutant-induced changes in stomatal aperture will influence the exchange of other gases between the interior of the leaf and the atmosphere. For example, carbon-dioxide flux and therefore photosynthetic rates are strongly influenced by stomatal behavior. In addition, a change in the flux of carbon dioxide and water vapor will be accompanied by a modification of the uptake of gaseous pollutants by the plant, and the transport of the pollutant to the sensitive metabolic sites in the mesophyll. It has been well-documented that plants exhibit greater damage from pollutants when they are exposed with stomata open. Therefore a general assumption can be made that the major pathway of gaseous pollutant uptake by plants is through the stomatal pore, although pollutant uptake may occur through the cuticle, especially if it is weathered or damaged. Control of pollutant uptake is largely effected by the stomata, but may be influenced by other resistances to flow in the pathway into the plant, such as the aerodynamic and internal resistances. Therefore not only are inherent stomatal responses to environmental conditions important in determining the rate of plant physiological processes, but any action of a pollutant on stomatal behavior will have important effects on gas exchange, with consequences for growth and development.

Most of the early work on the effects of pollutants on plant metabolism examined the severe depressions of photosynthesis and the appearance of visible injury that resulted when plants were exposed to very high concentrations of SO$_2$. Changes in photosynthesis and in carbon-dioxide concentration within the leaf were thought to be responsible for the observed reductions in transpirational water loss that were associated with the visible damage. However, a series of investigations in the 1970's has shown that SO$_2$ can have a direct effect on stomatal behavior that is independent of changes in photosynthesis or the appearance of visible injury (Biscoe, Unsworth, and Pinckney, 1973; Beckerson and Hofstra, 1979a,b; Black and Unsworth, 1979c). This work was stimulated by the findings of Mansfield and Majernik (1970) and Majernik and Mansfield (1970, 1972), who demonstrated not only that SO$_2$ increased stomatal opening in beans, but that this response occurred even when pollutant concentrations were relatively low. They also observed that polluted stomata still retained their capacity to respond to other environmental conditions, such as light and carbon dioxide.

Unfortunately, conflicting conclusions were drawn from investigations aimed at defining stomatal responses to SO$_2$. For example, stomata were reported to exhibit either increased opening in response to SO$_2$ (Majernik and Mansfield, 1970; Unsworth, Biscoe, and Pinckney, 1972), or increased closure (Bonté, 1975). These contradictory findings led Mansfield (1976) to write, as a postscript to a review article on the effects of pollutants on stomata, "The dearth of satisfactory information from well-designed, critical experiments is obvious in this commentary." Recently, however, it has become accepted that there is no simple pattern of stomatal response to SO$_2$, and that many of these apparent inconsistencies in results may be explained by the complexity of SO$_2$/stomatal interrelationships, in which stomatal responses are modified by many factors. For example, concentration and duration of pollutant exposure may determine stomatal responses; prevailing environmental conditions may influence whether stomata show enhanced opening or closure in response to exposure; and not only do species differ in their response, but even varieties, and individuals grown under different conditions, may differ in their stomatal sensitivity to SO$_2$. A more detailed account of these factors will be given later in this review.

Many researchers have realized that inadequate exposure techniques and methods of analysis of stomatal behavior may lead to conflicting results. Similarly, it is now accepted that physiological or biochemical effects may correlate less with the concentration of pollutant to which plants are exposed than with the flux or accumulated flux of pollutant into the plant. These modifications in thinking during the last few years have permitted advances in the definition of stomatal responses to SO$_2$, and in the elucidation of the sites and mechanisms of these responses to the pollutant.

SO$_2$-Induced Changes in Transpiration and Stomatal Conductance

I give here, as an overview of our current knowledge, a selection of papers that illustrate the range of stomatal responses to SO$_2$ and the extent to which the mechanisms of responses have been elucidated. For an exhaustive survey of the literature, see: Mansfield, 1976; Hällgren, 1978; Heath, 1980; Unsworth and Black, 1981; Black, 1982; Mansfield and Freer-Smith, 1984.

Direction of stomatal responses. Stomata may exhibit increases in either stomatal opening or stomatal closure when exposed to SO$_2$. For example, Majernik and Mansfield (1970) reported increased stomatal opening in *Vicia faba* plants exposed to 0.25 ppm (10.4 μmol m^{-3}) of SO_2. A similar observation was made by Unsworth, Biscoe, and Pinckney (1972) and by Biscoe, Unsworth, and Pinckney (1973), who also reported that stomata of *Zea mays* plants responded similarly and could be stimulated to open even if plants were exposed to SO_2 at night with stomata closed. In contrast, increased stomatal closure or depressed transpiration rates have been reported for polluted tobacco and pinto-bean plants (Menser and Heggestad, 1966; Sij and Swanson, 1974). Furthermore, Bonté, DeCormis, and Louguet (1977) described a large stomatal closure in *Pelargonium* within a few minutes of the commencement of SO_2 exposure.

Similar contrasting responses have been demonstrated in many investigations of various species. For example, stimulated stomatal opening has been reported for pine (Farrar, Relton, and Rutter, 1977), *Phaseolus* (Ashenden, 1978; Rist and Davis, 1979; Black and Unsworth, 1980), pea and corn (Unsworth, Biscoe, and Pinckney, 1972; Klein *et al.*, 1978), radish, soybean, cucumber, navy bean (Beckerson and Hofstra, 1979a,b), radish, sunflower, tobacco, field beans (Black and Unsworth, 1980), *Atriplex* (Winner and Mooney, 1980a,b,c), broad bean (Taylor, Reid, and Pharis, 1981), silver maple and black locust (Suwannapinunt and Kozlowski, 1980), and birch (Biggs and Davis, 1980). Insensitivity of stomata to SO_2 has been reported in pines (Hällgren *et al.*, 1982), and ryegrass (Cowling and Lockyer, 1976).

In contrast, SO_2-induced reductions in stomatal conductance and transpiration rates have been observed in broad beans (Mansfield and Majernik, 1970), pinto beans (Sij and Swanson, 1974), geranium (Bonté, 1975; Bonté, DeCormis, and Louguet, 1977), pine (Caput *et al.*, 1978), peanut, tomato, radish, perilla, and spinach (Kondo and Sugahara, 1978), *Diplacus* and *Heteromeles* (Winner and Mooney, 1980a,b,c), castor oil, swiss chard, rice, poplar, plane, sunflower, cucumber (Furukawa *et al.*, 1980), alday, wheat, corn, sorghum and bean (Kondo, Maruta, and Sugahara, 1980), apple (Schertz, Kender, and Musselman, 1980b), birch (Biggs and Davis, 1980; Mansfield and Freer-Smith, 1984), beans (Black and Unsworth, 1980; Taylor, Reid, and Pharis, 1981), peas (Olszyk and Tibbitts, 1981a,b), and soybeans (Brymer, 1982).

Factors influencing stomatal responses. These reports indicate the existence of different stomatal responses to SO_2. Not only do particular species respond differently, but also both opening and closing responses may be induced in the same species. For example, Black and Unsworth (1980) observed stimulated stomatal opening in response to SO_2 in tobacco and sunflower, whereas the opposite effect was reported by Kondo, Maruta, and Sugahara (1980) and by Furakawa *et al.* (1980). In addition, Mansfield and Majernik (1970), Black and Black (1979a,b), Black and Unsworth (1979c, 1980), and Taylor, Reid, and Pharis (1981) have reported both opening and closing responses when they exposed *Vicia faba* stomata to SO_2 or a solution of buffered sulfurous acid. An analysis of the exposure

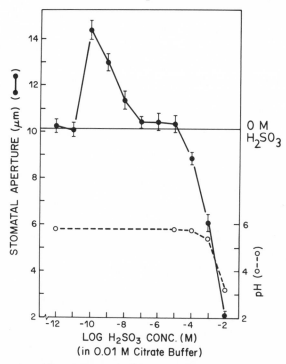

Fig. 6.1. The effect of sulfurous acid buffered in 10 mM sodium citrate buffer (pH 5.8) on stomatal aperture in isolated epidermal strips of *Vicia faba*. The strips were floated on the solutions and incubated for 4 h at 30°C with a light intensity of 46 W m⁻² prior to measurement of the stomates. The mean width (μm) of 180 stomates ± SE is given for each concentration of sulfurous acid. Reproduced from Taylor, Reid, and Pharis (1981).

conditions used in these experiments has identified several factors that may be responsible for this variation in stomatal responses to SO_2. Both Black and Unsworth (1980) and Taylor, Reid, and Pharis (1981) have observed that the pattern of stomatal response was influenced by the concentration of SO_2 or sulfurous acid (Fig. 6.1) to which plants were exposed. Although stomata may show increased opening in response to low pollutant concentrations, closure may result from exposure to higher concentrations. This variation of response with concentration has also been observed in birch stomata (Biggs and Davis, 1980). Indeed, most of the stomatal closing responses reported have been observed in plants exposed to high concentrations (sometimes as high as several ppm), whereas when plants are exposed to relatively low SO_2 concentrations, stomata generally exhibit either no response or increases in stomatal conductances or transpiration rates.

One exception to this generalization is the work of Bonté, DeCormis, and Louguet (1977), who reported a very marked and rapid stomatal closure when geranium plants were treated with a very low concentration of SO_2 (<.03 ppm; < 1.25 μmol m^{-3}). In contrast, using a similar exposure concentration, Black and Unsworth (1980) observed stomatal opening in a different species, field beans. This disparity indicates that stomatal sensitivity does differ from species to species. Indeed, Bonté, DeCormis, and Louguet (1975) have reported a variation in stomatal response from species to species, and even between individuals of the same clone. Kimmerer and Kozlowski (1981) observed clonal differences in stomatal sensitivity in *Populus* clones; the diffusive conductance of a susceptible clone decreased after exposure to 0.2 ppm (8.3 μmol m^{-3}) of SO_2 for eight hours, whereas the stomata of other clones were not affected. Species differences in plants polluted with relatively low concentrations of SO_2 were observed by Biggs and Davis (1980), who found that white birch stomata exhibited increased apertures after exposure to 0.3 ppm (12.5 μmol m^{-3}) of SO_2, whereas gray birch showed no response to this treatment.

Beckerson and Hofstra (1979a,b), who found that several species of plants opened their stomata in response to SO_2, also reported a great variation in stomatal behavior, not only between species, but also between upper and lower surfaces of the same leaf. For example, they observed that stomata on the adaxial surface of navy bean leaves were more responsive to SO_2 than those on the abaxial surface, whereas field-bean stomata were more sensitive on the lower surface (Black and Black, 1979a,b). A difference in sensitivity was observed by Amundsen and Weinstein (1981), who found that high concentrations of SO_2 (~2 ppm; 83 μmol m^{-3}) caused a greater increase of stomatal closure on the upper surface than on the lower. Species also vary in the magnitude and speed of response. Kondo, Maruta, and Sugahara (1980) examined the responsiveness of stomata to 2 ppm (83 μmol m^{-3}) of SO_2 in a range of species, and found great variation (Fig. 6.2). For example, in beans, a slight transient opening was followed by closure, whereas a slower closing response was observed in wheat plants; but a large rapid closing response, similar to that observed in *Pelargonium* by Bonté, DeCormis, and Louguet (1977), occurred in rice plants. However, the mechanism to explain the differences in sensitivity between species, clones, individuals, or leaf surfaces has not been elucidated.

Other factors may also influence stomatal reactions to SO_2. Olszyk and Tibbitts (1981a,b) found in pea plants exposed to SO_2 that stomata of expanding leaves closed to a greater degree than the stomata of expanded leaves. The stomatal reaction observed may also depend on the time of day during which plants were exposed, or the period in which stomatal measurements were made. Hällgren *et al.* (1982) suggested that stomata of *Pinus sylvestris* L. may be more responsive in the afternoon, whereas Olszyk and Tibbitts (1981a,b) found that stomatal responses of peas were greater when plants were treated early or late in the photoperiod than in the middle of the photoperiod (Fig. 6.3). Other workers have reported that stomatal responses may be observed in the dark period (Mansfield and Majernik, 1970; Unsworth, Biscoe, and Pinckney, 1972; Black and

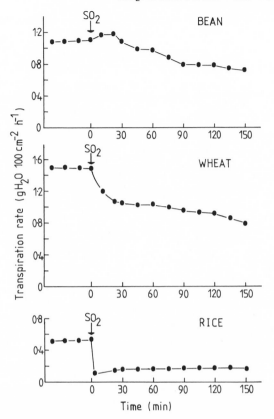

Fig. 6.2. Changes in transpiration rate of bean, wheat, and rice with SO$_2$ fumigation. Plants were preconditioned for about 2 h in the growth cabinet prior to 2.0 ppm (83 μmol m^{-3}) fumigation. SO$_2$ gas was introduced into the cabinet at the time 0, indicated by the arrow. Adapted from Kondo, Maruta, and Sugahara (1980).

Unsworth, 1980), or greater closure of stomata after the exposure period (Amundsen and Weinstein, 1981). Beckerson and Hofstra (1979a,b) and Brymer (1982) also observed changes in response of stomata if exposures were repeated for several days. Thus stomatal responses to SO$_2$ are very dynamic, and cannot be elucidated from occasional measurements of stomatal behavior.

   Another explanation for the variation observed in the responsiveness of a particular species or individual is that prevailing environmental conditions also may modify the behavior of stomata in response to SO$_2$. Water stress may either enhance or depress stomatal sensitivity to SO$_2$. Unsworth, Biscoe, and Pinckney (1972) reported that water-stressed bean plants exhibited a greater enhancement of stomatal apertures, but Olszyk and Tibbitts (1981a,b) reported that *Pisum sativum* plants grown under water-stressed conditions exhibited more stomatal clo-

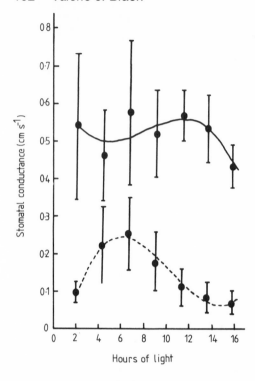

Fig. 6.3. Stomatal conductance of pea plants exposed for 2-h intervals at seven successive times during a 16-h light period, to 1.76 μl l^{-1} (73.2 μmol m^{-3}) of SO$_2$ (dashed line), compared to control plants (solid line). For exposed plants, $y = 0.130315 + 0.146795h - 0.017488h^2 + 0.000570h^3$, $r = 0.729$, $S_{yh} = 0.066$, $n = 56$. For control plants, $y = 0.632510 - 0.67441h + 0.010211h^2 - 0.000427h^3$, $r = 0.333$, $S_{yh} = 0.125$, $n = 56$. Values are averages of eight single-plant replicates with standard deviations indicated by bars. Adapted from Olszyk and Tibbitts (1981b).

sure in response to SO$_2$ than plants exposed to SO$_2$ but maintained at field capacity. A similar increase in closure has been reported also in plants exposed to ozone (Rich and Turner, 1972).

The water-vapor content of the atmosphere in which plants are exposed also influences the effect of SO$_2$ on stomatal behavior. Mansfield and Majernik (1970) observed that when broad-bean plants were exposed to low concentrations of SO$_2$ in conditions of low humidity, stomata exhibited reduced conductances relative to those of unpolluted plants. This was in contrast to the enhanced conductances observed when plants were exposed when relative humidity was high. This work was extended by Black and Unsworth (1980), who showed that these humidity-dependent responses were not unique, but also occurred in field beans, tobacco, and sunflower, though not in other species tested (Fig. 6.4). This humidity response occurred only in species able to exert some control over water loss by reducing stomatal apertures as evaporative demand increased. Perhaps the factors controlling stomatal responses to humidity interact with the effects caused by SO$_2$, to result in increased stomatal closure at low humidities. Increased stomatal closure in birch trees exposed to SO$_2$ under conditions of low humidity has been observed also by Mansfield and Freer-Smith (1984), whereas

Rosen, Musselman, and Kender (1978) suggest that the closure they observed in grape leaves might be partially dependent on the 50% relative humidity at which all fumigations were run. Other environmental factors during exposure may indirectly induce changes in stomatal aperture. No direct influence of light or carbon-dioxide concentration on SO$_2$-induced changes in stomatal behavior has been observed. However, both these factors, by their effect on stomatal apertures, will modify the flux of pollutant into the plant and thus the physiological and biochemical consequences.

Permanence of stomatal responses. Stomatal responses to SO$_2$ thus appear to be very complex. However, it is important to gauge whether these responses are transient or are maintained after exposure, and whether these SO$_2$-induced effects cause permanent malfunctioning of stomata and alterations in transpiratory water loss of plants. Unfortunately, few reports are available in which the permanence or reversibility of stomatal responses to SO$_2$ has been examined. Stomatal responses are often inferred from measurements of transpiration in uncontrolled and ill-defined conditions, and are rarely monitored throughout and following the exposure period. In addition, the influence of repeated dosages has been studied only occasionally. Part of the reason for this gap in our understanding of stomatal responses to pollutants lies in the difficulties of assessing whether a response is

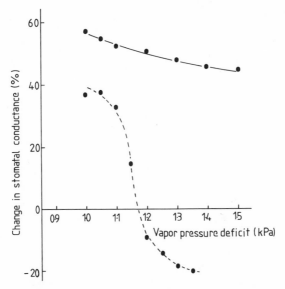

Fig. 6.4. Variations of the percentage change in stomatal conductance, $[(x - y)/y] \times 100$, where x = conductance (SO$_2$) and y = conductance (control), with vapor-pressure deficit in plants exposed to 0.035 ppm (1.5 μmol m^{-3}) of SO$_2$; solid line *Phaseolus vulgaris*; dashed line *Helianthus annuus*. Data from Black and Unsworth (1980).

reversible. Stomata respond very sensitively to many environmental factors or combinations of factors, and to the functioning of other physiological and bio-chemical processes within the plant itself. These factors, therefore, must either be kept constant during and following the exposure period, or be accounted for in the assessment of the observed stomatal responses. For example, following exposure, the effect of SO_2 may become progressively more severe, with several different symptoms developing, such as impairment of photosynthesis or damage to mesophyll cells. These effects would lead to changes in carbon-dioxide concentration or in leaf-water potential within the leaf, and these would alter stomatal behavior. These nonpollutant-induced changes would be very difficult to separate from stomatal effects arising from the removal of the pollutant from the atmosphere. In addition, the plant itself may modify its stomatal aperture to compensate for pollutant-induced changes in water loss or reductions in photosynthesis, thus amplifying or masking stomatal responses to the removal of the pollutant.

Because of these problems, the literature has reported few attempts to elucidate the permanence of these changes in stomatal aperture. Suwannapinunt and Koslowski (1980) observed that transpiration rates of *Robinia* and *Acer* plants increased progressively during exposure to SO_2, but these stimulating effects disappeared as fumigation was prolonged. Recovery to unpolluted conductances during exposure has been suggested by Bell, Rutter, and Relton (1979) and reported by Winner and Mooney (1980a,b,c). Bonté, DeCormis, and Louguet (1977) found, however, that the rapid stomatal closure observed in *Pelargonium* was not maintained following exposure. Similar responses were reported by Taniyama (1972), Unsworth, Biscoe, and Pinckney (1972), and Sij and Swanson (1974), although Black and Unsworth (1980) found that stomatal effects were maintained for several days following exposure, and Majernik and Mansfield (1970) and Ashenden (1979a,b) reported that recovery took place only several hours or days following exposure. Beckerson and Hofstra (1979a,b) and Brymer (1982) treated various species with SO_2 for several hours per day for several days, and showed opening and closing responses, respectively (Fig. 6.5). Both sets of workers observed that these changes in stomatal conductance could be maintained until the following day, and that stomata did not seem to become insensitive to these repeated exposures.

There is, therefore, very little conclusive information to elucidate the permanence of these SO_2-induced effects on stomata, or to explain the mechanism of these responses. It is necessary, therefore, not only to carry out repeated exposures and detailed, frequent observations of stomatal responses, but also to monitor simultaneously other factors, such as photosynthetic rates, water status, and appearance of visible injury. Only then will it be possible to separate direct effects of SO_2 on stomata from the indirect responses to other environmental and plant parameters. Significant advances have been made during the last few years in describing stomatal responses to SO_2, but we are still not able to make accu-

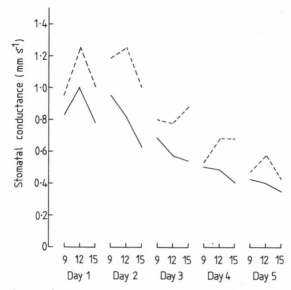

Fig. 6.5. Variation with time of the stomatal conductance of adaxial surface of primary leaves of *Phaseolus vulgaris* exposed in clean air (full line) or 0.150 ppm (6.2 μmol m^{-3}) of SO$_2$ (dashed line). Data from Beckerson and Hofstra (1979b).

rate predictions of stomatal responses of most plant species exposed to pollutants. In summary, there seems to be no unique stomatal response to SO$_2$; there are differences related to time of exposure, species stage of development of the plant, leaf age and prehistory, and environmental conditions, as well as to characteristics of the pollutant exposure itself, such as duration and concentration.

Mechanisms of SO$_2$ Action on Stomatal Behavior

What is the basis for these differences in stomatal sensitivity not only between species and between stomata on a given leaf, but also with prevailing environmental conditions? Are we in a position to identify sites of action of SO$_2$ and its derivatives, or the mechanisms to explain these changes in stomatal behavior induced by SO$_2$? There are several hypotheses emerging as our understanding of stomatal mechanisms increases, and as new techniques for use in investigating the functioning of stomata are developed.

Action of high SO$_2$ concentrations. Several mechanisms have been shown or postulated to be affected by SO$_2$, and may be responsible for the stomatal behavior observed. For example, the large reductions in aperture observed when plants are exposed to high pollutant concentrations may be the result of drastic impairment of photosynthesis and water status, and the development of visible injury, leading subsequently to reduced stomatal conductances. In addition, damage to cells of the stomatal complex induced by high concentrations of SO_2 could lead to closure. For example, Black and Black (1979a,b) have observed reductions in the viability of guard cells and the integrity of guard-cell chloroplasts in response to high concentrations of SO_2. Since stomatal apertures are determined by the balance between the turgor of these guard cells and that of adjacent epidermal or specialized subsidiary cells, any factor that alters this balance will cause changes in stomatal conductance. For example, a reduction in guard-cell turgor will increase stomatal closure. A similar role for SO_2 in damaging guard-cell metabolism, or the interaction between guard cells and subjacent mesophyll cells, has been suggested by Kimmerer and Kozlowski (1981).

Action of low SO_2 concentrations. It has proved more difficult to provide an explanation for observed stomatal responses to low concentrations of sulfur dioxide. Not only can stomata be stimulated by SO_2 to open or close, but these responses may be reversible, and stomata can still retain the capacity to respond to other environmental factors such as light, humidity, and carbon-dioxide concentration. Investigations of the direct effects of SO_2 on stomatal cells have allowed several hypotheses to be put forward to explain the action of SO_2.

In 1973, Biscoe, Unsworth, and Pinckney postulated that the turgor of epidermal cells adjacent to guard cells must be preferentially affected by SO_2, to explain the increased stomatal opening they observed in polluted bean plants. They thought this explanation more likely than the alternative possibility that SO_2 induced enhanced osmotic potentials and turgor of the guard cells. This view has been supported by Black and Black (1979a,b), Black and Unsworth (1980), and Suwannapinunt and Kozlowski (1980). Black and Black (1979a,b) investigated the turgor relations of cells of the stomatal complex using scanning electron microscopy, and showed that exposure to SO_2 for even 2 hours would cause collapse of adjacent epidermal cells, an effect that was maintained for over 24 hours. When they examined the viability of epidermal cells in epidermal strips taken from polluted plants, they also found that viability of these cells was reduced compared with that of controls. This observation does not indicate conclusively that the proportion of dead cells increases *in situ* on a leaf exposed to SO_2, but it does show that SO_2 causes some cellular weakening, possibly to membranes, which is increased when epidermal strips are taken and tested for viability. This preferential damage to epidermal cells was also postulated by Suwannapinunt and Kozlowski (1980), who suggested that preferential uptake of SO_2 by subsidiary cells would increase membrane permeability, decrease cell turgor, and therefore increase stomatal opening. As the exposure concentration is increased, the induction of stomatal opening subsides, and eventually an increase in stomatal closure results (Black and Black 1979a,b; Black and Unsworth 1980; Suwannapinunt and Kozlowski 1980; Taylor, Reid, and Pharis 1981). Suwannapinunt and Kozlowski believe that this closure is associated with increased SO_2 concentrations in the substomatal cavity and with inhibition of photosynthesis, rather than with the guard-cell damage postulated by Black and Black (1979a,b), Black and Unsworth (1980), and Taylor, Reid, and Pharis (1981). However, these hypotheses are not mutually exclusive, and stomatal closure may result from SO_2 action on both photosynthesis and guard cells.

Preferential damage and reductions in the turgor of epidermal cells caused by low concentrations of the pollutant, along with damage to guard cells when concentrations are higher, would help explain the different opening and closing responses observed with different pollutant treatments. If this explanation is correct, why do some species exhibit only stomatal closure even when concentrations of pollutant are very low? Mansfield and Freer-Smith (1984) have proposed that the guard cells of species that exhibit a closing response may be more sensitive to SO_2 and its derivatives than species that exhibit only an opening response. They suggest that this sensitivity could result if the guard cells of these species lacked chloroplasts, and thus were not able to convert SO_2 to hydrogen sulfide by using energy from photosynthetic electron transport. However, the absence of chloroplasts cannot be the sole explanation for these different responses, since *Pelargonium*, which Bonté, DeCormis, and Louguet (1977) have shown to exhibit very marked and rapid stomatal closure when exposed to very low concentrations of SO_2, does possess guard-cell chloroplasts. This observation does not preclude the existence of different capacities for detoxification in the guard-cell chloroplasts of different species.

In contrast, Heath (1980) hypothesized that the responses to low concentrations of SO_2 result not from reduced turgor of epidermal or subsidiary cells, but from a build up of sulfite concentrations in the guard cells that leads to an increase of osmotic pressure. At high concentrations, guard-cell photosynthesis would be inhibited, osmotic pressure would drop, and stomatal closure would ensue. There is no evidence now available, however, to confirm this hypothesis.

Bonté, DeCormis, and Louguet (1977), however, rejected the hypothesis that responses to low concentrations of SO_2 were purely passive. Their experiments with *Pelargonium* indicated that effects of SO_2 were active, and were perhaps due to an interference with the respiratory formation of the ATP necessary to drive fluxes of potassium and hydrogen ions across the plasmalemma membranes of cells of the stomatal complex. Taylor, Reid, and Pharis (1981) also believe that SO_2 must interfere with the mechanisms of movement of monovalent ions, in particular, of potassium, and with the synthesis of certain organic ions such as malate. Their results do indicate that high concentrations of SO_2 interfere with the basic processes involved in the production and movement of these ions. In addition, these workers and Kondo, Maruta, and Sugahara (1980) have observed an interaction between SO_2, stomatal behavior, and amounts of abscisic acid (ABA) in stomatal cells. Taylor, Reid, and Pharis (1981) suggested that there was an antagonistic relationship between SO_2 and ABA, and that stomatal response depended on which compound was present in greater concentration. For

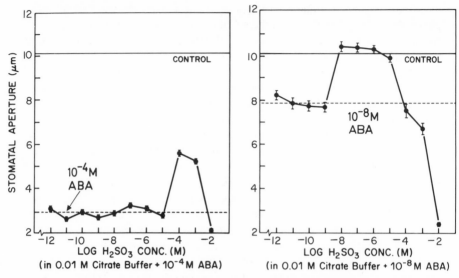

Fig. 6.6. The effect of sulfurous acid in 10 mM sodium citrate buffer (*p*H 5.8) on sto-matal aperture in *Vicia faba* epidermal strips in the presence of ABA. Incubation was as in Fig. 6.1. The mean width of 180 stomates ± SE is given for each concentration of sulfurous acid. The control lines represent stomatal response in buffer alone without ABA or sulfurous acid. *Left:* sulfurous acid plus 10^{-4} M ABA. *Right:* sulfurous acid plus 10^{-8} M ABA. (Reproduced from Taylor, Reid, and Pharis, 1981.)

example, if ABA was low in the plant during fumigation, stomatal opening would increase, whereas the presence of high amounts of ABA would counteract this opening, and stomatal closure would result (Fig. 6.6). Kondo and Sugahara (1978) and Kondo, Maruta, and Sugahara (1980) indeed have found that species that contain inherent high concentrations of ABA exhibited a larger rapid reduc-tion in transpiration rates on exposure to SO_2, whereas in those species with very low amounts of ABA, transpiration rates initially increased slightly, then ex-hibited a small decrease. They suggest that SO_2 may lower the *p*H of stomatal cells, and thus reduce the effective ABA concentration necessary to cause closure.

Although the SO_2 concentrations used by Kondo *et al.* were very high (~2 ppm, or 83 μmol m^{-3}), the interaction they observed between SO_2 and ABA concentrations does suggest that SO_2 might act by interfering with the flux of important ions or the integrity of membranes of the stomatal cells. Olszyk and Tibbitts (1981a,b) have expressed similar views. From the evidence that ozone can increase the flux of potassium ions across membranes, they suggest that pol-lutant-induced efflux of ions from guard-cell membranes would likely lead to a decrease in the osmotic pressure of guard cells, followed by a loss of water and a decline in guard-cell turgidity. This chain of events would lead to increased clo-sure, but does not explain the opening responses observed when some species are exposed to SO_2. They do point out, however, that the membranes of subsidiary cells could also be important sites of pollutant action, since reduced turgor of these cells could affect stomatal aperture. Several other possible mechanisms for stomatal responses to SO_2 are summarized by Olszyk and Tibbitts (1981a,b). For example, the decreased pH of cell cytoplasm as a result of uptake of SO_2 may affect hydrogen pumps or change the structure of membranes in ways that alter potassium flux from cell to cell. Alternatively, by interfering with important metabolic processes within stomatal cells, the pollutant may alter the amounts of metabolites that regulate the permeability of guard membranes.

Although the specific mechanism (or mechanisms) to explain the complex responses of stomata to SO_2 has not been identified, the evidence available does indicate that SO_2 is likely to interfere with the production or movement of ions that drive stomatal opening or closure. To define whether these are the primary sites of action, it is necessary to investigate how SO_2 interferes with the complex reactions that occur within epidermal and guard cells. Techniques developed to study stomatal responses to other environmental factors must now be extended and developed for use in understanding responses to pollutants such as SO_2. Such an approach is the one most likely to clarify the diverse responses to SO_2 and the interrelationship between these and environmental conditions, plant status, and stage of development.

Techniques for the Study of Stomatal Responses to SO_2

Stomata are extremely responsive to many environmental factors and to the rate of other processes operating within the plant; so it is often difficult to separate the direct effects of SO_2 on stomatal responses from those of other internal and external influences. A wide range of techniques has been used to investigate pollutant-induced effects on stomatal behavior and functioning. However, each technique is usually suitable only for answering a specific type of question. For example, a technique used to monitor the behavior of a single stoma will not permit results to be extrapolated to define the responses of whole plants growing in the natural environment. Similarly, the assessment of changes in water loss by whole plants by weighing will not help identify the mechanism by which SO_2 affects these stomatal changes. In addition, each technique has its own limitations and inherent problems.

Measurements of growth. Perhaps the most unsatisfactory method for defining stomatal responses to SO_2 is making assumptions about stomatal behavior from growth measurements. For example, Roberts (1975) proposed that the differences in growth he observed when he exposed tree seedlings to SO_2 indicated that stomata differed in their sensitivity to SO_2, but no physiological measurements were made to substantiate the proposal. Although this hypothesis was confirmed by Biggs and Davis (1980), they did state that these different stomatal responses were probably limited to relatively low concentrations of SO_2.

Measurements of water loss. Assessments of stomatal behavior are commonly inferred from measurements of water loss from plants. These losses may be quantified either by weighing (Kondo, Maruta, and Sugahara, 1980) or by measuring the addition of water vapor into an air stream, using techniques such as infrared gas analysis or dew-point hygrometry (Black and Unsworth, 1979c; Winner and Mooney, 1980a,b,c). However, in order to make quantitative measurements of stomatal behavior from water-loss measurements, the magnitude of nonstomatal water loss and the influence of external factors that affect rates of water loss have to be measured simultaneously. For example, the magnitude of the leaf boundary layer, which is influenced by wind speed, will modify transpiration rates. Similarly, pathways of water loss through leaf cuticle may be large if the cuticle is damaged. In addition, rates of water loss are influenced strongly by the concentration gradient of water vapor between the leaf and the atmosphere. However, if all these variables can be quantified, then stomatal conductances can be calculated.

Measurements of water loss can be made from one or more plants or leaves contained in chambers or cuvettes. A variety of cuvette designs are available for monitoring stomatal behavior under environmentally controlled conditions; these have been discussed by Jarvis *et al*. (1971). However, additional factors in chamber design and experimental procedure need to be considered when plant material is exposed to pollutants and physiological responses are monitored. SO_2 in particular is a very reactive gas; it is readily absorbed and adsorbed by water and most surfaces. Uptake by tubing and cuvette walls, especially if wet or dirty, may significantly reduce the concentration of SO_2 to which plants are exposed (Unsworth and Mansfield, 1980). Cuvettes should be constructed of materials such as teflon that exhibit minimal sorptive properties; or if there is unavoidable sorption of SO_2, care should be taken to account for this loss before physiological responses are related to exposure concentrations or to SO_2 uptake by the plant.

The environmental conditions within exposure chambers should be known and controlled (Legge, Savage, and Walker, 1979). Stomata respond to a large number of climatic variables, e.g., light, temperature, CO_2 concentration, relative humidity, and wind speed. Each of these variables will influence not only stomatal behavior directly, and indirectly via changes in rates of photosynthesis, but also the response of stomata to the pollutant. For example, relative humidity may determine whether stomata open or close in response to SO_2 in a number of species (Mansfield and Majernik, 1970; Black and Unsworth, 1979c, 1980). Rates of pollutant uptake are influenced by changes in stomatal and boundary-layer conductance that are induced by environmental conditions. For example, airflow characteristics of cuvettes should maximize boundary-layer conductance, but without causing plants to become water-stressed. Ashenden and Mansfield

(1977) and Black and Unsworth (1979c) showed that the aerodynamic conditions within exposure chambers may influence the sensitivity of plant response to SO$_2$.

It must be recognized that the environment within a cuvette may only approximate that of the natural environment; stress, or even abnormal behavior or growth, may arise when, for example, the quality or quantity of irradiance is inadequate or cuvette temperatures rise several degrees above ambient. Plants whose roots have been placed in polythene bags to restrict water and CO$_2$ loss from the soil may exhibit epinasty and display changes in stomatal behavior independent of the pollution treatment. Care must also be taken to ensure that plants grown in growth rooms and in the natural environment respond similarly to pollutant exposure. Plants grown in these two environments may, for example, exhibit dissimilar stomatal and cuticular characteristics.

Specific chamber designs are suited to particular purposes. Large chambers such as the Continuous Stirred Tank Reactor (C.S.T.R.) developed by Heck, Philbeck, and Dunning (1978) allow several whole plants to be exposed to pollutants, and their physiological responses to be monitored throughout plant growth and development. These reactors have been used successfully by Rogers *et al.* (1977) and Brymer (1982) to investigate pollutant uptake by plants and physiological responses induced by pollutants. Although rates of photosynthesis and transpiration can be measured successfully using this system, these chambers are not ideal for measuring subtle changes in stomatal behavior.

Smaller whole-plant chambers have been used (Ashenden, 1979a,b; Black and Unsworth, 1979c). These chambers present less major problems in the accurate air conditioning of the volumes of air necessary to give an adequate number of air changes per minute within the chambers. Thus they permit the simultaneous monitoring not only of photosynthesis and transpiration, but also of stomatal conductances continuously for several days. Physiological responses to pollutants can therefore be assessed in plants exposed in a range of environmental conditions. In addition, assessments of uptake of SO$_2$ by plants can be made, permitting physiological responses to be related to pollutant uptake rather than to exposure concentrations or doses.

This type of chamber does not permit, however, the separation of responses of individual leaves or leaf surfaces. Such investigations can be carried out in smaller cuvettes, in which single leaves or surfaces can be exposed to pollutants and their physiological responses monitored, including small changes in stomatal behavior (Legge, Savage, and Walker, 1979; Winner and Mooney, 1980a,b,c). However, care must be exercised to ensure that the rest of the plant is exposed to environmental conditions similar to those of the leaf in the cuvette, or spurious results may be obtained. Growth-room environments, for example, tend to have a high CO$_2$ concentration when inhabited by the experimenter. Ideally, a two-chamber system should be operated, in which the whole shoot is surrounded by a large, air-conditioned chamber, within which are small cuvettes attached to individual leaves. Such a system, however, may be complex to operate and maintain.

Exposure of single leaves to pollutants does permit accurate control of the environmental conditions within the chambers, rapid introduction of pollutant, assessment of pollutant uptake by the plant, and sensitive and accurate measurement of stomatal responses to SO_2. In contrast to the whole-plant chamber system, these measurements of the responses of a particular leaf will not reflect the great variation in the responses of the different leaves on the plant. The use of any type of enclosure does have the advantage that concurrent measurements of photosynthetic rates can be made, thus permitting the assessment of the indirect responses of stomata that arise from changes in photosynthesis and in CO_2 concentrations in the substomatal cavity.

Porometry. Stomatal responses to pollutants can be monitored by porometry, a technique of great practical use on plants exposed to pollutants in both laboratory experiments and in the natural environment. The instrument is portable and can be used to examine water loss from many leaves, from individual surfaces, or from different parts of the same leaf with great rapidity. It is a nondestructive technique and can therefore be used on plants exposed to pollutants in large chambers; and it allows simultaneous measurements of photosynthesis, growth, and pollutant flux to be carried out on the same plants for a long period of time. Porometry has been used widely, therefore, in air-pollution research (Majernik and Mansfield, 1970; Unsworth, Biscoe, and Pinckney, 1972; Beckerson and Hofstra, 1979a,b; Black and Black, 1979a,b; Rist and Davis, 1979; Amundsen and Weinstein, 1981; Kimmerer and Kozlowski, 1981; Olszyk and Tibbitts, 1981a,b).

The various types of porometers have been reviewed by Meidner (1981). In essence, the porometric technique operates by placing a small enclosure, or porometer cup, over part of a leaf and assessing either the water-vapor loss from this part of the leaf or the pressure required to bring about a flow of gas from one surface of the leaf to the other, i.e., diffusion or viscous-flow porometry. A viscous-flow porometer permits a continuous record of stomatal responses to SO_2 to be made, since the porometer cup can be left successfully on the leaf for long periods (Majernik and Mansfield, 1970). However, it can be used only on hypostomatous leaves, and gives an estimate of the average stomatal conductance of both leaf surfaces. In contrast, a diffusion porometer may be used on both amphistomatous and hypostomatous leaves, and allows water loss from individual leaf surfaces to be measured. This is an important characteristic, since stomata on upper and lower leaf surfaces may respond differently to SO_2. A diffusion porometer may be used only to make intermittent measurements of water loss, since it cannot be left permanently attached to a leaf. Environmental conditions within the cup attached to the leaf do not reflect those of the ambient environment, in particular light, CO_2, and water-vapor concentration, factors that greatly influence stomatal behavior. In addition, the stomata contained within the porometer cup are no longer in contact with the pollutant.

The porometric technique, however, provides information about the total water loss from only the area of leaf covered by the enclosure. Measurements will include, therefore, water loss via both the stomatal and cuticular pathways. Usually cuticular water loss is minimal compared to the loss through open stomata, and may be ignored. However, when leaves are weathered or damaged, significant water loss may occur via the cuticular pathways, and it must then be taken into consideration. This technique has made it possible to identify how such factors as stomatal position on a leaf, leaf age, stage of photoperiod, and duration of exposure affect stomatal response to SO_2.

Microscopic examination of stomatal apertures. Changes in stomatal behavior may be assessed by the microscopic examination of stomatal apertures. Meidner (1981) has advocated the use of this technique in stomatal physiology, since, although it is time-consuming, it does give precise information about the size of the standard pore. Stomata may be examined by direct examination, or stomatal impressions may be made of stomatal pores, and these measured using a microscope. Both methods have inherent difficulties, since the width of the stomatal pore varies at different depths; so it is difficult to ensure that measurements are made at a similar depth in each stomatal pore. The impression technique is of particular use when many simultaneous measurements are necessary. Impressions can be made rapidly, and measurements carried out at a later date. These techniques, however, have not been used widely in the study of stomatal responses to SO_2.

The behavior of stomata can be examined in epidermal strips. Lange *et al.* (1971) were able to use this technique to demonstrate the responsiveness of the stomata of certain plant species to relative humidity. The technique of isolating the layer of epidermal cells from the underlying mesophyll is sometimes difficult to carry out successfully. Cells may be killed if care is not taken, and indeed relatively few plant species can be stripped successfully. Where it can be carried out proficiently, it is a very useful tool for examining the responses of single stomata, and elucidating the mechanisms of stomatal response to SO_2. Either strips can be taken from plants previously exposed to SO_2 and their responses examined, or alternatively strips from unpolluted plants can be treated with sulfite or bisulfite in solution, and stomatal responses observed under the microscope. Taylor, Reid, and Pharis (1981) state that "there is some justification for assuming that the physiological effects produced by exposure to buffered H_2SO_3 are the same as those which occur following fumigation." Indeed, these workers have used this technique successfully, since it has several obvious experimental advantages over exposure of plants to gaseous SO_2. However, although it is a very useful tool in the definition of stomatal responses to SO_2, caution is needed in extrapolating from results obtained in these artificial pollution treatments to mechanisms that would explain the action of gaseous SO_2 in intact plants.

The viability of cells and the integrity of cellular organelles can be assessed by using various kinds of microscopes. Changes in the chloroplasts of guard cells of polluted plants have been observed under the light microscope (Black and Black, 1979a,b), and the transmission electron microscope was used to show that SO_2 could cause changes in the fine structure of chloroplasts (Wellburn, Majernik, and Wellburn, 1972). Scanning electron microscopy permits observation of the effects of SO_2 on leaf surface structure as long as care is taken to avoid preparation artifacts.

Many techniques are available, therefore, to investigate the different aspects of SO_2-induced effects on stomata. However, wrong techniques were often used in the past to fulfil specific aims, and prerequisites for successful use of these techniques, such as environmental control, have often been ignored. Assumptions have been made from insufficient data, or results obtained from an inappropriate technique. For the successful study of stomatal responses to SO_2, several criteria need to be satisfied. Not only do techniques need to be used with care and for the correct purpose, but a rigorous experimental protocol needs to be adhered to. This includes adequate rearing and handling of plants; sufficient control of environmental conditions and pollutant concentrations; relating effects to pollutant uptake rather than to external concentration; ensuring adequate measurements of stomatal behavior; measurement of the physiological processes that may indirectly affect stomata; and recognition of the great variability in stomatal responses to SO_2.

Implications of Changes in Stomatal Behavior for Growth and Yield

It is known that stomata do respond to short-term fumigations with SO_2, and mechanisms to explain these responses are partly understood. However, are these responses important in the long term, perhaps resulting in significant reductions in growth or dry-matter production? It has proved very difficult to assess these long-term implications, since stomatal responses to SO_2 are so diverse and dependent on many factors. For example, within a forest or crop canopy, stomatal responses to SO_2 will vary spatially and with time, in concert with variation in pollutant concentrations, environmental conditions, and biological state of each particular plant part. The long-term effects of stomatal responses to SO_2 can be postulated only if certain assumptions are made. For example, Unsworth, Biscoe, and Pinckney (1972) suggest that the increased stomatal opening they observed could lead to "increased transpiration rates leading to early and more severe water stress in spells of dry weather, and therefore a restriction of growth." These consequences may not result, however, if responses to SO_2 are not maintained for a significant time period, or if water is plentiful and available to the crop. However, in times of water stress, this reduced ability to restrict water loss could lead to changes in water potential and in photosynthesis, and to early senescense. In contrast, any stomatal closure that results from SO_2 exposure would ensure a beneficial restriction of water loss under these conditions.

Of greater significance are the consequences of altered stomatal conductances for carbon-dioxide exchange and SO_2 uptake. An increase in conductance will increase CO_2 exchange and potential photosynthetic rates. However, these advantages may be outweighed by the associated increase in flux of SO_2 to the sensitive metabolic sites within the plant. Closure in response to SO_2 could be regarded as having the opposite effect, reducing not only SO_2 flux but also photosynthesis. Unlike the opening responses, increased closure could form the basis for a useful resistance or avoidance mechanism, as long as these responses occur rapidly on exposure to pollutants, and are temporary, with apertures returning to normal following exposure. This type of beneficial response was observed in *Petunia* cultivars exposed to ozone (Butler and Tibbitts, 1979). However, if stomatal functioning is permanently impaired, the advantages of reduced pollutant flux conferred by closure may be negated by consequent reductions in photosynthesis, especially if pollutant exposures are infrequent, of short duration, or in concentrations unlikely to result in significant metabolic damage.

There is considerable evidence to support the hypothesis that stomata do have a role in determining plant sensitivity to SO_2. Kondo and Sugahara (1978) concluded that the number and size of stomata and the rapidity of stomatal closure on exposure to pollutant determined the resistance to SO_2. However, Mansfield (1976) pointed out that the plant sensitivity to SO_2 may be determined not by stomata alone, but also by the magnitude of other pathways of pollutant uptake and by the existence of physical and biochemical resistances to pollutant flux from the atmosphere to the sensitive metabolic sites within cells. Indeed, Bressan, Wilson, and Filner (1978) observed that the relative absorption of pollutant gas via the stomatal pores did partly explain the different sensitivities of species and varieties of the Cucurbitaceae, but suggested that a biochemically based resistance mechanism must also function after SO_2 has entered the leaf. It is also important to define the responsiveness of stomata at particularly sensitive periods of growth and development, and to find out whether preconditioning or frequent exposures to SO_2 change the sensitivity of the stomata to the pollutant. Only then can rational judgments be made about the importance of stomatal behavior in determining pollutant sensitivity.

It is known that stomata respond to SO_2 under laboratory exposures, but do these responses occur in the natural environment, and if so, do they lead to changes in subsequent growth and development of plants? Definitions of stomatal responses to SO_2 are difficult in the laboratory under controlled conditions; in the natural environment this task is even more problematical. Hällgren *et al.* (1982) observed a lack of stomatal responses in pine needles growing out of doors exposed either to "natural" fumigations or to additional pollutant concentrations. He does point out, however, there are no true controls in the natural environment with which to compare responses. This is true of most field fumigations, since the presence of low concentrations of SO_2 are common even in rural sites in Britain (Martin and Barber, 1981). In addition, stomata respond not only to SO_2 but also to the natural fluctuations in environmental conditions in the atmosphere and soil. Thus direct stomatal responses to SO_2 are difficult to identify, unless changes in aperture are large and relatively permanent. Definition of effects will also be difficult if stomata become less responsive with repeated exposures. Nevertheless, there is a little circumstantial evidence that some species show altered stomatal behavior in "naturally" polluted environments. For example, Godzik and Piskornik (1964) observed large increases in water loss from large chestnut leaves growing in a polluted area near a coke-producing plant. As in most environments, pollutants rarely exist singly but more often in combination; so these effects cannot be attributed specifically to SO_2.

Once stomatal responses to long-term exposures to pollutants have been defined, selection or breeding of resistant varieties of major agricultural and horticultural crops may be possible. However, such a program would have to be based on the proviso that other factors, such as yield production, quality, or marketability, were not impaired by the inclusion of these stomatal characteristics. Unfortunately, this approach cannot be extended to any great degree toward the protection of natural ecosystems. Perhaps after a prolonged history of exposure to SO_2, natural communities may have already evolved resistance mechanisms that involve modifications of stomatal behavior, especially if these confer advantage to the survival of the plant. There is, however, no evidence that any such evolution has occurred. Indeed, Bell and Mudd (1976) reported that S35 Ryegrass, a variety known to have evolved resistance to SO_2, does not exhibit stomatal behavior that in itself would confer resistance.

Directions for Future Research

Research carried out during the last few years has increased our understanding of stomatal responses to SO_2. The degree of opening or closing of stomata reflects an integrated response to a number of environmental factors and important physiological and biochemical processes occurring within the plant. The net stomatal response will be determined by the balance between factors that stimulate increased opening, and those that stimulate stomatal closure. SO_2 may act as an additional factor involved in determining this balance. When factors are biased toward stomatal opening and SO_2 concentrations are low, the action of SO_2 increases this opening. However, if conditions are such that stomatal closure is favored, the balance of responses to these factors will outweigh any SO_2-induced opening; indeed, under these conditions SO_2 may act as an additional stress factor causing increased closure. This may occur, for example, when other stress factors, such as lack of water and low humidity, operate on the stomatal complex, or when pollutant concentrations are very high. The interference of SO_2

with this balance could explain why polluted stomata still retain the ability to respond to changing environmental factors, albeit slightly out of phase.

Several areas of research are developing that promise to clarify these responses. Techniques previously unique to studies of pure stomatal physiology for examining the physiology and biochemistry of cells of the stomatal complex are now being used in air-pollution research. These should yield results that will not only aid our understanding of the mechanisms of stomatal responses to SO$_2$, but also be useful for developing criteria for the selection of resistant varieties. For example, the sensitivity of guard cells to SO$_2$ may determine stomatal behavior and ultimately the productivity of the plant itself.

Investigations of stomatal responses to SO$_2$ must also be extended and developed; we especially need a better understanding of how responses differ with species and stage of development, the magnitude and longevity of responses, the influence of prevailing environmental conditions during exposure, and the relationship between effects and pollutant uptake. Recent advances, in chamber design, sophisticated monitoring facilities, and environmental control, will facilitate studies of these subtle stomatal responses to SO$_2$. From such a knowledge of responses and the mechanisms responsible, an accurate prediction of stomatal behavior of plants exposed to SO$_2$ should be possible.

Although these investigations require different experimental approaches, they should be regarded not as discrete projects, but rather as complementary lines of research carried out in association with studies of other plant responses to SO$_2$. Similarly, since pollutants rarely occur singly in the atmosphere, responses to mixtures of pollutants must be investigated. In addition, we need to consider whether the effects elucidated in the laboratory occur in the natural environment and result in significant effects on growth and yield. It is, at present, difficult to envisage a satisfactory approach to investigating this complex problem. However, in order to assess the significance of SO$_2$-induced stomatal responses, field investigations must be given high priority. To achieve this aim, we must adopt an interdisciplinary approach in which the physical, chemical, physiological, biochemical, and growth aspects are investigated concurrently. Only then will it be possible to answer the many questions still perplexing air-pollution scientists.

7

Measuring and Assessing SO$_2$ Effects on Photosynthesis and Plant Growth

William E. Winner, Harold A. Mooney, Kimberlyn Williams, and Susanne von Caemmerer

Photosynthesis may respond to low SO$_2$ concentrations during even short time periods. For example, Black and Unsworth (1979a) found that photosynthesis of *Vicia faba* was reduced 10% within ten minutes of exposure to SO$_2$ at 0.07 ppm (2.9 μmol m^{-3}). This SO$_2$ concentration, which did not cause visible foliar injury, is lower than the U.S. Environmental Protection Agency secondary SO$_2$ standard, and is not unlike SO$_2$ concentrations periodically measured in ambient air at sites throughout Europe and North America. Thus it seems likely that SO$_2$ may cause small, temporary reductions in photosynthesis for many plants in farms and forests. The main goals of this paper are to discuss the biological importance of SO$_2$-caused changes in photosynthesis and to describe the methods used for monitoring photosynthetic responses to SO$_2$.

What is the physiological and biochemical basis for SO$_2$-caused changes in photosynthesis? Understanding these mechanisms will help define whether SO$_2$ exposures, such as described for *Vicia faba*, result in short-term or long-term photosynthetic changes. A few temporary, small depressions in photosynthesis may only slightly alter productivity during a plant's growing season, whereas long-term depressions or a few large depressions in photosynthesis may have a significant impact on plant growth, development, and reproduction. The underlying process by which SO$_2$ induces changes in photosynthesis is difficult to discover, because the photosynthetic rate of a leaf at any time is governed by several physiological factors, such as stomatal conductance, biochemical integrity of organelles, membranes, and enzymes in the leaf mesophyll cells, and leaf nutrient content. Although other plant processes are obviously related to photosynthesis, these physiological and biochemical factors provide a starting point for discussing the mechanisms of SO$_2$-caused changes in photosynthesis.

How is photosynthesis linked to plant productivity? It seems unlikely that a permanent 10% loss of photosynthesis of a fumigated *Vicia faba* plant would result in a 10% loss in either foliage or fruit production, because we would expect compensatory changes in allocation of plant resources. Although photosynthesis is an important component of plant growth, other processes, such as photosynthate translocation, mobilization of storage reserves, reallocation of nutrients between old and new leaves, and the switches that regulate the distribution of plant resources between roots, shoots, and reproductive plant parts, also influence patterns of plant growth. Here we will relate photosynthesis to plant productivity by discussing the effects of SO$_2$ on allocation and by demonstrating the interaction of photosynthesis and allocation in determining patterns of plant growth.

What are the instruments, techniques, and calculations needed to monitor the photosynthetic responses of plants to SO$_2$? Several technical advancements have improved our ability to treat plants with controlled doses of SO$_2$. These advances are reflected in gas-handling systems, cuvette design, and data-acquisition systems. In addition to improved experimental systems, recent advances have also been made in calculating rates of photosynthesis, transpiration, and SO$_2$ uptake from measurements of gas-exchange rates between leaves and air. In this paper we describe a gas-exchange system and the conceptual approaches that can be used to continuously monitor the photosynthetic responses of a leaf to SO$_2$.

SO$_2$-Caused Changes in Photosynthesis

Gas exchange by leaves. Leaf cells are bounded by a waxy cuticle, which provides a barrier to the exchange of SO$_2$ and water vapor between the leaf and air. The exchange of these gases occurs primarily through stomata, which respond to environmental or biochemical cues. The diffusion rate of gases through the stomata depends on (1) the diffusivity of the gases, (2) the magnitude of the concentration gradient for a specific gas, and (3) the degree to which stomata are open. In general, heavier molecules have lower diffusivities, and therefore diffuse more slowly than lighter molecules. The term "conductance" is used to describe the ease with which gas diffuses through the stomata. Conductance, the inverse of resistance, contains terms to account for stomatal aperture, density, and diffusivity of the gas. Conductance values, therefore, are specific for each gas, decrease as stomata close, and increase as stomata open. Plants can alter gas fluxes either by changing stomatal conductance or by changing their internal metabolism of the gas, which in turn alters the concentration gradient across the stomata.

Mechanisms of photosynthetic response to SO$_2$. SO$_2$ can affect photosynthesis by altering stomatal conductance or by changing the metabolic capacity of the mesophyll cells (Ziegler, 1975; Hällgren, 1978). SO$_2$ has been shown to cause either an increase (Black and Unsworth, 1980) or a decrease (Winner, Koch, and Mooney, 1982) in conductance. These changes in conductance will, in turn, change the rates of transpiration, SO$_2$ absorption, and CO$_2$ assimilation.

SO$_2$ that is absorbed into the leaf may directly alter the ability of mesophyll cells to fix CO_2. Absorbed SO_2 can oxidize cell and organelle membranes (Murray and Bradbeer, 1971), oxidize disulfide bonds of enzymes, destroying their tertiary structure (Anderson and Duggan, 1977), oxidize chlorophyll (Malhotra, 1977; Lauenroth and Dodd, 1981a), and compete with CO_2 for binding sites on the carbon-fixing enzymes for both C_3 and C_4 types of photosynthetic systems (Ziegler, 1972, 1973). Mechanisms by which SO_2 interferes with thylakoid function are discussed in Chapter 8, by Wellburn, in this volume, and the phytotoxic, free-radical reaction products are discussed in Chapter 9, by Peiser and Yang.

CO_2 assimilation and stomatal conductance are often closely correlated (Wong, Cowan, and Farquhar, 1979). This link presumably plays an important role in balancing carbon gain and water loss for plants growing where environmental factors continually change. This close correlation between conductance and photosynthesis may be lost after leaves are exposed to SO_2. More specifically, where SO_2 stimulates conductance, increases in photosynthesis rarely occur or are short-lived (Winner and Mooney, 1980c; Winner, Koch, and Mooney, 1982), because SO_2 decreases the photosynthetic capacity of the mesophyll cells. Photosynthetic decline is likely hastened by increased rates of SO_2 absorption caused by increased conductance. On the other hand, SO_2-caused decreases in conductance also reduce photosynthesis. Thus, whether SO_2 induces stomatal opening or stomatal closure, it generally has a negative influence on photosynthesis. In addition to directly influencing productivity, these SO_2-caused metabolic changes may also alter the capacity of plants to cope with other environmental factors, such as drought and competition.

We have developed a technique for partitioning SO_2-caused changes in photosynthesis between stomatal and nonstomatal components (Winner and Mooney, 1980b). Once plants are fumigated and SO_2 is removed from the airstream, ambient concentrations of CO_2 are increased in the cuvette. This increases the CO_2 concentration in the mesophyll even though stomata have closed. If photosynthetic reduction is related to stomatal closure, then photosynthesis will increase as internal CO_2 concentrations increase. On the other hand, if photosynthesis remains low while CO_2 internal concentration increases, then SO_2 decreases the CO_2 assimilation process by decreasing the ability of the mesophyll to fix CO_2.

Our results suggest that stomatal closure can occur within minutes after SO_2 exposure, and that, at least initially, photosynthetic decline is related to reduced conductance. As the fumigation continues, greater quantities of SO_2 are absorbed into the leaf, and the proportion of photosynthetic inhibition due to nonstomatal factors increases. Even though conductance and photosynthesis both decline, processes of stomatal function and CO_2 fixation are uncoupled; the stomata no longer respond to increases in CO_2 concentration in the leaf mesophyll that are experimentally induced, and photosynthetic decline may be much greater than would be expected from the magnitude of the SO_2-caused change in conductance (Winner and Mooney, 1980b,c).

Atmospheric Pollutants and Plant Growth

The effects of atmospheric pollutants, particularly SO$_2$, on photosynthetic capacity, allocation, and growth of plants have been well documented. However, the interactions between these processes have not been treated quantitatively. Here we will discuss each of these processes separately, and then attempt to interrelate them.

Photosynthetic capacity. As discussed above, it has now been demonstrated amply that various atmospheric pollutants, such as SO$_2$, have a direct effect on the photosynthetic capacity of leaves even at very low concentrations. As noted earlier, the principal pathway of pollutants into the leaf is through the stomatal pores. It is thus important to know what controls stomatal conductance. Recent work has demonstrated that photosynthetic capacity and conductance are closely correlated (Wong, Cowan and Farquhar, 1979) (Fig. 7.1). This fact enables us to make predictions, within certain limits, about the sensitivity of the photosynthetic process of different plant species to atmospheric pollutants. That is, leaves with the highest conductance, or with the least stomatal resistance to the entry of atmospheric gases, will receive the greatest amount of pollutants to the mesophyll, per unit time. The reduction in photosynthetic capacity of leaves is largely, though not entirely, dose-dependent at the mesophyll site (Winner, Koch, and Mooney, 1982); hence any predictor of photosynthetic capacity is also a predictor of stomatal conductance and thus of potential sensitivity of the leaves to damage by such pollutants as SO$_2$.

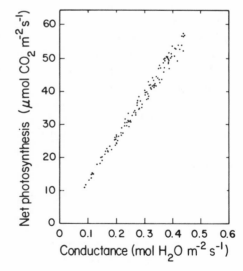

Fig. 7.1. Net photosynthesis versus conductance to water vapor in *Zea mays* leaves. Each point represents measurements from a different plant. Photosynthetic capacity was varied by altering nitrogen nutrition (from S. C. Wong, 1979).

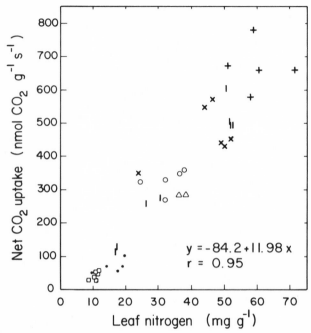

Fig. 7.2. Relationship between light-saturated net photosynthetic rates and leaf nitrogen contents. Values represent measurements on many different species (from Mooney and Field, unpublished).

Leaf nitrogen correlates strongly with light-saturated photosynthetic capacity (Fig. 7.2) and hence should be a rough indicator of potential sensitivity of leaves to pollutants, at least in the short term. Leaves of plants grown at high nitrogen availability and thus presumably having high conductance are relatively more sensitive to atmospheric pollutants than those grown at low nitrogen (Ormrod, Adedipe, and Hofstra, 1973). In these experiments, both stomatal conductance and the capacity of the leaf mesophyll to fix carbon were affected by the air pollutant. We do not have good information on the time course of this interactive process, or about how the carboxylating process adjusts itself to pollutants, if it does. We must add that even though Ormrod's experiments involved O_3 fumigations, most likely the trends reported and discussed here are applicable to SO_2 fumigations.

Carbon allocation and implications for growth. Each species has a carbon allocation schedule (Fig. 7.3), which may change during the life of the plant as well as with different environmental conditions (Fig. 7.4). A generalized model has been developed by Davidson (1969) to predict how allocation patterns shift with changing carbon or nitrogen capacities (Fig. 7.5). This model shows that

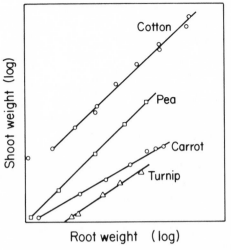

Fig. 7.3. Relationship between shoot and root weights from a variety of crops (from Russell, 1977, after Pearsall).

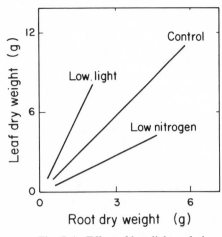

Fig. 7.4. Effect of low light and nitrogen on the proportional allocation to roots and shoots of *Phaseolus vulgaris* (from Russell, 1977, after Brouwer and de Wit).

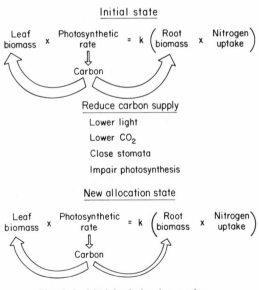

Fig. 7.5. Model relating interaction between root and shoot allocation (after Davidson, 1969).

factors that inhibit carbon fixation increase the fractional allocation to new leaves. Conversely, the model also shows that the reduction of nitrate uptake by plants caused by lowering either temperature or nitrate concentrations will result in a proportionately greater allocation to roots. Since atmospheric pollutants generally reduce photosynthetic capacity, it would be predicted that there would be an enhancement of allocation to leaf, and this effect has been noted (Oshima *et al.*, 1979; Reinert and Gray, 1981).

The growth of a plant results from the accumulation of allocated carbon through time. This accumulation is determined by the carbon surplus (net photosynthesis) and the allocation of carbon to leaf versus nonleaf tissue. A simple formulation of these relations given by Monsi (1968) illustrates the importance of photosynthetic capacity on the one hand and of the allocation pattern on the other (Fig. 7.6). Comparable dry-matter accumulations can occur at different combinations

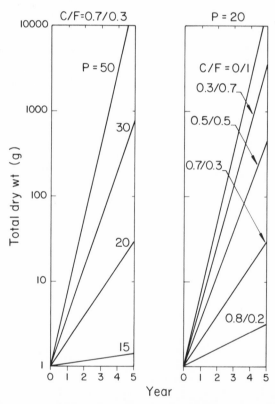

Fig. 7.6. *Left*: Growth rates of plants having the same allocation ratio (0.7) to nonphotosynthetic, *C*, and 0.3 to photosynthetic tissue, *F*, but dissimilar photosynthetic rates, *P* (photosynthetic rates in g dw per g dw leaf per year). *Right*: Growth rates of plants having the same photosynthetic rates but differing allocation ratios (from a simulation model of Monsi, 1968).

Week 5

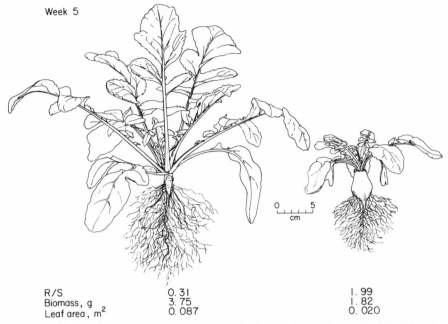

R/S	0. 31	1. 99
Biomass, g	3. 75	1. 82
Leaf area, m^2	0. 087	0. 020

Fig. 7.7. Leaf area, total biomass, and "root"/shoot ratios of five-week-old wild and cultivated (cherry-belle) radishes. Both types have comparable photosynthetic capacities (Mooney and Chu, unpublished).

of net photosynthesis and allocation schedules. This model assumes constant photosynthetic capacity during the life of the plant.

The importance of allocation for growth is further illustrated by the comparative dry-matter accumulation of wild versus cultivated radish plants (Fig. 7.7). Both radish plants start from seeds of similar weights, and the leaves of both plants have comparable photosynthetic rates. However, the two types of radish plants have very different patterns of carbon allocation. The wild radish allocates more carbon and nutrients to leaves than the commercial radish; and in 5 weeks from planting, it has twice the dry weight of the cultivated form (Fig. 7.7).

As we have noted, a change in the environment can change plant allocation patterns. Photosynthetic capacity itself will also change with the altered environment; so both must be considered simultaneously in predicting growth patterns. For example, an increase in available nitrogen will decrease allocation to roots (Davidson, 1969), but will also increase leaf-nitrogen content and hence photosynthetic capacity (Fig. 7.2).

Growth models derived for a wide variety of crops incorporate the environmental influences on net photosynthesis (Legg, 1981). Controls of allocation are derived either empirically or from hypotheses about controls on allocation. These models promise to be useful for predicting the impact of atmospheric pollutants on growth, as well as for understanding the possible indirect mechanisms that result in yield changes. For example, low concentrations of pollutants can sometimes actually increase plant production (Bennett, Resh, and Runeckles, 1974). This effect could occur, even though photosynthesis is reduced, if the allocation pattern to leaves is shifted in a way that increases production. For example, a plant with a photosynthetic capacity of 20 g g^{-1} y^{-1} may have higher biomass production than a plant with a higher photosynthetic capacity but greater allocation to nonphotosynthetic tissue (Fig. 7.6).

However, in natural systems plants probably have a genetically controlled allocation balance matched to their average pattern of resource availability (light, water, nutrients). For example, any reduction in allocation to roots because atmospheric pollutants have reduced photosynthesis would also result in a reduction in nutrient uptake, which is probably limiting to growth. However, in cultivated plants that are not nutrient-limited, a shift in allocation from roots to leaves could conceivably increase growth even if photosynthesis is reduced.

Techniques for Measuring SO$_2$-Caused Changes in Photosynthesis

Gas-exchange systems. C^{14} (Puckett *et al.*, 1973; Hällgren and Huss, 1975) and Warburg respirometry techniques (Beekley and Hoffman, 1981) have been used to assess SO$_2$ effects on photosynthesis. These techniques are useful for calculating the sum of SO$_2$-caused changes in photosynthesis during defined experimental periods. In addition, recently developed flow-through gas-exchange systems allow photosynthesis and conductance of intact leaves to be continuously monitored during exposure to SO$_2$. These systems commonly include an infrared gas analyzer (IRGA) for measuring CO$_2$ flux rates as well as instrumentation for analyzing transpiration and SO$_2$ uptake rates during fumigation. These systems show the dynamic responses of photosynthesis to SO$_2$, but are expensive to build.

Fig. 7.8 is a schematic diagram showing a flow-through gas-exchange system that operates in the differential mode for the analysis of CO$_2$ and water-vapor fluxes. The air stream going to a cuvette is preconditioned to attain the desired temperature, flow speed, and concentrations of CO$_2$, water vapor, and SO$_2$. The air stream reaches a "T" in the tubing before it reaches the cuvette. One branch of the "T," referred to as the sample line, goes to the cuvette and subsequently to analyzers to measure CO$_2$, water vapor, and SO$_2$ contents. The other branch of the "T," referred to as the reference line, goes directly to the analyzers. The flow rates in the sample and reference lines are usually kept equal, so that differences in gas concentrations between the two gas lines reflect gas exchange by the leaf.

IRGAs commonly are manufactured with two channels for differential analysis. Since capacitance-type humidity sensors are relatively inexpensive, two can

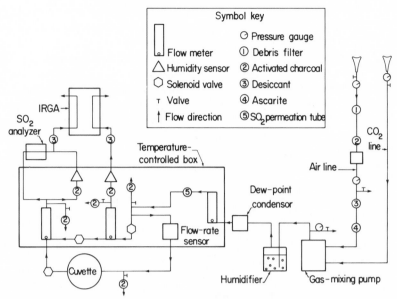

Fig. 7.8. Schematic diagram of a flow-through, differential gas-exchange system modified for SO_2 fumigations.

be purchased for simultaneous measurements of water-vapor concentrations in the reference and sample gas lines. A single SO_2 analyzer in the sample line is necessary to measure SO_2 concentrations leaving the cuvette. This should be done first with no leaf in the cuvette, before an experiment begins. Once the leaf is in the cuvette, SO_2 measurements of the cuvette-exit sample line are measurements of SO_2 concentrations around the leaf, since the cuvette is well-stirred with a fan. These ambient concentrations, along with stomatal conductance values, are used to calculate flux rates of SO_2 into the leaf. The difference between the SO_2 concentrations in the sample line with and without the leaf represents total SO_2 uptake (adsorption + absorption). The use of two SO_2 analyzers would allow constant measurements of SO_2 concentrations entering and leaving the cuvette and improve the capacity to analyze SO_2 fluxes.

The cuvette of a flow-through gas-exchange system has several requirements. It must have a design that promotes the mixing of gases and eliminates, or greatly reduces, the boundary layer. A fan positioned in the cuvette below the leaf accomplishes both purposes. The boundary layer of the leaf should be minimized because it provides resistance to gas exchange between the leaf and air. The cuvette must have a light sensor that measures radiation at the plane of the leaf. The temperature of the cuvette and the leaf must be monitored, and the cuvette should be temperature-controlled in order to manipulate leaf temperature. Leaf-temperature measurements have to be monitored accurately to calculate the saturated water-vapor pressure at leaf temperature. From the vapor-pressure difference between leaf and air, and measurements of the transpiration rate, leaf conductance to water vapor is calculated.

Gas-exchange systems used for SO_2 fumigations require monitoring of several gas analyzers, humidity sensors, light sensors, flow-rate sensors, and thermocouples. Some systems have computerized data-acquisition systems that allow all these sensors to be read at nearly the same instant. Data is processed in these systems within a few seconds, and can then be displayed at the CRT screen of a desk-top computer. Data-acquisition systems add greatly to the ability to analyze experiments that are in progress, and will become more common as new gas-exchange systems are built and the components for computer systems become less expensive.

SO_2 can be supplied in sufficient quantities for single-leaf gas-exchange experiments from either calibrated bottles of SO_2 in N_2 (available from most commercial suppliers of scientific gases) or from permeation tubes (Metronics Corp., Santa Clara, Calif.). Permeation tubes contain SO_2 in liquid phase; it degases at a rate dependent on the length of the tube and its temperature.

SO_2 is a corrosive gas that adheres to many types of surfaces; so introducing SO_2 into a gas-exchange system warrants several precautions. All tubing, valves, and flowmeter parts that contact SO_2 should be made of either stainless steel, glass, or teflon. If other materials are used, SO_2 will adsorb and/or degas, making control of SO_2 concentrations in experiments nearly impossible. FEP teflon film, which is clear and has adhesive on one side (Saunders Corp., Los Angeles, Calif.), can be used to line cuvettes made of other materials to reduce the SO_2 sink capacity of the cuvette interior.

Measuring total SO_2 uptake by leaves. With data from such a flow-through gas-exchange system, it is possible to make simultaneous calculations of photosynthetic rate, conductance, and SO_2 absorption. Two methods for calculating SO_2 movement into the leaf are possible using a flow-through gas-exchange system. One is to calculate it from the leaf conductance and concentration gradient of SO_2 into the leaf, as discussed in what follows, as well as in Chapter 6, by Black, in this volume. One can also measure total SO_2 flux to the leaf by using a flow-through gas-exchange system, as has been mentioned, but one must then find out what proportion of that flow is reaching the interior of the leaf in order to relate it to any metabolic changes observed.

Total uptake of SO_2 by leaves consists of several components. SO_2 may be deposited on the leaf surface, diffuse through the cuticle, or enter the leaf by diffusion through stomata (Unsworth, Biscoe, and Black, 1976). The major fluxes of CO_2 and water-vapor exchange are primarily via stomata. Cuticular fluxes are considered low, and can often be ignored. Adsorption of SO_2 by the leaf surface and cuticular diffusion, on the other hand, can represent between 10 and 50% of the total flux, depending on conditions and plant species (Black and Unsworth, 1979b; Taylor and Tingey, 1981).

Martin and Juniper (1970) review the factors contributing to the penetrability of a cuticle by substances, and conclude that major factors include the water- and lipoid solubility of the penetrating substance and the relative proportions of wax and cutin, a polar substance, in the cuticle concerned. The permeability of homogeneous membranes is proportional to the diffusivity and solubility of the diffusing substance in the membrane. Although the structure and heterogeneous nature of cuticles has hindered efforts to use this relationship for calculating permeabilities of cuticles, it serves as the conceptual basis for models of cuticular permeability (Schönherr, 1982). Since SO$_2$ is more soluble than CO$_2$ in a variety of substances, including such materials as rubber (van Amerongen, 1946) and aqueous solutions (Kruis, 1976), significant cuticular conductance is more probable for SO$_2$ than for CO$_2$. It should be noted that increased relative humidity has been shown to increase the flux of SO$_2$ to surfaces (Spedding, 1969), as well as increase the absorption of many substances through the cuticle, presumably because of the hydration and swelling of cutin (Martin and Juniper, 1970).

Measuring SO$_2$ absorption through stomata. It is experimentally difficult to distinguish between adsorption to the leaf surface and cuticular diffusion. In most experiments, total cuticular and surface flux (J_c) has been measured in the dark, when stomata were presumed closed. Stomatal flux (J_s) may then be estimated as $J_{total} - J_c$.

The stomatal flux, J_s (mol m^{-2} s^{-1}), of gas entering a leaf via stomata is then calculated as

$$J_s = g \, (s_a - s_i) \tag{1}$$

where g is conductance to diffusion of SO$_2$ through the stomata and surrounding boundary layer, and s_a and s_i are the mole fractions of SO$_2$ in the ambient air and stomatal cavities (Cowan, 1977). The conductance to SO$_2$ can be estimated from measurements of stomatal ($g_{s(H_2O)}$) and boundary-layer conductance ($g_{b(H_2O)}$) to water vapor from

$$g_s = g_{s(H_2O)} \times \frac{D_{(SO_2/air)}}{D_{(H_2O/air)}} \tag{2}$$

and

$$g_b = g_{b(H_2O)} \times \frac{D_{(SO_2/air)}}{D_{(H_2O/air)}}^{2/3} \tag{3}$$

and the relationship

$$1/g = 1/g_s + 1/g_b.$$

$D_{(SO_2/air)}$ and $D_{(H_2O/air)}$ are the binary diffusivities for SO$_2$ in air and for water vapor in air, respectively. The ⅔-power relationship results from the Pollhauser analysis of heat and mass transfer in laminar flow (Kays, 1966).

It is commonly believed that the diffusivity of gases is inversely proportional to the square root of the molecular weight (Graham's law). Farquhar, Wetselaar, and Weir (1982) pointed out that Graham's law applies only to diffusion in a vacuum and that the measured diffusivity for ammonia in air differs greatly from that calculated from Graham's law. Binary diffusion coefficients must, therefore, be measured or, in the absence of empirical data, calculated from an equation that accounts for the various molecular interactions in gas phase (e.g., Treybal, 1980, p. 31). $D_{(SO_2/air)}$ has been measured to be 0.122 cm^2 s^{-1} at 20°C (Andrew, 1955). This yields a ratio of $D_{(SO_2/air)}/D_{(H_2O/air)}$ as somewhere between 0.506 and 0.485; the uncertainty arises from the range of values measured for $D_{(H_2O/air)}$ (Andrussow, 1969).

Equation (1) can be used to estimate the flux of SO_2 through stomata if both conductance to SO_2 and the concentration gradient of SO_2 across the stomata are known. Unfortunately, although both conductance and ambient SO_2 concentration can be measured or calculated with relative ease, SO_2 concentrations inside the leaf airspaces are difficult to measure. Some researchers (e.g., Winner and Mooney, 1980a,b,c) have assumed s_i to be close to zero and use this equation to calculate flux.

Some attempts have been made to calculate s_i. Equation (1) can be used to calculate s_i if all other parameters are known. This approach is difficult, since it depends not only on accurate water-vapor exchange measurements, but also on accurate estimates of nonstomatal fluxes as well as total fluxes. Among the few attempts to use this approach, Taylor and Tingey (1983), working with *Geranium carolinianum* L., and our group (Williams, Koch, Winner, and Mooney, unpublished data), working with *Heteromeles arbutifolia* M. Roem., both calculate negative values for s_i. Taylor and Tingey suggest that this is evidence of a shorter pathlength of diffusion for SO_2 than for water vapor through the stomata. This explanation is open to question, since the average pathlength of water vapor is considered by some to be extremely short (e.g., Tyree and Yianoulis, 1980; Meidner, 1975). Sharkey *et al.* (1982) examined the accuracy of the gas-exchange equations in calculating intercellular CO_2 concentrations. They found that equations based on the assumption that CO_2 and water vapor had equal pathlengths gave values in good agreement with experimental results. The possibility of pathlength differences between SO_2 and water vapor, and its implications for calculating SO_2 flux, have not been adequately investigated, however.

Black and Unsworth (1979b) used an approach that did not depend on experimental measurement of cuticular flux to examine parameters affecting SO_2 flux. Their data for *Vicia faba* L. indicate that s_i is close to zero. From Eq. (1) and the relationship $J_t = J_s + J_c$, they derived

$$J_t/s_a = (1 - s_i/s_a) g + J_c/s_a. \tag{4}$$

Black and Unsworth found J_t/s_a to be linearly related to stomatal conductance, with a slope of approximately 1. This would indicate that s_i was negligible, according to Eq. (4).

A note should be made on the assumptions implicit in the use of resistance analogies in modeling fluxes. Resistance analogies assume that flux is directly proportional to the concentration gradient. Although this assumption holds true for some physical processes (e.g., diffusional processes), it does not necessarily hold true for other processes (e.g., enzymatic reactions or any other process in which the forward and backward rate constants are not identical). For this reason, there are limits to the extent to which resistance analogues may be carried. However, they are extremely useful for cases in which flux can be shown to be proportional to the concentration gradient, and for cases in which one merely wishes to show that there is some sort of hindrance to the flow of a compound between two points.

Effects of transpiration on SO₂ fluxes. An additional refinement of gas-exchange equations for CO_2 and H_2O exchange has been proposed by Jarman (1974). Jarman considered a ternary system of gases (water vapor, CO_2, and air) where there is no net molar flux of air. His treatment takes into account not only collisions between water vapor and air and between CO_2 and air, but also between CO_2 and water vapor. Von Caemmerer and Farquhar (1981) combined these equations with those of Cowan (1977) to calculate their gas-exchange parameters. Similarly, if the ternary system of gases, SO_2, H_2O, and air is considered, Eq. (1) is modified to

$$J_s = g_s (s_a - s_i) - \bar{s}E, \tag{5}$$

where E is the transpiration rate and $\bar{s} = (s_a + s_i)/2$. Jarman's equations can be simplified to Eq. (5) because the diffusivity of SO_2 in air is approximately equal to that of SO_2 in water vapor (0.124 cm^2 s^{-1} at $24.8°C$ as measured by Kimpton and Wall, 1952). The last term in the equation is the correction for the effect of effluxing water vapor. If one assumes that $s_i = 0$, evaluation of this equation shows that, under moderate evaporative conditions (e.g., around 50% relative humidity, leaf and air temperatures between 20°C and 30°C), this correction can be expected to reduce the calculated stomatal SO_2 flux by 2 or 3% of the uncorrected value. Under very extreme evaporative conditions, this correction may reduce the calculated SO_2 flux by up to 10% of the uncorrected value.

Directions for Future Research

Obviously, further studies of the effects of SO_2 on photosynthesis and stomatal conductance are needed. SO_2-caused decreases in conductance increase resistance to CO_2 flux and thereby reduce photosynthesis. In addition, SO_2-caused reductions in photosynthesis due to biochemical changes in the mesophyll will cause internal partial pressures of CO_2 ($CO_{2\ int.}$) to increase. One effect of increasing $CO_{2\ int.}$ is to induce stomatal closure. Future research with SO_2 may help develop our understanding of the relationships between stomatal physiology and photosynthesis. In addition, future studies will help define which of the many processes involved in photosynthesis are most susceptible to SO_2, as well as how long these processes need to recover from SO_2 exposures.

There is much to learn about the links between the photosynthetic capacity of plants and their growth. As we described, loss of carbon gain because of SO_2-caused decreases in photosynthetic capacity may be offset by changes in carbon-allocation patterns between shoots and roots; such changes usually favor growth of aboveground plant parts at the expense of belowground plant parts. SO_2-caused changes in allocations may therefore change the capacity of plants to tolerate environmental stresses such as drought, induce nutrient deficiencies, or make plants more susceptible to plant diseases. Much work on the physiological implications of SO_2-caused changes in carbon-allocation patterns needs to be done.

One of the major advances in gas-exchange technology in recent years has been the development of portable systems for measuring photosynthesis and stomatal conductance (e.g., Field, Berry, and Mooney, 1982). The portability and low electrical-power requirements of these systems make it possible to study gas exchange of plants in the field. Portable gas-exchange systems can also be used in laboratory experiments, and offer a wide range of experimental latitude. Responses of plants exposed to pollutants in controlled laboratory experiments can be compared with responses of plants growing in natural habitats with SO_2. In addition, it may be possible to modify existing portable gas-enclosure systems to control and monitor SO_2 concentrations. These modifications would be useful in comparing the metabolic responses of field-grown plants and plants grown in the laboratory (greenhouse and growth cabinet) to SO_2 exposures.

Additional research will improve our understanding of SO_2 absorption through stomata. The extent to which factors in the leaf mesophyll affect SO_2 flux is largely unknown. The extent to which SO_2 gas can exist, and increase in concentration, within mesophyll tissues is of great interest. SO_2 gas in the mesophyll would decrease net SO_2 flux rates through stomata. The extent to which sulfate and sulfite ions in solutions within the mesophyll may also influence SO_2 flux rates is unknown. Techniques exist, such as the approach used by Sharkey *et al.* (1982) to confirm $CO_{2\,int.}$ partial pressures, to help clarify questions concerning SO_2 flux rates. Use of these techniques will bring new information on the principles that govern SO_2 absorption into leaves.

8

SO$_2$ Effects on Stromal and Thylakoid Function

Alan R. Wellburn

Inhibition of photosynthesis is frequently thought to be one of the first effects of SO$_2$ on plants; so the chloroplast is regarded as the primary site of many of the disturbances caused by SO$_2$ or its products in aqueous solution (Ziegler, 1975; Hällgren, 1978). The pH of the chloroplast stroma is generally much greater than pH 7, and this favors the formation of sulfite ions at the expense of bisulfite when SO$_2$ ionizes in solution. As a consequence the effects of sulfite are often considered to reflect the action of SO$_2$ within chloroplasts, although this may not be strictly correct (Petering, 1977).

Several ultrastructural studies of the effect of SO$_2$ fumigation on plants have been undertaken. Structural disturbances within plastids have already been reported as the first signs of damage or disturbance (Wellburn, Majernik, and Wellburn, 1972; Fischer, Kramer, and Ziegler, 1973; Godzik and Sassen, 1974; Wong, Klein, and Jäger, 1977). Such a disturbance is usually observed as a swelling of the lumen spaces within the thylakoids, as can be seen by comparing a SO$_2$-polluted (0.25 ppm or 10.4 μmol m^{-3} for 2 h) bean plastid (Fig. 8.1a) with a chloroplast from an unfumigated plant (Fig. 8.1b). Initially this swelling is reversible, although the time needed for recovery is proportional to the dosage. Such swelling is indicative of the ionic disturbances and acidification that have been directly implicated in the toxicity of SO$_2$ to photosynthesis (Spedding *et al.*, 1980).

Fig. 8.2 schematizes some of the more important features of photosynthetic events as they occur within the chloroplast; these include the biochemistry of CO$_2$ fixation and of sulfur and nitrogen metabolism. Such events depend on the electron flow from water induced by the photosystems in the presence of light. Electron transport also creates a proton gradient across the thylakoid membrane, causing the stroma to become more alkaline and the lumen more acidic. This alkalinization increases the activity of enzymes within the Calvin cycle, and any decrease in pH caused by additional ions from SO$_2$ pollution will probably reduce the efficiency of CO$_2$ fixation. The proton gradient or ΔpH across the membrane enables the chloroplast coupling factor complex (CF$_0$ + CF$_1$) to form ATP (Mitchell, 1966). Fig. 8.2 emphasizes only a few of the many biosynthetic de-

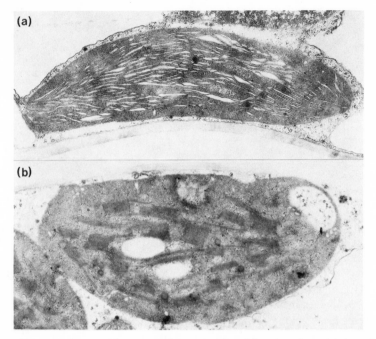

Fig. 8.1. Electron micrographs of bean tissue, showing chloroplasts from (a) tissue fumigated for 2 h with 0.25 ppm (10 μmol m^{-3}) of SO$_2$ as compared to (b) control tissue, taken from the experiments described by Wellburn, Majernik, and Wellburn (1972).

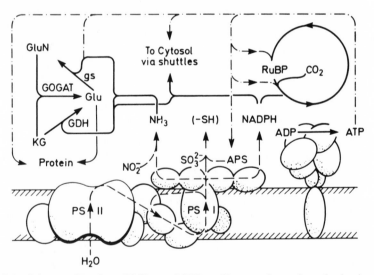

Fig. 8.2. Scheme taken from Wellburn (1982) to illustrate the various destinations for electrons from the photosynthetic membrane (dashed lines), some of the uses of the reduced products, and the involvement of ATP in various processes that depend on photosynthesis (alternate dots and dashes). As discussed in the text, gaseous pollutants are thought simultaneously to affect several of the processes outlined in this scheme.

mands placed on available ATP. Any detrimental effect that SO$_2$ or sulfite has on any of the processes leading to photophosphorylation (i.e., the formation of ATP from ADP and orthophosphate by light)—such as impairing the photosystems, interfering with electron flow, causing the membrane to leak protons, interfering with use of the proton gradient, or inhibiting the site of phosphorylation on chloroplast coupling factor (CF$_1$) particles—will reduce the rates of many biosynthetic activities, not the least of which would be protein and starch synthesis and CO$_2$ fixation.

In the following discussion, several likely sites of action by SO$_2$ and/or sulfite are considered, together with their mode of entry through the envelopes and the overall bioenergetic status. The probability that SO$_2$ or sulfite may induce defects in the different processes involved in the photosynthetic membranes is assessed for each case, as is the possibility that SO$_2$ and sulfite may have direct effects on CO$_2$ fixation, on nitrite reduction, or on assimilation of ammonia into amino acids and proteins.

The Effect of SO$_2$ on Plastids

Mechanisms of entry. From studies of the rates of uptake of orthophosphate, sulfite, and sulfate into chloroplasts, Hampp and Ziegler (1977) and Mourioux and Douce (1978) suggested that the transport of sulfite and sulfate is associated, at least in part, with the phosphate translocator of plastid envelopes. However, the rates of exchange for both sulfur anions are much lower than those for orthophosphate; so there seems to be no competition between orthophosphate and sulfite or sulfate. On the contrary, it would appear from the results of Hampp and Ziegler (1977) that orthophosphate increases the influx of sulfur anions. Influx is further increased if the anions are exchanged for photosynthetically synthesized triose phosphates; such exchange provides a mechanism to ensure optimum assimilatory sulfate reduction through the phosphorylating steps (Ziegler and Hampp, 1977). These uptake rates are greatly increased by reduced pH levels; this fact emphasizes how important the availability of protons is in influencing the toxicity of SO$_2$ within the plant cell. Nonionized sulfur species like SO$_2 \cdot$ H$_2$O, by contrast, appear to diffuse readily across the plastid envelopes, and are not under metabolic control (Spedding *et al.*, 1980).

Changes in the size of the nucleotide pool. The process of photophosphorylation is the major mechanism by which plants are provided with ATP. Their bioenergetic status depends on how available ATP is in relation to other nucleotides. A popular parameter to define the bioenergetic state of living systems has been the energy-charge ratio, but assessment of the phosphorylation potential may actually be more valuable. Endogenous amounts of ATP, ADP, and orthophosphate change greatly during transitions from light to dark, and they may temporarily change the cellular energy-charge ratios or phosphorylation potentials; however, they are usually restored quite rapidly. There is quite a variation in the sizes of pools of different adenine nucleotides across the different cellular compartments (Hampp, 1980). Surprisingly large amounts of ATP and high energy-charge ratios appear in the "cytosol" (i.e., cytoplasm plus vacuole plus broken nuclei); they are somewhat lower in plastids, and lowest of all in mitochondria.

Several field and fumigation studies have compared the amounts of adenine nucleotides in polluted and unpolluted plant tissues. Fumigation with 0.3 ppm (12 μmol m^{-3}) of SO_2 significantly reduces the amounts of ATP and other adenylates (Hoffman, Pahlich, and Steubing, 1976), but there is no significant change in the energy-charge ratio. Even at lower concentrations of SO_2 (0.05 to 0.2 ppm, or 2.1 to 8.3 μmol m^{-3}) there is an inverse linear relationship between ATP content and SO_2 concentration (Harvey and Legge, 1979) in field foliage, as well as in foliage that is field-acclimated and then laboratory-fumigated. Interestingly, this acclimation requirement may be linked to a slight deficiency in phosphate nutrition due to soil acidification by SO_2. Bisulfite derivatives are also known to reduce the cellular amounts of ATP (Lüttge *et al.*, 1972). *Lolium perenne* laminae exposed to 0.25 ppm (10 μmol m^{-3}) of NO_2, by contrast, have high amounts of ATP, but these amounts are much reduced if both SO_2 and NO_2 are present (Wellburn *et al.*, 1981).

Effects on photosystems 1 and 2. The direct effects of SO_2 and sulfite, etc., on the photosystems within thylakoids are unclear. If plants are fumigated with very high concentrations of SO_2 (more than 1 ppm, or 40 μmol m^{-3}), then the oxygen-evolution functions of photosystem 2 (PS 2) as measured polarographically are inhibited, but electron flow through PS 1 is unaffected (Shimazaki and Sugahara, 1979). Similarly, if thylakoid preparations are treated with either sulfite, bisulfite, or SO_2 at concentrations greater than 1 mol m^{-3}, there are reductions in equivalent PS-2 functions (Silvius, Ingle, and Baer, 1975). By contrast, many other investigations (see Hällgren, 1978) have failed to show that sulfur ions directly inhibit the rate of photosynthetic electron transport, partly because they have used additives in excess, which may not reflect the *in vivo* conditions. Furthermore, most published work has not been replicated statistically to isolate the effects of treatment from those of variations between individual preparations; the latter may be large. For studies where this has been done, no significant differences were detected by polarographic comparisons between PS-2 activities in preparations from nonfumigated grasses and activities in preparations from grasses treated with 0.25 ppm (10 μmol m^{-3}) of SO_2 and/or NO_2 for 11 days (Wellburn *et al.*, 1981). Furthermore, no significant differences were found by comparing control preparations of oat plastids (Robinson and Wellburn, unpublished results) and those treated with different concentrations of sulfite and/or nitrite up to 1 mol m^{-3}.

Intact plastid studies. In studies using isolated chloroplasts, there are certain measurable functions that are totally dependent on the presence of the envelopes that retain the stromal contents around the thylakoid membranes. For this reason, measurement of rates of oxygen evolution that depend on bicarbonate, ribulose-1,5-*bis*-phosphate, or 3-phosphoglycerate produces results quite different from those polarographic measurements of electron flow from water, or from artificial donors of electrons, that use preparations of detached thylakoids. Effective measurement of bicarbonate- or 3-phosphoglycerate-dependent O$_2$ evolution therefore relies on a variety of functions that encompass all aspects of photosynthesis within the intact plastids, from water splitting to CO$_2$ fixation. Not the least of these are those mechanisms by which ATP is formed by the thylakoids and consumed by various enzymic events within the Calvin cycle.

When the literature about the effects of SO$_2$, sulfite, and sulfate upon 3-phosphoglycerate-dependent O$_2$ evolution is compared at 1 mol m^{-3} (Table 8.1), a clear picture (relative to measurements of the effects of pollutants on electron flow) emerges. Sulfite and SO$_2$ are more detrimental than sulfate, but the magnitude of inhibition they cause is inversely proportional to the amount of orthophosphate present: in other words, the lower the orthophosphate concentration, the greater the reduction of 3-phosphoglycerate-dependent O$_2$ evolution. Silvius, Ingle, and Baer (1975) in their experiments unfortunately employed pyrophosphate rather than orthophosphate, but the similar relationships they obtained in

TABLE 8.1
Summary of the Literature on the Effect of 1 mol m^{-3} Sulfite, Sulfate, or SO$_2$ on the Rates of 3-Phosphoglycerate-dependent Oxygen Evolution by Isolated Intact Chloroplast Preparations

Additive (1 mol m^{-3}); and plant	Phosphate conc. (mol m^{-3})	Percent differences from controls[a]	Source
Sulfate			
Spinach	0	−60%	Baldry, Cockburn, and Walker (1968)
Spinach	0.2	−40	*Ibid.*
Spinach	0[b]	−35	Silvius, Ingle, and Baer (1975)
Spinach	0[c]	− 3	*Ibid.*
Sulfite			
Spinach	0[b]	−35%	*Ibid.*
Spinach	0[c]	−10	*Ibid.*
Pea	0.05	−70	Plesničar and Kalezić (1980)
Pea	0.4	−54	*Ibid.*
Pea	0.6	−33	*Ibid.*
Pea	1	−23	*Ibid.*
Pea	0.5	−43**	Previously unpublished
Pea	2	−20*	Previously unpublished
SO$_2$			
Spinach	0[b]	−43%	Silvius, Ingle, and Baer (1975)
Spinach	0[c]	−18	*Ibid.*

[a] Some of these reductions have been obtained by extrapolation.
[b] No orthophosphate, but 4 mol m^{-3} pyrophosphate instead.
[c] No orthophosphate, but 40 mol m^{-3} pyrophosphate instead.
*$p < 0.05$, **$p < 0.01$.

their results seem to show that proportional amounts of orthophosphate were present after endogenous partial hydrolysis of the supplied pyrophosphate. Table 8.1 also shows some previously unpublished results of our own on pea chloroplasts isolated and assayed by methods like those of Plesničar and Kalezić (1980) and with similar results. These Yugoslav workers have gone further by studying the kinetics of the inhibition of 3-phosphoglycerate-dependent O_2 evolution and by varying the amounts of sulfite in relation to orthophosphate concentrations. They have established that there is a competitive inhibition of sulfite on O_2 evolution associated with photosynthetic carbon assimilation, and have measured an inhibition constant (K_i) of 0.8 mol m^{-3} for sulfite competition with orthophosphate, which they ascribed to events associated with photophosphorylation.

Direct effects on photophosphorylation. There is now much literature on the effects of different sulfur compounds on photophosphorylation. Previously there was a tendency to use the phrase "cyclic (or noncyclic) photophosphorylation" to describe polarographic or spectrophotometric measurements of electron flow rather than of the actual formation of ATP or (at the very least) the disappearance of orthophosphate. Table 8.2 summarizes the literature on only the latter two forms of measurement, using 1 mol m^{-3} for comparison in each case.

Sulfate, sulfite, and SO_2 consistently depress light-induced ATP formation. The amounts of orthophosphate in the assay media have been included, but, on the whole, any orthophosphate interaction is concealed by the wide variety of electron donor/acceptor systems that have been used. Phenazine methosulfate (PMS) is known to react directly with P-700$^+$ to support cyclic electron flow at high rates with almost no light saturation (Witt, Rumberg, and Junge, 1968), whereas ascorbate plus diaminodurene (DAD) uses some of the natural electron-transport intermediates while generating proton gradients. Similarly, electron flow from water to ferricyanide also uses other natural intermediates, and supports the lowest rates of ATP formation of all three methods. Table 8.2 also shows some previously unpublished measurements of photophosphorylation supported by electron flow from ascorbate plus DAD to methylviologen (MV) by both oat and pepper preparations, using procedures described in Wellburn *et al.* (1981). For sulfite the more orthophosphate that was present, the less were the rates of ATP formation reduced. Very similar results were described by Plesničar and Kalezić (1980) for the rates of 3-phosphoglycerate-dependent O_2 evolution by pea chloroplasts.

Recently Cerović, Kalezić, and Plesničar (1983) have investigated the effect of SO_2 on the rates of ATP formation by pea envelope-free chloroplasts. The rates of photophosphorylation were depressed by increasing amounts of SO_2, especially for low orthophosphate concentrations. When these results are replotted using a Dixon plot (Fig. 8.3), a competitive inhibition of photophosphorylation by SO_2 is revealed. Moreover, the intersect reveals an inhibition constant (K_i) of 0.8 mol m^{-3}, precisely the same as for 3-phosphoglycerate-dependent O_2

TABLE 8.2

Summary of the Literature on the Effect of 1 mol m^{-3} Sulfate, Sulfite, or SO_2 on the Rates of Light-Induced ATP Formation by Isolated Thylakoid Preparations

Additive (1 mol m^{-3}); and donor/acceptor system[a]	Phosphate conc. (mol m^{-3})	Percent differences from controls[b]	Source
Sulfate			
Pyocyanine	5	−10%	Ryrie and Jagendorf (1971)
PMS	5	−10	Ibid.
PMS	2.5	−24	Hall and Telfer (1969)
PMS	1.3	−21	Asada, Deura, and Kasai (1968)
PMS	2[c]	−20	Silvius, Ingle, and Baer (1975)
H₂O/ferricyanide	2[c]	− 9	Ibid.
H₂O/ferricyanide	1.3	−11	Asada, Deura, and Kasai (1968)
H₂O/ferricyanide	2.5	−31	Hall and Telfer (1969)
H₂O/ferricyanide	5	−10	Ryrie and Jagendorf (1971)
Ascorbate/DAD/MV	5	−10	Ibid.
Sulfite			
PMS	2[c]	−27%	Silvius, Ingle, and Baer (1975)
H₂O/ferricyanide	2[c]	− 8	Ibid.
H₂O/ferricyanide	2.5	−33	Hall and Telfer (1969)
Ascorbate/DAD/MV (peppers)	2.5	−13**	Previously unpublished
Ascorbate/DAD/MV (oats)	0.2	−48**	Previously unpublished
Ascorbate/DAD/MV (oats)	1.2	−22**	Previously unpublished
SO₂			
PMS	2[c]	−25%	Silvius, Ingle, and Baer (1975)
H₂O/ferricyanide	2[c]	−14	Ibid.
H₂O/ferricyanide (peas)	1.7	−19	Cerović, Kalezić, and Plesničar (1983)

[a] Unless otherwise indicated, the plant source for the preparations was spinach.
[b] Some of these reductions have been obtained by extrapolation.
[c] Pyrophosphate instead of orthophosphate.
** $p < 0.01$.

evolution (Plesničar and Kalezić, 1980). This is strong evidence that the two assays are actually concerned with the same phenomenon, and that SO_2 or sulfite competes with orthophosphate for a binding site, probably on the chloroplast coupling-factor (CF₁) particles. This inhibition of photophosphorylation by SO_2 has also been shown by Cerović, Kalezić, and Plesničar (1983) to be fully reversible. By using an ATP-regenerating system involving phosphocreatine, they have also demonstrated that their previously reported depression of 3-phosphoglycerate-dependent O_2 evolution by sulfite was caused solely by inhibited photophosphorylation.

Effective pH gradients. In a chemiosmotic explanation of ATP formation, electron flow creates a proton gradient that, in turn, enables the coupling-factor complex (CF₁ + CF₀) to form ATP (Mitchell, 1966). At the same time, electrons

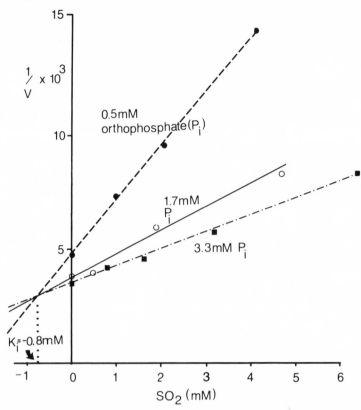

Fig. 8.3. Reciprocal values of the rates of formation of ATP at different orthophosphate concentrations, plotted as a function of SO_2 concentration. Redrawn from Cerović, Kalezić, and Plesničar (1983).

reduce nitrite or NAD^+ to form either ammonia or NADPH, which are utilized by the glutamine synthetase/glutamate synthase (GOGAT) and Calvin cycles, respectively. Hitherto, with the possible exception of the studies of Coulson and Heath (1975) on the effects of ozone on photoinduced amine-supported swelling of plastids, there have been no reports of experiments to find out how pollutants and their products directly affect the proton gradients across the thylakoids.

There are various methods for measuring the ΔpH generated across photosynthetic membranes in the light and the dark, but one of the most convenient, based on the uptake of fluorescent amines, was suggested by Schuldiner, Rottenberg, and Avron (1972). This uptake depends on the dissociation constant of the amine and the number of ionizable amines in the probe. The properties of 9-aminoacridine (9-AA) are highly suitable for these measurements and it has recently been widely adopted. Fluorescence of 9-AA is quenched by illumination of thylakoid suspensions, partly because of the redistribution of the probe in response to the proton gradient, partly because of increased binding of the probe to thylakoid membranes (Haraux and Kouchovsky, 1979). One can correct for binding of the probe by measuring light-induced quenching as a function of chlorophyll concentration, and thus arrive at meaningful assessments of ΔpH (Slovacek and Hind, 1981), but, leaving that possibility aside, the technique has the advantage for comparative purposes that the relative percentage change in light-induced quenching quickly indicates differences in the proton gradient across thylakoid membranes caused by different treatments.

Using a fluorescence apparatus very similar to that described by Mills, Slovacek, and Hind (1978) with identical filters and light intensities, we have followed the procedures and calculations of Slovacek and Hind (1981) in order to employ the technique of 9-AA light-induced fluorescence quenching to measure the effects of ozone, sulfite, sulfate, nitrite, and nitrate, singly and in combination, on thylakoid preparation from oats (Robinson and Wellburn, 1983).

Fig. 8.4 shows examples of the traces from untreated preparations and preparations treated with sulfite, nitrite, or sulfite plus nitrite. Typically the 9-AA fluorescence of 1.5-ml samples of thylakoid preparations (10 to 20 μg of chlo-

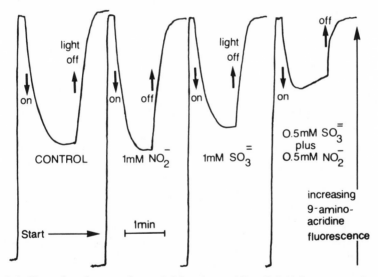

Fig. 8.4. Examples of traces of recorded 9-amino-acridine (9-AA) fluorescence from oat thylakoid preparations, in the dark and then illuminated with red light in the presence of (left to right): no addition (control); 1 mol m^{-3} of nitrite; 1 mol m^{-3} of sulfite; and a mixture of 0.5 mol m^{-3} of sulfite with 0.5 mol m^{-3} of nitrite. From Robinson and Wellburn (1983).

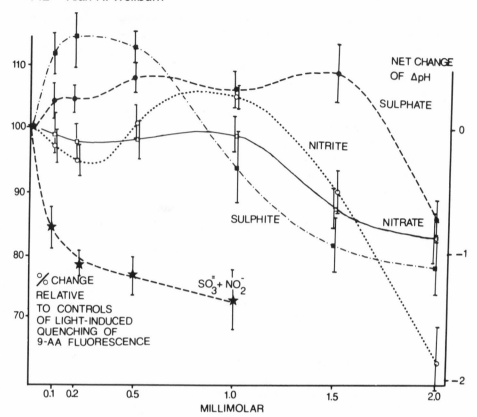

Fig. 8.5. Accumulated measurements of relative change in 9-AA light-induced quench in the presence of different individual concentrations of sulfate, sulfite, nitrate, nitrite, and mixtures of sulfite and nitrite, as compared directly to untreated control responses. The error bars represent the standard deviation of eight different measurements. Approximate ranges of net changes in ΔpH caused by the different ions are shown on the right-hand side. From Robinson and Wellburn (1983).

rophyll) in TRIS-Tricine buffer (pH 8.1) in the dark is considerably above instrumental zero (left-hand side of the traces), and is taken to be 100% relative fluorescence. Upon illumination with red light (>630 nm), the 9-AA fluorescence of the samples declines in proportion to the formation of ΔpH across the thylakoids. When the samples are again in the dark, the signal gradually increases as the proton gradient decays. Repeated signals with little or no change in properties could be obtained from the same sample by fresh illumination cycles.

Fig. 8.5 shows the means and ranges of maximum quenching of samples treated with different concentrations of sulfate, sulfite, nitrate, and nitrite relative to quenching of 9-AA fluorescence in untreated controls. With single treatments of any ion, no significant decline in light-induced 9-AA fluorescence quenching occurred at concentrations of 1 mol m^{-3} or less. Sometimes, especially with 0.1 to 0.5 mol m^{-3} of sulfite, a noticeable increase in quench was observed. This phenomenon is not understood, but is thought to arise because the binding between the membranes and 9-AA is altered by sulfite and, to a lesser extent, sulfate or nitrite. We do not know of any relationship between this response and the increase in growth sometimes reported for fumigations of whole plants with very low concentrations of SO$_2$.

Above 1 mol m^{-3}, nitrite appeared to be the most (and sulfate the least) effective ion in promoting a decline in light-induced 9-AA fluorescence quenching by the thylakoid samples. However, the most interesting observations were obtained with equal mixtures of sulfite and nitrite. Even at lower concentrations of both (0.1 to 0.5 mol m^{-3}) highly significant ($p < 0.001$) differences in the relative decline of quenching from controls, and also from individual additions of sulfite or nitrite, were obtained. Analysis of variance showed that there were "more than additive" interactions ($p < 0.01$) over the range 0.1 to 1 mol m^{-3}, relative to separate additions of 0.1 to 1 mol m^{-3} of sulfite or nitrite. However, in an analysis of this form, the sum concentration of the mixture is double that of the individual treatments. It may be more appropriate to test 0.1 mol m^{-3} of sulfite plus 0.1 mol m^{-3} of nitrite against 0.2 mol m^{-3} of sulfite alone or 0.2 mol m^{-3} of nitrite alone. When such analyses were undertaken for the range 0.1 to 0.5 mol m^{-3} of mixtures, highly significant ($p < 0.01$) interactions were again detected, but not for concentrations of 1 mol m^{-3}. An example of a trace obtained with a mixture of 0.5 mol m^{-3} of sulfite plus 0.5 mol m^{-3} of nitrite is included in Fig. 8.5.

We can estimate the ΔpH values for different light-induced fluorescence measurements for the quenching of 9-AA, but these are subject to corrections to allow for binding errors. On the right-hand side of Fig. 8.5 are shown comparative changes in ΔpH calculated in the manner described by Slovacek and Hind (1981), using their binding corrections and a thylakoid volume of 4.9 μl/mg chlorophyll. Estimates in ΔpH of control preparations under our conditions are in the range 3.5 to 4.5 pH units, but because of the uncertainty over binding corrections, which might also be changed by the different treatments, the equivalent net changes of ΔpH shown on the right-hand side of Fig. 8.5 should be regarded as only approximate.

In Fig. 8.4 preparations from plants treated with both sulfite and nitrite also show another feature that was also exhibited with studies using ozone. The decay of the 9-AA fluorescence is increased in the light, and darkness effects a partial repair, so that when the light is initially turned on again, most of the quenching recovers, but then decays even more in the light. This pattern is thought to be due to free-radical events elicited by the mixture of sulfite and nitrite, like those shown by ozone that cause ozonolysis of lipids and oxidation of amino-acid residues (Mudd, 1982), increased leakiness of the thylakoid membrane to protons, and thereby reduced effectiveness of the phosphorylation mechanisms associated with the membrane.

Carbon fixation. Most studies of pollutant effects on stromal reactions have focused on key enzymes and reactions, notably those involved in CO_2 fixation and nitrate reduction. Libera, Ziegler, and Ziegler (1975) demonstrated that, at concentrations greater than 1 mol m^{-3} of sulfite, the rate of $^{14}CO_2$ fixation declined rapidly, giving reduced amounts of radioactivity in phosphoglycerate and sugar phosphates. Measurements of ribulose-1,5-*bis*-phosphate carboxylase (RuBPC) activity after fumigation of peas for 6 days with various concentrations of SO_2 showed that only very high concentrations of SO_2 affected rates of enzymic activity (Horsman and Wellburn, 1975). With improved methods of *in vitro* activation of RuBPC devised by Lorimer, Badger, and Andrews (1976), the activity of the enzyme was reexamined by Miszalski and Ziegler (1980) and by Hällgren and Gezelius (1982). They suggested that lower concentrations of SO_2 did have an inhibitory effect, but the latter also showed that this inhibition was caused by a reduction in the amount of active RuBPC, not by a change in specific activity.

In vitro studies have shown that sulfite inhibits RuBPC assimilation of bicarbonate competitively ($K_i = 3$ mol m^{-3}), but only inhibits assimilation in the presence of Mg^{2+} non-competitively, suggesting that sulfite binds to the enzyme in the same manner as the bicarbonate (Ziegler, 1972). By contrast, Gezelius and Hällgren (1980) did not find sulfite to have such strong inhibitory effects on bicarbonate fixation ($K_i = 9$ to 13 mol m^{-3}), and showed that sulfate was about as effective as sulfite. These discrepancies may again be attributed to differences in activation of the RuBPC before assay, because Parry and Gutteridge (personal communication) have been able to show, by following the activation closely, that both the pattern of inhibition by sulfite and the kinetic constants change ($K_i = 2.5$ mol m^{-3}, increasing to 9 mol m^{-3}) with time as the RuBPC is consumed.

Fructose-1,6-*bis*-phosphatase and sedoheptulose-1,7-*bis*-phosphatase undergo light activation *in vitro*. Alsher-Herman (1982) has found that light activation can be inhibited by 1 mol m^{-3} of sulfite in SO_2-susceptible soybeans, but only by 5 mol m^{-3} of sulfite in those cultivars resistant to SO_2.

Nitrite reduction. Cereals and grasses appear to respond to low concentrations of NO_2 (less than 10 μmol m^{-3} or 0.250 ppm) by increasing nitrite-reductase activities. SO_2 by itself has no effect. However, if the two pollutants (NO_2 plus SO_2) are present together, the SO_2 appears to cancel the ability of the NO_2 to stimulate fresh increases in nitrite-reductase activity (Wellburn *et al.*, 1981). This inhibition of a potential detoxification mechanism of nitrite is believed to be one reason why the $SO_2 + NO_2$ combination exhibits such drastic synergistic effects on the growth of grasses. This important point has been discussed at length elsewhere (Wellburn, 1982, 1983).

Directions for Future Research

Numerous biophysical and chemical techniques are available for the study of chloroplasts and the impact of atmospheric pollution on their structure and function. Several have been used for both, and some have been exploited for the study of basic principles, but have yet to be used to measure the disturbance caused by the products of SO$_2$, etc. Those wishing to study effects of pollutants on chloroplast biochemistry are faced with some early decisions that predetermine some of the analytical approaches. They can either start with simpler systems or models (e.g., an enzyme) to follow the effect of the pollutants and then attempt to project the observations to increasing levels of complexity, or attempt to dissect out specific effects of pollution on cell and plastid alike. If the latter course is pursued, then the next decision is either to fumigate the whole plant and then attempt to compare preparations from polluted and untreated tissues or to adopt the *in vitro* approach.

Many *in vitro* experiments have studied the effects of the likely ionic products of SO$_2$ (Petering, 1977) on organelles (principally chloroplasts), detached membrane systems (usually thylakoids), simpler organisms (algae), model systems (artificial micelles), or enzymes. These have revealed a host of likely sites of action of the toxic products.

Internal concentrations of toxic anions. It has been evident for some time that between these two approaches to the study of the effects of atmospheric pollutants on plants (i.e., field and fumigation versus *in vitro*), there is a gap that hampers and threatens further progress. Finding out which event or function has significance depends on the concentration of SO$_2$ or sulfite chosen, the period of exposure (*in vitro* it is necessarily short), the events that remove or add to the toxic concentrations, and other critical factors, such as pH and buffering capacity. The problem is best put by quoting directly from two of the best reviews. Ziegler (1975) states that, "The actual concentration of sulfite within a cell or a cell compartment after fumigation is still completely unknown; however, this is a question that merits central attentions and demands resolution." Hällgren (1978) asks, "What are the actual concentrations inside the mesophyll cells and in the chloroplasts, after fumigation with SO$_2$, at different dosages? There are several good reasons that warrant an answer to this question."

How this might be accomplished is difficult to predict. Certainly there is good reason to believe that a starting point might be the isolation of protoplasts; techniques for doing so are well-developed for certain species. The source of protoplasts could be either fumigated or nonfumigated tissue; alternatively, some protoplasts from the latter could be exposed to SO$_2$ or sulfite, and then rapid separation of protoplasts into their constituent compartments could be undertaken by using the procedures similar to those of Hampp (1980). Various combined methods based on filtering silicone-oil centrifugation, electron microscopy, and light scattering are available to measure volumes such as the stromal space, but the actual measurement of ionic species within such small volumes would be extremely difficult. Ion-selective electrodes may be suitable for giant cells, but estimation of plastid contents must be undertaken by indirect methods. The emerging technique of high-performance ion chromatography (HPIC) may be capable of the resolution necessary for the separation and measurement of the ions of interest within these extremely small samples.

Buffering capacity and the use of NMR. Sulfite and bisulfite, the major ions produced by SO_2 fumigation, are buffered in the plastid by existing stromal proteins, free aminoacids (such as cysteine), and phosphate-containing compounds, but some acidification inevitably takes place if they rise to higher levels. A reduction in buffer capacity of plants has been shown in leaves exposed to SO_2 (Grill, Esterbauer, and Beck, 1975; Klein and Jäger, 1976). Furthermore, those lichens which are most sensitive to SO_2 pollution are generally those with the lowest buffering capacities (Nieboer, MacFarlane, and Richardson, 1983). However, this may not be the only criterion for lichen susceptibility. Several indirect methods for measuring interplastidic pH values have been employed (e.g., penetration of (^{14}C)-methylamine) with reasonable success (Heldt *et al.*, 1973), although the weak acid 5,5-dimethyl-(2-^{14}C)-oxazolidine-2,4-dione (DMO) may be a better alternative, since it is not metabolized or adsorbed onto proteins.

Nevertheless, for a rigorous study of pH changes, a direct method is required, preferably one that is rapid, noninvasive, and nondestructive. Intracellular pH in erythrocytes and yeast (Moon and Richards, 1973; Navon *et al.*, 1972) has been measured by means of nuclear magnetic resonance (NMR), as well as establishing cytoplasmic and vacuolar pH levels in plant cells (Roberts *et al.*, 1980; Roberts, Wade-Jardetzky, and Jardetzky, 1981). Nieboer, MacFarlane, and Richardson (1983) point out that similar techniques, when perfected, may also be extremely useful in pollution studies to measure stromal pH changes caused by SO_2 and NO_x exposure.

Electron-spin resonance studies. The oxidation of water to molecular O_2 in photosynthesis involves manganese in some manner (Radmer and Cheniae, 1977), and there appear to be at least three pools of manganese in thylakoids: a loosely bound Mn^{2+} associated with the photolysis mechanism, tightly bound manganese associated with light-harvesting chlorophyll-protein complexes, and a very loosely bound pool unrelated to O_2 evolution (Kanna *et al.*, 1981).

Manganese has a characteristic electron-spin resonance (ESR) spectrum at both room and cryogenic temperatures. Unfortunately, functional manganese in chloroplast membranes is ESR-silent, and can only be detected when it has been liberated from its functional site. This release is very dependent upon the immediate environmental conditions around the membrane, which suggests that ESR may well be a suitable probe for measuring the effects of atmospheric pollutants on photosynthetic membranes and indicating their influence on bioenergetic functions. Many studies using ESR to detect changes in the membranes of animals caused by pollutants have already been undertaken (Rowlands, Allen-Rowlands, and Gause, 1977).

In a series of preliminary studies, ESR has been used as a probe to see whether one can detect effects of the products of atmospheric pollution (e.g., sulfite) on photosynthetic membranes (Wellburn, 1983). Only single ionic species have been investigated so far, but studies of the mixed pollutant situation (i.e., sulfite plus nitrite) may well be more rewarding, as the 9-AA fluorescence-quenching studies have already indicated.

Leaf fluorescence. In vivo leaf fluorescence is a nondestructive technique that may be able to provide information on some of the electron-flow characteristics of photosynthetic membranes. The kinetics of the Kautsky effect comprise both fast and slow events, some of which are better understood than others. Arndt (1974) first used the technique to study effects of sulfite; he detected slight stimulations at low concentrations, and reduced signal at concentrations greater than 1 mol m^{-3}. Hällgren (1978) reinvestigated the phenomenon and found it to be pH-dependent. At pH 8 the fluorescence yield was increased by 1 mol m^{-3} sulfite, but was correspondingly reduced at pH 6. We have found that variations in signal strength that depend on where the probe is placed on the leaf can be larger than differences caused by fumigation with either SO$_2$ or SO$_2$ plus NO$_2$. Consequently we have devised methods with a microcomputer to allow us to average many signals from different locations of the probe. The only consistency in our results using this modification so far appears to be a significant reduction of the much later S peak, which is thought to be due to changes in the redox poising of plastoquinone. We have noted similar changes in leaves subjected to water stress, fungal infection, or aphid attack; so the technique may be rather unspecific as a monitor of SO$_2$ pollution, and rather better as a means of measuring general environmental stress.

In conclusion it is apparent that existing techniques may be supplemented by other possibilities (NMR, HPIC, ESR, etc.) to enable us to answer some important questions. Of those that merit special attention, the need to establish intra-plastidic concentrations of active ions such as sulfite ranks high. We also need to learn how to discriminate between general pH effects and lesions created by specific ions, and how to find out what interactions take place between sulfite and nitrite (or other combinations) within photosynthetic membranes. In this context, a greater understanding of free-radical events, and how the chloroplast combats their detrimental effects by various antioxidation devices (such as superoxide dismutase), would be valuable and might lead to improved selection of SO$_2$-tolerant cultivars of plants.

9

Biochemical and Physiological Effects of SO_2 on Nonphotosynthetic Processes in Plants

Galen Peiser and Shang Fa Yang

Sulfur dioxide has been known for many years to damage plants (Thomas, 1951); yet we still do not have a clear understanding at the chemical and biochemical level of how it adversely affects plants. Here we examine the chemistry and biochemistry of sulfur dioxide and how these reactions may be related to injury.

Gaseous SO_2 is highly soluble in water, and is ionized in a manner depending on the pH of the solution. At pH 7 it exists largely in the form of bisulfite and sulfite ions (Schroeter, 1966):

$$SO_2(g) \longleftrightarrow SO_2(aq) \underset{pK = 1.76}{\xrightarrow{H_2O \; H^+}} HSO_3^- \underset{pK = 7.20}{\xrightarrow{H^+}} SO_3^{2-}$$

Unless otherwise noted, sulfite will be used to denote the total of SO_2, bisulfite, and sulfite. It has been found that the Henry's partition coefficient ($[SO_2]aq/[SO_2]g$) at 25°C is 30 (Hales and Sutter, 1973; Johnstone and Leppla, 1934). Thus, an atmosphere of 4.1 μmol m^{-3} (0.1 μl l^{-1}; 262 μg m^{-3}) of SO_2 will give at equilibrium 0.12 μmol l^{-1} of SO_2 (aq), 16 mmol l^{-1} of HSO_3^-, and 10 mmol l^{-1} of SO_3^{2-}. However, if sulfite is removed by oxidation or other reactions in the cells faster than the rate of the transport or the solubilization and ionization of $SO_2(g)$ into bisulfite, an equilibrium will never be reached.

Sulfite as a Sulfonation Reagent

Sulfite is a nucleophile, and can attack several biological substrates by nucleophilic substitution or nucleophilic addition. The species responsible for sulfonation may be either sulfite or bisulfite ion. Disulfide (Cecil, 1963) or thiamine (Fry, Ingraham, and Westheimer, 1957) can be attacked by sulfite through sulfitolysis (nucleophilic substitution):

$$RSSR + SO_3^{2-} \leftrightarrow RS^- + RSSO_3^-$$

Addition of sulfite to aldehydes, ketones (including sugars), conjugated alkenes, quinones, coumarines (Gilbert, 1965), the pyridine ring of NAD (Swoboda and Massey, 1966), the pyrazine ring of folic acid (Vonderschmitt *et al.*, 1967) and the isoalloxazine ring of flavin coenzymes (Muller and Massey, 1969) has been demonstrated. The products of sulfite addition to aldehydes or ketones are α-hydroxylsulfonate, and the reactions are reversible (Gilbert, 1965):

$$R-CHO + HSO_3^- \leftrightarrow R-\underset{\underset{OH}{|}}{CH}-SO_3^-$$

Sulfite has been reported to possess mutagenic activity (Shapiro, Servis, and Welcher, 1970; Hayatsu, Wataya, and Kai, 1970). By addition-elimination reactions, sulfite catalyzes the specific conversion of cytosine derivatives under physiological conditions to uracil derivatives:

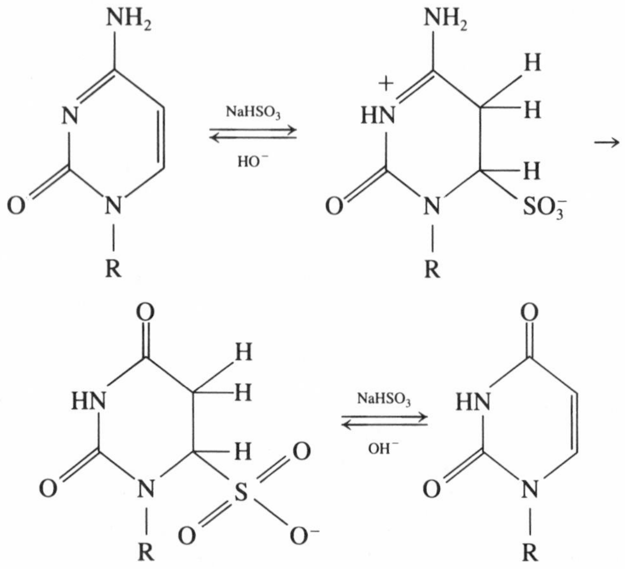

Biochemistry of Sulfite Oxidation

Sulfite is a metabolic intermediate both in the oxidation of organic sulfur to sulfate by mammals and in the reduction of sulfate to organic sulfur by autotrophic organisms. In plant tissues, SO_2 is mainly oxidized to sulfate (Weigl and Ziegler, 1962; de Cormis, 1969; Garsed and Read, 1977b). There are two mechanisms by which sulfite can be oxidized to sulfate. One is aerobic oxidation through a free-radical mechanism; the other is mediated by sulfite oxidase. The oxidation of sulfite to sulfate through a free-radical chain mechanism can be initiated by metal ions (Abel, 1951), ultraviolet irradiation (Hayon, Treinin, and Wolfe, 1972; Backstrom, 1927), illuminated dyes (Fridovich and Handler, 1960; Asada and Kiso, 1973; Peiser and Yang, 1977), or enzymatic reaction (Fridovich and Handler, 1958; Yang, 1967; Nakamura, 1970), all of which can generate free radicals. The spontaneous oxidation of sulfite can be induced by superoxide radical (O_2^-), since the reaction is effectively inhibited by superoxide dismutase, which catalyzes the disproportionation of O_2^- radical (McCord and Fridovich, 1969a,b). The available data are in good agreement with the view that O_2^-, $OH\cdot$, and SO_3^- radicals are generated during the aerobic oxidation of sulfite, and that these radicals are, in turn, responsible for the propagation of the sulfite-oxygen chain reaction as described by the following schemes (Hayon, Treinin, and Wolfe, 1972; Yang, 1970):

Initiation: $SO_3^{2-} + O_2 \rightarrow SO_3 \cdot^- + O_2^-$ (1)

Propagation: $SO_3 \cdot^- + O_2 \rightarrow SO_5 \cdot^-$ (2)

$SO_5 \cdot^- + SO_3^{2-} \rightarrow SO_4 \cdot^- + SO_4^{2-}$ (3)

$SO_4 \cdot^- + SO_3^{2-} \rightarrow SO_4^{2-} + SO_3 \cdot^-$ (4)

$SO_4 \cdot^- + OH^- \rightarrow SO_4^{2-} + OH \cdot$ (5)

$OH \cdot + SO_3^{2-} \rightarrow OH^- + SO_3 \cdot^-$ (6)

$O_2^- + SO_3^{2-} + 2H^+ \rightarrow 2OH \cdot + SO_3 \cdot^-$ (7)

Termination: $OH \cdot + SO_3 \cdot^- \rightarrow OH^- + SO_3$ (8)

$SO_3 \cdot^- + O_2^- + 2H^+ \rightarrow SO_3 + H_2O_2$ (9)

$(H_2O_2 + SO_3^{2-} \rightarrow H_2O + SO_4^{2-}$

$SO_3 + H_2O \rightarrow SO_4^{2-} + 2H^+)$

Once Eq. (1) is initiated by metal ions, UV light, or enzymic reactions, chain-propagation reactions—Eqs. (2) through (7)—are maintained, with the concomitant formation of sulfate. The chain length of the reaction has been estimated to be 30,000 mol per mol O_2^- in the xanthine-xanthine oxidase system (Fridovich and Handler, 1958), and 300 mol in isolated chloroplasts under illumination (Asada and Kiso, 1973). The metal-initiated autooxidation of sulfite is inhibited by many reagents, including organic acids, alcohol, thiols, amines, and proteins, which are abundant in the cells, and thus may serve as radical scavengers (Schroeter, 1966; McCord and Fridovich, 1968). However, in the presence of enzymes such as peroxidase, the chain-initiated sulfite oxidation is not inhibited by these scavengers to a significant extent (Yang, 1970), suggesting that such reactions *in vivo* must be catalyzed by enzymes. It is important to note, especially in relation to ambient SO_2 concentrations, that sulfite reacts primarily by an ionic mechanism at high concentrations in the mol l^{-1} range (Shapiro, Servis, and Welcher, 1970; Schroeter, 1966), but it undergoes free-radical oxidation at low concentrations, in the mmol l^{-1} range or less (Inoue and Hayatsu, 1971; Yang, 1970; Kudo, Miura, and Hayatsu, 1977).

It has been demonstrated *in vitro* that many physiologically important compounds are destroyed by free radicals produced during the aerobic oxidation of sulfite. The amino acids methionine and tryptophan (Yang, 1970, 1973; Inoue and Hayatsu, 1971), as well as the plant hormone indole-3-acetic acid (Yang and Saleh, 1973), are destroyed when coupled to the oxidation of sulfite. The $OH \cdot$ and O_2^- radicals were implicated as the oxidation reagents in the destruction of these compounds (Yang, 1973; Horng and Yang, 1973, 1975). NADH and NADPH are destroyed during the oxidation of sulfite, and it was suggested that the sulfite radical served as the oxidant (Tuazon and Johnson, 1977; Klebanoff, 1961):

$SO_3^{2-} + \text{initiator} \rightarrow SO_3^- \cdot$

$NADH + SO_3^- \cdot \rightarrow NAD \cdot + SO_3^{2-} + H^+$

$NAD \cdot + O_2 \rightarrow NAD^+ + O_2^-$

$SO_3^{2-} + O_2^- + 2H^+ \rightarrow SO_3^- \cdot + H_2O_2$

Sulfite also reacts by a free-radical chain mechanism with certain DNA and RNA molecules (Inoue, Hayatsu, and Tanooka, 1972; Kitamura and Hayatsu, 1974). Although the exact nature of such reactions is not known, evidence suggested that the free radicals generated during the autooxidation of bisulfite were responsible for the glycosidic bond cleavage of pyrimidine nucleosides.

Kharasch, May, and Mayo (1938) reported the oxygen-mediated addition of bisulfite to olefins. Since the reaction is inhibited by radical scavengers, they concluded that the sulfite radical is the reactive species in this reaction. Sulfonation of 4-thiouridine by such a mechanism has been recognized (Hayatsu and Inoue, 1971):

$$RSH + SO_3^- \cdot \rightarrow RS \cdot + HSO_3^-$$
$$RS \cdot + SO_3^- \cdot \rightarrow RSSO_3^-$$

Also chlorophyll (Peiser and Yang, 1977, 1978), β-carotene (Peiser and Yang, 1979b), and linoleic and linolenic acid (Lizada and Yang, 1981) are destroyed during the free-radical oxidation of sulfite, as will be discussed below.

The other mechanism of sulfite oxidation is catalyzed by sulfite oxidase. The existence of this enzyme has been observed in animals (MacLeod *et al.*, 1961; Cohen, Fridovich, and Rajagopalan, 1971; Cohen *et al.*, 1973), bacteria (Lyric and Suzuki, 1970), and plants (Tager and Rautanen, 1956; Fromageot, Vallant, and Perez-Milan, 1960; Ballantyne, 1977; Kondo *et al.*, 1980). In well-characterized systems, sulfite-oxidase activity is located in the mitochondria (Cohen and Fridovich, 1971; Ballantyne, 1977; Shibuya and Horie, 1980). The oxidation of sulfite by purified sulfite oxidase can use either cytochrome c or O_2 as an electron acceptor (Cohen and Fridovich, 1971). The reaction with cytochrome c, when coupled to the mitochondrial respiratory chain, would reduce molecular oxygen to water, whereas the product of oxygen reduction by sulfite oxidase is H_2O_2 (Cohen and Fridovich, 1971; Oshino and Chance, 1975):

$$SO_3^{2-} \rangle \langle \; 2 \text{ Cyt } c \text{ (Fe}^{3+}) \rangle \langle \; H_2O$$
$$SO_4^{2-} \rangle \langle \; 2 \text{ Cyt } c \text{ (Fe}^{2+}) \rangle \langle \; \tfrac{1}{2}O_2$$
$$SO_3^{2-} + H_2O + O_2 \rightarrow SO_4^{2-} + H_2O_2$$

Kondo *et al.* (1980) examined the ability of extracts from several plants to reduce cytochrome c in the presence of sulfite (a common assay for sulfite oxidase) in relation to the ability of these plants to tolerate SO_2 exposure. In general, they observed that a higher sulfite oxidase activity was correlated with tolerance to SO_2.

Reduction of Sulfite to Hydrogen Sulfide

In addition to the enzymatic oxidation of sulfite to sulfate, sulfite or SO_2 can be reduced to H_2S. Plants exposed to SO_2 release H_2S (de Cormis, 1969; Hällgren and Fredricksson, 1982) and recently Sekiya, Wilson, and Filner (1982) have examined the relationship between the ability of young and old cucumber leaves to produce H$_2$S and their resistance to SO$_2$. They observed that young leaves, which are resistant to SO$_2$, converted approximately 10% of the absorbed SO$_2$ to H$_2$S, but mature leaves, which are susceptible to SO$_2$, converted only about 0.2% less of the absorbed SO$_2$ to H$_2$S. The conversion of SO$_2$ to H$_2$S was primarily light-dependent, as is consistent with work on spinach sulfite reductase, where ferrodoxin was found to serve as the electron donor for reduction (Tamura, Hosoi, and Aketagawa, 1978; Aketagawa and Tamura, 1980).

Free Radical Involvement in Chlorophyll and β-Carotene Destruction

Leaf chlorosis is one of the primary visible symptoms of SO$_2$ damage (Thomas and Hill, 1935; Hällgren, 1978). The participation of free radicals, produced during the aerobic oxidation of sulfite, in the destruction of chlorophyll was examined in two *in vitro* systems (Peiser and Yang, 1977). In the light, and in the dark with added Mn^{2+}, a rapid destruction of chlorophyll occurred in the presence of sulfite (Fig. 9.1). Both of these systems required sulfite and O$_2$ (Table

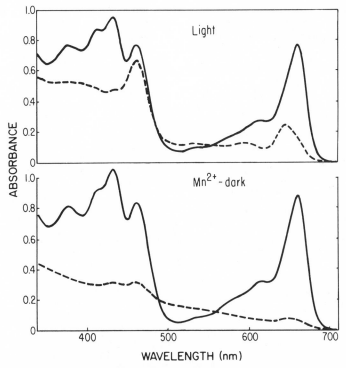

Fig. 9.1. Visible absorption spectra of chlorophyll in the absence (solid line) and presence (broken line) of bisulfite for the light and Mn^{2+}-dark systems. Illumination time for the light system was 3 min and incubation time for the Mn^{2+}-dark system was 1 min. Reproduced from Peiser and Yang (1977), p. 278.

9.1), and there was a concomitant loss of chlorophyll and oxidation of sulfite. In the light system chlorophyll served as a photosensitizer, to initiate sulfite oxidation; this function could be, however, replaced by Mn^{2+} in the dark system. The free-radical scavengers hydroquinone, α-tocopherol, and tiron (1,2-dihydroxybenzene-3,5-disulfonic acid) were effective at inhibiting chlorophyll destruction. β-Carotene was destroyed in a similar system (Fig. 9.2), with sulfite using Mn^{2+} as the initiator for sulfite oxidation (Peiser and Yang, 1979b). As with chlorophyll, sulfite and O_2 were required, and α-tocopherol and tiron effectively inhibited β-carotene destruction.

TABLE 9.1
Requirements for the Destruction of Chlorophyll in the Light or Mn^{2+}-dark Systems

Light system	Percent Chl destroyed	Mn^{2+}-dark system	Percent Chl destroyed
Complete	83%	Complete	89%
Minus HSO_3^-	12	Minus HSO_3^-	0.3
Minus light	4	Minus Mn^{2+}	0.4
Minus O_2	3	Minus O_2	0.9
		Minus glycine	0.6

SOURCE: Peiser and Yang (1977).
NOTE: Illumination time was 3 min for the light system, and incubation time for the Mn^{2+}-dark system was 30 s.

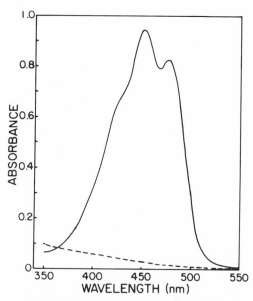

Fig. 9.2. Visible absorption spectrum of β-carotene after incubation for 1 min with bisulfite in the absence (solid line) or presence (broken line) of Mn^{2+}. Adapted from Peiser and Yang (1979b), p. 447.

TABLE 9.2
Effects of Tiron, Ascorbate, and Hydroquinone on the
Destruction of Chlorophyll a
in SO$_2$-fumigated Leaves of Spinach

Additions (mM)	Percent chlorophyll a destroyed after illumination time of	
	3 h	6 h
None	100%	100%
Tiron		
1	27	84
10	2	29
Ascorbate		
1	73	56
10	27	16
Hydroquinone		
0.1	0	39
1	−12	10

SOURCE: Adapted from Shimayaki et al. (1980).
NOTE: SO$_2$ fumigation was performed for 2 h at 2.0 ppm (82 μmol m^{-3}). The chlorophyll a destruction in the reference samples (None) were about 30% of total chlorophyll a for 3-h illumination and 60% of it for 6-h illumination. The reagents were added to the leaf disks under vacuum infiltration.

Shimazaki et al. (1980) have examined chlorophyll destruction in spinach leaves exposed to 82 μmol m^{-3} (2 μl l^{-1}) of SO$_2$. They observed that light was necessary for chlorophyll destruction, as in the light system described above (Peiser and Yang, 1977). After the spinach plants were exposed to SO$_2$, discs were taken from the leaves and infiltrated with free-radical scavengers. Tiron, ascorbate, and hydroquinone effectively inhibited the chlorophyll destruction (Table 9.2). The involvement of free radicals, specifically, superoxide radical, was further demonstrated by the use of superoxide dismutase, which dismutates superoxide radical to oxygen and hydrogen peroxide. Superoxide dismutase was added to a homogenate obtained from SO$_2$-fumigated leaves, and was found to inhibit chlorophyll destruction when the homogenate was exposed to light. Also, the endogenous superoxide dismutase activity was found to decrease in leaves during exposure to SO$_2$, suggesting that the observed damage could have partially resulted from the loss of the protective effect of superoxide dismutase (Shimazaki et al., 1980).

Lipid Peroxidation and Membrane Damage

Membrane damage occurs during SO$_2$ injury, although the primary cause of this damage is not clear. Leakage of solutes occurred when lichens were exposed to SO$_2$ (Puckett et al., 1977). Wellburn, Majernik, and Wellburn (1972) reported that swelling of thylakoids in chloroplasts was the first ultrastructural change observed in leaves exposed to SO_2, and that this occurred with no visible injury. Also, this thylakoid swelling was reversed when SO_2 was removed. SO_2 or sulfite can react with proteins (Yokoyama, Yoder, and Frank, 1971) and cause disruption of proteins by cleaving disulfide bridges (Cecil and Wake, 1962) and by inactivating enzymes without disulfide bridges (Puckett *et al.*, 1974). The other component of membranes, fatty acids, can also be damaged by sulfite (Kaplan, McJilton, and Luchtel, 1975; Lizada and Yang, 1981). Lizada and Yang (1981) examined the peroxidation of linoleic and linolenic acids in the presence of bisulfite. The time course shows that the formation of peroxidized products of linoleic acid correlated with the disappearance of bisulfite (Fig. 9.3). Peroxidation and sulfite oxidation were inhibited by low concentrations of the free-radical scavengers butylated hydroxytoluene, α-tocopherol, and hydroquinone. The authors concluded that free radicals generated from the free-radical oxidation of sulfite were responsible for the peroxidation of linoleic and linolenic acids. Lipid per-

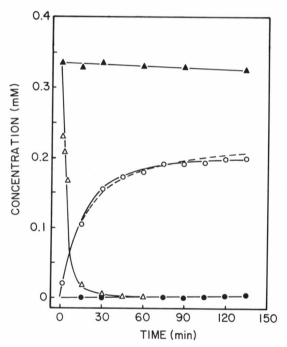

Fig. 9.3. Time course of conjugated diene (CD) and linoleic acid hydroperoxide (LOOH) formation and sulfite consumption. CD concentrations in the complete reaction mixture (open circles) and control (closed circles, no sulfite added). Sulfite concentrations in the complete reaction mixture (open triangles) and blank (closed triangles, no linoleic acid). The broken line represents LOOH concentrations as measured by the ferrous thiocyanate assay carried out on complete reaction mixtures in which more than 90% of the sulfite had been oxidized. Reproduced from Lizada and Yang (1981), p. 190.

TABLE 9.3
Ethane Formation from Broken and Intact Chloroplasts

Chloroplasts	Ethane formed (after 1 h in light)	
	pmol	percent
Broken		
+SO$_3{}^{2-}$	491	100%
+SO$_4{}^{2-}$	20	4
Intact		
+SO$_3{}^{2-}$	46	9
+SO$_4{}^{2-}$	3	1

SOURCE: Peiser, Lizada, and Yang (1982).

oxidation, determined by ethane formation (Riely, Cohen, and Lieberman, 1974; Dillard and Tappel, 1979), has been measured in plants exposed to SO$_2$ (Peiser and Yang, 1979a; Bressan *et al.*, 1979). However, in these two studies, it was not determined whether lipid peroxidation was a primary cause or only a result of injury. In relation to this question, we have recently examined the peroxidation of chloroplast membranes in the presence of sulfite and the role of free radicals in this peroxidation (Peiser, Lizada, and Yang, 1982). Isolated spinach chloroplasts were used in this study, and ethane evolution was used as a measure of lipid peroxidation. In broken and intact chloroplasts, sulfite stimulated ethane formation much more than the sulfate controls (Table 9.3). Sulfite caused approximately ten times more increase in ethane production in intact chloroplasts than the sulfate controls, and in broken chloroplasts the effect of sulfite was even more marked. Sulfite oxidation occurred along with ethane formation, and this is an important characteristic of systems involving the free-radical oxidation of sulfite (Abel, 1951; Yang, 1970, 1973; Peiser and Yang, 1977, 1978; Lizada and Yang, 1981). Light was required for both ethane formation and sulfite oxidation. Earlier, Asada and Kiso (1973) studied the oxidation of sulfite by illuminated chloroplasts, and observed that photosynthetic electron transport was involved. They concluded that the free-radical oxidation of sulfite was initiated by superoxide radical, which was formed on the reducing side of photosystem I. However, they did not examine any deleterious effects of the sulfite oxidation. In our system, we observed that photosynthetic electron transport was required for both sulfite oxidation and ethane formation (Table 9.4). DCMU [3(3,4-dichlorophenyl)-1,1-dimethylurea], which is an inhibitor of photosynthetic electron transport, phenazine methosulfate (PMS), which promotes cyclic electron flow, and the free-radical scavengers tiron and ascorbate effectively inhibited both sulfite oxidation and ethane formation. Specific involvement of superoxide radical was clearly indicated, since superoxide dismutase inhibited ethane formation. An outline of these results is presented in Fig. 9.4. Photosynthetic electron transport provides electrons to reduce O$_2$ to O$_2{}^-$, which initiates sulfite oxidation. Radicals produced from sulfite oxidation then lead to the peroxidation of membrane lipids, which

TABLE 9.4

Effect of DCMU, Phenazine Methosulfate, and Methyl Viologen (MV)
on Ethane Formation and Sulfite Oxidation from Broken Chloroplasts

Addition	Percent ethane formed (after 1 h in light)	Percent O_2 uptake[a]
Sulfite	100%	100%
Sulfite + DCMU, 10 μM	17	12
Sulfite + PMS, 20 μM	20	18
Sulfite + MV, 100 μM	685	
Sulfate + MV, 100 μM	8	

SOURCE: Peiser, Lizada, and Yang (1982).
[a] Rate calculated from linear portion of uptake curve, with the small rate in the dark subtracted from the rate in the light.

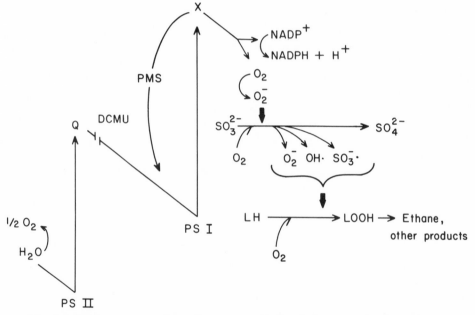

Fig. 9.4. Proposed scheme for sulfite-induced ethane formation in illuminated chloroplasts. The bold arrows indicate initiation reactions. Reproduced from Peiser, Lizada, and Yang (1982), p. 997.

result in the formation of ethane. These results show that free radicals generated from the aerobic oxidation of bisulfite can cause damage to isolated chloroplasts.

There is other work that carries this idea of free-radical involvement in SO_2 damage further, to the *in vivo* situation. Using leaf discs prepared from SO_2-fumigated spinach plants, Shimazaki *et al.* (1980) observed a marked inhibition of malondialdehyde formation, a product of lipid peroxidation, when the radical scavenger tiron was infiltrated into the discs. In experiments on curcurbit culti-vars (Bressan, Wilson, and Filner, 1978; Bressan *et al.*, 1979) and poplar (Ta-naka and Sugahara, 1980), researchers found that young leaves are more re-sistant to SO_2 than old leaves, even though young leaves absorb more SO_2 than old leaves (Bressan *et al.*, 1979). Tanaka and Sugahara (1980) found higher en-dogenous superoxide-dismutase activity in young poplar leaves than in old pop-lar leaves, and this activity was inversely correlated with greater chlorophyll de-struction and malondialdehyde formation in old leaves than in young leaves when plants were exposed to SO_2 (Fig. 9.5). Furthermore, they observed that super-oxide-dismutase activity could be induced by low concentrations of SO_2, which caused no visible injury. In young leaves, superoxide-dismutase activity in-creased fourfold when plants were exposed to 0.1 μl l^{-1} (4.1 μmol m^{-3}) of SO_2 for 20 days. When plants both with and without the 0.1 μl l^{-1} prefumigation were treated with 2 μl l^{-1} (82 μmol m^{-3}) of SO_2, those with the prefumigation showed less chlorophyll destruction than those without the prefumigation. The work of Shimazaki *et al.* (1980) and Tanaka and Sugahara (1980) provides evi-dence that free radicals play a role in the *in vivo* SO_2 damage to plants.

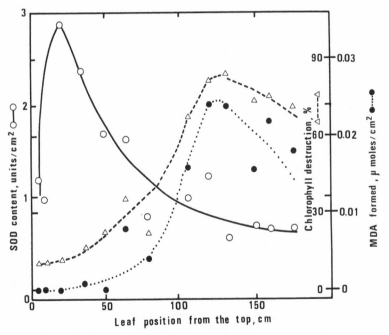

Fig. 9.5. Relationship between the content of superoxide dismutase (SOD) and the resis-tance to SO_2 in poplar leaves at different ages. SOD activity has been measured before SO_2 fumigation. After 82 μmol m^{-3} (2 μl l^{-1}) fumigation for 22 h, the formation of malon-dialdehyde and the destruction of chlorophyll were measured. Reproduced from Tanaka and Sugahara (1980), p. 604.

Active Species in Injury

In nearly all biochemical studies, bisulfite or sulfite is used instead of SO_2 gas, partly because sulfite is easier to use than SO_2 gas, but also, and more importantly, because it is generally accepted that sulfite is the active agent in damage to plants (Thomas, Hendricks, and Hill, 1950; Ziegler, 1975). However, there is evidence that sometimes SO_2 may be the culprit. Inglis and Hill (1974) examined the effect of sulfite at various pH's on $H^{14}CO_3^-$ fixation in several mosses. They observed that the greatest inhibition by sulfite occurred at low pH (pH 3 to 5). When the concentration of each species (SO_2, bisulfite, and sulfite) was calculated at each pH and compared to the amount of inhibition observed, the best correlation was with SO_2. Ziegler (1972) reported that sulfite competitively inhibited ribulose bisphosphate carboxylase with respect to HCO_3^-. Since this enzyme fixes CO_2 rather than HCO_3^- (Cooper *et al.*, 1969), it is most probable that it is SO_2, rather than sulfite, that inhibits ribulose bisphosphate carboxylase. Similarly, inhibition of moss HCO_3^- fixation by SO_2 (observed by Inglis and Hill, 1974) may have been at the site of ribulose bisphosphate carboxylase. In our laboratory the effect of sulfite on spore germination of *Cladosporium* and *Botrytis* at various pH's was examined (L. Strand and S. F. Yang, unpublished). We found that inhibition of spore germination correlated with the calculated concentration of SO_2, but not with that of bisulfite or sulfite. However, here as elsewhere, it may be that biochemical injury was in fact caused by bisulfite or sulfite ion; yet the limiting step of the injury process may have been passage across the cell membrane, which favors the passage of the nonionic form (SO_2) rather than the ions sulfite or bisulfite. Thus plant injury depends on the entry of SO_2 into the leaf, passage across the cell membrane, and biochemical injury within the cell; the form of sulfur that passes across cell membranes may not be the same one that causes biochemical injury.

Discussion

As outlined in the preceding sections, sulfite can undergo sulfonation and free-radical reactions that may be deleterious to the plant. Most of the work presented has been from *in vitro* experiments; yet the work of Shimazaki *et al.* (1980) and of Tanaka and Sugahara (1980) is consistent with the view that free radicals play a role in SO_2 damage *in vivo*. A more specific consideration about the importance of free-radical reactions relates to the processes involved in acute versus chronic injury. Garsed (1981) has pointed out that the processes by which long-term low concentrations of SO_2 reduce growth may be different from those by which short-term high concentrations of SO_2 cause acute damage. Horsman *et al.* (1979) subjected 28 clones of perennial ryegrass to acute and chronic exposures of SO_2, 41 μmol m^{-3} (1 μl l^{-1}) for two weeks and 10.3 μmol m^{-3} (0.25 μl l^{-1}) for 8 weeks, respectively, and observed that there was no correlation between visible injury from acute exposure and reduction in yield (dry matter) from chronic exposure. The results of Shimazaki *et al.* (1980) are for acute injury, since they used 22 μmol m^{-3} of SO$_2$ for 2 hours. The results of Tanaka and Sugahara (1980), however, showed that superoxide-dismutase activity increased fourfold after a chronic exposure of 4.1 μmol m^{-3} for 20 days; they concluded that increased superoxide-dismutase activity results from an increase in free radicals, specifically O$_2^-$, in the leaves. More conclusive evidence is needed, using whole plants exposed to long-term low concentrations of SO$_2$, to discover the importance of free-radical reactions under these conditions.

Miller and Xerikos (1979) found that the residence time of sulfite in eight cultivars of soybean with varying degrees of tolerance to SO$_2$ was shortest in those cultivars with the greatest tolerance. Presumably the sulfite was oxidized to sulfate; yet when sulfite was measured, no consistent differences between tolerant and sensitive cultivars were found. They concluded that a plant's ability to remove sulfite may influence its tolerance to SO$_2$. They did not speculate about the mechanism of sulfite removal, although there are at least two possibilities. Higher activities of sulfite oxidase in the more tolerant cultivars would explain these results. Also, Sekiya, Wilson, and Filner (1982) found that young (tolerant) cucumber leaves were more able to convert SO$_2$ to H$_2$S than old (sensitive) leaves, and Miller and Xerikos (1979) did notice the odor of H$_2$S from the sulfite-treated plants.

Conclusion

Sulfite can react with organic compounds via sulfonation reactions that could interfere with cellular metabolism. Also, sulfite can undergo aerobic oxidation to sulfate, with the concomitant formation of free radicals, which have been shown to destroy many cellular components. Both of these reactions may therefore be involved in the deleterious effects of SO$_2$ on plants.

Plants can resist or reduce damage from SO$_2$ by means of at least three biochemical detoxification mechanisms: reducing the free-radical oxidation of sulfite by endogenous superoxide dismutase and antioxidants; converting sulfite to sulfate via sulfite oxidase; and converting sulfite to H$_2$S.

10

The Effect of SO_2 on CO_2 Metabolism: Response to SO_2 in Combination with Other Air Contaminants

Roger W. Carlson

This paper focuses on how SO_2, in combination with other common air contaminants, affects carbon metabolism, defined as all the processes by which CO_2 enters the leaf, is fixed by photosynthesis, and is released by respiration. "Other air contaminants" will include oxides of nitrogen (NO and NO_2, collectively referred to as NO_x), ozone (O_3), peroxyacetyl nitrate (PAN), fluoride (F), and particulates.

To discuss the interaction of SO_2 and other air pollutants, one must first understand the effects of single pollutants. Therefore, I shall begin by summarizing the following data for each pollutant: (1) physical characteristics, (2) suggested mode(s) of action, (3) reactive chemical species, (4) plant functions and/or components affected, and (5) mechanisms assumed to be important in conferring resistance. Certain patterns will emerge that will be helpful for interpreting plant response to combinations of air contaminants. In the last two sections I will then discuss the results of treatments that contain one or more pollutants in addition to SO_2, identify gaps in the data, and suggest profitable avenues for future research.

Physical Properties and Effects of Individual Pollutants

A detailed critique of original research on the mechanisms of pollutant injury is not within the scope of this paper; therefore I have relied heavily on previous summary and review papers for background information. The papers contained in Naegle (1973), Mudd and Kozlowski (1975), N.I.E.S. (1980), and Unsworth and Ormrod (1982) have been especially helpful. Heath (1980) brought together known responses of plants to various pollutants, and constructed a cohesive theory for the early effects of major pollutants on plants. However, he neglected to consider one important aspect of early plant response to air pollution, that of the initiation of repair processes. There is much evidence to suggest that certain kinds of compensatory mechanisms begin to operate as soon as the plant is exposed to an air contaminant (Sutton and Ting, 1977; Tingey and Taylor, 1982).

Although repair activity helps the plant endure pollutant exposure, it also complicates the analysis of pollutant effects. Another complicating factor is the variation in pollutant chemistry. Hocking and Hocking (1977) suggested that physical environmental factors may provide enough variation in the chemistry of SO$_2$ in solution to account for many of the differences observed in the effect of this pollutant on plants.

Sulfur dioxide. The solubility of SO$_2$ in water is very high (1.6 mol l^{-1}, or 1.6 × 10^3 mol m^{-3}, at 0°C). This is equivalent to dissolving 80 volumes of SO$_2$ in one volume of water. SO$_2$ is also quite soluble in some organic solvents. The partition coefficient for H$_2$O/chloroform is 0.72, and for H$_2$O/benzene is 0.44 (Mudd, 1975a). SO$_2$ hydrates in aqueous systems very rapidly, forming a weak acid that at physiological *p*H is present as a bisulfite anion. The equilibrium for dissolved SO$_2$ strongly favors the presence of HSO$_3^-$ at *p*H 2 to 5, but SO$_3^=$ is the predominant anion at *p*H greater than 5 (Fig. 10.1).

Conductance of SO$_2$ to a plane water surface is shown in Fig. 10.1. SO$_2$ uptake by water is affected by several factors. For example, temperature and SO$_2$ concentration in air and water will influence SO$_2$ uptake by aqueous solutions. In addition, the efficiency with which an aqueous solution absorbs SO$_2$ increases with *p*H, to nearly 100% at *p*H greater than 5 (Nash, 1975).

The increased acidity of cell surface water caused by SO$_2$ in solution could lead to increased anion influx into cells (Heath, 1980). This could in turn increase the osmotic potential of the cell and induce water flow into the cell. Increased acidity can also cause the loss of Mg from chlorophyll and uncouple phosphorylation in chloroplasts. This suggests that plants with a large *p*H buffering capacity, or the ability to mobilize such a buffering potential, would be less susceptible to injury by SO$_2$ than plants lacking such capacity. Species with an inherently high cell-sap *p*H have been found to be less sensitive to SO$_2$ than species with a lower *p*H (Shuwen *et al.*, 1981).

Sulfite has been shown to inactivate and/or compete with carbonate for sites on the CO$_2$-fixing enzymes of ribulose-bisphosphate (RuBP) carboxylase and phosphoenol pyruvate (Ziegler, 1972, 1973, 1975). However, Hällgren and Gezelius (1982) have suggested that reduced RuBP carboxylase activity in response to fumigation with SO$_2$ may be a consequence of reduced photosynthetic activity rather than a cause of reduced CO$_2$ fixation, since (1) high sulfate concentrations are required *in vitro* to reduce RuBP activity (Ziegler, 1972; Gezelius and Hällgren, 1980), and (2) SO$_2$ concentrations from 0.11 to 0.17 ppm (4.5 to 5.9 μmol m^{-3}) *in vivo* reduce CO$_2$ fixation much more than they reduce RuBP carboxylase activity (Hällgren and Gezelius, 1982). Sulfite can also compete with phosphate for sites on the phosphate transporter and thereby gain entrance into the chloroplast (Heath, 1980).

Sulfur-dioxide fumigation of *Helianthus annuus* (sunflower) leaf disks may

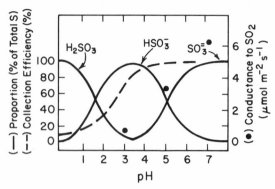

Fig. 10.1. Relationship between *p*H and the proportion of dissolved SO_2 as H_2SO_3, HSO_3^-, and SO_3^{2-} for wheat leaves (Shuwen *et al.*, 1981); the resistance to SO_2 sorption by a plane water surface (Petit, LeDoux, and Trinite, 1977); and the collection efficiency of SO_2 by water (Nash, 1975).

cause large reductions in chlorophyll a (but not chlorophyll b), in total carotenoids, and in superoxide dismutase (SOD) activity (Shimazaki *et al.*, 1980). The inhibition of SOD activity may be particularly important, since most of the O_2^- formed by illuminated chloroplasts is believed to be scavenged by SOD. O_2^- is believed to be a major source for other forms of reactive oxygen species, including singlet oxygen, peroxide, and hydroxyl radicals (Asada, 1980; Tanaka and Sugahara, 1980).

SO_2 at low concentration may cause stomata to open, but at high SO_2 stomata are induced to close. The effect of SO_2 on stomatal conductance for some species varies with the water-vapor-pressure deficit (vpd) of the air (Mudd, 1975b; Black and Unsworth, 1980). Stomatal conductance of *Vicia faba* (broad bean), *H. annuus*, and *Nicotiana tobacum* (tobacco) was increased by SO_2 at low vpd and reduced at high vpd relative to that of control plants. In contrast, stomatal conductance increased in response to SO_2 at both high and low vpd for *Raphanus sativus* (radish) and *Phaseolus vulgaris* (bean). Photosynthesis may also show a temporary stimulation at low SO_2, but with continued exposure or at high SO_2, photosynthesis declines (Black, 1982; Takemoto and Noble, 1982). Increased stomatal opening at low SO_2 in broad bean was correlated with extensive injury of adjacent epidermal cells (Black and Black, 1979a,b). However, at higher SO_2 concentrations, guard-cell viability was also reduced. Reduction in photosynthesis depends on SO_2 concentration, and is reversible only at low to moderate SO_2 concentrations.

Stomatal closure due to SO_2 is usually not enough to account for the observed simultaneous reduction in photosynthesis (Muller, Miller, and Sprugel, 1979; Barton, McLaughlin, and McCanathy, 1980; Furukawa, Natori, and Totsuka, 1980; Carlson, 1983a,b). In fact, those studies that have suggested that a large reduction in stomatal conductance causes an observed large reduction in photosynthesis may need to be reexamined. If the relationship between stomatal conductance and the rate of photosynthesis was assumed to be linear, then the importance of stomatal closure was probably overestimated. This conclusion arises from an analysis of the curvilinear response of photosynthesis to CO_2 concentration. Although the concentration of ribulose bisphosphate may be limiting at low CO_2 availability, at high CO_2 photosynthesis is limited by the capacity of the leaf to regenerate ribulose bisphosphate (Farquhar and Sharkey, 1982). Therefore the importance of stomatal control is demonstrated only when both stomatal conductance and internal CO_2 concentration change in the same direction as the rate of photosynthesis.

Measurements of respiratory response to SO_2 indicate a wide variety of effects, including no effect, inhibition, and stimulation (Black and Unsworth, 1979a; Furukawa, Natori, and Toksuka, 1980; Black, 1982). Similar results have been found for both dark respiration and photorespiration.

Sulfur is an essential plant nutrient. It is absorbed by roots as sulfate, transported to leaves, and reduced before it is assimilated into complex organic molecules. Sulfate-reduction enzymes are found in leaves. Consequently, sulfate usually accumulates in plants exposed to SO_2, and H_2S may be given off from foliage during or after exposure to SO_2 or bisulfite (Dodrill, 1976; Wilson, Bressan, and Filner, 1978; Winner et al., 1981; Hällgren and Fredriksson, 1982; Sekiya, Wilson, and Filner, 1982). The reduction of sulfate and subsequent release of H_2S could be an important detoxifying mechanism for plants exposed to SO_2. The sulfur derived from SO_2 when it is oxidized to sulfate can enter the sulfur pool and be converted to sulfur-containing metabolites. The ability of the plant to produce organic derivatives by metabolizing absorbed SO_2 may be an important way in which the plant avoids injury by SO_2 (Pahlich, 1975). Under certain conditions, sulfur from SO_2 can supply up to 40% of the sulfur required by plants (McCune, 1973).

Other studies indicate that sensitivity to SO_2 in certain plants is correlated with: (1) the rate at which sulfate is converted to sulfite (Miller and Xerikos, 1979; Kondo et al., 1980), (2) differences in leaf thickness, and (3) the rate of SO_2 uptake (Furukawa et al., 1980). Stomatal closure occurs more quickly in response to SO_2 in species that have a high concentration of endogenous abscisic acid (ABA) than in plants with low ABA (Kondo and Sugahara, 1978). However, increased plant resistance to an air pollutant is not always correlated with reduced pollutant uptake (Taylor and Tingey, 1981).

Oxides of nitrogen. NO is only sparingly soluble in water (0.002 mol l^{-1}, or 2 mol m^{-3}, at 20°C). In solution, NO acts as a dissolved gas in accordance with Henry's law, existing at a concentration that is proportional to the concentration in the atmosphere above the solution. NO_2 decomposes in water to form a mixture of nitrous and nitric acids. High concentrations of both strong and weak acids can be formed from NO_2 in the cell (Heath, 1980). Other reactive moieties include free radicals, formed when NO_x reacts with unsaturated compounds, and nitrosamines, formed when nitrite reacts with secondary amines (Mudd, 1973).

Exposure of plants to low NO concentration causes a rapid reduction in photosynthesis to a new steady state. Photosynthesis increases rapidly back to prefumigation rates after fumigation ends (Taylor *et al.*, 1975). The rapidity of the response may be partially due to the behavior of NO as a dissolved gas. The dissolved products of pollutants that do not follow Henry's law—e.g., bisulfite (for SO_2 uptake) and nitrite (for NO_2 uptake)—tend to build up and cause longer-term impairment during postfumigation periods (Bennett and Hill, 1975a,b).

Stomatal conductance may be decreased by exposure to NO_2, but usually not enough to account for the observed simultaneous reductions in photosynthesis (Srivastava, Jolliffe, and Runeckles, 1975a,b). NO_2 inhibits transpiration and photosynthesis much less than comparable ambient concentrations of SO_2 (Carlson, 1983b).

The primary detoxifying mechanism for NO_x appears to be reduction to ammonia. This process is light-dependent and requires NADPH (nicotinamide dinucleotide phosphate). Higher irradiance results in less plant damage by NO_x. Any diversion of NADPH from normal uses may be expected to reduce other NADPH-requiring processes, including photosynthesis. Under certain conditions, NO_x may be incorporated into nitrogen-containing compounds and transported throughout the plant (Anderson and Mansfield, 1979; Kaji *et al.*, 1980). A study using $^{15}NO_2$ has shown that virtually all the absorbed NO_2 taken up by *Phaseolus vulgaris* (bean) was metabolized (Rogers, Campbell, and Volk, 1979). Ambient NO_2 concentrations in this study ranged from 0.01 to 0.33 ppm (0.42 to 14 μmol m^{-3}), with total uptake being linearly correlated with concentration.

Ozone. The solubility of O_3 in water is 0.012 mol l^{-1} (12 mol m^{-3}). The O_3 molecule is very reactive, having a standard redox potential of $+2.1$ volts. It decomposes rapidly in alkaline media, releasing molecular O_2, but tends to be stable in acidic media (Heath, 1975). O_3 reacts in aqueous solution to form hydroxyl, hydroperoxyl, superoxide anion, and other free radicals (Tingey and Taylor, 1982). These species may be more reactive to biological constituents than O_3, and may also increase the *p*H of cellular solutions (Heath, 1980). Many researchers believe that the plasmalemma membrane is the initial site of O_3 action. In fact, evidence is accumulating to suggest that specific sites, such as those associated with osmoregulatory processes, are most sensitive (Heath, 1980). The results of other studies suggest that O_3, or its reaction products, can pass through the plasma membrane and affect subcellular organelles and processes (Mudd, 1982).

Membrane integrity is required to maintain electron-transport processes in the chloroplast. An agent that even partially disassociates chloroplast membranes would also uncouple parts of the electron-transport system (Coulson and Heath, 1974). The ionic balance of the cell solution would also be altered. Photosynthesis and respiration would be impaired, and the ability to accept electrons from water would be reduced. This could lead to a surplus of free radicals and subsequent chloroplast degradation. Such events would help explain why the first microscopically observable damage on exposure to O$_3$ occurs in the chloroplast grana and not at the cell membrane (Heath, 1980).

Stomata of certain species are sensitive to O$_3$ and may begin to close shortly after the initiation of fumigation (Olszyk and Tibbitts, 1981a,b). Reduced stomatal conductance would help to protect the plant from further injury by this and other gaseous pollutants. *Phaseolus vulgaris* (bean) plants pretreated with low concentrations of O$_3$ are less sensitive to subsequent fumigation with higher O$_3$ concentrations (Runeckles and Rosen, 1974). This suggests that plants may be able to acclimate to O$_3$ when exposed to low concentrations for extended periods of time.

SOD activity is also correlated with O$_3$ sensitivity in plants. Leaves with low SOD activity exhibited much more leaf injury than leaves with high SOD in an ozone-sensitive variety of *Phaseolus vulgaris* (bean) (Lee and Bennett, 1982). SOD is an enzyme for the reaction

$$2O_2^- + 2H^+ \rightarrow H_2O_2 + O_2$$

and plays an important role in protecting cells against oxy-radical reaction products. The fact that biological systems differ in their capacity to buffer themselves against free radicals suggests a way to explain changes in O$_3$ sensitivity with age as well as differences between plants.

Fluoride. Fluoride is the most phytotoxic of the common air pollutants, with susceptible plants being injured at less than 0.001 ppm, or 0.042 μmol m^{-3} (Weinstein and Alscher-Herman, 1982). The typical rate of fluoride uptake by plants is much more rapid than that of SO$_2$, NO$_x$, or O$_3$ (Bennett and Hill, 1973a). Soluble forms of fluoride may be taken up by plants through the roots, and gaseous fluoride (HF, SiF$_4$) may enter through both the leaf cuticle and the stomata (Weinstein and Alscher-Herman, 1982). HF penetrates the cuticle because it dissolves readily in aqueous media (18 mol l^{-1}, or 18 $\times$ 10^3 mol m^{-3}) forming a weak acid.

Fluoride moves in the transpiration stream as an inorganic ion, and tends to accumulate at the tips and margins of leaves. Fluoride may also be found in high concentrations in and around the chloroplasts of certain species treated with fluoride. This localization correlates with the reported sensitivity of photosynthesis to fluoride (Bennett and Hill, 1973a,b). Reductions in photosynthesis caused by fluoride at low atmospheric concentrations are reversible. In isolated chloroplasts, Mg was found to stoichiometrically neutralize the inhibitory effect of KF on the Hill reaction (Ballantyne, 1972).

Both Ca and Mg appear to influence the effect of fluoride on plant processes (Weinstein and Alscher-Herman, 1982). Increased Ca appears to both stimulate fluoride uptake, by selectively altering membrane permeability, and reduce fluoride toxicity. Ca may be involved in sequestering fluoride, because of the low solubility of CaF (0.2 mmol l^{-1}, or 200 mol m^{-3}; Weast, 1972). Liming the soil around plants is an effective way to reduce fluoride injury (Weinstein and Alscher-Herman, 1982).

Peroxyacetyl nitrate. PAN degrades stoichiometrically upon reaction with water, into acetate, nitrite, and molecular O_2 (Mudd, 1975b). Its half-life in solution decreases from 21.5 min at pH 6.0 to 6.5 min at pH 8.0. The toxicity of PAN cannot be ascribed solely to nitrite, since plant-injury symptoms are very different from those observed for NO_2. PAN appears to be a highly specific oxidizing agent. Reaction with sulfhydryl groups may produce a mixture of disulfide and acetylated compounds, with PAN being more toxic when leaf-sulfhydryl content is high. However, this observation was made in a study in which irradiance may have confounded the results (Mudd, 1975b). Illumination before, during, and after fumigation is required for exposure to PAN to result in injury (Taylor *et al.*, 1961). An interaction with light quality suggests that PAN may also interfere with the red/far-red response of phytochrome (Mudd, 1975b).

The reaction of PAN with specific biochemicals has been examined (Mudd, 1973, 1975b). Epoxides are formed when PAN reacts with ethylenic double bonds. Indoleacetic acid is oxidized by PAN, but tryptophan is resistant. Enzymes of cellulose synthesis may be inhibited. The toxicity of PAN analogs increases as the length of the alkyl chain lengthens. Nicotinamide derivatives are reversibly oxidized by PAN, with the reaction proceeding more rapidly at higher pH and at higher ionic strengths (Mudd and Dugger, 1963). These results indicate that availability of a reductant is important in dissipating the oxidizing power of an air contaminant. However, by reacting with NADP, PAN diminishes the availability of an essential reductant for photosynthetic CO_2 fixation.

Particulates. Particulate contamination of industrial origin is usually a conglomerate of chemically heterogeneous substances. The exact composition differs between industries, between similar kinds of manufacturing processes within an industry, and perhaps even at different times at a location close to a particular manufacturing plant. Cement-kiln dusts are high in alkaline metals, causing water suspensions of these dusts to be alkaline in pH. Particulate matter near aluminum-reduction plants may be high in fluoride. Deposited particulates near highways are high in lead, and moistened soot particulates may be acidic. In general, particulates seem to affect leaf surfaces only when sufficient water is also present, as dew, fog, or light mist (Lerman and Darley, 1975). A heavy rain tends to remove particulates from leaf surfaces (Carlson, Bazzaz, and Stukel, 1976) and to remove soluble compounds from leaves.

Leaf damage from wetted particulates appears to be related both to high KCl (salt damage) and to high alkalinity if Ca compounds are present (Lerman and

TABLE 10.1
Known Primary Effects of Air Pollutants on Plants

Effect	Pollutant						
	SO$_2$	O$_3$	NO$_2$	NO	HF	PAN	Particulates
Changes in stomatal conductance:							
Directly	X	X	X				X
Because of VPD	X						
Erodes leaf surface							X
Increases acidity of cell solution	X		X	X			
Forms injurious concentrations of							
anions in cells	X		X		X		
Forms free radicals	X	X					
Affects chlorophyll function	X	X	X		X	X	
Increases sensitivity to SOD activity	X	X					
Reacts with dinucleotides		X				X	
Activates detoxifying mechanisms	X	X	X		X		
Affects mineral nutrition	X		X				
Causes injury if quantity and quality of							
light allow						X	

Darley, 1975). Alkaline solutions saponify the cuticle and permit entry of the solution into the mesophyll. Acidic particles can also corrode the leaf surface, thus reducing the impenetrability of the cuticle and epidermis.

Deposited particulates interfere with stomatal function, not by reducing conductance, but by lodging in the stomatal pore and preventing it from closing (Ricks and Williams, 1974). This could have an important effect if other gaseous pollutants are present. Catalase enzyme activity, chlorophyll content, and respiration all decreased for *Helianthus annuus* (sunflower) leaves experimentally treated with cement dust (Borka, 1980). A reduction of 5 to 20% in each parameter was observed after 100 days of treatment with dust applications of 30 g m^{-2} of leaf area per month.

The known primary responses of plants to the common air pollutants are summarized in Table 10.1.

A general scenario can be developed for the action of air pollutants on plants. The pollutant first makes contact with the boundary layer surrounding the leaf. The amount available for reaction with the leaf depends on the concentration of the pollutant in the bulk air and the conductance of the boundary layer. Boundary-layer conductance is a function of windspeed and leaf-surface characteristics, but usually does not limit uptake except where the rate of air movement is very low (Carlson and Bazzaz, this volume). A portion is adsorbed on the leaf surface, and the remainder passes through the stomatal pore. The rate of pollutant uptake into the leaf depends on stomatal conductance, which varies as a function of light intensity and plant water potential. Hence, air pollutant uptake is much less during the night, when stomata are generally closed, than during the day for the same ambient concentration, except when the pollutant itself induces stomatal opening or when lodged particulates prevent stomatal closure. Either of these last two circumstances would expose internal leaf surfaces to increased concentrations of both the pollutant responsible for causing stomatal opening and any other pollutants that might be simultaneously present in the air. At high pollutant concentrations, or for long exposure times, stomatal closure is the dominant response.

When stomatal conductance decreases, the influx of all gases into the leaf is reduced. The consequence then could be to reduce injury, possibly so much that pollutants that occur together could cause less injury than they would induce alone at the same concentrations. To find out if this is the case, one must compare the amount of pollutant absorbed into the leaf during the combined treatment with the amounts taken up during exposure to the individual pollutants. Uptake must be expressed in moles in order to allow comparison of relative reactive potential.

Once inside the stomatal cavity, the pollutant may or may not be readily absorbed by the water adhering to interior cell walls. SO_2, NO_2, and HF will be readily absorbed and form bisulfite, nitrite, and fluoride anions, respectively. These chemical changes result in an increase of hydrogen-ion concentration in the cell-wall solution. Injury that can be attributed strictly to a decrease in pH is not likely to differ whether it is due to the absorption of one, or more than one, pollutant. However, the biochemical pathways, reaction products, organelle sensitivity, and detoxification mechanisms are different for each of the anions. Therefore, plant injury from a combination of pollutants might be induced simultaneously along very dissimilar pathways.

Response to Pollutant Combinations

The terms additive, antagonistic, synergistic, and interaction will be used in describing plant response to pollutant combinations. *Additive* will indicate a response to combined pollutant exposure equal to the sum of the effects observed for treatments with individual pollutants at the same concentration. *Antagonistic* will indicate a combined pollutant effect that is less than the sum of the effects for individual pollutants at the same concentration. *Synergistic* will refer to a response to a combined pollutant treatment that is greater than additive. The term *interaction* will be used in a statistical sense to refer to treatments containing two or more independent variables.

Exchange of water vapor. The cuticle is believed to be an almost impermeable barrier to water vapor, CO_2, and gaseous air pollutants (except HF); therefore, the primary port of entry for gaseous substances into the plant is through the stomata. Consequently, a direct effect on transpiration and/or stomatal opening is an important aspect of the response of plants to polluted air.

A direct effect on stomatal opening by a combination of air pollutants was reported by Menser and Heggestad (1966). They exposed a sensitive variety of *Nicotiana tobaccum* (tobacco) to 0.03 ppm (1.3 μmol m^{-3}) of O$_3$, 0.25 ppm (10 μmol m^{-3}) of SO$_2$, and a combination of the two pollutants together. Stomatal width was reduced by the pollutant combination, but not by exposure to each pollutant alone. Stomatal response agreed closely with the development of visible symptoms.

Leaf conductance of *Pinus strobus* (white pine) and *P. ponderosa* (ponderosa pine) was altered by SO$_2$ and O$_3$ in combination (Boone, 1978). Synergistic response was concentration dependent, and only appeared under certain conditions. In general, stomatal conductance was decreased by exposure to the pollutant combination only as much as by exposure to O$_3$ alone.

The transpiration rate of *Phaseolus vulgaris* (bean), as measured by daily plant plus pot weighing, increased with exposure to 0.10 ppm (4.2 μmol m^{-3}) SO$_2$ or NO$_2$, but decreased upon exposure to the combination of SO$_2$ + NO$_2$ at 4.2 μmol m^{-3}, as compared with that observed for plants in clean air (Ashenden, 1979a). Statistical significance was shown early in the five-day experiment for the single pollutant treatments, and late in the experiment for the bipollutant treatment. It was suggested that the decreased rate of transpiration in the combined treatment may have been due to physiological injury within the leaf. Stimulation of stomatal opening at low concentrations of SO$_2$ has been attributed to a decrease in the turgidity of epidermal cells surrounding the guard cells (Black and Black, 1979a,b).

Stomatal conductance was measured for three different cultivars of *Petunia hybrida* (petunia) exposed to 0.8 ppm (33 μmol m^{-3}) of SO$_2$ and/or 0.4 ppm (17 μmol m^{-3}) O$_3$ for up to 4 hours (Elkiey and Ormrod, 1979a). Stomatal conductance decreased in response to the O$_3$ treatment, but not in response to the SO$_2$ treatment. Although the combined pollutant treatment caused a more rapid decrease in stomatal conductance, it did not decrease beyond the point observed for treatment with O$_3$ alone.

The effects of SO$_2$ and O$_3$ alone or together at 0.15 ppm (6.2 μmol m^{-3}) on stomatal conductance of several agricultural species were reported by Beckerson and Hofstra (1979a,b). The plants were exposed for 6 hours on each of five consecutive days, with measurements of stomatal conductance being taken immediately before, in the middle of, and at the end of the exposure period. Stomatal conductance of *Phaseolus vulgaris* (bean), *Raphanus sativus* (radish), *Cucumis sativus* (cucumber), and *Glycine max* (soybean) was lower for the pollutant combination than for each pollutant alone.

Stomatal conductance was measured with a diffusive resistance porometer for *Pisum sativum* (pea) after a 4-hour exposure to 0.42 ppm (17 μmol m^{-3}) of SO$_2$, 1.2 ppm (50 μmol m^{-3}) of SO$_2$, 0.13 ppm (5.4 μmol m^{-3}) of O$_3$, or the two pollutants together (Olszyk and Tibbitts, 1981a). SO$_2$ alone decreased stomatal conductance, but O_3 alone increased conductance. The combined pollutant treatments caused conductance to decrease as much as that observed for SO_2 alone. This appears to be a case where SO_2 was the pollutant of overriding importance.

Pollutant uptake. Elkiey and Ormrod (1980b, 1981b) compared the uptake of various pollutants alone and in mixture. The 1980 paper deals with SO_2 and O_3 uptake by three different cultivars of petunia, and the 1981 paper with the uptake of SO_2, O_3, and NO_2 alone or as a tripollutant combination by nine different cultivars of *Poa pratensis* (Kentucky bluegrass). Both studies indicate that uptake into the leaf for pollutants in combination was less than for each pollutant alone. Leaf-surface adsorption was the same for each pollutant in combination as it was in the single pollutant treatment. In general, the rate of pollutant absorption into the leaf was less for the less-sensitive cultivars.

In another study, pollutant uptake was measured for *H. annuus* fumigated with NO_2 and/or O_3 (Omasa *et al.*, 1980b). Measurements were made of total plant uptake over a 5-hour period at 1.0, 2.0, and 4.0 ppm (42, 83, 170 μmol m^{-3}) of NO_2 with or without 0.2 ppm (8.3 μmol m^{-3}) of O_3, and at 0.2, 0.45, and 0.6 ppm (8.3, 19, 25 μmol m^{-3}) of O_3, with or without 1.0 ppm (42 μmol m^{-3}) of NO_2. Transpiration declined with increasing time of exposure at the higher concentration treatments for individual pollutants and for all combined pollutant treatments. The rate of pollutant uptake in all treatments was closely correlated with the rate of transpiration, increased with pollutant concentration, and, for each pollutant alone, was not different from that observed for the combined pollutant treatment.

The uptake of air pollutants into leaves should theoretically be the best indicator of the dose that the plant has been subjected to. However, it is difficult to experimentally quantify pollutant uptake into a leaf because adsorption also occurs onto chamber walls. Leaf-surface adsorption is frequently taken as the amount of pollutant removed from leaves in the dark, adjusted for adsorption on chamber walls and other surfaces in the measurement system. The assumption is made (and of course must be verified in each experiment) that the stomata are completely closed during surface adsorption measurements. Other environmental variables should be identical for the light and dark measurements. Relative humidity is especially important, since even small differences in humidity can cause a large difference in the amount adsorbed on chamber walls for soluble air pollutants (Omasa *et al.*, 1980a; Taylor and Tingey, 1981).

Pollutant uptake into leaves can be estimated by using measurements of transpiration to calculate stomatal conductance (Unsworth, Biscoe, and Black, 1976; Unsworth, 1981). The primary requirements are (1) similarity in the pathway between water-vapor efflux from the leaf and pollutant influx, and (2) a known efficiency of absorption for the pollutant at the sites of absorption within the leaf. If it is suspected that either of these two requirements is not being met, one must experimentally or otherwise analyze any differences. The literature detailing the differences between CO_2 and water-vapor efflux should be consulted for help in pathway analysis. (Papers of particular importance include Cowan, 1977; Rand, 1977, 1978; Tyree and Yianopolis, 1980; Farquhar and Sharkey, 1982.) The error in pollutant uptake because of the distribution of absorption within the leaf is likely to be small for highly soluble air pollutants. However, the absorption of less soluble pollutants, such as O_3 and NO, may be significantly affected, and it may not be possible to consider internal leaf surfaces as a perfect sink (O'Dell, Taheri, and Kabel, 1977; Tingey and Taylor, 1982). In such cases quantification of internal conductance to the pollutant is necessary. For a useful technique to distinguish between leaf surface adsorption and pollutant absorption into the leaf through the stomata, see Black and Unsworth (1979b) and Unsworth (1981).

Photosynthesis. The photosynthetic response of a higher plant to a pollutant combination was first reported by Hill and Bennett (1970). They found that combination of NO plus NO_2 caused an additive reduction in photosynthesis.

Bull and Mansfield (1974) found an additive response to SO_2 plus NO_2 in a 4 by 4 factorial experiment with *Pisum sativum* (pea). Pollutant concentrations ranged from zero to 0.25 ppm (10 μmol m^{-3}). This was the first fully balanced pollutant combination experiment on photosynthesis, that is, all concentrations of one pollutant were paired with all concentrations of the other pollutant.

SO_2 in combination with either NO_2 or HF caused a synergistic decrease in photosynthesis of *Medicago sativa*, alfalfa (White, Hill, and Bennett, 1974; Bennett and Hill, 1974). The concentrations in ppm (μmol m^{-3}) included in these experiments were: $0.15 + 0.15$ (6.2), $0.25 + 0.25$ (10), and $0.50 + 0.40$ (21 + 17) for $SO_2 + NO_2$; $0.30 + 0.10$ (12 + 4.2), and $0.30 + 0.20$ (12 + 8.3) for $SO_2 + O_3$; and $0.25 + 0.03$ (10 + 1.3) for $SO_2 + HF$; respectively. Treatments were not included that contained a high concentration of one pollutant paired with a low concentration of the second; so we cannot fully describe the kinetics of the effects of these bipollutant combinations on photosynthesis. However, a synergistic response was evident at low pollutant concentrations, i.e., concentrations that caused a barely measurable photosynthetic reduction in single pollutant exposures. An additive response was found at pollutant concentrations that were above the threshold for photosynthetic reduction with single pollutants. In subsequent experiments, this research group found that the inhibition of photosynthesis by SO_2 and NO_2 could be reduced by half if the CO_2 concentration were raised from 315 ppm (13.1 mmol m^{-3}) to 645 ppm (26.8 mmol m^{-3}) (Hou, Hill, and Soleimani, 1977). Photosynthesis was reduced by SO_2 and NO_2 much more slowly at elevated CO_2 concentration than at 315 ppm (13.1 mmol m^{-3}).

A study of the effects of a pollutant combination on photosynthesis in tree species was reported by Carlson (1979). *Acer saccharum* (sugar maple) and *Fraxinus americana* (white ash), but not *Quercus velutina* (black oak), showed a synergistic response to $SO_2 + O_3$. The synergism persisted for a week for *A. sac-*

charum. In this study only one concentration, 0.50 ppm (21 μmol m^{-3}), of each pollutant was studied. Photosynthesis was reduced more at low light intensity and at high light intensity than it was at intermediate intensities. This suggests that photo-induced repair processes were able to keep pace with the injury caused by the combination of SO_2 and O_3 only at certain intermediate light intensities.

Further evidence that the type of plant response may depend on pollutant concentration was shown by the effect of SO_2 and O_3 on photosynthesis in *Vicia faba* (broad bean) (Ormrod, Black, and Unsworth, 1981). A comparison was made between the response to O_3 (varied from 0.0 to 0.25 ppm; 10 μmol m^{-3}) with or without SO_2 added at a concentration of 0.40 ppm (17 μmol m^{-3}). The results indicate a synergistic response to concentrations of O_3 below 0.11 ppm (4.5 μmol m^{-3}), and an additive or antagonistic response to O_3 concentrations above 0.11 ppm (4.5 μmol m^{-3}). The authors suggested that a synergistic response is most likely to occur above the concentration at which the presence of one pollutant is not influenced by the presence of a second, and below the concentration at which response to one pollutant overrides the effect caused by the second. To test this hypothesis, one must do a fully balanced experiment, in which the various concentrations of SO_2 are paired with all combinations of a similar set of concentrations of O_3.

A study in which *Helianthus annuus* (sunflower) was fumigated with both bi-pollutant and tripollutant mixtures of SO_2, NO_2, and O_3 was reported by Furukawa and Totsuka (1979). They found a synergistic decrease in photosynthesis for $SO_2 + O_3$, and for $NO_2 + O_3$, at 0.20 ppm (8.3 μmol m^{-3}) of each pollutant, relative to that observed for single pollutant exposures. Exposure to 1.0 ppm (42 μmol m^{-3}) of NO_2 and 0.20 ppm (8.3 μmol m^{-3}) of SO_2 (but not 8.3 μmol m^{-3} of each) also caused a synergistic reduction in photosynthesis. Adding SO_2 at 8.3 μmol m^{-3} to either of the two $NO_2 + O_3$ treatments (42 μmol m^{-3} of NO_2 and 8.3 μmol m^{-3} of O_3, or 8.3 μmol m^{-3} of each) did not significantly reduce photosynthesis below the value obtained at the end of the 2-hour exposure to the bipollutant. However, photosynthesis in the tripollutant treatment decreased more rapidly than with either of the bipollutant exposures that did not include SO_2. Treatments containing both SO_2 and O_3 caused a rapid curvilinear reduction in photosynthesis, whereas treatments with either SO_2 and NO_2 or O_3 and NO_2 reduced photosynthesis in a linear fashion. This result suggests that the response to treatments with $SO_2 + O_3$ differs from the response to treatments not containing this pair of pollutants.

In a recent study several photosynthetic parameters were measured for *Glycine max* (soybean) in response to combinations of SO_2 and NO_2 (Carlson, 1983b). Individual plants were exposed for 2-hour periods to these pollutants and photosynthesis was measured before, during, and after fumigation. At the end of fumigation, measurements were taken of dark respiration, apparent quantum yield, photorespiration, and photosynthetic saturation for both CO_2 concentration and irradiance. Leaf conductance to CO_2 was partitioned into its various components: boundary layer, stomatal, mesophyll, and carboxylation conductance. The experiment included treatments at all combinations of SO_2, NO_2, and SO_2 + NO_2, at concentrations of 0, 0.20, 0.40, and 0.60 ppm (8.3, 17, and 25 μmol m^{-3}). The results were analyzed statistically using a stepwise multiple-regression analysis of variance in order to (1) identify the type of response for each parameter, (2) calculate the amount of variance accounted for by each pollutant alone and/or in combination, and (3) produce a simple mathematical expression for each parameter in terms of pollutant concentration that could be used for predictive purposes.

The results show that, by the end of the 2-hour period of fumigation, photosynthesis and residual conductance (the inverse sum of mesophyll and carboxylation resistance) were both reduced synergistically by the combination of SO_2 and NO_2. Stomatal conductance, however, was reduced more by SO_2 alone than by either NO_2 or the combination of the two pollutants. Carboxylation was found to be a small portion of the total residual resistance, with mesophyll resistance accounting for approximately half the total leaf resistance to CO_2. Both photorespiration (measured as an increase in photosynthesis in response to the reduction of the O_2 concentration to less than 3%) and dark respiration were relatively insensitive to the pollutant treatments. There was a slight tendency for the observed reduction in respiratory rates to be associated with the concentration of NO_2. Apparent quantum yield (measured as the slope of the curve relating photosynthesis to light intensity at low irradiance) was reduced slightly in an additive fashion by SO_2 and NO_2. The data obtained in this study are not only valuable from a mechanistic standpoint, but also may be readily used in the development of ambient-air quality criteria.

An example of how data from such a study might be used to construct an air-quality criterion for a combination of pollutants is shown in Fig. 10.2. This figure was derived from a regression equation:

$$p = 20.75 - 0.0134(sn) - 0.553(s) + 0.0115\,(s^2),$$

where photosynthesis (p) is expressed as a function of SO_2 (s) and NO_2 (n) concentration in μmol m^{-3} (Carlson, 1983b). The equation was rearranged to express n in terms of p and s. Different values of photosynthetic reduction (10, 20, 30, and 40%) were entered for p and the equation solved for n at different values of s. A plot of the relationship between the three variables can be used to indicate the concentration of each pollutant in combination that would be expected to cause a specific reduction in photosynthesis. A similar graph could be constructed for plant growth and yield, for combinations of other pollutants, or for combinations of more than two pollutants. The only requirement is to have enough observations to cover the desired response surface. High, medium, and low concentrations of one pollutant must be paired in all combinations with a similar set of concentrations for the second and third pollutants.

The experiments discussed above, as well as the many that deal with plant

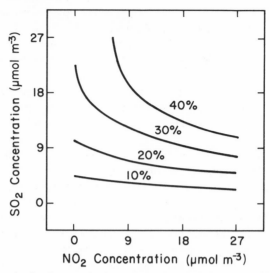

Fig. 10.2. Reduction in photosynthesis expressed as a percentage for different concentrations of SO_2 and NO_2 in combination. From Carlson (1983b).

growth and biochemistry, indicate that plant response to pollutant mixtures depends on both the type and the concentration of the pollutants. Ambient pollutant concentrations may vary both absolutely and relative to one another. It is therefore important that experiments be designed in such a way that all combinations of each pollutant (at reasonable concentrations) be included. Since response is to be measured as a function of more than one independent variable, it is necessary to construct the experiment in such a way that it is amenable to proper statistical analysis. Only then will it be possible to describe sufficiently the "interaction" between air pollutants.

Directions for Future Research

An understanding of how individual pollutants affect the metabolism of CO_2 is beginning to develop. The evidence suggests some similarities in the ways that various species respond to specific air contaminants. However, we probably still do not have enough data to be able to formulate a mechanistic model for any individual pollutant, even for a single species. Exceptions might be the extensive work with SO_2 by Unsworth, his associates, and others on *Vicia faba*, and a similar body of information that is developing for *Glycine max*. Enough information is available for these species such that formulation of a general mechanistic model should be attempted. The principles outlined by Tingey and Taylor (1982) for a conceptual model of physiological response to O_3 would provide a very useful guide. A model developed for SO_2 should include variables derived from both the chemistry of SO$_2$ and the physiology of plant response. This would tend to make the model general enough to be easily modified for application to the study of other air contaminants and other species. A model would also serve to focus experimental design toward the quantification of specific sensitive and/or less-well-known parameters.

The data available on physiological response to combinations of pollutants does not permit the development of anything more than the most general scenario. A useful model must include details on (1) the chemistry of, (2) the plant uptake for, and (3) physiological response to each contaminant by itself and in combination with the co-contaminant(s). Gas phase and solution chemistry differ greatly for the common air pollutants. Of major importance is the rate of absorption by external and internal leaf surfaces, and the chemical species formed by reaction with the absorbing medium. The salient features of a response curve similar to Fig. 10.1 for SO$_2$ must be appreciated for each contaminant in the combination that is under consideration. The residence time in solution for each reactive moiety is also important, since this parameter, in part, controls the rate of pollutant uptake and is indicative of the rate of reaction with cell constituents. Physiological response to the individual pollutant must be examined concomitantly with response to the pollutant combination. A range of concentrations must be included in the experiment, and the design must be properly statistical. Only then will it be possible to obtain the desired response "surface" for a pollutant combination and identify the type of interaction. An additional benefit of such an experimental design is that a complete response surface will allow better predictions than discrete concentration combinations would for formulating air-quality criteria.

11

Intraspecies Variability in Metabolic Responses to SO_2

David T. Tingey and David M. Olszyk

Vegetation responses to air pollutants may range from subtle changes in cellular processes to modifications of whole-plant physiology, visual injury, or altered productivity. For any plant organizational level for which the response is measured, phytotoxicity originates in molecular events that culminate in perturbed cellular structure and function. Different plant responses to a given SO_2 concentration are related to differences in environmental conditions and genic expression. The interaction between the genome and the environment is the ultimate source of differences in plant response to SO_2, whereas the immediate cause is differing physiological states.

Differences in plant response to SO_2 are common, and occur at various levels of plant organization. Differences within a species or between cultivars in sensitivity to SO_2, as measured by foliar injury responses, frequently have been used to make inferences about metabolic and growth processes. The injury symptoms result from a series of physiological events that occur before, during, and after exposure, and are influenced by genic and environmental factors. Many reports documenting differences in foliar injury for a wide range of crops, other herbaceous plants, and trees are summarized in Table 11.1.

A wide range of response was observed in most species, even when only two or three plants were examined. In species such as *Ilex aquifolium* (Brennan and Leone, 1970), where differences were not observed, the inclusion of additional plants or higher SO_2 concentrations may have produced responses large enough for reliable detection. In some species, such as *Lolium perenne* (Horsman, Roberts, and Bradshaw, 1978; Ayazloo and Bell, 1981) or *Zea mays* (Grzesiak, 1979), there were significant differences between cultivars in growth or yield without apparent differences in foliar injury. To reduce the possible confounding influences of different environmental conditions or developmental phases, all plants to be compared must be exposed simultaneously under the same conditions. Unless this condition is satisfied, one cannot attribute differences in responses to underlying genic characteristics.

To illustrate why plants exhibited different responses to O_3, Tingey and Taylor (1982) developed a conceptual model of its effects on plant metabolic processes.

TABLE 11.1
Intraspecies Differences in Foliar Injury Response to SO_2

Species	Conc.[a] (μmol m^{-3})	Time[b]	Comparison[c]	Results[d]	Source no.
AGRONOMIC CROPS					
Glycine max (soybean)	83	1.75 h	19 cv	d**	1
	83	4 h	3 cv	d	2
Hordeum vulgare (barley)	62	1 h	2 cv	nd	3
Humulus lupulus (hops)	8–17	4 h (4 h d^{-1})	4 cv	d	4
Medicago sativa (alfalfa)	56	4 h	5 cv	nd*	5
Nicotiana tabacum (tobacco)	83	1 h	4 cv	d	6
	62	3 h	6 cv	d	7
	6, 12	50–57 d	2 cv	d	8
Oryza sativa (rice)	21–83	1.5–7.5 h	303 cv	d	9
	25–29	5 h (1 h d^{-1})	22 cv	d	10
	83	6, 8, 12 h	4, 5 cv	d	11
Triticum aestivum (wheat)	8–25	30–100 h	4 cv	d	12
Zea mays (maize)	4–21	6–100 h	7 cv	nd	12
	21	20 d (5 h d^{-1})	7 lines, 2 cv	nd	13
GRASSES AND NATIVE HERBACEOUS PLANTS					
Dactylis glomerata (orchard grass)	83	6 h	2 pop	d***	14
Festuca rubra (red fescue)	31–75	6 h	2 cv	d	15
	6	10 d (6 h d^{-1})	2 cv	d	16
	82	6 h	2 pop	d	14
Geranium carolinianum (wild geranium)	33	12 h	6 pop	d	17
Holcus lanatus (velvet grass)	83	6 h	2 pop	d**	14
Lolium perenne (perennial ryegrass)	31–75	6 h	2 cv	d	15
	83	6 h	2 pop	nd	14
Muhlenbergia asperifolia	41–456	8 h	2 pop	nd	18
Poa pratensis (Kentucky bluegrass)	31–75	6 h	2 cv	d	15
	8	2 h	17 cv	d*	19
	6	10 d (6 h d^{-1})	12 cv	d**	16
Zoysia japonica (Japanese lawngrass)	31–75	6 h	2 cv	nd	15
HORTICULTURAL CROPS					
Antirrhinum majus (snapdragon)	166	2 h	3 cv	nd	20
Begonia × hiemalis (Eliatior begonia)	21	4 d (4 h d^{-1})	5 cv	d	21
B. semperflorens (begonia)	17–91	3, 4 h	6, 16 cv	d	22
	166	2 h	3 cv	nd	20
Chrysanthemum morifolium (chrysanthemum)	104, 187	3, 4 h	16 cv	d	23
Coleus spp. (coleus)	4–124	4 h	15 cv	d*	24
Cucumis sativus (cucumber)	17–37	6–24 h	2 cv	d	25

TABLE 11.1—cont.

Species	Conc.[a] (μmol m^{-3})	Time[b]	Comparison[c]	Results[d]	Source no.
Cucurbita pepo (squash, pumpkin)	17–37	6–24 h	2 cv	d	25
Daucus carota (carrot)	10–36	2 h	4 cv	nd	5
Euphorbia pulcherrima (poinsettia)	4–124	1–3 h	8 cv	d**	26
Fragaria × ananassa (strawberry)	4–42	4 d (6 h d^{-1})	6 cv	nd	27
Hibiscus syriacus (hibiscus)	124, 2,073	1 h	2 cv	d	28
Ilex aquifolium (English holly)	29–166	6 d (7 h d^{-1})	2 cv	nd	29
Lycopersicon esculentum (tomato)	41, 83	2, 4 h	26 cv	d*	30
Malus domestica (apple)	17, 83	4 h	3 cv	d**	31
Pelargonium spp. (cultivated geranium)	456	1 h	11 cv	d	32
Petunia hybrida (petunia)	104, 207, 415	1 h	14 cv	d	33
	166	2 h	4 cv	d	20
	33	4 d (4 h d^{-1})	3 cv	d	34
	116	16 h	5 cv	d*	35
Phaseolus vulgaris (bean)	42	5 d (6 h)	32 cv	d	36
Rhododendron spp. (azalea)	124	1–3 d (3–6 h d^{-1})	19 cv	d	37
	10	6 h (3 h d^{-1})	8 cv	nd	38
Tagetes spp. (marigold)	42, 83	2, 4 h	39 cv	d*	39
T. patula (marigold)	166	2 h	2 cv	nd	20
Vitis labrusca (grape)	3–11	seasonal	6 cv	d	40
	17, 33	4 h	2 cv	d**	41
TREES					
Acer platanoides (Norway maple)	42	7.5 h	8 cv	d	42
A. rubrum (red maple)	42	7.5 h	4 cv	d	42
A. saccharum (sugar maple)	42	7.5 h	3 cv	nd	42
Fraxinus americana (white ash)	42	7.5 h	10 pr, 50 hsf	d**	43
F. pennsylvanica (green ash)	42	7.5 h	2 cv	d	42
	42	7.5 h	16 pr, 59 hsf	d**	43
Ginkgo biloba (ginkgo)	42	7.5 h	4 cv	d	42
Gleditsia tricanthos inermis (thornless honeylocust)	42	7.5	6 cv	d	42
Larix spp. (larch)	14–30	20–50 h	12 cl	d	44
	829	10 × (12 mx^{-1})	18 cl	d	45
	30	20–50 h	54 cl	d	46
Picea abies (Norway spruce)	8–207	6–14 d (12 h d^{-1})	26 cl	d	47
	8–33	10–28 d	36 cl	d	48
	field	field	8 cl	d	49

TABLE 11.1—*cont.*

Species	Conc.[a] (μmol m^{-3})	Time[b]	Comparison[c]	Results[d]	Source no.
Pinus contorta (lodgepole pine)	829	10 × (12 mx^{-1})	19 pr	d**	50
	58–66	109 h	4 pr	d**	51
	field	field	14 pr	d	52
Pinus strobus (eastern white pine)	4	20–40 d (4–8 h d^{-1})	2 cl	d	53
	1–19	6 h	30 cl	d	54
	field	field	600 sdg	d	55
	1	6 h	10 cl	d	56
	37–83	6 h	8 cl	d	57
P. sylvestris (Scotch pine)	8–207	6–14 d (12 h d^{-1})	17 cl	d	47
	83	6 h	5 pr	d	57
	42–83	—	23 cl, 4 seed	d**	58
	42–83	24 h	17 cl	d**	59
	83	2 h	3 sel	d*	60
Plantanus occidentalis (American sycamore)	83	3 h	3 pr	d	61
P. orientalis	83	3 h	4 pr	d	61
Populus spp. (poplar)	field	field	22 cv	d	62
	17–124	2 h	9 cv	d	
	166	3–12 h	4 cl	d	63
	10	3 d (8 h d^{-1})	2 cl	d	64
	207	1.5–6 h	2 cl	d	64
P. tremuloides (quaking aspen)	15–27	3 h	5 cl	d	65
	21	3 h	10 cl	d**	65
Ulmus americana (American elm)	83	3 h	4 pr	d	61

SOURCES: 1, Miller, Howell, and Caldwell (1974). 2, Amundson and Weinstein (1981). 3, Baba and Sakai (1976). 4, Fiala (1966). 5, Tibbitts *et al.* (1982). 6, Hsieh (1973). 7, Menser, Hodges, and McKee (1973). 8, Miyake and Uno (1973). 9, Omura, Sato, and Sugahara (1978). 10, Valenzona, Saladagu, and Silva (1978). 11, Omura *et al.* (1980). 12, Laurence (1979). 13, Grzesiak (1979). 14, Ayazloo and Bell (1981). 15, Brennan and Halisky (1970). 16, Elkiey and Ormrod (1980a). 17, Taylor and Murdy (1975). 18, Campbell (1972). 19, Murray, Howell, and Wilton (1975). 20, Adedipe, Barrett, and Ormrod (1972). 21, Reinert and Nelson (1980). 22, Leone and Brennan (1972a). 23, Brennan and Leone (1972). 24, Semeniuk and Heggestad (1981). 25, Bressan, Wilson, and Filner (1978). 26, Heggestad, Tuthill, and Stewart (1973). 27, Rajput, Ormrod, and Evans (1977). 28, Ahn and Yeam (1977). 29, Brennan and Leone (1970). 30, Howe and Woltz (1982a). 31, Schertz, Kender, and Musselman (1980b). 32, Bonte *et al.* (1977). 33, Feder *et al.* (1969). 34, Elkiey and Ormrod (1979d). 35, Mikkelsen *et al.* (1981). 36, Beckerson, Hofstra, and Wukasch (1979). 37, Kunishige and Hirata (1975). 38, Sanders and Reinert (1982). 39, Howe and Woltz (1982b). 40, Fujiwara (1970). 41, Schertz, Kender, and Musselman (1980a). 42, Karnosky (1981a). 43, Karnosky and Steiner (1981). 44, Schönbach *et al.* (1964). 45, Börtitz and Vogl (1965). 46, Enderlein, Kastner, and Heidrich (1966, 1967). 47, Rohmeder, Merz, and Schönborn (1962). 48, Rohmeder and Schönborn (1965, 1967). 49, Fer *et al.* (1972). 50, Tzschachsch, Vogl, and Thummler (1969). 51, Lang, Neumann, and Schutt (1971). 52, Tzschachsch (1972). 53, Dochinger and Seliskar (1970). 54, Houston (1970). 55, Berry (1973). 56, Houston and Stairs (1973). 57, Genys and Heggestad (1978). 58, Mejnartowicz, Bialobok, and Karolewski (1978). 59, Smith and Davis (1978). 60, Bialobok, Karolewski, and Oleksyn (1980). 61, Santamour (1969). 62, Lampadius, Pelz, and Pohl (1970). 63, Dochinger *et al.* (1972). 64, Dochinger and Jensen (1975). 65, Karnosky (1976). 66, Karnosky (1977).

[a] Concentration × 0.02413 = ppm.

[b] Abbreviations: h (hour), d (day), m (minutes).

[c] Abbreviations: cl (clones), cv (cultivars), hsf (half-sib families), pop (populations), pr (provenances), sdg (seedlings), seed (seed sources), sel (selections).

[d] Abbreviations: d (difference between comparisons), nd (no difference between comparisons), ***, **, and * (statistical significance at 0.1, 1, and 5% levels, respectively), no indication (no statistical test reported).

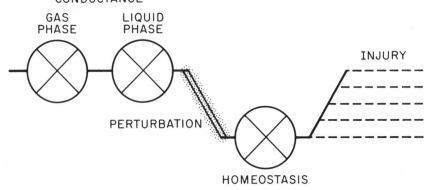

Fig. 11.1. Conceptual model of metabolic processes that determine plant response to SO_2; major processes regulating plant variation are gas- and liquid-phase conductance and homeostasis. Adapted from Tingey and Taylor (1982), p. 114.

The basic concepts are also applicable to SO_2; therefore the model is used here to assist in summarizing and interpreting data (Fig. 11.1). Plant responses to SO_2 may be viewed as the consequence of a series of biochemical and physiological events, beginning with SO_2 uptake and ending with foliar injury and/or effects on growth and yield. The biochemical and physiological processes may be influenced by environmental or genic conditions. The proposed model includes pollutant uptake, perturbation, homeostasis, and injury, which may be equated with or result in foliar injury. The model can be summarized as follows.

Leaf conductance. Sulfur-dioxide phytotoxicity results from biochemical reactions within the leaf interior. A primary factor controlling injury, therefore, is the rate at which SO_2 diffuses from the ambient air into the cellular perturbation sites. Following sorption onto the moist cell surfaces, the pollutant distribution within the leaf and cell is influenced by liquid-phase reactions, including diffusion, formation of derivatives, and reactions with scavenging systems.

Perturbation. Consists of alterations induced by SO_2 in cell structure or function. The initial phytotoxic response probably results from the reaction of SO_2 or its derivatives with macromolecules that are instrumental for the maintenance of cell function.

Homeostasis. The recovery processes that occur after the perturbation event. These cellular processes may be either repair or compensatory responses.

Injury. Cellular changes that result after pollutant exposure. The injury is controlled by the interplay between leaf conductance, perturbation, and the extent to which homeostatic processes are able to repair, restore, or compensate for cellular dysfunctions.

Leaf Conductance

Sulfur-dioxide uptake (flux) is controlled by the physical and chemical properties of the pathway. Physical properties are more important for the initial gas-phase flux into the leaf; chemical properties are more important for subsequent liquid-phase flux into cells (Fig. 11.2). Sulfur-dioxide flux is usually modeled with the same analog model, assumptions, and approach used to describe CO_2 uptake and H_2O vapor loss. SO_2 flux (J_{SO_2}) into the leaf is therefore proportional to the SO_2 concentration gradient between the atmosphere (C_a) and the leaf interior (C_i), as well as to the pathway conductance for SO_2, g, where

$$J_{SO_2} = (C_a - C_i) \cdot g. \tag{1}$$

This relationship between conductance (resistance^{-1}) and flux has frequently been used to model air-pollution fluxes into leaves (e.g., Tingey and Taylor, 1982; Black and Unsworth, 1979b; Unsworth, Biscoe, and Black, 1976; Winner and Mooney, 1980a).

Conductance can be partitioned into various components, i.e., boundary layer (g_a), stomatal (g_s), and residual (g_r, also called mesophyll, internal, or chemical, among other terms). Boundary-layer conductance influences the fraction of the gas-phase flux diffusing to the leaf surfaces ($J_{surface}$). Stomatal conductance determines the fraction of the total gas-phase flux that passes through the stomatal pore into the substomatal cavity. Residual conductance depends in part on diffusion distance and chemical reactivity, and it affects the liquid-phase flux from the substomatal cavity into the mesophyll cells. The stomatal and residual conductance jointly determine flux to the active sites within cells ($J_{internal}$).

Gas-phase conductance. Many studies have focused on differences in gas-phase conductance as the primary mechanism causing differences in plant response to SO_2. However, these studies produced conflicting results: some indicated that differences in injury responses were correlated with the gas-phase conductance of SO_2; no such correlation was observed in others (Table 11.2). These divergent conclusions may, in part, result from actual differences in gas-phase conductance within or between species, or may also result from errors in basic assumptions and in the experimental procedures used to measure gas-phase conductance and SO_2 uptake.

Because experimental methodology has a major effect on data interpretation, the various approaches will be briefly examined. Plant uptake of SO_2 can be measured either by finding out how much sulfur has been removed from the atmosphere or, antithetically, by finding out how much more sulfur the plant has acquired during exposure. Both approaches involve some inherent difficulties in experimental procedures and interpretation of results. However, with careful delineation of the implicit assumptions and experimental methods, both approaches can provide useful information. Because the two types of approach use different techniques, we will consider their methodologies separately.

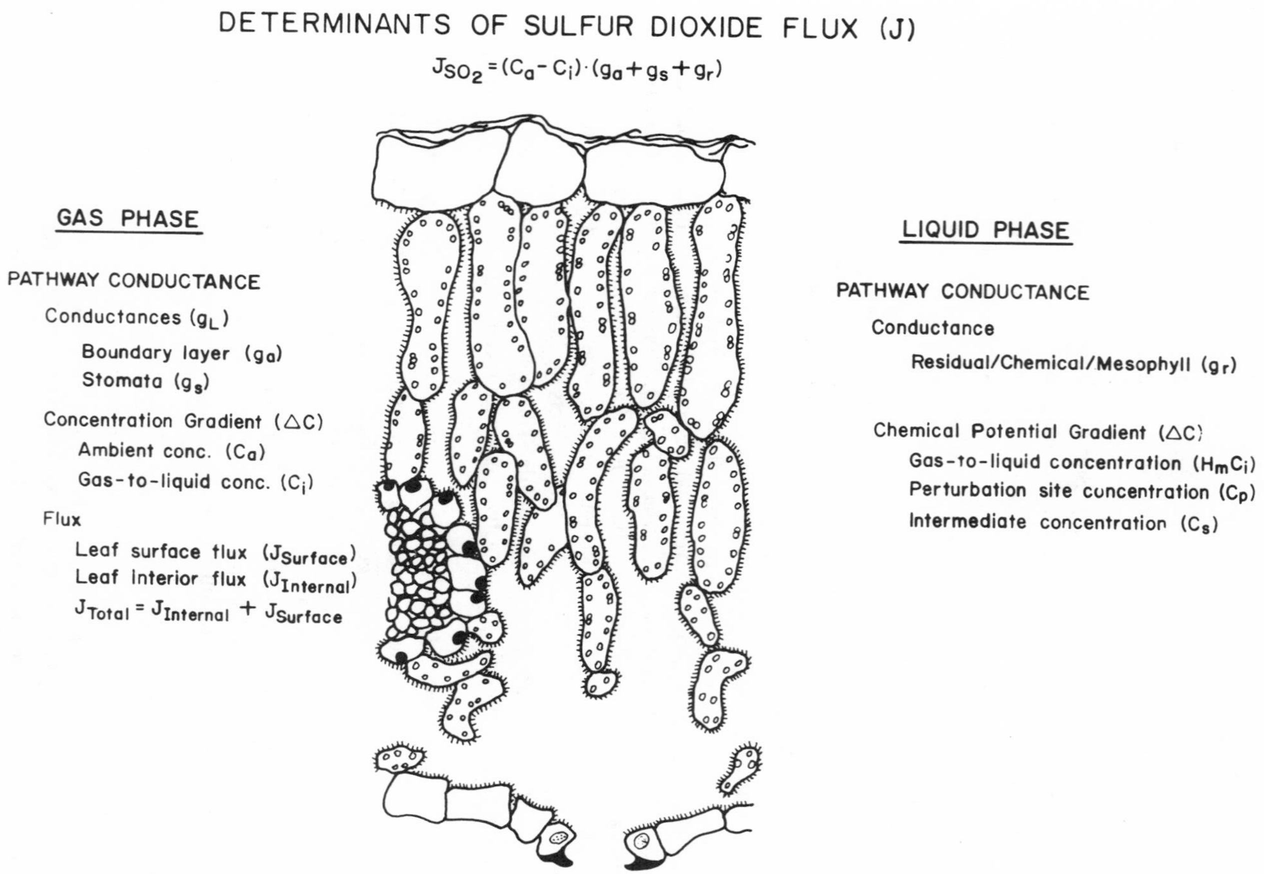

DETERMINANTS OF SULFUR DIOXIDE FLUX (J)

$$J_{SO_2} = (C_a - C_i) \cdot (g_a + g_s + g_r)$$

GAS PHASE

PATHWAY CONDUCTANCE

Conductances (g_L)

    Boundary layer (g_a)

    Stomata (g_s)

Concentration Gradient $(\triangle C)$

    Ambient conc. (C_a)

    Gas-to-liquid conc. (C_i)

Flux

    Leaf surface flux $(J_{Surface})$

    Leaf Interior flux $(J_{Internal})$

    $J_{Total} = J_{Internal} + J_{Surface}$

LIQUID PHASE

PATHWAY CONDUCTANCE

Conductance

    Residual/Chemical/Mesophyll (g_r)

Chemical Potential Gradient $(\triangle C)$

    Gas-to-liquid concentration $(H_m C_i)$

    Perturbation site concentration (C_p)

    Intermediate concentration (C_s)

Fig. 11.2. Analog model and representative leaf cross section describing the relationships between SO_2 flux and leaf conductances (g).

TABLE 11.2
Intraspecies Differences in Gas-Phase Conductance

Species	Comparison[a]	Parameter[b]	Source no.
NO CORRELATION WITH PLANT RESISTANCE			
Cucumis sativus (cucumber)	2 cv	Stom. density	1
Dactylis glomerata (orchard grass)	2 pop	Stom. density, length	2
		Stom. cond. SO_2	
Geranium carolinianum (wild geranium)	2 pop	Stom. cond. SO_2	3, 4
		SO_2 $J_{surface}$	4
		SO_2 $J_{internal}$	
Glycine max (soybeans)	3 cv	Stom. cond. air	5
Holcus lanatus (velvet grass)	2 pop	Stom. density, length	2
		Stom. cond. SO_2	
Lolium perenne (perennial ryegrass)	2 pop, 2 cl	Stom. density, length	6, 2
		Stom. cond. SO_2	
		Stom. cond. air	6
Malus domestica (apple)	3 cv	Stom. cond. air	7
		Stom. cond. SO_2	
Nicotiana tabacum (tobacco)	6 cv	Stom. density, width air	8
Pelargonium spp. (cultivated geranium)	11 cv	Stom. density	9
		Stom. dimensions air	
	2 cv	Stom. cond. air	
Petunia hybrida (petunia)	3 cv	Stom. cond. air	10
		Stom. cond. SO_2	11
	5 cv	SO_2 J_{total}	12
Poa pratensis (Kentucky bluegrass)	9 cv	SO_2 $J_{surface}$	13
Populus tremuloides (quaking aspen)	5 cv	Stom. cond. SO_2	14
Vitis lambrusca (grape)	2 cv	Stom. cond. air	15
		Stom. cond. SO_2	
CORRELATION WITH PLANT RESISTANCE			
Cucumis sativus (cucumber)	2 cv	SO_2 J_{total}	1
Cucurbita pepo (squash, pumpkin)	2 cv	Stom. density	1
		SO_2 J_{total}	
Dactylis glomerata (orchard grass)	2 pop	Stom. cond. air	2
Euphorbia pulcherrima (poinsettia)	5 cv	Leaf surface	16
		Stom. cond. air	
Glycine max (soybean)	3 cv	Stom. cond. SO_2	5
Holcus lanatus (velvet grass)	2 pop	Stom. cond. air	2
Lolium perenne (perennial ryegrass)	2 pop	Stom. cond. air	2
Pelargonium spp. (cultivated geranium)	2 cv	Stom. cond. SO_2	9
Petunia hybrida (petunia)	3 cv	SO_2 $J_{surface}$	17
		SO_2 $J_{internal}$	17, 18
		Leaf surface morphology	10
Picea abies (Norway spruce)	30 cl	Needle morphology	19
		Gas permeability	
		Stom. cond. SO_2	19, 20
		Stom. cond. air	19
Poa pratensis (Kentucky bluegrass)	9 cv	SO_2 $J_{internal}$	13
Populus tremuloides (quaking aspen)	5 cv	Stom. cond. air	14

SOURCES: 1, Bressan, Wilson, and Filner (1978). 2, Ayazloo, Garsed, and Bell (1982). 3, Taylor (1976). 4, Taylor and Tingey (1981). 5, Amundson and Weinstein (1981). 6, Bell and Mudd (1976). 7, Schertz, Kender, and Musselman (1980b). 8, Menser, Hodges, and McKee (1973). 9, Bonté *et al.* (1977). 10, Elkiey, Ormrod, and Pelletier (1979). 11, Elkiey and Ormrod (1979a). 12, Mikkelsen *et al.* (1981). 13, Elkiey and Ormrod (1981b). 14, Kimmerer and Kozlowski (1981). 15, Schertz, Kender, and Musselman (1980a). 16, Krizek, Wergin, and Semeniuk (1982). 17, Elkiey and Ormrod (1980b). 18, Elkiey and Ormrod (1981a). 19, Braun (1977c). 20, Braun (1977d).

[a] Abbreviations: cl (clones), cv (cultivars), pop (populations).
[b] Abbreviations: Stom. (Stomatal), cond. (conductance), air (no SO_2), SO_2 (during exposure to SO_2).

Methodology for gas-phase studies. Sulfur-dioxide flux either may be measured directly or may be calculated from the concentration gradient and conductance measurements for SO_2 or H_2O, with appropriate adjustments made for differences in diffusivity according to Eq. 1.

Direct measurement of SO_2 flux by the mass-balance approach (e.g., Winner and Mooney, 1980a; Taylor and Tingey, 1981) uses the equation

$$J_{SO_2} = ([SO_2]_{in} - [SO_2]_{out}) F \cdot A^{-1}. \qquad (2)$$

The required measurements include (a) SO_2 concentration in the air before and after it passes over the plant, (b) air-flow rate (F) through the chamber, and (c) amount of plant material (A), usually leaf area, sometimes dry weight. The measured difference between the SO_2 concentration in and out of the gas-exchange chamber represents the SO_2 sorbed by the plant and the experimental system. Accurate measurement of SO_2 loss to the system (primarily to the exposure chamber or cuvette) is therefore required for calculation of actual SO_2 flux to the plant. Losses to the system are influenced by factors such as dewpoint (Taylor and Tingey, 1979), type of chamber surfaces, and the amount of time SO_2 has been in the system. Calculation of SO_2 losses to the system must therefore be made under conditions identical to those when a plant is present. For example, to measure chamber sorption accurately, the dewpoint of the air in the chamber must be kept the same as when the plant is present and actively transpiring. If system losses are not properly measured, caution is required when evaluating SO_2 flux to the plant.

Total SO_2 flux (J_{total}) to the plant represents the sum of the SO_2 adsorbed onto the leaf surface ($J_{surface}$) and that absorbed into the leaf interior ($J_{internal}$). Generally, it has been assumed that $J_{surface}$ has no physiological role, and that only $J_{internal}$ induces plant responses (Runeckles, 1974; Winner and Mooney, 1980a). However, both $J_{surface}$ and $J_{internal}$ are important for evaluating gas-phase flux to plants.

If $J_{surface}$ is large or boundary-layer conductance is small, the effective SO_2 concentration at the stomatal pore will be less than that of the bulk air. The measured SO_2 concentration in the air therefore may not accurately reflect either the concentration gradient through stomata into the leaf or the concentration affecting cells inside the leaf, which is of prime interest in metabolic studies (Runeckles, 1974; Unsworth, 1982). This may be a significant problem as estimates of $J_{surface}$ approach 10 to 60% of the total SO_2 sorbed by the plant (Spedding, 1969; Garsed and Read, 1974; Fowler, 1978; Elkiey and Ormrod, 1980b, 1981b; Taylor and Tingey, 1981).

Calculation of $J_{surface}$ is either (a) by direct measurement of J_{total} when stomata are assumed to be closed and $J_{internal}$ to be negligible, or (b) by subtraction of $J_{internal}$ from J_{total}, $J_{internal}$ usually being derived by analogy to H_2O vapor. Direct measurement of J_{total} is usually carried out by measuring SO_2 uptake by plants in the dark (Elkiey and Ormrod, 1980b; Tingey and Taylor, 1982); however, problems may arise if the stomata are not closed or if they open in response to SO_2 (Mansfield and Majernik, 1970; Majernik and Mansfield, 1970; Elkiey, Ormrod, and Marie, 1982). The possibility of stomatal opening in the dark may be assessed by comparison to nonexposed plants or to flux measurements in the light. Procedures to insure stomatal closure, such as elevated atmospheric CO_2 levels (Raschke, 1975a,b) or ABA treatments (Mittelheuser and van Steveninck, 1971; Raschke, 1975b), also may be used. Calculation of $J_{surface}$ by subtracting $J_{internal}$ from J_{total} assumes that SO_2 diffusion is controlled by the same processes and follows the same diffusion pathway as H_2O vapor, because it uses the diffusion of H_2O vapor as a surrogate for SO_2 (Winner and Mooney, 1980a). The only required adjustment is a correction for differences in diffusivity. In addition, all water-vapor flux from the plant during light is assumed to be through the stomata, and water-vapor flux through the cuticle is generally neglected, even though cuticular transpiration may be large.

Water-vapor flux from leaves has been widely used as a surrogate for the flux of SO_2 and other gases into leaves. For example, water-vapor flux was used to make inferences about SO_2 uptake in nearly all the studies listed in Table 11.2. Water-vapor flux may be measured by using a gas-exchange chamber or by porometry, where a humidity sensor is placed near the leaf. Both procedures measure the amount of water vapor diffusing from a leaf or plant to calculate stomatal conductance to water vapor, which may represent conductance for SO_2 once differences in gas diffusivity for the two gases are reconciled. The adjusted stomatal-conductance values and the ambient concentration of SO_2 are used in Eq. 1 to calculate SO_2 flux. The use of water-vapor flux to estimate SO_2 uptake has several problems, including: (a) the possibility that SO_2 does not follow the same diffusive pathway as H_2O vapor, (b) presence of chemical or other internal (residual) factors not present for H_2O vapor that may affect conductance for SO_2, (c) the failure to consider boundary-layer conductance if porometry is used, and (d) the need to estimate $J_{surface}$ for SO_2.

The presence of residual conductance factors that influence SO_2 but not water vapor merits additional comment. The influence of residual factors was not considered in most SO_2 flux studies (e.g., Winner and Mooney, 1980a; Kimmerer and Kozlowski, 1981), even though these factors were suggested by other authors (Black and Unsworth, 1979b; Bressan, Wilson and Filner, 1978; Hällgren, 1978; Heath, 1980). Some of the studies suggest that these residual factors decreased the SO_2 flux to less than that expected for gas-phase (primarily stomatal) conductance (Hällgren et al., 1982). Decreased conductance associated with residual factors has been described for other pollutants, e.g., O_3 (Taylor and Tingey, 1982), and H_2S (Cope and Spedding, 1982). Other data suggest that SO_2 conductance is higher than expected from gas-phase conductance values (Klein et al., 1978; Taylor and Tingey, 1983).

Given the preceding considerations and questions about methodology, very few studies appear to provide a critical investigation of differences in gas-phase flux within species. Additional studies with better experimental controls are needed if one is to accurately calculate surface loss and residual conductance to insure the same $J_{internal}$ for subsequent studies on physiological and biochemical processes.

Variability in boundary-layer conductance and $J_{surface}$. Available data suggest that $J_{surface}$ may vary within some species but not others (Table 11.2); however, it has actually been measured in only a few studies (Elkiey and Ormrod, 1980b, 1981b; Taylor and Tingey, 1981). Elkiey and Ormrod reported variability in $J_{surface}$ for SO_2 in *Petunia hybrida* (1980b) and *Poa pratensis* (1981b) cultivars, but it was correlated with leaf injury only in *Petunia hybrida* (Table 11.3). The most SO_2-tolerant *P. hybrida* cultivar, 'Capri,' exhibited the highest $J_{surface}$ and had the roughest cuticle and largest number of trichomes (Elkiey, Ormrod, and Pelletier, 1979). These cuticular characteristics could decrease boundary-layer conductance; this decrease, in combination with SO_2 removal by the leaf surface, would lead to a lower SO_2 concentration at the stomatal pore, reducing $J_{internal}$ compared to that in other cultivars. A potential problem with Elkiey and Ormrod's studies was the lack of information about how the chamber sorption for SO_2 was measured. If chamber sorption was not measured at the same dewpoint as when plants were present, SO_2 loss to the chamber system would have been underestimated, producing errors in $J_{surface}$ and $J_{internal}$. A more serious problem would have occurred if stomatal closure in the dark had not been checked by porometry and stomatal impressions. If conductance varied in the dark, then the estimations of $J_{surface}$ would have been inaccurate, and $J_{internal}$ could not have been reliably estimated (Elkiey, Ormrod, and Marie, 1982).

Resistant populations of *Geranium carolinianum* had a slightly higher $J_{surface}$ than sensitive populations (Table 11.3), with the difference increasing with the SO_2 concentration (Taylor and Tingey, 1981). However, in this study, $J_{surface}$ values were not correlated with SO_2 resistance, since $J_{internal}$ was similar for both populations. For the studies with *G. carolinianum*, the sorptive losses of SO_2 to the chamber were measured at the same dewpoint as when the plants were present. In addition, the leaf conductance (combination of boundary-layer and stomatal conductance) values in the dark were low, indicating that the stomata for both populations were probably closed (Taylor and Tingey, 1981).

Morphological data also suggest differences in $J_{surface}$ between SO_2-tolerant and SO_2-sensitive plants. In *Picea abies*, needles from resistant clones were longer and wider than needles from sensitive clones (Braun, 1977c). *Euphorbia pulcherrima* leaves from resistant cultivars had a greater trichome density than leaves from sensitive cultivars (Krizek, Wergin, and Seminiuk, 1982). However, since $J_{surface}$ was not measured in either species, it is uncertain whether these features actually influenced SO_2 uptake. Furthermore, in both species additional factors also appeared to influence SO_2 resistance (Braun, 1977a; Krizek, Wergin, and Seminiuk, 1982).

TABLE 11.3
Intraspecies Differences in Rates of SO_2 Flux

Species; and cultivar	Sensitivity	Exposure (μmol m^{-3})	$J_{surface}$ (nmol m^{-2} s^{-1})	$J_{internal}$
Geranium carolinianum (wild geranium)[a]				
(Population)	Resistant	25–0.4 h	67.3	72.9
(Population)	Sensitive		66.1	81.6
Petunia hybrida (petunia)[b]				
Capri	Resistant	33–0.75 h	10.1	7.5
White Magic	Intermediate		7.9	9.7
White Cascade	Sensitive		1.1	9.6
Poa pratensis (Kentucky bluegrass)[c]				
Fylking	Resistant	17–4 h	4.9	7.8
Touchdown	Resistant		5.9	9.3
Plush	Resistant		4.5	7.9
Merion	Intermediate		8.1	8.1
Skofti	Intermediate		6.6	11.1
Cheri	Sensitive		9.3	10.0
Baron	Sensitive		4.6	10.2

[a] Adapted from Taylor and Tingey (1981).
[b] Adapted from Elkiey and Ormrod (1980b).
[c] Adapted from Elkiey and Ormrod (1981a,b), using data only for day 1 from cultivars where SO_2 sensitivity was indicated.

Differences in stomatal conductance. Many studies have focused on stomatal conductance, morphology, or density as the principal factor controlling differences in SO_2 injury between plants (Table 11.2), because the stomata are the principal pathway by which gases diffuse into the leaf. In most of these studies, it was assumed that stomatal conductance to water vapor was proportional to $J_{internal}$ for SO_2. Differences in stomatal conductance were observed for many species, both as a specific response to SO_2 exposure and as a preexisting condition that may determine how much injury plants of a species may suffer. Resistant cultivars of *Glycine max* (Amundson and Weinstein, 1981) and *Pelargonium* spp. (Bonté *et al.*, 1977) exhibited greater stomatal closure than sensitive ones in response to SO_2 exposures, suggesting a reduction in $J_{internal}$ for SO_2. For both species, all cultivars showed similar conductance in SO_2-free air. Resistant clones of *Picea abies* exhibited more stomatal closure during SO_2 exposures, and also had less stomatal conductance in SO_2-free air than sensitive clones (Braun, 1977d).

In other species, different plant responses to SO_2 were not clearly associated with differences in stomatal conductance. Resistant populations of *Dactylis glomerata*, *Holcus lanatus*, and *Lolium perenne* (Ayazloo, Garsed, and Bell, 1982) all had less conductance in SO_2-free air than sensitive populations. However, in response to an SO_2 exposure, the leaf conductance of tolerant populations of *D. glomerata* and *H. lanatus* decreased, but there was no change in *L. perenne*.

Populus tremuloides clones tolerant to SO_2 had less leaf conductance than sensitive clones in SO_2-free air; whereas, in response to SO_2, the leaf conductance decreased in sensitive clones, but did not change in tolerant ones (Kimmerer and Kozlowski, 1981). Data on the stomatal conductance of *L. perenne* clones in SO_2-free air are contradictory; Ayazloo, Garsed, and Bell (1982) reported a lower conductance for resistant populations, but Bell and Mudd (1976) found no difference between clones. Resistant cultivars of *Euphorbia pulcherrima* exhibited less stomatal conductance than sensitive cultivars in SO_2-free air (Krizek, Wergin, and Seminiuk, 1982).

No difference in stomatal conductance, either in response to SO_2 or in SO_2-free air, was observed among cultivars of other species. The two cultivars of *Vitis lambrusca* exhibited similar patterns of stomatal opening (Schertz, Kender, and Musselman, 1980a), whereas all cultivars of *Malus domestica* exhibited stomatal closure in response to SO_2 exposure (Schertz, Kender, and Musselman, 1980b). Cultivars of *Petunia hybrida* (Elkiey and Ormrod, 1979a) and populations of *Geranium carolinianum* (Taylor, 1976; Taylor and Tingey, 1981) had similar stomatal conductances in SO_2-free air, and showed little change in conductance during SO_2 exposure.

In summary, even though stomatal conductance differs within species, the importance of such differences for resistance to SO_2 injury depends on the species, and interpretation is influenced by the inherent problems with the methodology, as described earlier. All previous studies (except possibly that of Braun, 1977d) used water vapor as a surrogate for SO_2, and most did not consider the effects of boundary-layer conductance and residual factors on SO_2 flux.

Differences in J_{total} and $J_{internal}$. Intraspecific differences in SO_2 flux, both J_{total} and $J_{internal}$, have been compared, for plants that show different injuries, in only a few species (Table 11.2). Total SO_2 flux differed in populations of *Cucumis sativum* and *Cucurbita pepo*, with resistant cultivars exhibiting 40 and 13% lower J_{total}, respectively, than corresponding sensitive cultivars (Bressan, Wilson, and Filner, 1978). In *Petunia hybrida*, J_{total} differed 27% between cultivars (Mikkelsen *et al.*, 1981); however, variability within cultivars was of the same magnitude, and there was no correlation between J_{total} and resistance to SO_2. This lack of correlation between J_{total} and injury in *P. hybrida*, as reported by Mikkelsen *et al.* (1981), may be due to differences in $J_{surface}$ as documented by Elkiey and Ormrod (1980a), but use of different cultivars of *P. hybrida* by the two research groups may also have contributed to the difference in results.

Elkiey and Ormrod found large intraspecies differences in $J_{internal}$ for *Poa pratensis* (1981b) and *Petunia hybrida* (1981b, 1981a) cultivars that differed in foliar injury (Table 11.3). In *Poa pratensis*, $J_{internal}$ was 22% lower in three resistant cultivars than in two sensitive cultivars. For cultivars of intermediate sensitivity, $J_{internal}$ was similar to both resistant and sensitive cultivars. In *Petunia hybrida*, $J_{internal}$ was approximately 30% lower in the resistant cultivar than in the sensitive cultivar. For the intermediate-sensitivity cultivar, $J_{internal}$ was similar to that in the sensitive cultivar.

Not only do the chamber-loss considerations described earlier apply here, but the data (Elkiey and Ormrod, 1980b, 1981a,b) were not evaluated to see if differences in g_a could account for the large differences in $J_{surface}$. An increase in g_a would decrease the SO_2 concentration at the stomatal pore, thereby reducing the SO_2 concentration gradient between the pore and the leaf interior, thereby producing a lower $J_{internal}$ than would be expected from the atmospheric SO_2 concentration. This concern may be especially important for *P. hybrida*, since the resistant cultivar had a much larger $J_{surface}$ than the sensitive cultivar, and the resulting decrease in cuticular SO_2 concentration may have been large enough to account for the lower $J_{internal}$ for the resistant cultivar.

Taylor and Tingey (1981) reported only a slight difference in $J_{internal}$ between two populations of *Geranium carolinianum* (Table 11.3). The resistant population exhibited a $J_{internal}$ that averaged 5 to 11% lower than that of the sensitive population over a range of SO_2 concentrations, compared to a much larger (9 to 68%) variability in J_{total} or $J_{surface}$ within each population. For *G. carolinianum*, $J_{surface}$ was similar for both populations, and the cuticular SO_2 concentration was adjusted for boundary-layer conductance.

Sulfur enrichment. Gas-phase conductance can be estimated by measuring how much additional sulfur a plant acquires during exposure, with sulfur content evaluated as [35]sulfur or as nonradioactive total sulfur or sulfate-sulfur (Table 11.4).

Acquisition of [35]sulfur did not differ between resistant and sensitive populations of *Dactylis glomerata* or *Holcus lanatus* (Ayazloo, Garsed, and Bell, 1982), or between resistant and sensitive populations and clones of *Lolium perenne* (*ibid.*; Garsed and Read, 1977c). For each species, similar [35]sulfur concentrations were washed from leaves that had been fumigated with [35]SO_2; so $J_{surface}$ was similar for resistant and sensitive plants. In addition, [35]sulfur concentrations in internal fractions (methanol : chloroform : H_2O + N-ethylmaleimide soluble and insoluble) indicated that [35]sulfur concentrations were similar. For *Festuca rubra*, the resistant population had 43% less [35]sulfur in the leaf-wash solutions than the sensitive population (Ayazloo, Garsed, and Bell, 1982). However, [35]sulfur was essentially the same in the internal fractions. These data illustrate the [35]sulfur uptake method (Garsed and Read, 1974, 1977c) for calculating $J_{surface}$ and $J_{internal}$ that avoids the problems inherent in measuring differential SO_2 concentrations.

Nonradioactive total sulfur content of plant tissues following an SO_2 exposure has been evaluated for plant selections within a number of species. Four herbaceous plants, *Lolium perenne* (Bell and Mudd, 1976), *Lycopersion esculentum* (Howe and Woltz, 1981), *Nicotiana tabacum* (Miyake and Uno, 1973), and *Petunia hybrida* (Elkiey and Ormrod, 1981c), exhibited little difference in total sul-

TABLE 11.4
Intraspecies Differences in Increased Sulfur

Species	Comparison[a]	Method	Source No.
NO CORRELATION WITH RESISTANCE			
Dactylis glomerata (orchard grass)	2 pop	^{35} sulfur	1
Festuca rubra (red fescue)	2 pop	^{35}sulfur	1
Holcus lanatus (velvet grass)	2 pop	^{35}sulfur	1
Lolium perenne (perennial ryegrass)	2 cl	^{35}sulfur	2
	2 pop	^{35}sulfur	1
	2 cl	total sulfur	3
Lycopersicon esculentum (tomato)	10 cv	total sulfur	4
Muhlenbergia asperifolia	2 pop	total sulfur	5
Nicotiana tabacum (tobacco)	2 cv	total sulfur	6
Petunia hybrida (petunia)	3 cv	total sulfur	7
Picea abies (Norway spruce)	18 cl	total sulfur	8
Zea mays (maize)	3 cv	total sulfur	9
CORRELATION WITH RESISTANCE			
Petunia hybrida (petunia)	3 cv	sulfate-sulfur	7
Picea abies (Norway spruce)	rst. vs. sens.	sulfate-sulfur	10
Pinus strobus (eastern white pine)	2 cl	total sulfur	11
P. sylvestris (Scotch pine)	rst. vs. sens.	sulfate-sulfur	10
Populus tremuloides (quaking aspen)	5 cl	total sulfur	12

SOURCES: 1, Ayazloo, Garsed, and Bell (1982). 2, Garsed and Read (1977c). 3, Bell and Mudd (1976). 4, Howe and Woltz (1981). 5, Campbell (1972). 6, Miyake and Uno (1973). 7, Elkiey and Ormrod (1981c). 8, Braun (1977c). 9, Laurence (1979). 10, Rohmeder, Merz, and Schönborn (1962). 11, Roberts (1976). 12, Kimmerer and Kozlowski (1981).

[a] Abbreviations: pop (populations), cl (clones), cv (cultivars), rst. vs. sens. (resistant versus sensitive).

fur content, even though plant selections within these species differed in injury or growth. *Zea mays* cultivars showed no differences in total sulfur content (Laurence, 1979). There was no difference in total sulfur content between resistant and sensitive clones of *Picea abies* (Braun, 1977e). *Pinus strobus* (Roberts, 1976) exhibited a higher total sulfur content in resistant than in sensitive clones, whereas *Populus tremuloides* (Kimmerer and Kozlowski, 1981) exhibited a lower total sulfur content in resistant than in sensitive clones following exposure to SO_2.

Total sulfur content in *P. tremuloides* paralleled stomatal conductance; the lowest total sulfur content was observed in resistant clones, which also had the lowest stomatal conductance (Kimmerer and Kozlowski, 1981). The authors assumed that the net increase in total sulfur content after SO_2 exposure represented $J_{internal}$, and thus concluded that SO_2 resistance in *P. tremuloides* was determined by gas-phase conductance. However, the total sulfur content actually represents J_{total}. In contrast, other researchers found no correlation between total sulfur content and stomatal conductance (Biggs and Davis, 1981c; Elkiey and Ormrod, 1979a, 1980b, 1981a,c). Also, the total sulfur content does not consider compartmentalization of sulfur into different fractions, or translocation of sulfur to different organs of plants, factors that may be important in determining SO_2 resistance (Garsed and Read, 1974, 1977c).

Sulfate-sulfur was correlated with resistance to SO_2 in *Picea abies* and *Pinus strobus*, with resistant clones exhibiting a lower sulfate-sulfur content than sensitive clones (Rohmeder, Merz, and Schönborn, 1962). In *Petunia hybrida*, the cultivars differed in sulfate-sulfur content when plants were exposed at a high relative humidity, whereas all cultivars had similar sulfate-sulfur contents when plants were exposed at a low relative humidity (Elkiey and Ormrod, 1981c). Sulfate-sulfur was not correlated with resistance to injury at either humidity level (Elkiey and Ormrod, 1979d). Sulfate-sulfur may be a more appropriate indicator of increased sulfur than total sulfur, since sulfate is the major product of SO_2 metabolism in plant tissue (Weigl and Ziegler, 1962; DeCormis, 1969).

Liquid-phase conductance. Thomas, Hendricks, and Hill (1950) proposed that phytotoxicity resulted from the diffusion of SO_2 through the stomata into the mesophyll, where toxic metabolites were formed. When SO_2 uptake outran the detoxification reactions in the mesophyll, toxic metabolites would accumulate, and when they exceeded certain cellular concentration thresholds, injury would result. The lack of a consistent correlation between SO_2 uptake and plant response (Table 11.2) supports this concept that gas-phase conductance alone is insufficient to explain intraspecies differences in injury. Several studies have suggested that plants have "avoidance" (i.e., pollutant exclusion) and "tolerance" (i.e., control of toxicity) mechanisms that contribute to interspecies and intraspecies differences in injury (Taylor, 1978b; Ayazloo, Garsed, and Bell, 1982). For example, studies on resistance to SO_2 within populations of several grass species found no correlation between the degree of acute injury and the degree of chronic injury; hence Ayazloo and Bell (1981) have suggested these plants may have several independent mechanisms for tolerance. Alterations in the liquid-phase conductance of SO_2 may contribute, in part, to intraspecies differences in injury.

After SO_2 has diffused through the stomata, some additional pathways for SO_2 transfer may constitute internal hindrances to SO_2 conductance. These hindrances may include internal components of SO_2 metabolism, such as solubilization in the H_2O film on the cell surface, conversion into additional compounds, and transport (Jäger and Klein, 1980).

After SO_2 dissolves into H_2O in and on the cell, it is rapidly hydrated, forming bisulfite and sulfite (Peiser and Yang, this volume). At typical cellular pH values, more bisulfite is present, but the proportion of sulfite increases with pH (Ziegler, 1975; Peiser and Yang, this volume). Both bisulfite and sulfite have been implicated in SO_2 toxicity, being approximately 30 times more toxic than sulfate (Thomas, 1961).

Sulfite and bisulfite may be detoxified by either oxidative or reductive reactions (Fig. 11.3). Depending on the redox conditions that prevail within the cells, sulfite may undergo autooxidation or possibly oxidation by a mitochondrial sulfite oxidase (Ziegler, 1975; Hällgren, 1978). Sulfite may also be photooxi-

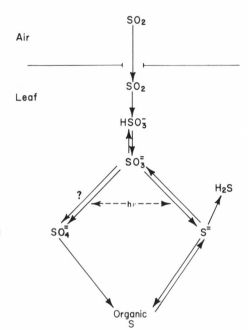

Fig. 11.3. Pathways for SO$_2$ metabolism and potential detoxification. The question mark indicates that a non-photochemical process that oxidizes sulfite to sulfate is not certainly known to exist.

dized to sulfate by a free-radical chain process that is initiated by superoxide anion produced via the photosynthetic electron-transport system in the chloroplast (Asada, 1980; Peiser, Lizada, and Yang, 1982). Sulfite may also be photoreduced to sulfide and subsequently volatilized as H$_2$S (Silvius *et al.*, 1976).

The oxidation of sulfite to sulfate has been proposed as a potential detoxification mechanism for SO$_2$. Differences in oxidation rate and in concentration of sulfite in tissues have been associated with different plant responses to SO$_2$. Exogenous sulfite was metabolized more rapidly in *Glycine max* cultivars resistant to foliar SO$_2$ injury than in sensitive cultivars (Miller and Xerikos, 1979). Studies with *Lycopersicon esculentum* plants (Howe and Woltz, 1981) showed that foliar sulfite levels increased with the SO$_2$ exposure concentration, illustrating the association between uptake and cellular metabolic concentrations. In addition, foliar injury was highly correlated with tissue sulfite concentrations, confirming the observations of Miller and Xerikos (1979).

Kondo *et al.* (1980) reported that SO$_2$ tolerance in various species was correlated with sulfite oxidase activity. Ayazloo, Garsed, and Bell (1982) found that a sensitive genotype of *Lolium*, S23, had a higher proportion of inorganic sulfur as sulfite than the tolerant genotypes did after both light and dark incubations in sulfite solutions. They also reported that, of the populations of several species exhibiting differences in sensitivity to SO$_2$, only the more tolerant population of *Holcus lanatus* had a higher rate of sulfite oxidation after SO$_2$ exposure; no significant differences in sulfite oxidation rates were observed between genotypes of two other species.

The photoreduction of sulfite to sulfide may also be a detoxification mechanism. Plants exposed to SO_2 emit H_2S in the light (DeCormis, 1968; Silvius *et al.*, 1976; Hällgren and Fredriksson, 1982). Hydrogen sulfide was detected within 15 to 30 minutes after the start of exposure, with emissions reaching a maximum within 45 minutes (DeCormis, 1968; Taylor and Tingey, 1983). A $J_{internal}$ of at least 48 μmol of SO_2 m^{-2} was required to stimulate emissions, suggesting that the cellular concentration had to exceed a threshold before emissions occurred (Taylor and Tingey, 1983). Dodrill (1976) suggested that the emission of H_2S was a method of detoxifying SO_2, since H_2S emissions were stimulated at lower SO_2 concentrations than would cause foliar injury. Young leaves of *Cucumis sativus* plants were resistant to acute exposures of SO_2, but older leaves were sensitive even though they absorbed less SO_2 (Sekiya, Wilson, and Filner, 1982). The authors suggested that the differences in sensitivity between different-aged leaves of *C. sativus* cultivars occurred because the young leaves were more able to photoreduce SO_2 to H_2S. In contrast, H_2S emissions were similar from plants of populations of *Geranium carolinianum* that differed in SO_2 sensitivity (G. Taylor, personal communication). In some species, emission of H_2S may provide a means of detoxifying SO_2 if the sulfide is formed by the photoreduction of sulfite. In some plant systems, cysteine may also be metabolized into H_2S (Rennenberg *et al.*, 1982; Sekiya *et al.*, 1982), but this would apparently not contribute to detoxification.

Studies on the inheritance of resistance to SO_2 injury in inbred lines and crosses of *C. sativus* led to the conclusion that the plasma membrane was a barrier to the uptake and transport of both SO_2 and sulfite (Bressan *et al.*, 1981). The authors suggested that this barrier contributed to the differences in sensitivities of plants to SO_2. They supported this conclusion by the observation that the relative resistance to SO_2 or bisulfite was the same as for a variety of other stresses likely to affect membrane properties.

The capacity of plant tissue to buffer hydrogen ions was decreased by SO_2 exposures without an accompanying decrease in cellular pH (Thomas, Hendricks, and Hill, 1950; Jäger and Klein, 1977). Although buffering capacity has been implicated in plant resistance to SO_2, the few investigations of differences in buffering capacity within species produced conflicting results. Resistant clones of *Picea abies* had a higher buffering capacity than sensitive clones (Scholz and Reck, 1976; Braun, 1977e). No difference in buffering capacity was found between two populations of *Muhlenbergia asperifolia* (Campbell, 1972).

In summary, both gas-phase and liquid-phase conductance pathways control SO_2 flux to cellular perturbation sites within leaves. Interspecies differences in gas-phase components g_a, g_s, $J_{surface}$, and $J_{internal}$ occur in a wide range of species. However, these differences were not consistently associated with differences in plant injury. Much study will be needed to understand gas-phase conductance of SO_2, and it will require rigorous evaluation of methodology and documentation of components that indicate metabolically active SO_2, e.g., $J_{internal}$. Liquid-phase conductance is now poorly understood. However, recent evidence suggests that mesophyll detoxification reactions are important in controlling plant response to SO_2, and highlights the need for critical study of components of liquid-phase conductance.

Perturbation

Perturbation, as defined here, is the initial SO_2-induced change in cell structure or function. Phytotoxicity results from the spontaneous reaction of bisulfite, sulfite, or the free radicals produced by their oxidation with the macromolecules needed to maintain cellular function. Evidence from a variety of sources suggests that the primary SO_2 toxic effects occur within the chloroplasts. The first ultrastructural change that results from SO_2 exposure is swelling of the thylakoids, without other effects being observed; swelling is reversible if the SO_2 is removed (Wellburn, Majernik, and Wellburn, 1972; Wong, Klein, and Jäger, 1977).

When plants are exposed to radioactive SO_2, the sulfite formed within the cytoplasm is incorporated into the chloroplasts, mitochondria, and peroxisomes. The sulfur is uniformly distributed in the protein moieties of the mitochondria and peroxisomes, but in the chloroplasts there is a preferential incorporation into the thylakoid membranes (Ziegler, 1977). In the light, superoxide-anion free radicals are produced on the surface of the thylakoid membranes as a result of electron transport, and these free radicals initiate the chain oxidation of sulfite to sulfate (Asada and Kiso, 1973; Asada, 1980; Peiser, Lizada, and Yang, 1982). Concurrent with the free-radical chain oxidation of sulfite to sulfate is the production of increased amounts of hydroxyl, bisulfite, and additional superoxide-anion free radicals, which are cytotoxins that can oxidize chloroplast membranes (Asada, 1980; Peiser, Lizada, and Yang, 1982). Sulfur dioxide may also form free radicals indirectly in the chloroplast; e.g., conditions such as stomatal closure that would reduce the amount of CO_2 for photoreduction can also cause an increased superoxide-anion production (Asada, 1980). Free-radical disruption of chloroplast membranes also would affect electron transport, and could thus contribute to photooxidative injury in the cells. This is consistent with the suggestion of Noack (1929) that SO_2 interfered with the catalytic properties of the chloroplast and induced photooxidative processes that promoted chlorophyll decomposition and cell death.

Several types of evidence suggest that free-radical formation is one mechanism of SO_2 phytotoxicity. Foliar applications of free-radical scavengers, such as α-tocopherol, diphenylamine, or propyl gallate, reduced foliar injury induced by SO_2 (Chang *et al.*, 1981). The enzyme superoxide dismutase (SOD) disproportions free radicals and forms hydrogen peroxide, which is metabolized by catalase to H_2O and oxygen. Low concentrations of SO_2 have been reported to induce SOD in *Populus euramericana* leaves, imparting protection against subsequent exposures to a higher concentration (Tanaka and Sugahara, 1980). The same authors showed that a copper chelator that reduced SOD activity increased

SO_2 toxicity. They also attributed the fact that young *P. euramericana* leaves tolerate SO_2 better than old ones to a higher SOD activity. Asada (1980) proposed that SOD was a tolerance mechanism for SO_2 injury because it inhibited the chain oxidation of sulfite to sulfate (Asada, 1980; Ziegler, 1975). If the SOD scavenged the superoxide anion used to initiate sulfite oxidation, as suggested by Asada (1980), it would prevent the chain oxidation of sulfite and the resultant free radicals, but sulfite and bisulfite, also phytotoxic, would remain. Given the increased free-radical production during oxidation of sulfite to sulfate, it is not clear that this is a reliable detoxification mechanism.

Most plants close their stomata in the dark, and therefore are assumed to be much more tolerant of SO_2 in the dark than in the light. However, in species where the stomata may not close (e.g., potato), the leaves are more sensitive in the dark than light (Nielsen, 1938). Individual alfalfa leaves that were observed to have open stomata in the dark displayed about as much SO_2 sensitivity as in light exposures (Thomas, 1961). The phytotoxicity observed in the dark suggests that other SO_2 metabolites, such as bisulfite or sulfite, were also phytotoxic, as were the photochemical-induced free radicals. Bisulfite has been shown to alter the selective permeability of plant membranes in both the light and the dark (Lüttge *et al.*, 1972), and it may disrupt disulfide bonds in proteins, altering structure and activity.

There have been no studies of intraspecies differences in the initial perturbation from SO_2. Limited studies, with mixed results, have been conducted on the possible effects of SO_2 on the permeability of the plasmalemma. The relationship between membrane permeability and resistance to SO_2 injury was evaluated in six species by using potassium leakage or conductivity as an indicator of membrane integrity. No difference in potassium leakage was observed between cultivars of *Petunia hybrida* (Elkiey and Ormrod, 1979b) or clones of *Picea abies* (Braun, 1977c). No consistent pattern in membrane permeability was observed in sensitive and resistant populations of four grass species; resistant and sensitive populations had similar potassium-leakage rates after SO_2 exposure (Ayazloo, Garsed, and Bell, 1982). The lack of consistent results in plasmalemma permeability is not surprising, since the primary effects of SO_2 seem to occur within the chloroplast. Substantial metabolic changes may occur before there are substantial changes in the outer limiting membrane.

Homeostasis

After perturbations, living systems begin reestablishing their normal metabolic state. Recovery may be accomplished by repair of or compensation for the perturbation and its consequences (Levitt, 1972). Although the final state (i.e., recovery) of these processes is largely the same, their underlying mechanisms differ. The repair processes rectify the perturbation, thereby eliminating its physiological dysfunction. Compensation is somewhat different from repair, in that

TABLE 11.5
Intraspecies Differences in Parameters Indicating SO_2 Injury

Species	Comparison[a]	Parameter	Resistance correlation	Source no.
CARBON DIOXIDE EXCHANGE				
Dactylis glomerata (orchard grass)	2 pop	Photosynthesis	Yes	1
Festuca rubra (red fescue)	2 pop	Photosynthesis	Yes	1
Holcus lanatus (velvet grass)	2 pop	Photosynthesis	Yes	1
Hordeum vulgare (barley)	16 cv	Photosynthesis	—	2
Lolium perenne (perennial ryegrass)	2 pop, 3 cl	Photosynthesis	Yes	1
Muhlenbergia asperifolia	2 pop	Photosynthesis	Yes	3
Picea abies (Norway spruce)	rst. vs. sens.	Photosynthesis	Yes	4
Pinus strobus (eastern white pine)	63 cl	Photosynthesis	Yes	5
P. sylvestris (Scotch pine)	3 cl	Photosynthesis	Yes	6
	6 sdg	Photosynthesis Photorespiration CO_2 compensation point	Yes	7
Zea mays (maize)	7 lines, 2 cv	Photosynthesis respiration	Yes	8
ENZYME ACTIVITY AND ISOENZYMES				
Lolium perenne (perennial ryegrass)	3 cl	Glutamate synthetase activity	Yes	9
	2, 4 cl	Nitrite reductase activity	No	
Picea abies (Norway spruce)	18 cl	Protein bands	No	10
Pinus strobus (eastern white pine)	63 cl	Acid phosphatase activity	Yes	5
		Acid phosphatase isoenzymes	Yes	
	36 cl	Esterase isoenzymes	No	11
	36 cl	Peroxidase activity	No	11
	36 cl	Peroxidase isoenzymes	Yes	11
	63 cl	Peroxidase isoenzymes	No	5
		Phenoloxidase isoenzymes	No	
P. sylvestris (Scotch pine)	23 cl, 4 sdg	L-leucine amino-pipidase isoenzymes	No	12
	2 pop	GPT, GOT activity	Yes	13
		RuDPC activity	Yes	
		Peroxidase activity	Yes	

TABLE 11.5—*cont.*

Species	Comparison[a]	Parameter	Resistance correlation	Source no.
		HYDROCARBON EMISSIONS		
Cucumis sativus (cucumber)	2 cv	Ethane[b]	Yes	14
Cucurbita pepo (squash, pumpkin)	2 cv	Ethane[b]	Yes	14
Hordeum vulgare (barley)	2 cv	Ethylene	No	15
		PIGMENT CONTENT		
Euphorbia pulcher-rima (poinsettia)	17 cl	Flavonol	No	16
		Anthocyanins	No	
Lolium perenne (perennial ryegrass)	2 cl	Chlorophyll	No	17
Picea abies (Norway spruce)	rst. vs. sens.	Chlorophyll	Yes	18
	12 cl	Chlorophyll	No	10
		OTHER PARAMETERS		
Lolium perenne (perennial ryegrass)	3 cl	ATP formation	Yes	9
Muhlenbergia asperifolia	2 pop	Sulfur assimilation	Yes	3
Petunia hybrida (petunia)	3 cv	Leaf water potential	No	19
Picea abies (Norway spruce)	rst. vs. sens.	Protein synthesis	No	18
		Thioester content	Yes	
		Sulfhydryl compound content	Yes	
	18 cl	Protein synthesis	No	10
		Sulfur distribution	No	
		Organic sulfur	Yes	
		Sulfhydryl content	Yes	
	16 cl	Water potential	Yes	20
		Water content	Yes	
Pinus strobus (eastern white pine)	59 cl	Phenol content	Yes	5
		Polyphenol content	Yes	
		Monoterpene content	No	
	36 cl	Phenol content	Yes	11
P. sylvestris (Scotch pine)	9 cl	Specific gravity	No	21

SOURCES: 1, Ayazloo, Garsed, and Bell (1982). 2, Brough, Parry, and Kendall (1975). 3, Campbell (1972). 4, Pelz (1962). 5, Eckert (1978). 6, Börtitz and Vogl (1965). 7, Lorenc-Plucinska (1982). 8, Grzesiak (1979). 9, Wellburn *et al.* (1981). 10, Braun (1977e). 11, Houston (1970); Houston and Stairs (1972). 12, Mejnartowicz, Bialobok, and Karolewski (1978). 13, Horsman and Wellburn (1977). 14, Bressan *et al.* (1979). 15, Baba and Sakai (1976). 16, Krizek and Semeniuk (1981). 17, Bell and Mudd (1976). 18, Schindlbeck (1977). 19, Elkiey and Ormrod (1979c). 20, Braun (1977d, 1978). 21, Patton (1981).

[a] Abbreviations: cl (clones), cv (cultivars), pop (populations), sdg (seedlings), rst. vs. sens. (resistant versus sensitive).

[b] Based on sensitivity to bisulfite with the assumption that bisulfite sensitivity paralleled SO_2 sensitivity.

the perturbation and its physiological consequences persist, but the cell responds with processes that counter the detrimental effects of the perturbation. Because the repair of or compensation for the cellular perturbation may range from partial to complete, the plant response will vary similarly.

There have been very few studies on homeostatic processes in plants after SO_2 exposure; however, such processes would be expected to occur, just as they do for other pollutants, such as O_3. Repair of O_3 injury involves the reestablishment of normal membrane processes, and apparently is an active process requiring energy. Differences in plant sensitivity to ozone have been attributed to differences in endogenous repair rates (Taylor and Tingey, 1982).

One early visual symptom of acute SO_2 injury is the appearance of H_2O-soaked interveinal areas on leaves. If these areas are not subjected to extensive drying by either wind or sunshine, the H_2O may be reabsorbed into the cells, and thus result in little or no visual injury (Thomas, Hendricks, and Hill, 1950); the initial stages of injury therefore seem to be reversible. Wellburn, Majernik, and Wellburn (1972) reported that the SO_2-induced swelling of thylakoids was reversible if polluted air was replaced by clean air. This response also suggests that homeostatic processes were operative.

The SO_2-induced depression of photosynthesis (e.g., Winner and Mooney, 1980a,b; Black and Unsworth, 1979a; Bennett and Hill, 1973b) is a commonly observed response (Table 11.5), and probably results from the disruption of the chloroplast membranes. When SO_2 is removed, the photosynthetic rate rapidly returns to the prefumigation level unless acute injury has occurred (e.g., Black and Unsworth, 1979a; Bennett and Hill, 1973b). This recovery is probably a consequence of homeostatic processes repairing the altered chloroplast membranes, unless the depressed photosynthesis was a result of only a decreased leaf conductance to CO_2. The recovery from the SO_2-induced depression of photosynthesis may provide an entry point for investigation of homeostatic processes.

Chronic SO_2 exposures can increase the polyamine content of pea plants (Priebe, Klein, and Jäger, 1978). These authors suggested that one important metabolic function of polyamines was the regulation of cellular pH, because the polyamines can function as polyvalent cations. The relative cation deficiency induced by SO_2 exposures may therefore be partially compensated for by the synthesis of polyamines (*ibid.*). Polyamines have also been reported to stabilize cellular membranes (Naik and Srivastava, 1978), and may thus protect membrane integrity in the presence of SO_2.

In C_3 plants, photorespiration may be a mechanism for preventing the electron-transport chain, in the chloroplasts, from being oxidized and for increasing free-radical formation. From this perspective, photorespiration may also contribute to homeostatic processes in plants.

The repair and/or compensatory processes that are induced by SO_2 perturbation need additional research. As studies are conducted, they should consider the probable initial effects of SO_2. In addition, the physiological responses measured should assess the rate of repair. There also may be different homeostatic processes for acute and chronic perturbations.

Injury

Injury is the result of perturbations that are not repaired or compensated for. Because the initial SO_2-induced perturbations apparently occur at specific sites within the cell, such as the chloroplast, the initial metabolic dysfunctions probably originate in specific organelles or cellular regions. Biochemically, injury consists of changes in metabolism, including effects on enzyme activities and metabolite pools. The altered biochemical processes, when summed across numerous cells, are expressed as altered tissue and organ activities, which in turn may induce foliar injury symptoms, altered carbon allocation, and changes in plant growth, reproduction, and yield. These changes are all consequences of the SO_2 perturbation rather than part of the initial perturbation itself. The magnitude of the plant response or injury observed reflects the extent to which homeostatic processes are unable to correct the perturbation.

Intraspecies differences in plant responses to SO_2 have been reported for characteristics that range from enzyme activities and metabolite pools to plant pigments (Table 11.5). Many of the responses, such as altered photosynthesis, changes in plant pigments, and altered enzyme activities, are typical stress responses that have also been observed with other air pollutants and with various abiotic and biotic stresses. They represent the consequences of altered cellular function, but are not the cause of the perturbation. As our knowledge of homeostatic processes improves, we may find that certain of these metabolic changes are aspects of repair or compensatory processes that are correcting secondary or tertiary perturbations resulting from the initial SO_2 effect. For example, the reduction in photosynthesis (e.g., Börtitz and Vogl, 1965; Grzesiak, 1979; Lorenc-Plucinska, 1982) is a transient response unless there is extensive tissue injury. This reduction is a consequence of, but probably not part of, the primary SO_2 perturbation. The restoration of normal metabolic activity after SO_2 exposure results from the repair of altered membrane functions or enzyme activities that are necessary for normal metabolism, and illustrates the probability of homeostatic processes that can reverse these secondary perturbations.

Plants exposed to SO_2 have been shown to emit both ethylene and ethane (Peiser and Yang, 1979a; Bressan et al., 1979). Ethylene emissions began shortly after exposure, peaked within 1 to 8 hours, and were produced in greater quantity in the dark than in light, whereas ethane emissions were associated with visual injury, and production was stimulated by light (Peiser and Yang, 1979a). The release of ethylene is a general stress response that is elicited by air pollutants and many other stresses. In contrast, ethane emissions were not stimulated by O_3, various metals, or organic compounds that induced ethylene production, and emissions were similar in the light and dark (Tingey, unpublished data). Ethane is apparently formed by lipid peroxidation (Peiser, Lizada, and Yang, 1982), initiated by free radicals formed during the aerobic oxidation of sulfite to sulfate. The lipid peroxidation may be associated with, or a consequence of, the initial SO_2 perturbation (*ibid.*; Shuwen *et al.*, 1982). However, it is not clear if the lipid peroxidation is closely associated with the initial perturbation, because it takes several hours after exposure for emissions to increase. In either case, the emission of ethane as an indicator of lipid peroxidation may be a useful indicator of perturbation, provided the limitations of the data are considered.

Given the many physiological mechanisms responsible for different plant responses to SO_2, we suggest a common progression of events, beginning with SO_2 conductance (gas and liquid phase), continuing through perturbation and homeostasis, finally terminating with injury, the consequence of these events. Many studies have attempted to document the significance of gas-phase conductance; far fewer studies have focused on liquid-phase conductance. The importance of perturbation and homeostasis has not been well-documented.

The association between gas-phase conductance and injury is still unclear. From the conceptual model, it is obvious that gas-phase conductance would be the decisive factor controlling plant response only if it were the limiting factor. Physiologically significant processes that occur between gas-phase conductance and injury would decrease the association. Because plant response is not always correlated with gas-phase conductance (i.e., SO_2 uptake), it is self-evident that liquid-phase conductance and homeostasis are important factors influencing SO_2 response.

Importance of Metabolic Differences

Metabolic differences within species are important both for understanding the mechanism of plant response to SO_2 and as a way to screen plant populations for individuals that may exhibit resistance to SO_2 for use in genetic studies. Ideally, a physiological parameter that achieves both objectives would be most useful. However, in practice, certain physiological parameters may be more appropriate as indicators of the mechanism of resistance, whereas others may be more useful in providing information for genetic studies.

The source of differences in foliar injury to plants by SO_2 is not clear. Only a few reports are available that describe differences within species in SO_2 flux into leaves. In studying plant responses to SO_2, it is important to know the internal exposure, since SO_2 exerts its metabolic effects within the leaf. This exposure must be defined and quantified if we are to understand plant response, since it is the dose reaching the perturbation sites, and not the ambient concentration, that is important. Only when studies have assured that the internal exposures are similar between treatments of a plant selection is it possible to accurately interpret plant-response data. In studies on the perturbation mechanism, it has not been possible to find out if the variable responses reflect differences between plants in the sensitivity of the particular site or only the fact that different amounts of SO_2 reached the active site. The same problem complicates the interpretation of homeostatic processes and metabolic changes that occur with injury. Differentially sensitive strains within species should provide a useful means for studying the perturbation mechanism if equality of cellular exposure is assured. The metabolic variability should be smaller within than between species, making it easier to detect changes in metabolites that may be related to sensitivity.

Only a few studies have attempted to study a broad range of responses that may contribute to differences in sensitivity within the same species (e.g., Ayazloo, Garsed, and Bell, 1982; Braun, 1977a; Elkiey and Ormrod, 1979a,b,c, 1980b, 1981c). Such comprehensive studies should be encouraged. Only by considering the full range of responses by a few plants and understanding the influence of cellular exposures will we be able to make progress in understanding the causes of plant response to SO_2.

For genetic studies, parameters that indicate plant injury, e.g., chlorophyll content (Schindlbeck, 1977), peroxidase activity (Wellburn et al., 1976), ethane emissions (Bressan et al., 1979), or visible necrosis, as in the German "Rapid Test" for trees (Börtitz and Vogl, 1965), may be used to evaluate the resistance of parent or progeny plants to SO_2. The parameter chosen should support the research aims and allow the quantitative screening of many plants within a short time.

Other parameters more directly linked to a mechanism for resistance to injury, such as internal SO_2 flux (Elkiey and Ormrod, 1981a; Taylor and Tingey, 1981) or sulfite metabolism (Howe and Woltz, 1981; Miller and Xerikos, 1979; Asada, 1980), may be effective indicators of plant response once the resistance to SO_2 has been established in various lines. Individual plants with these characteristics could then be used for further study to produce tolerant progeny. If possible, the parameter should be related physiologically to plant productivity; e.g., stomatal conductance influences pollutant flux, and thereby the photosynthetic rate, and hence yield. Characteristics not necessarily associated with either injury or physiology of resistance, such as isoenzyme patterns (Mejnartowicz, Bialobok, and Karolewski, 1978), may not be appropriate for breeding research.

Prospects and procedures for breeding plants resistant to SO_2 have been described for agricultural crops (Ryder, 1973; Guderian, 1977) and trees (Demeritt et al., 1972; Jensen et al., 1976; Bialobok, 1979, 1980; Scholz, 1981). Enough progress in breeding in some species, e.g., Larix spp., has been made that resistant strains are available for forest testing (Enderlein, Kastner, and Heidrich, 1967). The particular methods used in breeding of course depend on the species in question; but, in general, the process of producing plants resistant to SO_2 should be like the process of developing plants with disease resistance or other desirable characteristics.

The obvious differences in SO_2 sensitivity within species (Table 11.1) illustrate the heritability of SO_2 resistance, as has been documented by studies such as the clonal repeatability tests for *Pinus strobus* (Houston and Stairs, 1973) and *Populus tremuloides* (Karnosky, 1977). However, the genetic system for resistance to SO_2 has been evaluated in only a few species. Additive inheritance was found in *Geranium carolinianum* (Taylor, 1978a), but was not associated with leaf conductance parameters. In contrast, one or two genes appeared to control SO_2 resistance in *Cucumis sativus* (Bressan *et al.*, 1981) and *Oryza sativa* (Omura *et al.*, 1980); that is, a relatively simple physiological mechanism controlled injury. Studies of SO_2 flux and bisulfite sensitivity in *C. sativus* indicated that, although SO_2 flux was lower in resistant plants than in sensitive plants, injury was not necessarily dependent on stomatal conductance (Bressan, Wilson, and Filner, 1978; Bressan *et al.*, 1981). General inheritance was demonstrated when parents and progeny of *Pinus sylvestris* (Vogl, 1970) and *Larix* spp. (Enderlein, Kastner, and Heidrich, 1967) were examined for SO_2 resistance; however, in these studies the mechanism was not evaluated. This limited research seems to indicate that the genetic systems for resistance to SO_2 parallel those for O_3, where resistance is controlled sometimes by few, sometimes by many genes (Butler, Tibbitts, and Bliss, 1979; Hanson, Addis, and Thorne, 1976); the question of how metabolic differences between genetic lines are related to the apparent system of heritability in a species could provide a profitable area of research.

Direct genetic and reproductive effects of SO_2 (Shapiro, 1977) could have a significant effect on breeding for resistance. SO_2 decreased pollen germination and tube growth in *Pinus strobus* (Karnosky and Stairs, 1974) and *Tradescantia paludasa* (Ma and Khan, 1972); SO_2-increased chromosome aberrations also were found in *T. paludasa* pollen-tube cultures (Ma *et al.*, 1973). Reduced seed production and abortion of seeds were reported in conifers injured by ambient SO_2 (Hedgcock, 1912; Mamajev and Shkarlet, 1970).

Differences in reproductive response to SO_2 within species could be important both in breeding programs and in natural environments. Research by Murdy (1979) on *Lepidium virginicum* indicated that greater fertility in polluted environments was associated with resistance of populations to SO_2. Examination of the metabolic basis for this or other reproductive responses in other species would be a significant contribution to an understanding of the long-term effects of SO_2 on plant populations.

That it is possible to select for SO_2 resistance is evident from the evolution of resistant populations of plants in areas that have been subjected to SO_2 emissions for a long time (Bradshaw, 1976; Taylor and Murdy, 1975; Roose, Bradshaw, and Roberts, 1982). In these plants, resistance was documented by comparing the response of populations from affected and nonaffected areas when both groups were exposed to SO_2 under controlled laboratory conditions. For three pasture grasses, *Dactylis glomerata*, *Festuca rubra*, and *Holcus lanatus*, resistance was observed in terms of both reduced leaf injury and increased growth; in *Lolium perenne* resistance was indicated by enhanced growth but no difference in injury was found between the two populations (Ayazloo and Bell, 1981). Resistance in

Geranium carolinianum was evaluated only in terms of reduced visible injury (Taylor and Murdy, 1975), whereas in *Lepidium virginicum* resistance was evident as greater fertility (Murdy, 1979).

Similar evolution of resistance to pollutants has been demonstrated in species affected by heavy metals (Antonovics, Bradshaw, and Turner, 1971; Wu, Bradshaw, and Thurman, 1975), and possibly by oxidants (Dunn, 1959) and acid rain (Hodgkin and Briggs, 1981). Indeed, in areas where pollutants are significant environmental factors, it is not surprising that, given enough time, genetic mechanisms produce resistant populations (Roose, Bradshaw, and Roberts, 1982); so a species may adapt by either avoiding or tolerating the stress (Taylor, 1978b). Additional examples of evolution of resistance to SO_2 may be expected, especially if herbaceous species with a short generation time are evaluated (Roose, Bradshaw, and Roberts, 1982).

If breeding for resistance to SO_2 is to become successful, the general adaptability of the resistant plants to the environment in which they are to be grown must be considered. Recent evidence indicates a complex interaction between environment and SO_2 exposure. For example, low concentrations of SO_2 may sensitize plants to cold injury (Huttunen, 1978; Keller, 1981a; Baker, Unsworth, and Greenwood, 1982; Davison and Bailey, 1982), possibly by increasing H_2O stress (Huttunen, Havas, and Laine, 1981). Conversely, environmental factors may increase the SO_2 sensitivity of plants (Guderian, 1977), as shown by the increase in SO_2 injury with desiccation (Okoloko and Brewley, 1982). Thus, even if clones or cultivars are developed that are relatively resistant to SO_2, injury may occur if they are not fully adapted to their growing environment (Huttunen, 1978; Guderian and Küppers, 1980; Huttunen, Havas, and Laine, 1981).

Summary

Differences in metabolic responses to SO_2 are widespread in many plant species. Plant responses are a consequence of a series of biochemical and physiological events, beginning with SO_2 flux into plants progressing through perturbation and homeostasis, and ending with foliar injury or effects on growth and yield. Gas-phase and liquid-phase conductance are important determinants of SO_2 flux. However, the relationship between these factors and plant response is incompletely understood, at least in part because of difficulties with SO_2 gas-exchange methodology. Perturbation events that occur primarily in the chloroplast and homeostatic mechanisms have received only brief attention.

The entire sequence of plant response to SO_2 needs additional investigation, especially to identify components useful in genetic studies of SO_2 resistance. Evidence to date demonstrates the heritability of SO_2 resistance in both native and cultivated vegetation. Understanding the mechanisms that cause differences in plant response would improve success of breeding for resistance to SO_2 injury.

Part 3
SO$_2$ Effects on Plant Growth

12

SO$_2$ Effects on the Productivity of Grass Species

J. N. B. Bell

Grasslands are the primary source of nutrition for a substantial proportion of the world's grazing animals, both wild and domesticated. They range from native permanent pastures, often very old and subject to minimal husbandry, to high-yielding, intensively managed systems, sown with cultivars bred for particular uses and for specific environmental conditions. Grasslands may either be used for grazing, and thus subject to a more or less continuous process of defoliation, or be harvested at intervals and the grass stored for fodder. In addition, many grasslands are established for use as parks and sports pitches, where, although productivity is not important in itself, maintenance of a continuous sward depends on the vigor of the plants in providing tolerance to trampling and other wear.

Most grass species are perennial, and thus may experience SO$_2$ pollution for many years; so it is difficult to extrapolate to the field from short-term fumigation experiments. In temperate regions with oceanic climates, grasses overwinter in a green condition, and so are exposed to SO$_2$ when concentrations are at a maximum in most places. Park and sports grasslands, in particular, are likely to experience high concentrations of SO$_2$, because they are located mainly in urban areas. Grasslands differ from most other agricultural systems in that they are usually made up of several species instead of being established as monocultures. Within the sward there is intense competition within and between species; this permits grassland structure to change rapidly in response to changing environmental conditions.

These unique characteristics of grasslands also make it hard for us to understand the response of grass species to air pollution in the field. This paper is a review of some recent developments in studies of SO$_2$ effects on grass species. It will also provide some explanation of apparent discrepancies in the literature, and help explain how SO$_2$ affects grassland productivity in areas subject to low to moderate concentrations of air pollutants.

Review of Effects of SO_2 on Grass Productivity

In general, much less attention has been paid to the effects of SO_2 on grass species than to those on forest trees and arable crops (Bell, 1982). There are relatively few reports in the literature of obvious injury by air pollutants to grasslands, and, indeed, in areas subject to very severe pollution, such as near smelters of sulfide ores, grasslands may replace forests damaged by SO_2 (Smith, 1974). Until recently, most studies on the relative sensitivity to SO_2 of different grass species and cultivars have employed short-term, high-concentration fumigations, followed by measurement of the resulting foliar necrosis (e.g., Brennan and Halisky, 1970; Murray, Howell, and Wilton, 1975).

However, recent work has demonstrated unequivocally that plant responses to acute doses of SO_2 do not bear any relationship to chronic growth reductions produced by long-term exposure to lower concentrations of the gas. It is, of course, such long-term exposures to the concentrations that are characteristic of most polluted locations in the field. Ayazloo and Bell (1981) fumigated 20 populations of five grass species with either acute (5,300 μg m^{-3}, or 82.81 μmol m^{-3} for 6 hours) or chronic (250 to 500 μg m^{-3}, or 3.91 to 7.81 μmol m^{-3} for 60 to 180 days) doses of SO_2, and assessed injury in terms of percentage foliar necrosis and dry-weight reduction, respectively. In no case was there any indication of a significant correlation between the two types of injury on individual genotypes. Furthermore, there also appears to be no pattern to how species differ in their relative susceptibility to chronic and to acute SO_2 injury. In the same study, there was an almost exact reversal of the ranking of the species in their response to the two types of fumigation (Ayazloo, Garsed, and Bell, 1982).

Clearly, moderate SO_2 levels can reduce grass productivity, even when there are no other visible symptoms. Is there, then, any evidence that SO_2 reduces grass growth in the field in areas subject to such ambient concentrations?

Evidence for chronic SO_2 injury in the field. The effect of low to moderate concentrations of SO_2 on grass growth has been the subject of controversy in recent years. This conflict has arisen because different results have been obtained by different (or even the same) research groups (Bell, 1982). Despite this, two different types of firm experimental evidence indicate that substantial, but hitherto undetected, effects may be taking place at sites where annual mean SO_2 concentrations do not exceed 150 μg m^{-3} (2.34 μmol m^{-3}).

Several pollutant-exclusion experiments have been performed during the last decade in British cities. In these studies the growth of grasses has been compared in outdoor chambers that are ventilated with either ambient or charcoal-filtered air (Table 12.1). Experiments in a suburb of Sheffield showed remarkable improvements in the growth of *Lolium perenne*, *L. multiflorum*, and *Dactylis glomerata*, during periods ranging from 28 to 131 days, when air containing mean SO_2 concentrations between 38 and 70 μg m^{-3} (0.59 and 1.09 μmol m^{-3}) was

TABLE 12.1

Effects of Ambient Air on Shoot Dry Weight of Grass Species in Pollutant-Exclusion Experiments in British Cities, 1973–80

Species	Mean SO$_2$ concentration μmol m^{-3} (μg m^{-3})	Duration (d)	Reduction in ambient air (pct. clean-air control)	$p <$
Sheffield				
1. *Lolium perenne*	1.09 (70)	56	36%	0.001
2. *Lolium perenne*	0.92 (59)	131	20	0.01
3. *Lolium perenne*	1.08 (69)	86	25	0.001
4. *Lolium perenne*	0.98 (63)	116	26	0.001
5. *Lolium perenne*[a]	0.69 (44)	28	14	0.05
6. *Lolium perenne*[a]	0.59 (38)	28	25	0.05
7. *Lolium multiflorum*	1.05 (67)	56	36	0.001
8. *Lolium multiflorum*[a]	0.69 (44)	28	14	0.05
9. *Dactylis glomerata*	0.70 (45)	72	42	0.001
10. *Dactylis glomerata*[a]	0.69 (44)	28	39	0.01
St. Helens				
11. *Lolium perenne*	1.92 (123)	240	15	0.05
12. *Lolium perenne*	1.50 (96)	300	+16	N.S.[b]
13. *Lolium perenne*	1.09 (70)	310	+6	N.S.[b]

SOURCES: 1–4, Crittenden and Read (1978b). 5, 6, 8, and 10, Awang (1979). 7 and 9, Crittenden and Read, (1979). 11, Roberts *et al.* (1983). 12 and 13, Colvill *et al.* (1983).

[a] Total dry weight.
[b] Not significant at $p = 0.05$.

filtered (Crittenden and Read, 1978b, 1979; Awang, 1979). Studies in St. Helens (Colvill *et al.*, 1983; Roberts *et al.*, 1983), however, showed only a 15% reduction in shoot dry weight of *L. perenne* after 240 days in air containing a mean SO$_2$ concentration of 123 μg m^{-3} (1.92 μmol m^{-3}), but no significant effect was observed at 96 and 70 μg m^{-3} (1.50 and 1.09 μmol m^{-3}) after 300 and 310 days, respectively. Although some of these pollutant-exclusion experiments arrive at inconsistent results, the possible causes of which I will discuss in the section on assessment of techniques, they do indicate that urban air containing SO$_2$ can reduce the growth of grass species.

The second line of research that provides evidence for SO$_2$ injury at moderately polluted sites is the study of selection for SO$_2$ tolerance in the field. Ayazloo and Bell (1981) collected specimens of *Lolium perenne, Holcus lanatus, Phleum bertolonii, Festuca rubra,* and *Dactylis glomerata,* from three polluted sites in northern England, where annual mean SO$_2$ concentrations were less than 150 μg m^{-3} (2.34 μmol m^{-3}). These plants were then screened for tolerance to chronic and acute SO$_2$ injury alongside bred cultivars or populations of the same species collected from unpolluted areas. All five species showed evidence of the evolution of tolerance to acute and/or chronic injury at one or more of the polluted sites. Total dry weight was reduced by chronic SO$_2$ fumigations more in clean-site-bred cultivar populations than in the polluted-site populations for *L. perenne, D. glomerata, F. rubra,* and *P. bertolonii.* For all except the last,

TABLE 12.2
Examples of the Development of SO_2 Tolerance in Grass Populations at Polluted Sites in the United Kingdom

Species	Chronic injury: Reduction in total dry weight in SO_2 (pct. clean-air control)		Acute injury: Length of necrotic leaf (pct. total leaf length)	
	Polluted-site plants	Clean-site plants/ bred cultivars	Polluted-site plants	Clean-site plants/ bred cultivars
Lolium perenne	16%	23%	20.3%	33.5%
Dactylis glomerata	21	33	24.2	41.4
Festuca rubra	8	33	3.5	9.0
Holcus lanatus	—	—	0.8	3.0
Phleum bertolonii	12	29	—	—

SOURCE: Ayazloo and Bell (1981).
NOTE: Chronic SO_2 fumigations in outdoor chambers were 319 to 482 μg m^{-3} (4.98 to 7.53 μmol m^{-3}) for 72 to 180 days; acute SO_2 fumigations in a controlled environment cabinet were 5,320 μg m^{-3} (83.13 μmol m^{-3}) for 6 h, and 9,610 μg m^{-3} (150.16 μmol m^{-3}) for *D. glomerata* for 6 h.

populations from polluted sites showed less foliar injury in response to acute fumigations, this response also occurring in *H. lanatus* (Table 12.2).

This finding has important implications in that, if selection for SO_2 tolerance is taking place at sites where the annual mean is less than 150 μg m^{-3} (2.34 μmol m^{-3}), then the pollutant must be exerting some type of deleterious effect on sensitive genotypes at these locations. These effects may be due to SO_2 either on its own or in combination with another pollutant or environmental stress. Furthermore, the time-scale for such selection to take place has recently been shown to be as short as 4 years in the presence of a mean of about 120 μg m^{-3} (1.88 μmol m^{-3}) of SO_2 (Bell, Ayazloo, and Wilson, 1982). These authors reported the results of an investigation where tolerance to acute SO_2 injury was measured in populations of five grass species (*Lolium perenne, L. multiflorum, Phleum pratense, Festuca rubra*, and *Poa pratensis*) sampled annually from monoculture plots in a park in Manchester, U.K. In each case the amount of leaf necrosis produced by an acute SO_2 fumigation was compared with that on the same species grown from the seed batch that was used to establish the plots. During the first 3 years after sowing, no difference in tolerance was found between the two samples of each species. However, by the fourth year the *L. perenne* plants from Manchester showed significantly less injury than those grown from the original seed batch. This difference was maintained into the fifth year, when the development of tolerance was also detected in *Phleum pratense*, with significantly less injury appearing in response to SO_2 than on plants of the same species grown from the original seed (Table 12.3).

It is surprising that tolerance to *acute* injury should develop in the presence of a mean SO_2 concentration of about 120 μg m^{-3} (1.88 μmol m^{-3}), but this reinforces the argument that concentrations of less than 150 μg m^{-3} (2.34 μmol m^{-3}) can damage grass species in the field.

TABLE 12.3
Leaf Length Injured after Acute Fumigations of Grass Species Collected in 1976–80 from Plots Established in a Manchester Park, in Comparison with Plants Grown There from the Original Seed
(percent of total leaf length)

| Species | 1976 | | 1978 | |
	Original seed	Plots[a]	Original seed	Plots[a]
Lolium perenne	20.5%	18.5%	23.5%	20.1%
Lolium multiflorum	50.7	52.5	48.0	44.7
Phleum pratense	59.8	61.4	67.0	62.5
Poa pratensis	6.1	5.8	9.4	8.8
Festuca rubra	4.5	4.4	4.4	3.9

| Species | 1979 | | 1980 | |
	Original seed	Plots[a]	Original seed	Plots[a]
Lolium perenne	9.4%	4.3%**	42.7%	33.5%*
Lolium multiflorum	20.0	12.0	55.2	64.3
Phleum pratense	5.8	9.0	67.3	55.0*
Poa pratensis	3.2	5.1	44.5	35.0
Festuca rubra	4.1	8.1	24.5	20.2

SOURCE: Bell, Ayazloo, and Wilson (1982).
[a] No significant difference unless otherwise noted.
*$p = 0.05$, **$p = 0.01$.

Implications for grassland productivity of selection for SO$_2$ tolerance. This demonstration of widespread selection for tolerance to SO$_2$ in grass species at polluted sites is probably a reflection of the intense intraspecies competition that is far more characteristic of grasslands than of other agricultural systems. A grassland can thus blunt the adverse effect of SO$_2$ by selection for tolerance. Roose, Bradshaw, and Roberts (1982) have pointed out that the development of tolerance to SO$_2$ often involves some "costs," in that the tolerant plant has a poorer performance than sensitive individuals when growing in clean air. An examination of the yield in clean air of the pairs of tolerant and susceptible populations identified by Ayazloo and Bell (1981) reveals that half the latter plants grew faster (Table 12.4). This is not surprising, since these four susceptible populations were cultivars originally bred for high productivity. However, both tolerant and sensitive populations of *Phleum bertolonii* had originated from the same cultivar, but the former grew faster in clean air. In the remaining three examples, tolerant and susceptible wild populations of *Dactylis glomerata*, *Festuca rubra*, and *Holcus lanatus* (collected from polluted and clean sites, respectively) did not differ significantly in growth in clean air. Thus, slow growth is by no means a prerequisite for the development of tolerance to either chronic or acute SO$_2$ injury; so we cannot generalize about changes in grassland productivity brought about by shifts in tolerance to SO$_2$.

TABLE 12.4
Growth in Clean Air of Pairs of Grass Species Populations Differentially Tolerant to Chronic and/or Acute SO_2 Injury

Species	Type of SO_2 tolerance	Age of plants (d)	Mean total dry weight (g ± SEM)	
			Tolerant population	Susceptible population[d]
Lolium perenne	Chronic/acute	105	1.91 ± 0.07	2.81 ± 0.08**[a]
Lolium perenne	Acute	61	1.71 ± 0.05	2.06 ± 0.03*[a]
Dactylis glomerata	Chronic/acute	180	2.16 ± 0.09	3.00 ± 0.08**[a]
Dactylis glomerata	Acute	131	3.06 ± 0.24	2.98 ± 0.20[b]
Festuca rubra	Chronic/acute	105	1.14 ± 0.07	2.95 ± 0.08**[a]
Festuca rubra	Acute	131	2.16 ± 0.19	2.27 ± 0.14[b]
Holcus lanatus	Acute	131	3.04 ± 0.21	2.59 ± 0.28[b]
Phleum bertolonii[c]	Chronic	72	3.76 ± 0.10[a]	3.32 ± 0.14*[a]

SOURCES: Ayazloo (1979); Ayazloo and Bell (1981).

NOTE: Tolerant and susceptible populations were grown together in outdoor chambers ventilated with clean air.

[a] Bred cultivars.
[b] Wild populations from unpolluted sites.
[c] Tolerant population from polluted site originated from the susceptible bred cultivar.
[d] No significant difference unless otherwise noted.
*p = 0.005, **p = 0.01.

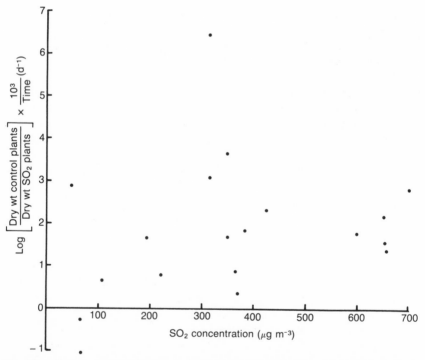

Fig. 12.1. Ratios of dry weights of spaced *Lolium perenne* plants from control and fumigation treatments, normalized for duration of exposure, in relation to mean SO_2 concentration. Data are from: Bell and Clough (1973); Bell, Rutter, and Relton (1979); Ayazloo and Bell (1981); Horsman, Roberts, and Bradshaw (1978, 1979); Horsman *et al.* (1979); and Ashenden and Mansfield (1977). Adapted from Bell (1982), p. 232.

Dose-response relationships for chronic SO$_2$ injury. Considerable difficulty has been experienced in obtaining a clear dose-response relationship and threshold for chronic SO$_2$ effects on grass species. *Lolium perenne* is the only species that has been subjected to enough comparable fumigation experiments to permit an attempt at the derivation of such a relationship. In Bell (1982), I examined all data published up to 1981 by three research groups who performed fumigations of spaced *L. perenne* plants with chronic SO$_2$ concentrations. I plotted the logarithm of the ratio of the dry weights from control and SO$_2$ treatments, divided by time, against the SO$_2$ concentrations (Fig. 12.1), in order to eliminate the effect of fumigation duration, which otherwise results in an increasing % growth reduction in time compared with the controls, when SO$_2$ depresses the relative growth rate by a constant amount. However, no significant correlation between SO$_2$ concentration and response was revealed, and I concluded that we still cannot demonstrate a dose-response relationship for chronic SO$_2$ injury. Subsequently, however, Mansfield and Freer-Smith (1981) reexamined the data, and found a significant positive correlation ($p < 0.02$) between SO$_2$ concentration and growth response when experiments of less than 40 days' or greater than 160

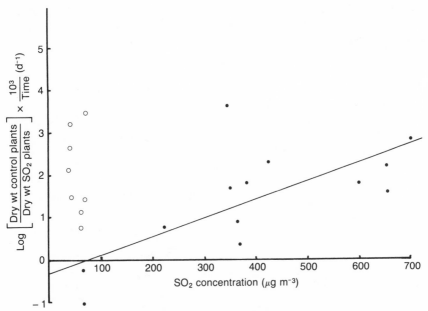

Fig. 12.2. Ratios of dry weights of spaced *Lolium perenne* plants from control and fumigation/pollutant-exclusion treatments, normalized for duration of exposure, in relation to mean SO$_2$ concentration. Experiments of <40 days and >160 days duration are excluded. The regression $y = -0.2298 + 0.00417x$ (for fumigations only) is significant at $p < 0.02$. Data from pollutant-exclusion experiments at Sheffield by Crittenden and Read (1978b) and Awang (1979) are also shown (open circles). Adapted from Mansfield and Freer-Smith (1981), p. 350.

days' duration were excluded. Their regression line (Fig. 12.2) shows a zero response threshold of about 75 μg m^{-3} (1.17 μmol m^{-3}) of SO$_2$.

It might be argued that the way Mansfield and Freer-Smith (1981) selected data was arbitrary, and that the results of short-term and very prolonged fumigations must be taken into account. However, there is evidence, outlined below, that fumigations of very young plants, and of more mature plants for longer than six months, are not comparable with those of intermediate duration.

Circumstantial evidence from both pollutant-exclusion (Crittenden and Read, 1979) and fumigation (Bell, Rutter, and Relton, 1979) experiments suggests that grass seedlings are much more susceptible than older plants to chronic SO$_2$ concentrations. Whitmore and Mansfield (1983) have recently investigated this possibility by the simultaneous fumigation with 177 μg m^{-3} (2.77 μmol m^{-3}) of SO$_2$ for 7 months overwinter of single, newly emerged seedlings and of 42-day-old plants (with three or four tillers) *Poa pratensis*, *Dactylis glomerata*, *Lolium perenne* cv. S23 and cv. S24, and *Phleum pratense* cv. S48 and cv. Eskimo. The plants were harvested in the spring, and it was found that *Poa pratensis*, *D. glomerata*, and *Phleum pratense* cv. S48 and cv. Eskimo were reduced in growth by SO$_2$ only if plants were fumigated from emergence (Table 12.5). However, the two *Lolium* cultivars showed the opposite effect, with significant reductions being limited to the older plants. The results of this experiment should be interpreted with caution, since there were no intermediate harvests, but there is some indication that fumigation during the first 40 days after emergence may be particularly critical in reducing growth.

Recent work by Whitmore and Freer-Smith (1982) has demonstrated dramatic changes with time in response to SO$_2$ during a prolonged fumigation. *Poa pratensis* plants were subjected to 177 μg m^{-3} (2.77 μmol m^{-3}) of SO$_2$ in outdoor chambers for 11 months, starting in October, with destructive harvests at intervals. Initially the pollutant had little effect, but from January onward there was

TABLE 12.5

Effects of Fumigation on Shoot Dry Weight of Six Grass Species and Cultivars Commencing at Emergence or 42 Days Old

		Reduction in dry weight in SO$_2$ (pct. clean-air control)	
Species		Fumigated from emergence[a]	Fumigated from 42 days old[a]
Poa pratensis		45***	N.S.
Dactylis glomerata		40***	N.S.
Phleum pratense	cv. Eskimo	42***	N.S.
Phleum pratense	cv. S48	20*	N.S.
Lolium perenne	cv. S23	N.S.	24**
Lolium perenne	cv. S24	N.S.	22**

SOURCE: Whitmore and Mansfield (1983).
NOTE: Fumigation was of 177 μg m^{-3} (2.77 μmol m^{-3}) of SO$_2$ for 7 months.
[a] N.S., no significant difference.
*$p = 0.05$, **$p = 0.01$, ***$p = 0.001$.

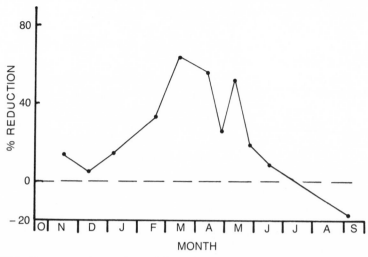

Fig. 12.3. Percentage changes in total dry weight of *Poa pratensis* in 177 μg m⁻³ (2.77 μmol m⁻³) SO₂ during 11 months compared with clean air controls. (Shoot dry weight only after May). Adapted from Whitmore and Freer-Smith (1982), p. 56.

an increasing reduction in dry weight, which reached a maximum of 64% in March (Fig. 12.3). During the summer the adverse effect of SO₂ declined progressively, and by the final harvest, in September, it had been transformed into a significant stimulation of 17% in comparison with the controls. This shows very clearly the different types of result that can be obtained for a given concentration of SO₂, when only a single harvest is made, but the fumigation is terminated after different times. SO₂ resulted in stimulatory, inhibitory, and zero effects after the fumigation exceeded four months. These results provide some justification for Mansfield and Freer-Smith's exclusion of experiments of more than 160 days duration in Fig. 12.2.

A further study of changes in the growth of a grass species in a long-term fumigation has been reported recently by Colvill *et al.* (1983). Swards of *Lolium perenne* were grown in open-top chambers ventilated with ambient rural air or with air containing SO₂ at concentrations adjusted seasonally to simulate those in the industrial town of St. Helens, in northern England. The fumigation started in October and terminated in the following August, with five consecutive harvests when the swards were clipped and allowed to regrow. Mean SO₂ concentrations between harvests ranged from 135 μg m⁻³ (2.11 μmol m⁻³) during the winter to 85 μg m⁻³ (1.33 μmol m⁻³) during the summer. At the end of the first six weeks, SO₂ had caused a 14% reduction in dry weight. During the winter and spring, this reduction increased to 30 to 35% of the controls (Fig. 12.4). During the summer, SO₂ stimulated growth, although this gain disappeared by the final harvest. Overall yield was reduced significantly by 12% during the 11 months of the experiment, during which the mean SO₂ concentration was 120 μg m⁻³ (1.88

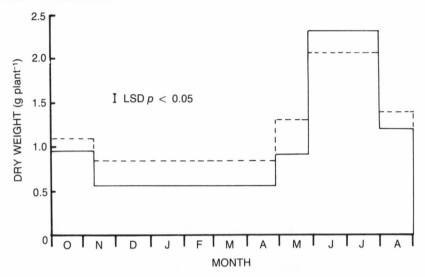

Fig. 12.4. Effects of SO_2 on dry weight per plant of *Lolium perenne* grown in open-top fumigation chambers for 11 months, with five consecutive harvests by clipping. Mean SO_2 concentrations in open-top chambers with unfiltered air (unbroken lines) were 116 μg m^{-3} (1·82 μmol l^{-1}) and with charcoal filtered air (broken lines) were 39 μg m^{-3} (0·61 μmol l^{-1}). Data from Colvill *et al.* (1983).

μmol m^{-3}). There is reasonably good agreement between the results of Colvill *et al.* (1983) and those of Whitmore and Freer-Smith (1982), despite very different experimental procedures. There is a need for further studies of this type to quantify the net effect of different concentrations of SO_2 during a long period, and to elucidate the importance of seasonal and phenological factors in modifying response.

Effect of plant density on SO_2 injury. There has been some controversy about whether fumigation of spaced plants or of dense swards is more relevant to prediction of SO_2 effects on grass species in the field. If effects on a newly sown grassland are being considered, then experiments with single plants are more appropriate, but this situation will prevail for only a very short period during the lifetime of any particular grassland, and not at all in permanent pastures. In Bell (1982), I reviewed the literature on chronic SO_2 injury to grass species, and concluded that the development of a sward would probably ameliorate the adverse effect of the pollutant. This amelioration could occur if SO_2 uptake per unit weight of plant is reduced as leaf area increases. In addition, intraspecies competition may select tolerant individuals at the expense of the more sensitive genotypes, further reducing the effects of pollutants. Hence it is reasonable to doubt the validity of experiments with spaced plants for predicting SO_2 effects on grasslands, except for newly reseeded areas. However, my conclusions were based on comparisons of experiments with spaced plants and with swards performed under different environmental conditions and/or with different types of fumigation system.

Whitmore and Mansfield (1983) have questioned the idea that swards are more tolerant to SO$_2$ than spaced plants. In a fumigation with 177 μg m^{-3} (2.77 μmol m^{-3}) of SO$_2$, both single seedlings and small swards (six seedlings in 7.5-cm pots) were grown from October until the following June. On harvesting, they found that the shoot dry weight of the swards was decreased in SO$_2$ by 27% compared with the controls, but there was no significant effect of the gas treatment on the single plants. This finding suggests that experiments with spaced plants might, in fact, underestimate the impact of SO$_2$ on a sward in the field. However, each "sward" in this experiment consisted of only six plants in a very small pot, and their exposure to the gas may have resembled that of spaced plants more than that of a sward. Also, that many more competing individuals are present in the field would provide a greater opportunity for amelioration of injury by selection for SO$_2$ tolerance. Further studies on this subject are needed: spaced plants should be used together with swards that are more representative of grasslands in the field.

Comparison of pollutant-exclusion and fumigation experiments. When the results of pollutant-exclusion and fumigation experiments are compared, it can be seen that the former usually show much more reduction in yield. This trend is based on mean SO$_2$ concentrations during the course of the experiment (Fig. 12.2). It has been suggested that this apparently greater toxic effect of ambient SO$_2$ may be caused by the presence of other pollutants in urban air and/or by fluctuating SO$_2$ concentrations in the field. In the United Kingdom, intermittent peak SO$_2$ levels may be concealed within the overall mean concentration (Mansfield and Freer-Smith, 1982; Bell, 1982).

In the past, all long-term fumigations with SO$_2$ have used constant concentrations, and thus have not simulated ambient pollution characteristics. Two recent publications, however, have compared the effect on a grass and a tree species of the same overall mean SO$_2$ concentration, given either as a constant concentration or as containing high peaks with intervening lower concentrations. Jones and Mansfield (1982b) grew *Phleum pratense* for 41 days in a wind-tunnel fumigation system, either in clean air or subjected to three SO$_2$ treatments that all had a mean of 160 μg m^{-3} (2.50 μmol m^{-3}). The SO$_2$ treatments consisted of a constant concentration of 160 μg m^{-3} (2.50 μmol m^{-3}), exposure to 2,140 μg m^{-3} (33.44 μmol m^{-3}) for 1.5 hours for five days per week, and exposure to 1,070 μg m^{-3} (16.72 μmol m^{-3}) for three hours for five days per week; 72 to 75 μg m^{-3} (1.13 to 1.17 μmol m^{-3}) of SO$_2$ was given between the peak exposures. SO$_2$ had no effect on shoot growth, but the root dry weight was reduced in all gas treatments, with a corresponding fall in the shoot:root weight ratio (Table 12.6). However, the two peak concentration treatments did not produce a significant

TABLE 12.6
Effects of Constant and Fluctuating SO_2 Treatments
on Growth of *Phleum pratense* During 41 days

Category	Shoot dry weight (g)	Root dry weight (g)	Shoot/root weight ratio
Controls	0.320	0.097	3.35
SO_2 treatment, $\mu g\ m^{-3}$ ($\mu mol\ m^{-3}$)			
Constant, 160 (2.50)	0.310	0.081[a]	3.94[a]
Peaks of 1,070 (16.72)	0.289	0.079[a]	3.89[a]
Peaks of 2,140 (33.44)	0.291	0.074[a]	4.11[a]
LSD at $p < 0.05$	0.044	0.014	0.38

SOURCE: Jones and Mansfield (1982b).
NOTE: Mean SO_2 concentration over 41 days for all treatments was 160 $\mu g\ m^{-3}$ (2.50 $\mu mol\ m^{-3}$).
[a] Significantly different from the controls at $p = 0.05$.

difference in either root dry weight or shoot:root weight ratio from the constant-concentration fumigation.

Some support for these findings is provided by Garsed, Mueller, and Rutter (1982) in a study on *Pinus sylvestris*: plants between 2 and 4 years old were fumigated in outdoor chambers for 650 days with 100 $\mu g\ m^{-3}$ (1.56 $\mu mol\ m^{-3}$) of SO_2 given either at a constant concentration or as 4 treatments, comprising factoral combinations of high (750 $\mu g\ m^{-3}$ or 11.72 $\mu mol\ m^{-3}$) and low (300 $\mu g\ m^{-3}$ or 4.69 $\mu mol\ m^{-3}$) peaks during short periods (5 hours) or long periods (21 hours). The constant treatment resulted in a 14% reduction in dry-weight increment, but peak concentrations did not produce any further growth reduction unless they were of 21 hours' duration. More work is needed to find out whether peak concentrations fail universally to modify chronic SO_2 injury, but this pattern has been observed in two very different types of plant species.

It now seems highly probable that the large growth reductions produced by ambient urban air containing low SO_2 concentrations is actually caused by the presence of other phytotoxic pollutants. In particular, the presence of nitrogen oxides (NO_x) has been shown as an inevitable accompaniment to SO_2 in urban areas, being produced in abundance by high-temperature combustion processes. It appears that mean NO_x concentrations are normally greater than mean SO_2 levels (volume/volume) at both rural and urban sites (Fowler and Cape, 1982), with the $NO:NO_2$ ratio being greater in or near cities. The direct effects of NO and NO_2 on plants have hitherto been largely neglected, with attention being focused on the role of NO_x in the formation of photochemical oxidant smogs. However, there is an increasing volume of literature reporting both marked synergistic and additive effects of SO_2 and NO_2 mixtures on grass species within the range 166 to 194 $\mu g\ m^{-3}$ (2.59 to 3.03 $\mu mol\ m^{-3}$) and 118 to 139 $\mu g\ m^{-3}$ (2.57 to 3.02 $\mu mol\ m^{-3}$), respectively (Ashenden and Mansfield, 1978; Whitmore and Freer-Smith, 1982; Whitmore and Mansfield, 1983).

There is now a good case for considering the combined effects of SO_2 and NO_2

in the field rather than treating them as independent phytotoxic agents (Well-burn, 1982). It would be of considerable interest to find out what the effects are on grass species of long-term fumigations with mixtures of SO$_2$ and NO$_x$ representative of the area of Sheffield where Awang (1979) and Crittenden and Read (1978a,b, 1979) recorded large growth reductions in their pollutant-exclusion experiments. In addition to NO$_x$, one must take the intermittent occurrence of phytotoxic levels of O$_3$ into account when interpreting the British pollutant-exclusion experiments. This pollutant has recently been identified for the first time as causing damage to crop species in the United Kingdom (Ashmore *et al.*, 1980).

Beneficial effects of SO$_2$ on grass species. The mechanisms of the growth stimulation in SO$_2$ observed by Colvill *et al.* (1983) and Whitmore and Freer-Smith (1982) are unknown. However, it is well known that low to moderate concentrations of SO$_2$ can increase the growth of sulfur-deficient grasses (Cowling, Jones, and Lockyer, 1973; Cowling and Lockyer, 1976, 1978). During the last 30 years there has been a substantial rise in the rate of application of nitrogen fertilizers to agricultural land in western Europe, accompanied by a progressive fall in the amount of sulfur used in this manner (Bache and Scott, 1979). A sulfur deficiency is likely to appear as the N:S ratio falls below 16:1 (Murphy, 1978), and it has been claimed that ambient SO$_2$ is preventing such problems in rural areas by acting as a nutrient source (Prince and Ross, 1972).

Until recently such claims were not clearly substantiated. Most of Britain receives an annual input of atmospheric sulfur by both wet and dry deposition of at least 15 kg ha^{-1}, and there is only limited evidence of sulfur deficiency in grasslands. However, the Republic of Ireland, a country with low SO$_2$ emissions and subject to prevailing westerly winds that bring clean air from the Atlantic, experiences total deposition of sulfur from the air of only about 10 to 12 kg ha^{-1} year^{-1}. Murphy (1979) has reported the results of a recent major field trial in Ireland, where sulfur was applied in the form of gypsum to 36 grassland sites, mainly dominated by *Lolium perenne*, scattered throughout the country. On 12 of these sites, the addition of sulfur increased grass yield by at least 10% and up to 45% during the growing season. Thus, when assessing the effects of SO$_2$ on grass performance in the field, one must take into account its nutritional aspects, which may be of importance under certain modern agricultural practices. However, the benefits gained in this manner could be gained as well by the addition of sulfur to nitrogen fertilizers, this being unlikely to represent more than a 10% increase in the cost of fertilizer production (Bache and Scott, 1979).

Assessment of Techniques Used to Measure the Effect of SO$_2$ on Grass Growth

The major problem in predicting the effect of SO$_2$ on the growth of grasses lies in the design of experiments in which pollutant dose, climate, and cultural conditions simulate conditions in polluted areas. There is little doubt that the difficulty in establishing a threshold SO_2 level for growth reductions in grasses is partly the result of environmental differences between the various experimental systems. In addition, the existence of seasonal or phenological effects on the nature of SO_2 response, demonstrated by Whitmore and Freer-Smith (1982) and Colvill *et al.* (1983), contributes further to this problem.

The experiments on the effects of SO_2 on grass productivity discussed in this paper were mainly performed in outdoor closed chambers (Crittenden and Read, 1978b, 1979; Awang, 1979; Bell, Rutter, and Relton, 1979; Ayazloo and Bell, 1981; Garsed, Mueller, and Rutter, 1982) or open-top chambers (Colvill *et al.*, 1983; Roberts *et al.*, 1983), but sometimes a controlled environment system was employed (Davies, 1980a,b; Jones and Mansfield, 1982a,b). Outdoor closed chambers inevitably modify ambient climatic conditions, particularly in summertime, when problems with overheating may occur. For example, in the experiments of Ayazloo and Bell (1981) temperatures inside the chamber often exceeded the outdoor temperature by 2 to 7°C in summer. Overheating still occurred even when chambers were shaded, so that incident radiation was reduced by 50% (Bell, Rutter, and Relton, 1979). Temperatures in the pollutant-exclusion chambers used in Sheffield generally exceeded ambient temperatures by 5 to 10°C (Crittenden and Read, 1978a).

In chambers with some degree of climatic control, there is usually no difficulty in maintaining realistic temperatures, but light regimes are often lower than those in the field. For example, in the wind-tunnel systems used for SO_2 fumigations of grasses by Horsman *et al.* (1979) and Davies (1980a), the maximum irradiance levels were 70 W m^{-2} and 130 W m^{-2}, respectively, which compare with a summertime maximum of about 800 W m^{-2} in England (Monteith, 1973).

Recent experiments on *Phleum pratense* by Davies (1980a) and Jones and Mansfield (1982a) leave little doubt that chronic SO_2 injury is increased when growth is reduced by low temperatures and low light intensities. In my (1982) paper, I reviewed all available data on the effects of SO_2 on individual organs of grass species, and pointed out some major discrepancies between the results of different experiments. These should now be reexamined in the light of recent developments in the understanding of SO_2/environment interactions on plant growth.

One apparently inexplicable feature I noted (1982) is the difference in how SO_2 may affect the rate of tillering, which can accompany reduction in shoot dry weight. When the results of recently published work are included with these data, some degree of consistency becomes apparent, according to the type of fumigation system employed. Out of seven experiments using wind tunnels with artificial illumination, six showed no effect on tiller number despite significant reductions in growth. On the other hand, in 15 fumigation and pollutant-exclusion experiments conducted in outdoor chambers, suppression of tillering accompanied growth reduction in all but one case. Without more detailed studies, it is probably futile to speculate on just what environmental factors may be responsible for these differences. However, the ability of grass species to produce tillers is known to be very sensitive to changes in light intensity, which in itself can affect the optimum temperature for tiller production (Langer, 1972). It is thus apparent that more work is required on how climatic conditions interact with chronic SO$_2$ injury to the whole plant if we are to understand the significance of air pollution for grassland productivity.

The concept that growth rate is important in modifying chronic SO$_2$ injury of grass species receives some support from an examination of the performance of the control plants in pollutant-exclusion and fumigation experiments (Bell, 1982, 1983). In the pollutant-exclusion experiments at Sheffield, where ambient air containing very low SO$_2$ concentrations produced marked reductions in yield, growth rates were often slow in the chamber ventilated with filtered air. For example, *Lolium perenne* plants showed a relative growth rate in clean air ranging between 0.006 and 0.045 gg^{-1} d^{-1} during a 45-week experiment, covering both summer and winter conditions (Crittenden and Read, 1978b). This rate compares with a relative growth rate of between 0.009 and 0.073 gg^{-1} d^{-1} in a winter experiment in outdoor chambers reported by Bell, Rutter, and Relton (1979). The importance of growth-rate in SO$_2$ injury will only be fully understood when chamber experiments are performed in which environmental parameters can be manipulated to simulate the conditions experienced by plants growing outdoors (see section on future research directions).

Open-top chambers are considered to be an important advance over closed systems in providing climatic conditions as close as possible to the ambient. For example, Roberts *et al.* (1983) found only a 1 or 2°C increase over low ambient temperatures in the open-top chambers used in the pollutant-exclusion experiments at St. Helens, although light intensity was reduced by 15 to 25%. Unfortunately, open-top chambers have other defects, notably a much lower filtration efficiency than closed systems have, because ambient air enters through the top, particularly under turbulent meteorological conditions. The pollutant-exclusion chambers used by Roberts *et al.* (1983) and Colvill *et al.* (1983) had an efficiency of only 50 to 70% for SO$_2$; this compares with 98 to 100% in most closed systems. Thus their "clean air" control plants in the open-top chambers were subjected to SO$_2$ levels within the range found in the ambient air chambers at Sheffield. This inefficient removal of SO$_2$ may account for the relatively small effects on growth recorded in the experiments at St. Helens. In addition, even though activated charcoal is effective at removing SO$_2$, NO$_2$, and O$_3$, it does not absorb much NO. Consequently, all chambers in urban experiments are likely to be subjected to NO, which may therefore influence plant growth in the control treatment.

All the experiments on the effects of SO$_2$ on grass productivity discussed in this paper and in Bell (1982) have used plants grown in various types of containers. This represents an artificial constraint on root growth that may modify

TABLE 12.7
Effect of Artificial Misting on Acute Injury on *Poa pratensis*
Cultivars after SO_2 Fumigation
(percent of leaf area destroyed)

	Misting	
Cultivar	None	10 minutes per day
Cheri	<7%	>15%
Merion	8–15	>15
Touchdown	0	>15

SOURCE: Elkiey and Ormrod (1981d).
  NOTE: Fumigation was at 400 μg m^{-3} (6.25 μmol m^{-3}) for 10 days.

the response of plants to SO_2 in comparison with plants growing in the ground. Some evidence for this has been provided by Davies (1980b), who fumigated *Phleum pratense* with 320 μg m^{-3} (5.00 μmol m^{-3}) SO_2 for 32 days and then repotted half the plants into larger containers. Harvests after a further 8 and 16 days showed that repotting delayed the senescence induced by SO_2 in comparison with the plants that remained in the smaller containers. Davies suggested that restriction of the root system aggravated the effect of SO_2 in promoting senescence, by reducing the nutrient supply from the soil. In Bell (1982), I concluded that accelerated senescence was not the principal mechanism by which SO_2 reduced grass growth, because large effects on yield are not always accompanied by a significant increase in leaf death. Perhaps increased senescence is caused by SO_2 in some experiments, but not others, because of the degree of constraint imposed by the container on the roots. An examination of the literature, however, does not reveal any consistent relationship between SO_2-induced senescence in grasses and the ratio of plant weight to container volume. Nevertheless, the findings of Davies (1980b) are worthy of further investigation, in view of their demonstration that container-grown plants may respond differently to SO_2 from plants in the field.

Recent work by Elkiey and Ormrod (1981d) has pinpointed another, hitherto neglected, factor that may modify substantially the response of grass species to SO_2. They carried out an experiment in which three *Poa pratensis* cultivars were fumigated with 400 μg m^{-3} (6.25 μmol m^{-3}) of SO_2 for 10 days, with half the plants being subjected to a deionized water-misting treatment for 5 minutes twice daily. In all cultivars, misting increased the amount of acute SO_2 injury on the foliage (Table 12.7). This raises serious questions about the nature of watering regimes employed in chamber experiments. Aerial application of water, which wets the foliage, may increase chronic SO_2 injury more than the subirrigation processes employed in many experiments. These effects of moisture on leaves could be of considerable importance for SO_2 injury in grasslands in areas with a high relative humidity and fairly uniform distribution of rainfall throughout the year. The foliage of vegetation in Britain is wet, because of dew or precipitation, about 30% of the time (Garland and Branson, 1977), which probably represents a much larger proportion of the growth period than is likely to result from the brief misting treatments of Elkiey and Ormrod (1981d).

Directions for Future Research

The complexities of the effects of SO_2 on grass species are becoming more apparent as more research data are published. The search for an overall long-term threshold concentration at which chronic injury always occurs appears increasingly futile. Attention should be switched to considering how dose-response relationships may be modified by environmental conditions, cultural practices, and the presence of other pollutants at locations of interest in the field.

SO_2 injury to plants is affected by artificial conditions imposed by the chamber systems, the size and shape of pots, and the use of unrealistic watering regimes. These problems are related to laboratory systems, and provide support for the employment of open-air fumigation systems in grasslands, where plants can be subjected in the field to fluctuating concentrations of pollutant mixtures. The recent development of such systems is obviously a welcome step forward toward understanding the real effect of SO_2 on grass growth under field conditions. Nevertheless, some problems that arise in field fumigation systems do not occur in chamber experiments. In particular, the presence of other phytotoxic pollutants, notably O_3 and acid precipitation, at many sites where SO_2 concentrations are generally low may confound the results of open-air fumigations.

More research is needed on the effects of ambient mixtures of SO_2 and other pollutants, particularly NO_2, NO, and O_3. Techniques currently available for pollutant-exclusion experiments are not entirely satisfactory, because efficient filtration and realistic climatic conditions apparently cannot be combined in the same system. An approach that might assist in overcoming these problems is the development of an open-air pollutant-exclusion system in which clean air is blown continually over the plants. Growth of plants in clean air would then be compared to plant growth in other plots subjected to the same treatment with ambient air. A further treatment could add back SO_2 to clean-air plots, with and without other pollutants, to mimic concentrations in the ambient-air plots. Despite advances in open-air fumigation technology, however, chamber systems will remain indispensable for carefully controlled experiments designed to find out how SO_2 injury is modified by climatic and edaphic factors.

Future work should be aimed at designing experiments that take into account the special features of both agricultural and leisure-use grasslands, as outlined at the beginning of this paper. In particular, continuous fumigations for several years are required in order to understand the importance of SO_2-induced changes in the species that make up a sward. In addition, long-term studies are needed to show the seasonal/phenological fluctuations in response of the type noted by

Colvill *et al.* (1983) and Whitmore and Freer-Smith (1982). Experiments should incorporate management practices for agricultural and leisure-use grasslands, including typical cropping regimes and fertilizer additions, as well as using realistic climatic conditions. Furthermore, studies are needed on the interactions of SO_2 with other environmental stresses peculiar to grasslands; little or nothing is known about the effects of grazing, mowing, or trampling (by animals or humans). Only when such neglected features of grassland management are considered will it be possible to make meaningful estimates of the current full economic impact of SO_2 and to predict the consequences of future changes in air quality.

13

SO$_2$ Effects on Dicot Crops:
Some Issues, Mechanisms, and Indicators

S. B. McLaughlin, Jr., and G. E. Taylor, Jr.

The issues involved in assessing the effects of SO$_2$ in the environment are as diverse as the landscapes on which this pollutant has been deposited during the many decades since people began intensive use of fossil fuels. In spite of many years of research, the principal issue, "How much should SO$_2$ be controlled?" has not been satisfactorily answered. Several major areas of uncertainty should be addressed if we are to achieve an integrated understanding of plant responses to SO$_2$ and define strategies for managing SO$_2$ emissions. These are as follows.

1. The economic or ecological consequences of the current ambient concentrations of SO$_2$ have not been quantified well enough to allow their costs to be weighed against the costs required for emission controls.

2. The distribution of SO$_2$ concentrations in rural areas has not been measured and/or statistically described well enough to provide meaningful estimates of short-term or long-term exposure dose.

3. The effect of a given concentration of SO$_2$ on productivity may be influenced by many diverse environmental variables, and hence will differ greatly between sites that differ in important environmental characteristics.

4. SO$_2$ is known to have both positive and negative effects on plant productivity, but thresholds for these effects are not well-defined.

5. Adequate data for describing dose-response relationships are not available for many species, and are not generally available for yield responses under field conditions.

6. Although physiological responses are commonly used as measurement endpoints, the mechanisms of whole-plant integration of physiological changes into yield responses are not well-understood.

7. Although diverse physiological responses to SO$_2$ have been identified, the kinetics of primary metabolic responses at ambient concentrations of SO$_2$ have not been clearly described.

It is typically easier to identify gaps in the information needed for large-scale assessments than to fill those gaps. Our approach in this paper will be to build on existing knowledge in order to offer generalizations about dicot responses to SO$_2$, to point out deficiencies in our data, and to define more specific research needs. Our focal points will be (1) evaluation of our present database on SO_2 effects on crop yield; (2) discussion of the significance of the distribution of major sources of SO_2 on regional air quality and its relationship to observed yield responses; (3) examination of mechanisms of plant response to SO_2 and some potentially useful physiological indicators of SO_2 effects on crop productivity under different environmental conditions; and (4) identification of additional information needs and some experimental approaches that might prove appropriate in meeting those needs.

The Data on Yield Effects

A summary of studies of field exposures of dicot crop plants to SO_2 is shown in Table 13.1. These data are derived from many experiments and represent a variety of sampling and exposure methodologies. Included among the experimental methodologies are the Zonal Air Pollution System (ZAPS), open-top chambers, linear-gradient exposure studies, and studies of plots located at different distances from a smelter emitting large amounts of SO_2. Also identified in Table 13.1 are datapoint number, species, average exposure concentration, exposure duration, total dose (concentration × time), exposure methodology, and references for dicot crops for which specified yield losses have been measured. Most of the datapoints (1 to 20) represent studies with soybeans (*Glycine max*) or snap beans (*Phaseolus vulgaris*), over a range of SO_2 concentrations. Most of the field data are derived from studies in which SO_2 was added either to naturally occurring concentrations of ozone in the ambient air or to charcoal-filtered air to which ozone was also added at near-ambient concentrations.

To find out if these data provided a useful basis for predicting yield losses, we evaluated several indices of pollutant exposure by regression against yield responses. Yield losses were plotted against average SO_2 concentration during exposure, total exposure dose in ppmh (or μmol m^{-3} h) (average SO_2 concentration × exposure duration in hours), and log-transformed dose in ppmh (or μmol m^{-3} h). Of the three regressions, log dose (Fig. 13.1) provided the most consistent description of yield loss potential with a regression of the form (yield loss = 28.9 (log dose) − 15.9) and a correlation coefficient (r^2) of 0.77; r^2 values for total dose and average concentration were 0.74 and 0.01, respectively. The log dose vs. yield regression provides an extrapolated threshold estimate of 3.5 ppmh (145.6 μmol m^{-3} h) below which yield is not reduced. This regression also indicates that dose-response relationships were generally consistent across these exposure methodologies. In Fig. 13.2 we show the relationship between this regression line and other sources of data. The latter include data primarily from two sources: (1) a study by Reich, Amundson, and Lassoie (1982) using a linear gradient system to expose soybeans during the interval from flowering until harvest; and (2) potato yields from two years of container studies at plots located at various distances from the iron smelters at Biersdorf, Germany (Guderian and

TABLE 13.1

Characteristics of Field Studies of Growth and Yield of Dicot Crops

Data point	Species	SO₂ concentration in ppm (µmol m⁻³)		Duration (h)	Dose (ppmh)	Log dose (ppmh)	Yield effects (percent)	Exposure mode
1	Soybean (cv Wells)	1.40	(58.23)	4.3	6.0	0.78	5%	Zonal Air Pollution System (SO₂ added to ambient air).
2	Soybean (cv Wells)	1.70	(70.71)	4.3	7.3	0.86	11	
3	Soybean (cv Wells)	2.00	(83.19)	4.3	8.6	0.93	15	
4	Soybean (cv Wells)	0.12	(4.99)	113	13.5	1.13	13%	Zonal Air Pollution System (SO₂ added to ambient air).
5	Soybean (cv Wells)	0.30	(12.48)	113	34	1.53	21	
6	Soybean (cv Wells)	0.79	(32.86)	113	89	1.95	45	
7	Soybean (cv Wells)	0.09	(3.74)	75.6	6.8	0.83	6%	Zonal Air Pollution System (SO₂ added to ambient air).
8	Soybean (cv Wells)	0.10	(4.16)	75.6	7.6	0.89	5	
9	Soybean (cv Wells)	0.19	(7.90)	75.6	14.4	1.16	12	
10	Soybean (cv Wells)	0.25	(10.40)	75.6	18.9	1.28	19	
11	Soybean (cv Wells)	0.36	(14.97)	75.6	27.2	1.44	16	
12	Snap beans, 3 vars.	0.06	(2.50)	144	8.6	0.93	9%	Open-top chambers. SO₂ added to ambient air. Three-species regression.
13	Snap beans, 3 vars.	0.12	(4.99)	144	17.2	1.24	18	
14	Snap beans, 3 vars.	0.30	(12.48)	144	43	1.63	43	
15	Soybean (cv Davis)	0.026	(1.082)	337	8.8	0.94	17%	Open-top chambers. SO₂ added to 0.055 ppm O₃ (41 ppmh) pts 15–17, and to 0.065 ppm O₃ (51 ppmh) pts 18–20.
16	Soybean (cv Davis)	0.085	(3.536)	337	28.7	1.45	14	
17	Soybean (cv Davis)	0.367	(15.266)	337	124	2.09	39	
18	Soybean (cv Davis)	0.026	(1.082)	337	8.8	0.94	18	
19	Soybean (cv Davis)	0.085	(3.536)	337	28.7	1.45	23	
20	Soybean (cv Davis)	0.367	(15.266)	337	124	2.09	51	
21	Soybean (cv Hark)	0.04	(1.66)	66	2.3	0.36	10%	1.9 ppmh O₃ added — Linear
22	Soybean (cv Hark)	0.09	(3.74)	66	5.2	0.72	28	2.5 ppmh O₃ added — Gradient
23	Soybean (cv Hark)	0.16	(6.66)	66	9.0	0.95	35	3.5 ppmh O₃ added — System
24	Potato	0.01	(0.42)	86	1.0	0.01	4%	Field plots with plants in containers at distances of 325–1,900 m from iron smelter. Assume 12 h d⁻¹ effective exposure (pts 24–28, 1959 data; pts 29–31, 1960 data).
25	Potato	0.02	(0.83)	130	2.8	0.44	17	
26	Potato	0.05	(2.08)	215	10.3	1.01	21	
27	Potato	0.09	(3.74)	279	24.5	1.38	31	
28	Potato	0.14	(5.82)	323	43.6	1.64	49	
29	Potato	0.05	(2.08)	166	8.0	0.90	12	
30	Potato	0.09	(3.74)	214	18.8	1.27	34	
31	Potato	0.14	(5.82)	248	33.5	1.52	64	
32	Soybean (cv Dare)	0.10	(4.16)	798	80	1.90	1%	Open-top chamber.
33	Alfalfa	0.06	(2.50)	1632	98	1.99	26	Pots in open-top chambers.

SOURCES: 1–3, 7–11, Miller et al. (1978). 4–6, Sprugel et al. (1980). 12–14, Heggestad and Bennett (1981). 15–20, Heagle et al. (1983). 21–23, Reich, Amundson, and Lassoie (1982). 24–31, Guderian and Stratmann (1962). 32, Heagle, Body, and Neely (1974). 33, Neeley, Tingey, and Wilhour (1977).

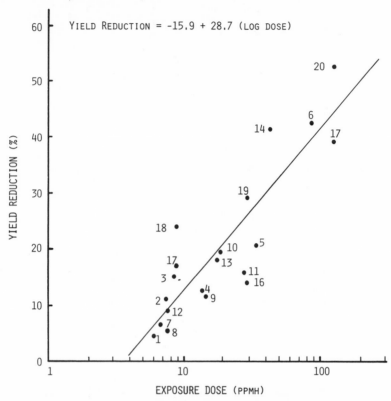

Fig. 13.1. Combined dose-response surface for yield responses of soybeans and snap beans to field exposure with SO_2. Data points are identified in Table 13.1.

Stratmann 1962). The study by Reich, Amundson, and Lassoie (1982) (points 21 to 23) is particularly interesting, in that it provides a lower estimate of the threshold dose for yield loss, 1.3 ppmh (54.1 μmol m^{-3} h), than was obtained from any of the other bean-derived data sets examined. This may be attributed to any of a number of factors, including growing conditions, soybean variety, or exposure methodology. Exposure conditions would have been particularly important in this study. Since the exposure occurred only during the sensitive flowering stage, both SO_2 and ozone concentrations were increased simultaneously in this study. However, average ozone concentrations in at least the lowest two SO_2 doses were still below O_3 levels measured in ambient air in the open-top chamber studies of Heggestad and Bennett (1981). In the latter experiments, SO_2 was added to ambient air.

The data on potato yields from Biersdorf (points 24 to 28 for 1959 and 29 to 31 for 1960) represent the only information from near a point source of SO_2, and at the same time constitute the most complete data characterizing SO_2 exposure conditions. Both the SO_2 exposure conditions and yield reductions are shown in

Table 13.1 in relationship to distances from the smelter. The yield data plotted in Fig. 13.2 show distinct differences from year to year. Dosage thresholds for yield effects were 0.7 ppmh (29.1 μmol m^{-3} h) for 1959 and 6.0 ppmh (249.6 μmol m^{-3} h) in 1960, and both response surfaces were described well by the yield:log-dose relationship.

Because of the completeness of the exposure data for these studies, we have examined relationships between potato yield and a wide variety of indices of SO₂ exposure. In Table 13.2, yield relationships to maximum 30-minute average, geometric and arithmetic means, distance from source, total dose (ppmh or μmol m^{-3} h), log dose, and percent exposure time have been plotted. Based on the 1959 data, a variety of indices of exposure level were found to be well-correlated with yield losses. The highest correlation coefficients were found with geometric and arithmetic means, percent exposure time, and total exposure dose ($r^2 = 0.94$ to 0.96). Close agreement between many of these two parameters was, of course, to be expected, because they are strongly correlated with each other. What

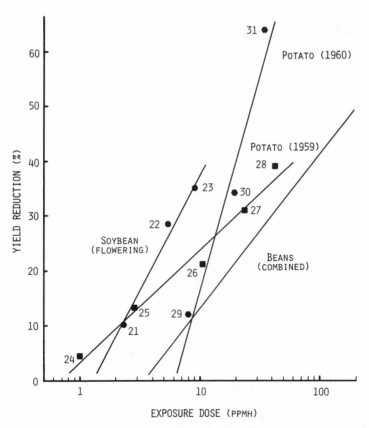

Fig. 13.2. Comparative dose-response surfaces for SO₂ effects on soybeans, snap beans, and potatoes for various studies identified in Table 13.1.

TABLE 13.2

Exposure Characteristics and Yield Responses of Potatoes from Studies near an Iron Smelter at Biersdorf, Germany

Parameter	Stations					Correlation with Yield (r^2)
	I	II	III	IV	V	
Distance from source (m)	325	600	725	1,350	1,900	0.84
Maximum 30 min. $\bar{x}$[a]	6.0	5.1	2.3	1.6	1.3	0.88
Geometric $\bar{x}$	0.45	0.34	0.24	0.18	0.15	0.96
Exposure time	30%	26%	20%	12%	8%	0.95
Arithmetic $\bar{x}$	0.14	0.09	0.05	0.02	0.01	0.95
Peak: mean[b]	13.3	15	9.6	8.9	8.7	0.60
Exposure time (h) [c]	323	279	215	130	86	0.90
Exposure dose (ppmh)[d]	43.6	24.5	10.3	2.8	1.0	0.94
Log exposure dose (ppmh)[d]	1.64	1.38	1.01	0.44	0.01	0.89
Yield reduction	49%	31%	21%	17%	4%	

SOURCE: Guderian and Stratmann (1962).

[a] All SO_2 concentrations in ppm represent average of 1959 and 1960.

[b] Thirty-minute maximum: geometric mean.

[c] Calculated for 1959 season based on 12 potentially effective exposure hours per day over an approximate 90-day growth period.

[d] 1 ppm dose can be converted to SI units where 1 ppmh = 42.0 μmol m^{-3}, as explained in the Note on Units at the beginning of this volume.

TABLE 13.3

Characteristics of Laboratory and Greenhouse Studies of Growth and Yield of Dicot Crops

Species	SO_2 concentration in ppm (μmol m^{-3})		Duration (h)	Dose (ppmh)	Growth or yield effects (percent)
1. Tobacco (Bell W-3)	0.05	(2.08)	100	5	22%
2. Alfalfa	0.05	(2.08)	100	5	26
3. Tobacco (Burley 21)	0.05	(2.08)	100	5	0
4. Soybean	0.05	(2.08)	720	36	0
5. Soybean	0.20	(8.32)	720	144	0
6. Pea	0.10	(4.16)	216	22	5
7. Pea	0.15	(6.24)	216	33	8
8. Pea	0.25	(10.40)	216	55	26
9. Bean	0.30	(12.48)	60	18	25
10. Sunflower	0.30	(12.48)	60	18	41
11. Bean	0.30	(12.48)	130	39	15
12. Sunflower	0.30	(12.48)	130	39	29
13. Tobacco	0.53	(22.06)			+44
14. Sunflower	0.53	(22.06)			+27
15. Sunflower	0.35	(14.56)			+44

SOURCES: 1–3, Tingey and Reinert (1975). 4–5, Tingey et al. (1973b). 6–8, Jäger and Klein (1977). 9–12, Markowski, Grzesiak, and Schramel (1975). 13–15, Faller (1970).

was surprising was the low correlation between the peak:mean ratio and yield responses.

In contrast to field studies, results obtained by controlled laboratory or greenhouse studies (Table 13.3) represent a broader range of species and exposure conditions and provided no indication of a consistent relationship between dose level and plant response. Yield responses indicated intermediate sensitivity generally, and fell between the regression lines delineated by the linear gradient and ZAPS systems. Growth-chamber experiments with two species, sunflower and tobacco, growing in a sulfur-deficient soil (Faller, 1970), resulted in positive effects (7 to 44% increase) of SO_2 on plant yield. Since sulfur is a plant nutrient, such effects should be anticipated wherever SO_2 is supplied at low atmospheric concentrations to plants growing in conditions where sulfur is limiting plant growth (Faller, 1970).

In summarizing the data on yield, the following points can be made.

1. The data from field experiments provide a much clearer picture of dose-response relationships than do laboratory data.

2. Yield responses for soybeans and snap beans exposed during the entire growing season show a general consistency in relationship to log dose in ppmh (or μmol m^{-3} h), in spite of the use of different varieties and different exposure methodologies.

3. Different species and even the same species in different years may show different dose thresholds and dose:response curves as a function of varietal sensitivity or environmental conditions; however, losses in yield appear to be rather well-correlated with log-transformed dose over a wide range of conditions.

The Distribution of SO_2 Sources and Concentrations

The major concerns about SO_2 effects on crops are centered around measuring the concentrations at which yield loss occurs and the size of the area in which loss occurs in the ambient environment. SO_2 is emitted primarily from point sources that burn either coal or oil; currently, coal- and oil-fired power plants in the United States contribute approximately two-thirds of the total 27 million tons of SO_2 annually emitted into the atmosphere of the United States (U.S. Environmental Protection Agency, 1983). Their distribution is distinctly nonuniform, with approximately 80% of these facilities being east of the Mississippi River. Frequent air stagnation and high soil and atmospheric moisture are also characteristic of this area, and combine to increase the likelihood that air pollutants will cause adverse effects to vegetation in this region (McLaughlin, 1981). A county-level map (Fig. 13.3) of the emissions of SO_2 from all sources (Shriner *et al.*, 1984) further emphasizes the high density of SO_2 emissions in the eastern United States.

The likelihood that SO_2 emissions will produce adverse effects should be considered from two perspectives: (1) local effects in the vicinity of individual point sources; and (2) regional effects attributable to chronic exposures to low concentrations of SO_2 from multiple sources. In either, SO_2 may act additively, synergistically, or antagonistically with other regional-scale pollutants, such as ozone or acid rain.

Point-source effects. Effects of SO_2 and other combustion-derived gases from electric-generating plants or industrial sources are primarily a function of local

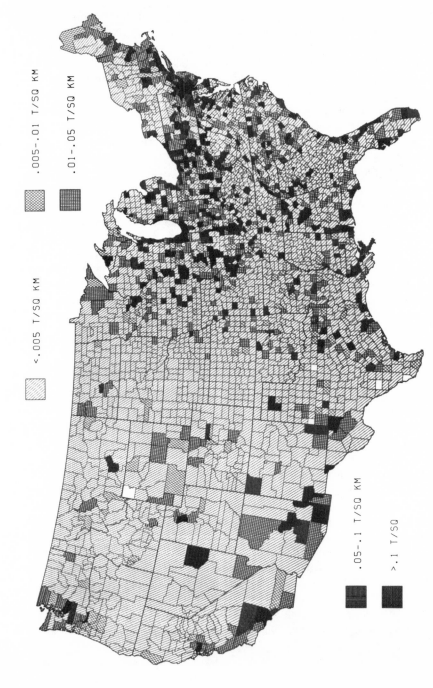

Fig. 13.3. 1978 county-level emissions of SO_2 from all sources, in tonnes/sq km, demonstrating regional differences in SO_2 emission sources within the United States.

.005–.01 T/SQ KM

.01–.05 T/SQ KM

<.005 T/SQ KM

.05–.1 T/SQ KM

>.1 T/SQ

meteorological conditions (wind speed and direction, and atmospheric stability), which generally bring the plume to ground level for varying periods of time. Under these conditions, exposure episodes with SO$_2$ typically occur with rapidly fluctuating concentrations characterized by peak:mean concentration ratios significantly greater than 1.0. It is under these conditions that the concept of exposure dose has its most diverse and significant dimensions. Because of the kinetics of typical SO$_2$ exposures from fossil-fuel power plants (McLaughlin, Jones, and Schorne, 1976), an evaluation of the type and severity of effects to be expected from point-source exposures to SO$_2$ must consider distribution of SO$_2$ both within and between exposure events.

Laboratory experiments by Zahn (1961, 1970) have strengthened the evidence for increased toxicity of short-term ($\leqslant$30 min), high-concentration exposures, and have emphasized the importance of the concentration and duration of exposure to SO$_2$ during "recovery periods" between higher-concentration episodes. From field experiments near smelters at Biersdorf, Germany, Stratmann (1963) identified several aspects of pollutant exposure that were important for describing the potential for plant damage. He characterized these as follows: (1) mean concentration, (2) total dose (integration of concentration over time), (3) variability of concentrations, and (4) frequency of concentration peaks.

Repeated SO$_2$ exposure episodes may increase plant sensitivity to subsequent exposures; so it may be important to document exposure history. Unfortunately, relatively few studies have concentrated on evaluating the significance of dose kinetics or dose sequence on plant response (McLaughlin *et al.*, 1979). Most controlled fumigation work has been done with steady-state concentration regimes, and evaluations of plant response have been based either on single-exposure episodes or on unrealistically constant SO$_2$ levels.

In the United States, Jones *et al.* (1979) have analyzed data from field studies in the vicinity of power plants operated by the Tennessee Valley Authority. This analysis was designed to provide a way to estimate probabilities of effects at various concentrations of SO$_2$. Their approach consisted of examining both the probability that a given SO$_2$ dose would occur and the probability that effects would be produced when that dose occurs.

From analysis of field-survey data on about 180 plant species, of which 86 were visibly injured one or more times during the six-year study interval, Jones *et al.* (1979) concluded that SO$_2$ thresholds for visible foliar symptoms to occur were 0.32 and 0.22 ppm (13.31 and 9.15 μmol m^{-3}) for 1- and 3-h averaging intervals, respectively. Soybean was among the most sensitive, economically valuable species found. The probability of soybean experiencing $\geqslant$5% leaf-area chlorosis at 3-h average SO$_2$ concentration of 0.50 ppm (20.8 μmol m^{-3}) was approximately 50%. Peak concentrations that were two or more times the 3-h average concentrations for the intervals during which they occurred were important in producing visible injury.

To find the probability that visible injury might occur around a large, properly

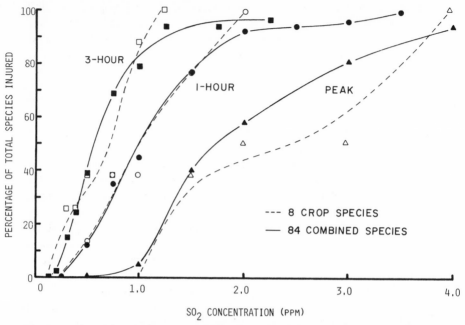

Fig. 13.4. Percentage of plant species on which visible injury was detected as a function of peak, 1-h avg, and 3-h avg SO_2 concentrations near a coal-fired power plant. Data for eight dicot crops (dashed lines) are compared with responses of the total 84 species affected. Data modified from McLaughlin (1981).

sited power plant, Jones *et al.* (1979) measured the frequency distribution of 1-h average SO_2 concentrations around the 2,600-megawatt Cumberland Steam Plant, a modern coal-fired plant with 1,000-ft stacks sited in low-profile topography. They found that the frequency of occurrence of SO_2 concentrations ≥ 0.20 ppm (8.3 μmol m^{-3}) for one hour was $\leq 2\%$. Thus with modern power plants in proper terrain, the probability of visible injury occurring may be very low.

As an example of a poorly-sited facility, TVA's Widows Creek Steam Plant provided a wealth of data on plant response during the early 1970's before stack elevation and partial scrubbing substantially reduced SO_2 pollution problems (McLaughlin and Lee, 1974; McLaughlin, 1981). Under these conditions, exposure to SO_2 (concentrations ≥ 0.1 ppm, or 4.16 μmol m^{-3}) occurred on approximately 40% of the days, 11% of daylight hours, and lasted approximately 2.3 hours on each occurrence. A review of the McLaughlin and Lee (1974) dose-response data showing the relationship of visible foliar injury to the 84 plant species to varying peak, 1-h and 3-h average SO_2 concentrations stressed the greater sensitivity of a wide variety of plants to short-term exposure characteristics (McLaughlin, 1981). When the eight agricultural crops were examined separately (Fig. 13.4), it was evident that response thresholds for visible injury were similar to those for the combined 86 species.

Visible injury *per se*, however, may mean little as a predictor of SO$_2$ effects on yield. A major question is whether yield loss occurs at the more frequently encountered lower concentrations of SO$_2$, which are experienced as a result of either regional background levels of SO$_2$ or local point-source additions to those levels.

Regional-scale exposure to SO$_2$. The SO$_2$ monitoring stations reported by EPA (1980c) are located predominantly in urban areas; however, data from these stations may be used to provide a very general index of the SO$_2$ concentrations within a region not directly associated with nearby point-source emissions. To provide an index of the total SO$_2$ dose experienced by vegetation within various air-quality regions, McLaughlin (1981) has used annual 90th-percentile arithmetic mean and maximum 24-h doses for valid reporting stations within the ten EPA Air Quality Regions (EPA, 1980c). When adjusted for the number of daylight hours (12) and length of typical growing seasons (6 months), these data provide a very rough estimate of the potential annual dose of SO$_2$ during "high-concentration" episodes. By this approach, high-concentration episodes are defined as the upper 10% of all exposure hours occurring during daylight hours within the growing season. Resultant total mean exposure doses (concentration times duration) ranged from a low of 0.9 ppmh (0.37 μmol m^{-3} h) for Region IX to 5.5 ppmh (229 μmol m^{-3} h) for Region VIII. For large cities in the eastern United States, e.g., Chicago, Cincinnati, St. Louis, and Philadelphia, an analysis of actual 90th-percentile SO$_2$ concentrations for 1977 (EPA, 1977) gave SO$_2$ levels of 0.06, 0.04, 0.10, and 0.08 ppm (2.5, 1.66, 4.16, and 3.33 μmol m^{-3}), respectively. Over a growing season these concentrations occurring 10% of the time during daylight hours would give dosages ranging from 6.6 to 14.5 ppmh (275 to 603 μmol m^{-3} h). Maximum concentrations for these cities ranged from 0.25 ppm (10.4 μmol m^{-3}) for Chicago to 0.67 ppm (27.87 μmol m^{-3}) for Cincinnati.

Estimating a SO$_2$-Response Threshold

The preceding calculations provide a rough estimate of higher-concentration SO$_2$ dosages in predominantly urban areas, and a basis for comparison of these doses with those shown to produce yield losses in Fig. 13.1. If these SO$_2$ doses are evaluated in terms of the regression line defined in Fig. 13.1, which is based primarily on soybean responses, such doses provide yield loss estimates of ≤15%. However, this approach to predicting SO$_2$-induced crop losses should be used with extreme caution, since the concentrations of SO$_2$ at which SO$_2$ effects change from beneficial or innocuous to harmful are unknown. Certainly exposure to SO$_2$ at all concentrations does not result in accumulative toxicity to plant processes, since sulfur is a plant macronutrient, and can be readily metabolized for amino-acid and protein synthesis. It is generally thought that SO$_2$ becomes toxic to plant metabolic processes when the flux rate of SO$_2$ into the plant exceeds the plant's capacity to convert SO_2 from the very toxic sulfite form to the less-toxic sulfate. Sulfate, though less toxic than sulfite, can also accumulate to toxic concentrations under prolonged exposure to SO_2 when sulfur is nonlimiting for plant growth (Thomas, Hendricks, and Hill, 1950).

The critical question in evaluating plant-yield responses to SO_2 as it occurs in the field, however, is, "At what minimum ambient concentration is SO_2 toxic?" Certainly there can be no easy answer to this question, since toxicity will ultimately depend on sulfur status of the plant-soil system, the activity of sulfur-metabolizing pathways, the pattern (continuous or intermittent) and frequency of exposure, and the status of other environmental modifiers of SO_2 uptake. The threshold will also depend on the process being measured. Plant responses to SO_2 may occur at many levels of biochemical and physiological resolution, and may constitute either primary or secondary reactions. To be ultimately useful in quantifying changes in plant productivity, these measurements should include either changes in growth of whole plants or plant organs, or changes in physiological indicators that can be related to plant growth. The latter responses will be discussed later in this chapter.

Evidence for changes in crop growth at very low concentrations of SO_2 in the field is sparse. Sprugel *et al.* (1980) and Miller *et al.* (1978) reported yield losses of about 5% for soybeans (cv Wells) following 18 exposures of about four hours each to SO_2 concentrations in the range of 0.09 to 0.10 ppm (3.74 to 4.16 μmol m^{-3}). Heagle, Body, and Neely (1974), on the other hand, found no significant effects on yield of soybeans (cv Dare) after 92 days of open-top field-chamber exposure to 0.10 ppm (4.16 μmol m^{-3}) of SO_2 for 6 h d^{-1}. Jones and Noggle (1980) found a 13% yield reduction of Essex soybeans exposed during the pod-fill stage for 65 hours to 1-h average SO_2 concentrations $\geqslant 0.1$ ppm (4.16 μmol m^{-3}; total dose = 16.7 ppmh, or 694 μmol m^{-3} h) near a coal-fired power plant. However, no effects were observed during the following two years to as many as 29 hours of exposure to 1-h average $\geqslant 0.10$ ppm (4.16 μmol m^{-3}; 9.8 ppmh, or 408 μmol m^{-3}, total dose). Heggestadt and Bennett (1981), on the other hand, developed a regression of yield of three varieties of snap beans versus SO_2 at three concentrations: 0.06, 0.10, and 0.30 ppm (2.50, 4.16, and 12.48 μmol m^{-3}) for 6 h d^{-1} for 24 days. Their results suggest a statistically significant inverse relationship between SO_2 concentration and yield at all concentrations including 0.06 ppm (2.50 μmol m^{-3}), which provided a total dose of 8.6 ppmh (358 μmol m^{-3} h). Yield losses at this concentration averaged across the three varieties were approximately 9%. In this experiment, in which SO_2 was added to nonfiltered air, a significant finding was that yield losses were not produced by ozone alone, but only when low concentrations of SO_2 were added to ambient air, and the effects of SO_2 and ambient air were much greater than when SO_2 was added to charcoal-filtered air. It is also significant to note that at the Beltsville, Maryland, field station where these studies were conducted, the average daily maximum O_3 value was 0.065 ppm (2.70 μmol m^{-3}), and ambient SO_2 concentrations averaged less than 0.002 ppm (0.083 μmol m^{-3}). Thus even in close proximity to two large cities (Washington and Baltimore), rural SO$_2$ concentrations were quite low. Another important aspect of these studies was that yield losses of one bean cultivar, Astro, which showed no visible foliar injury, were as large as those of the other two varieties, whose foliar symptoms were more typical of oxidant injury.

The Physiological Basis for SO$_2$ Effects

Plant responses to SO$_2$ may be measured at many organizational levels, each differing in relevance to plant productivity. Some of these responses are shown in Fig. 13.5, beginning with SO$_2$ uptake, and extending through effects on physiological processes that ultimately influence resource allocation. Each process can be used to measure or infer effects at higher levels of organization. Because of their unique utility as physiological indicators, the process of gas-exchange (CO$_2$ and SO$_2$) and carbon allocation are the focal points for this discussion.

Photosynthesis. As a diagnostic tool, net photosynthesis (P_s) is a valuable monitor of stress effects in plants. For gaseous air pollutants in general and SO$_2$ specifically, the use of P_s as an indicator of a plant's physiological state is particularly appealing (Taylor, Cardiff, and Mersereau, 1965), since cells of the leaf interior are initial sites of pollutant toxicity.

The threshold and nature of photosynthetic response as a function of SO$_2$ exposure is not uniform among all plant species and exposure conditions, and the occurrence of different patterns of response as a function of atmospheric SO$_2$ concentration makes a consensus about threshold levels of injury difficult (Black, 1982). The graph of photosynthetic inhibition as a function of exposure dose in the atmosphere may identify some common features of photosynthetic response to SO$_2$ for dicot crop species (Fig. 13.6). The data are predominantly from experiments in which individual leaves or whole plants were fumigated, and thus their applicability to field conditions may be limited. Most data provide

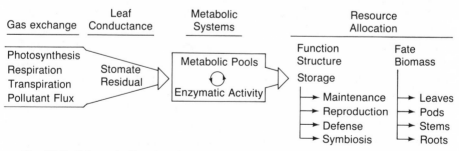

Fig. 13.5. Schematic diagram of some plant processes that have been used as indicators of SO$_2$ effects on plant systems.

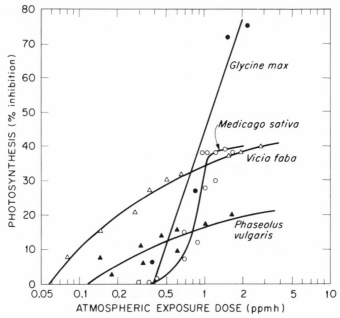

Fig. 13.6. Percentage inhibition of photosynthesis in dicotyledenous crop species exposed to SO_2. The abscissa (atmospheric exposure dose) is the product of concentration and time. The source of data for each species is as follows: *Glycine max* (Carlson, 1983a), *Medicago sativa* (Bennett and Hill, 1973b), *Vicia faba* (Black and Unsworth, 1979b), and *Phaseolus vulgaris* (Taylor, unpublished data).

only a few measurements from which linear (*via* least-squares regression) or curvilinear response lines were fitted; for the studies in which many coordinates were available (Black and Unsworth, 1979b; Bennett and Hill, 1973b), the relationship between photosynthesis and exposure dose was curvilinear. The extrapolated dosages below which P_s was not influenced were within the range of 0.05 to 0.4 ppmh (2.1 to 17 μmol m^{-3} h); the mean threshold for a P_s response among the reported studies was 0.25 ppmh (10 μmol m^{-3} h). This is equivalent to a 6-h exposure to approximately 0.04 ppm (1.66 μmol m^{-3}). Fowler and Cape (1982) estimated that the annual arithmetic mean SO_2 concentration of 0.04 ppm (1.66 μmol m^{-3}) occurs over 15.4 × 10^5 ha in western Europe (including Great Britain). Comparable data for North America are not available. The mean exposure dose to cause a 10% inhibition of P_s was 0.45 ppmh (18.7 μmol m^{-3} h), and the range was 0.1 to 0.7 ppmh (4.2 to 29 μmol m^{-3} h).

The exposure doses associated with a threshold effect (0.25 ppmh, or 10 μmol m^{-3} h) and 10% inhibition (0.45 ppmh, or 19 μmol m^{-3} h) of P_s are an order of magnitude less than the comparable values for threshold concentrations that cause yield effects under field conditions (Fig. 13.1). Thus a single exposure dose sufficient to reduce photosynthesis by 10% will not necessarily result in a quantitatively comparable effect on yield. However, repeated low-concentration exposures during the course of a growing season may impair plant processes enough cumulatively to cause a yield reduction.

Although the significance of differences between species in governing the relationship between exposure dose and photosynthesis is evident in Fig. 13.6, a modifying role for environmental factors is not specifically considered in these data. The concentration threshold for effects may shift markedly as the edaphic, climatic, and atmospheric conditions change. For example, SO_2 concentrations of 0.04 ppm (1.66 μmol m^{-3}) significantly depressed P_s in *Vicia faba*, and the addition of ozone caused an even greater decline in CO_2 fixation (Ormrod, Black, and Unsworth, 1981). Experiments with *Phaseolus vulgaris* indicate that SO_2 concentrations representative of those monitored near point sources were nearly twice as inhibitory of P_s if they coincided with increases in O_3 concentrations (Taylor, unpublished data). The greater toxicity of SO_2 in the presence of O_3 has also been recorded under field conditions for yield responses in *G. max* (Reich, Amundson, and Lassoie, 1982) and *P. vulgaris* (Heggestad and Bennett, 1981). A second example is the observation that P_s is more responsive to SO_2 in low vapor-pressure-deficit regimes than in high vapor-pressure-deficit regimes (Barton, McLaughlin, and McConathy, 1980), which is explained by differences in pollutant uptake into the leaf interior (McLaughlin and Taylor, 1981). The SO_2 threshold concentrations required to produce effects are therefore likely to be lower in regions experiencing the combination of increased O_3 concentrations and low vapor-pressure deficit.

SO₂ uptake. The exposure dose (ppmh or μmol m^{-3} h) is a characterization of atmospheric pollutant conditions (concentration × time) in the turbulent air above the plant canopy, and may differ from that in the intercellular spaces of the leaf interior. The atmospheric exposure dose may not be an accurate measure of the dose to which the plant responds, and a characterization of *effective pollutant dose* (Runeckles, 1974; Taylor, McLaughlin, and Shriner, 1982) might provide a more reliable criterion for evaluating threshold concentrations for physiological and yield effects. A most pressing need is to evaluate the usefulness of *effective pollutant dose* as a basis for understanding differences in plant response as a function of edaphic, climatic, and atmospheric conditions.

The effective dose of SO_2 is defined as the cumulative amount of SO_2 absorbed per unit leaf area during the exposure period (units of mg of SO_2 m^{-2}), and its relationship to plant response shows some promising features (Fig. 13.7). In *V. faba*, photosynthesis is related to effective dose in a curvilinear fashion (Black and Unsworth, 1979b), and the regression depicts an SO_2 threshold level of 0.9 mg m^{-2}. In *P. vulgaris* (Taylor, unpublished data) and *G. max* (Sprugel and Miller, 1980), photosynthesis is also well-correlated with effective SO_2 dose (Fig. 13.7), showing threshold levels of 1.0 and 1.5 mg m^{-2}, respectively. Photosynthesis was inhibited as the mean effective dose exceeded 1.1 mg m^{-2}. The

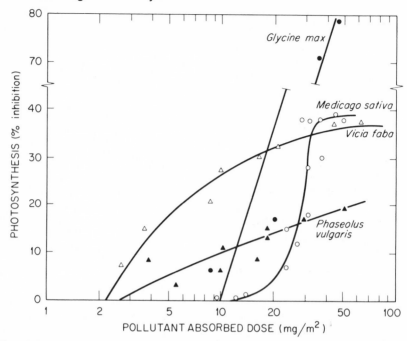

Fig. 13.7. Percentage inhibition of photosynthesis as a function of the amount of SO_2 that diffuses into the leaf interior (i.e., *effective SO_2 dose*); see text for description. The source of data for each species is as follows: *Glycine max* (Carlson, 1983a), *Vicia faba* (Black, 1982), *Phaseolus vulgaris* (Taylor, unpublished data), and native California species (Winner, Koch, and Mooney, 1982).

progressive inhibition of photosynthesis with increasing effective SO_2 dose was also observed for native California plant species (Fig. 13.7), although the extrapolation of the regression to a generalized no-effects threshold across all species is not justified. The data in Winner, Koch, and Mooney (1982) show several threshold responses for photosynthesis, each occurring at effective SO_2 doses exceeding 10 mg of SO_2 m^{-2}. This threshold is higher than the mean (1.1 mg of SO_2 m^{-2}) of the reported dicot species, and may reflect a generally lower susceptibility of xeric species.

Although the relationship between effective SO_2 dose and photosynthesis shows some promising features, the data are too incomplete to allow conclusive generalizations about SO_2 threshold concentrations for P_s response. Given the intensive research effort needed to provide such data, it is unlikely that a complete database will be developed soon. However, Fowler and Cape (1982) proposed an alternative inductive technique, assuming an analogy between leaf conductance to H_2O vapor and SO_2. This estimate, *pollutant absorbed dose*, is derived as follows:

pollutant absorbed dose = [exposure dose] × [H$_2$O conductance × 0.53]

$$= \left[\frac{\text{mg SO}_2}{\text{m}^3}\right] \times [\text{h}] \times \left[\frac{\text{m}}{\text{h}}\right]$$

$$= \text{mg SO}_2 \text{ m}^{-2}.$$

Conductance to H$_2$O is either measured experimentally (e.g., by porometry) or obtained from the existing literature (e.g., Körner, Scheel, and Bauer, 1979). As an example of this approach, we calculated a pollutant absorbed dose for the dicot crops in Fig. 13.6 based on the exposure doses for each study and leaf conductances to H$_2$O vapor in Körner, Scheel, and Bauer (1979). The mean pollutant absorbed dose for a threshold response was 7.5 mg m^{-2} and ranged from 2 to 12 mg m^{-2} (Fig. 13.8). This dose estimate is roughly 7.5 times higher than the previously derived dose ($\bar{x} = 1.1$ mg m^{-2}) for dicot crops. Thus, the two analy-

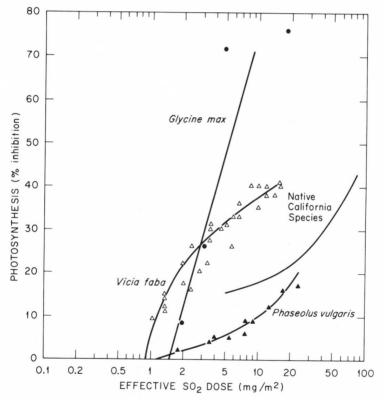

Fig. 13.8. Percentage inhibition of photosynthesis in dicotyledenous crop species as a function of the amount of SO$_2$ absorbed. This dose estimate is calculated by assuming an analogy between leaf conductance to H$_2$O and to SO$_2$ (Fowler and Cape, 1982). The dose estimates are derived from data in Fig. 13.6, and mean leaf conductances to H$_2$O for each species are as reported by Körner, Scheel, and Bauer (1979).

ses provide dissimilar estimates of threshold SO_2 doses, for reasons not fully resolved. One possible explanation is that the Körner, Scheel, and Bauer (1979) data may overestimate SO_2 uptake. An additional source of error may be the assumption of an analogy between leaf conductance to SO_2 and to H_2O vapor. The observation in *P. vulgaris* that SO_2 uptake is negatively influenced by increasing vapor-pressure deficit suggests that a close analogy between SO_2 and H_2O flux may not exist in all exposure conditions (McLaughlin and Taylor, 1981).

Carbon allocation. The allocation of photosynthate to plant growth is an integrative process controlled at many levels of physiological resolution. A wide variety of physiological processes determine both the total amount and the distribution of increases in plant biomass. Photosynthesis, respiration, and biomass distribution have been the most frequently used indicators of SO_2-induced stress on carbon allocation for plant growth processes. Another potentially useful indicator of incipient alteration of growth processes, and one that has not been widely used, is the measurement of altered rates and patterns of translocation of photosynthate. Evans (1975a) has stressed the importance of concurrent examination of translocation, photosynthesis, and respiration in studies of primary production of forest trees. The same can be said for higher plants in general; yet relatively few studies of air-pollution effects have focused on the critical allocation processes that link photosynthetic production to growth.

Several recent studies have indicated that the processes of carbohydrate translocation may be both susceptible to exposure to air pollutants and useful as general indicators of pollution-related stress. Noyes (1980) followed movement of ^{14}C products from $^{14}CO_2$-labeled bean leaves, and found loss rates (translocation and presumably respiration) to have been reduced 39, 44, and 66% by exposure to SO_2 at 0.1, 1.0, and 3.0 ppm (4.16, 41.6, and 124.8 μmol m^{-3}) for 3 h, respectively. Photosynthesis was unaffected at 0.1 ppm (4.16 μmol m^{-3}), and reduced 13 and 73% at SO_2 concentrations of 1.0 and 3.0 ppm (41.6 and 124.8 μmol m^{-3}) of SO_2. Teh and Swanson (1982) noted a 75% decrease in photosynthesis and a 45% decrease in translocation of bean (*Phaseolus vulgaris*) exposed to SO_2 for 2 h at 2.9 ppm (120.6 μmol m^{-3}).

Controlled field studies in Montana at the much lower SO_2 concentrations typical of field conditions (0.08 ppm, or 3.33 μmol m^{-3}) showed translocation of ^{14}C photosynthate of bluestem (*Agropyron smithii*) leaves to have been stimulated by 12% (Milchunas, Lauenroth, and Dodd, 1982). This stimulation occurred at a mean exposure level of 0.08 ppm (3.33 μmol m^{-3}) a value in the threshold range for beneficial SO_2 effects at this site. Jones and Mansfield (1982b), on the other hand, found both reduced root growth and ^{14}C translocation to roots of *Phleum pratense* exposed to 0.06 ppm (2.50 μmol m^{-3}) of SO_2 for six weeks. Field studies in the eastern United States by McLaughlin *et al.* (1982) have documented concurrent changes in photosynthesis, dark respiration, ^{14}C-photosynthate allocation, and growth of white pine trees chronically exposed to increased concentrations of ozone and other regionally transported pollutants. As in the study by Teh and Swanson (1982), photosynthate allocation appeared to be a more sensitive indicator of pollutant stress than was photosynthesis.

Recent studies in our laboratory have explored the usefulness of measuring retention of photosynthate by foliage of *Phaseolus vulgaris* following exposure to both SO$_2$ and O$_3$, and the relationship of altered translocation to allocation of photosynthate to biomass (McLaughlin and McConathy, 1984). In these studies the hypothesis that the maintenance requirements of foliage would be increased by air-pollution stress was tested. This hypothesis was based on previous studies that had indicated that foliage maintenance requirements (percentage of photosynthate retained in foliage) are rather high, $\geqslant 10$ to 15% of gross photosynthesis (McLaughlin and McConathy, 1979). Thus, foliar maintenance requirements were hypothesized to constitute a sensitive indicator of cellular damage to physiological systems of plant foliage. Changes in retention of photosynthate are known to be induced by foliar diseases, and appear to be related to increased sink strength of the affected leaves (McLaughlin and Shriner, 1980).

A series of laboratory exposures of two varieties of bush bean (*Phaseolus vulgaris*, Variety 274 and Variety 290) was conducted to measure the sensitivity of [^{14}C]-photosynthate allocation patterns to alteration by SO$_2$ and O$_3$. Experiments with the pollution-resistant 274 Variety demonstrated short-term changes in both ^{14}C and biomass allocation to roots of ^{14}CO$_2$-labeled plants, but no significant effect on yield by up to 40 h of exposure to SO$_2$ at 0.50 ppm (20.80 μmol m^{-3}) or 4 h of O$_3$ at 0.40 ppm (16.64 μmol m^{-3}). Subsequent experiments with the more sensitive 290 Variety demonstrated significant alteration of phytosynthesis, translocation, and partitioning of photosynthate between plant parts, including developing pods. Significant increases in foliar retention of photosynthate (+40%) occurred after eight hours of exposure to SO$_2$ at 0.75 ppm (31.20 μmol m^{-3}; 6.0 ppmh, or 250 μmol m^{-3} h) and 11 h of exposure to O$_3$ at 0.30 ppm (12.48 μmol m^{-3}; 3.3 ppmh, or 137 μmol m^{-3} h) (Fig. 13.9). Time-series sampling of labeled tissues after ^{14}CO$_2$ uptake showed that the disruption of translocation patterns persisted for at least one week after exposures ceased. Subsequent longer-term exposures at lower concentrations of both O$_3$ (0.0, 0.10, 0.15, and 0.20 ppm, or 4.16, 6.24, and 8.32 μmol m^{-3}) and SO$_2$ (0.0, 0.20, and 0.40 ppm, or 8.32 and 16.64 μmol m^{-3}) demonstrated that O$_3$ more effectively altered allocation than SO$_2$, that primary leaves were generally more sensitive than trifoliate leaves, and that responses of trifoliate leaves varied with plant growth stage. Altered rates of allocation of photosynthate to leaves were generally associated with alterations of similar magnitude and opposite direction in developing pods. Collectively, these experiments suggest that allocation patterns can provide sensitive indices of incipient growth responses of pollution-stressed vegetation, and that these patterns may prove useful for documenting threshold doses for plants exposed to SO$_2$ under field conditions. The changes in allocation of carbon to root systems noted in our experiments and those of Jones and

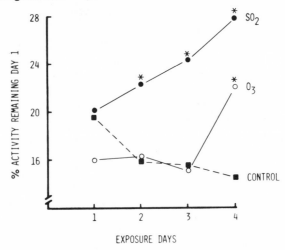

Fig. 13.9. Effects of 2 h/d SO_2 (0.75 ppm, or 31.20 μmol m^{-3}) or O_3 (0.30 ppm, or 12.48 μmol m^{-3}) on foliar retention of [^{14}C]-photosynthate by *P. vulgaris* (CV 290) after 1, 2, 3, and 4 d of exposure. An asterisk indicates a significant difference from the corresponding control value at $\geq 95\%$ confidence.

Mansfield (1982b) are particularly interesting, since they suggest that altered root : shoot ratios and altered root vigor may occur in response to SO_2 exposure. Such changes may be inconsequential where nutrients or water are nonlimiting; however, in other situations, secondary growth effects may occur. Thus, whole-plant responses, including gas exchange and partitioning of carbon, may ultimately be required to adequately understand how and when plant growth will be affected by SO_2.

Summary

Numerous indices of plant response to SO_2 are offered in the literature, and range in organizational level from subcellular processes to whole-plant physiology. We have examined some of these indices from the standpoint of describing dose-response relationships and estimating threshold doses for plant responses in dicot crop species, and the relevant limited data for various species or groups of species are summarized in Table 13.4. In general, physiological processes, such as photosyntheses and carbon allocation, are most sensitive to SO_2 exposure doses (≥ 0.05 ppmh, or 2.1 μmol m^{-3} h) followed in order by visible injury (0.25 to 0.50 ppmh, or 10.4 to 20.8 μmol m^{-3} h) and yield responses (0.7 to 6.5 ppmh, or 29 to 270 μmol m^{-3} h). There was good agreement between plant responses and the log-transformation of SO_2 dose in ppmh (or μmol m^{-3} h) for a variety of plant indices; however, significant differences in dose-response relationships generally occurred between different species or the same species in different years. The major exception was the relationship between dose and yield response in soybeans and snap beans. The most extensive data set available on SO_2 response in dicot crops existed for beans, and included studies from two varieties of soybeans exposed to SO_2 using three different exposure methodologies.

The use of exposure dose (as a product of concentration and time in general and the log transformation of this parameter specifically) to predict yield loss appears counterintuitive, because it is well-recognized that high concentrations for short time intervals can produce more visible injury (Zahn, 1970) or physiological change (McLaughlin *et al.*, 1979) than the same dose from lower concentrations for a longer time interval. Yet the available data on dicot crops demonstrated that dose (ppmh or μmol m^{-3} h) was a reliable parameter for predicting yield for a wide variety of field studies, including the Biersdorf experiments, where peak:mean ratios were high and differed greatly between stations. The question then becomes one of finding out how closely laboratory studies of visible leaf injury or physiological response describe yield losses of field-grown plants under seasonal exposure conditions.

Yield responses of field-grown crops to SO_2 are a consequence of numerous physiological changes integrated over time; yet not all plant responses to increased concentrations of SO_2 are negative, for there must be some concentration threshold that must be exceeded with some frequency or to some accumulated amount before yield loss ensues. Perhaps constant-SO_2-addition field experiments or studies near prominent point sources do not have a sufficient proportion of subthreshold concentrations to significantly "dilute" the total dose. However, chronic regional-scale exposures may consist principally of subthreshold SO_2 concentrations, and total dose under these conditions will have little meaning unless this nontoxic component is excluded from total dose calculations. Estimating such a threshold is further complicated in many rural areas by the occurrence of potentially more toxic concurrent concentrations of ozone.

The resolution of these issues will not be easy, but it may prove useful to combine established physiological methodologies (which have been restricted primarily to the laboratory) with the realism afforded by evolving field-exposure techniques. Systematic field studies such as the National Crop Loss Assessment Network (Heck and Heagle, this volume) offer the possibility of obtaining the crucial endpoint of yield in concert with a mechanistic basis for yield responses. To obtain this product, it is necessary to develop techniques of concurrently tracking exposure concentration and short-to-intermediate-term plant response. Measurements of photosyntheses, water and nutrient relations, and carbon allocation can be valuable tools in this effort.

There can be no universal dose-response relationship to characterize the SO_2 response of all plant species under all environmental conditions. As a consequence, the challenge becomes one not only of finding out what the final response was, but of defining its qualitative and quantitative physiological basis. By combining existing analytical capabilities with carefully planned experiments, we

TABLE 13.4

Summary of Estimates of Threshold SO_2 Exposure Dose and SO_2 Concentrations Associated with Responses of Dicot Crops

Response: and species	Dose (ppmh)[a]	Response threshold Concentration (ppm/time)[b]	Comments
Yield reduction			
Bean	3.5	0.06/144 h	Combined data, soybean and snap beans, SO_2 added to near-ambient O_3. Multiple methodologies.
Soybean	1.2	0.04/66 h	SO_2 added to low O_3 during flowering. Linear gradient.
Potato (1959)	0.7	0.15/geometric $\bar{x}$ }[a] 0.01/arithmetic $\bar{x}$ }[a]	Container plots near iron smelter. "a" produced 4% yield loss; not statistically significant. "b" resulted in 17% yield loss, which was significant. "a" and "b" from 1959 data.
Potato (1960)	6.5	0.24/geometric $\bar{x}$ }[b] 0.05/arithmetic $\bar{x}$ }[b]	
Foliar injury			
Crops	0.5–0.9	1.5/peak 0.5/1 h 0.3/3 h	Combined data for eight crop species from field surveys near a coal-fired power plant.
Photosynthesis			
Pea	0.05	0.02/2 h	Whole plant, vegetative stage.
Bush bean	0.12	0.15/1 h	Whole plant, vegetative stage.
Soybean	0.40	0.18/2 h	Whole plant, vegetative stage.
Alfalfa	0.30	0.16/2 h	Whole plant, vegetative stage.
Carbon allocation			
Bush bean	1.5–3.0	0.75/2–4 h	Whole-plant, time-series, reduced foliar retention of fixed $^{14}CO_2$. Variety 290, five weeks old.
Bush bean	≤0.2	0.10/2 h	Eleven-day-old seedlings of Variety Black Valentine. Single leaves still expanding; ^{14}C flux followed by geiger tube.

[a] Dose threshold for yield responses has been estimated from the no-effects intercept of the dose-response regression line. 1 ppmh dose can be converted to SI units where 1 ppmh = 42.0 μmol m^{-3}, as explained in the Note on Units at the beginning of this volume.

[b] Actual minimum SO_2 concentration and time associated with a response.

can achieve a more complete description of plant response to SO$_2$. The end product should provide a basis for extending the existing data to describe and predict potential plant responses to SO$_2$ under a broader range of exposure conditions.

Acknowledgment

Our research was sponsored by the U.S. Environmental Protection Agency under interagency agreement 40-740-78 and by the U.S. Department of Energy under contract W-7405-eng-26 with Union Carbide Corporation. Publication 1629, Environmental Sciences Division, Oak Ridge National Laboratory. This manuscript has not been internally reviewed by the Environmental Protection Agency and does not necessarily reflect the views of the agency, and no official endorsement should be inferred.

14

SO$_2$ Effects on Tree Growth

Theodor Keller

This paper focuses on the effects of SO$_2$ on forest trees and is not intended to survey the literature on this subject; for a literature survey, see Legge (1980) or Smith (1981). Because forests cover a large proportion of the Earth's surface even in the industrialized northern hemisphere, I begin with a general survey of forest functions, and then consider why forest trees are particularly sensitive to air pollutants such as SO$_2$. Since SO$_2$ affects forest growth, growth may be an important biological indicator of pollution effects. In this chapter, I present data on SO$_2$ effects on shoot growth, differentiating between the use of seedlings and the use of older branches of mature trees that can be cloned. I then present data on SO$_2$ effects on root growth, argue that the quality of tree growth should not be neglected for the sake of quantitative measurements, and close with some remarks on future research needs.

Forest Functions and Sensitivity to SO$_2$

Forests provide more than lumber and firewood. Additional functions include protecting the landscape from erosion, maintaining quality supplies of drinking water, providing recreational areas for people and habitats for wildlife, and playing an active role in the biogeochemical cycling of H$_2$O, CO$_2$, and elements such as N and S. Many of these functions cannot be valued in economic terms but can be influenced by SO$_2$-caused changes in forest productivity. Another forest function that is related to air-pollution sensitivity is the filtering of air via the uptake of pollutants by forest canopies. Tree species differ in their capacity to purify air, and this may be related to their sensitivity to air pollution. I would like to outline how conifers, deciduous trees, and crops compare in air-pollution tolerance and identify characteristics of trees that help explain how they might differ in their capacity to absorb SO$_2$.

The Sensitivity of Trees to SO$_2$

When we consider tree sensitivity, two questions immediately arise. (1) what should be investigated: foliage alone or the whole plant? (2) Why are trees so sensitive to SO$_2$?

Perhaps the first modern research in air-pollution effects on plants was carried out at the Forest Academy of Saxony at Tharandt, where Schroeder (1873) pointed out in short-term studies that even though conifer needles are more resistant to SO$_2$ than deciduous foliage or herbs, coniferous trees may be more sensitive because of their needle longevity. In general, deciduous trees completely renew their photosynthesizing apparatus each year, whereas conifers keep their needles for several years. Thus in fumigations with SO$_2$, deciduous trees may quickly show visible symptoms of injury, shed their affected foliage, form new leaves, and recuperate when brought to pure air. Conifers sometimes do not exhibit any visible symptoms of injury for a long time, until they suddenly drop their needles. O'Gara (1922, quoted from Thomas, 1964) disregarded this fact, studied responses of needles to air pollution, and found pines to be seven to fifteen times more resistant than alfalfa. Nevertheless, pines have long disappeared from the forests in the Ruhr area where alfalfa still grows well. The following points may help to explain aspects of tree physiology, morphology, and ecology that are related to their capacity to absorb air pollution.

1. *Crown volume*. For each unit of soil surface, tree crowns usually filter a much larger volume of air than do low-growing herbs. Thus the uptake of pollutants from air must be relatively high for trees. This includes adsorption of pollutants to leaf exteriors as well as absorption of pollutants into leaf mesophyll tissues. Adsorbed pollutants reside on leaf surfaces until they are washed into the understory by rain. Although adsorbed pollutants are thought to have little direct effect on tree physiology, absorbed pollutants may alter stomatal conductance, reduce photosynthesis, and have injurious effects on other metabolic processes and organelles. This may be illustrated by the leaf area index (foliage leaf area/ground area), which ranges from 3.7 to 7.4 for mountain beech forests, but is 4.8 for adjacent grasslands, according to Schoenenberger (personal communication). Many more examples are given by Larcher (1980, p. 151).

2. *Layers with increased wind speed*. High tree crowns usually extend into air layers where there is a greater wind speed than at ground level. Fig. 14.1, using data from the extensive work done by Geiger (1950, p. 329), clearly shows the increase of wind velocity with height above ground in a 65-year-old pine forest (upper curve). A very similar increase of air-polluting fluorides accumulating in spruce needles (μg F in 1 g needles, lower curve) was found by Knabe (1968). We may speculate that increased wind speed in tree canopies results in increased uptake of pollutants, including SO$_2$. In fact, studies of grasses by Ashenden, Mansfield, and Wellburn (1978) are relevant for trees: 4-week SO$_2$ fumigation with 4.8 μmol m^{-3} did not cause a growth depression at a wind speed of 0.6 km h^{-1} (0.16 m s^{-1}), but did depress growth at a wind speed of 1.5 km h^{-1} (0.42 m s^{-1}). These authors pointed out that this difference of wind velocity greatly affects the leaf boundary layer, and with it gas diffusion through stomata.

3. *Longevity and seasonal variation*. Whereas herbaceous crops normally have a short life, trees survive for many years, and may be exposed to many episodes of air pollution. In addition, evergreens may also suffer exposures during winter time, when SO$_2$ levels are increased over large areas. Winter is known

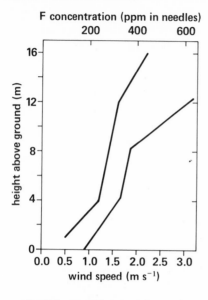

Fig. 14.1. Wind speed and F content in needles of *Picea abies* in relation to height above ground (tree height). Wind speed data from Geiger (1950), F contents from Knabe (1968).

as the "dormant" season for vegetation, during which it is considered to be more resistant to pollutants. It has been shown, however, that plant damage is sometimes most severe during winter (Davies, 1980a).

Needles may live for several years, although their physiological activity may decline with age, as is known, e.g., for CO_2 uptake. Experimental evidence indicates that the capacity of plants to assimilate absorbed SO_2 also decreases with age. It has been observed in several fumigations that older needles were affected by lower concentrations than young needles. Seasonal variation in sensitivity to SO_2 (see also Fig. 14.2) is an important factor for all perennials, including trees (Van Haut and Stratmann, 1970). The physiological mechanisms linking maturation and seasons to changing SO_2 sensitivity are poorly understood.

4. *Morphology and anatomy.* Within one and the same tree, these aspects of foliage differ, as does the light regime. There are thin shade leaves with thin epidermis and cuticle, and thick, leathery sun leaves, which are relatively well-protected against solar radiation and air pollutants. Their thick epidermis, however, contains more stomata; therefore they may take in more CO_2 than do shade leaves. However, shade leaves are less resistant to pollution; because they receive less light energy, their physiology (including detoxification) is more sluggish or less efficient. There are also significant differences in the morphology and anatomy of resistant versus susceptible spruce clones (Braun, 1977c). Morphological aspects are discussed by Kärenlampi and Soikkeli (1980) and by Winner and Mooney (1980b).

5. *Genetic variability.* In the forest each tree has its individual genetic make-up; thus a forest is genetically more heterogeneous than a field of crops, and there are wide differences in SO_2 sensitivity within species. These differences may be

genetically controlled (Heck and Brandt, 1977; Ormrod, 1983). Resistance to air pollution may be related to drought resistance (Braun, 1977e). Recently, the great differences in sensitivity in conifer populations have been pointed out once again by Garsed and Rutter (1982b). Because a forest is a complex ecosystem, differences among species in sensitivity to SO$_2$ must be evaluated.

6. *Supply of water and nutrients.* This factor is known to influence pollution resistance. Each species is thought to have highest resistance in its ecologically optimal niche. Over the centuries, however, the human population has restricted forests largely to areas that are unsuitable for agriculture or where improvements are often prevented by financial considerations. In addition, fertilization with macronutrients may cause a deficiency of micronutrients or water, and thus create new difficulties, although in several instances fertilization improves forest growth in polluted areas (Dässler, 1976; Olszowski, 1980).

The Importance of Biological Indications

By biological indications, I mean the physiological, biochemical, or ecological reactions of a plant to outside influences. These reactions may be related to the plant's susceptibility to abiotic (e.g., frost) or biotic (e.g., insects) influences. They may be short-lived and reversible (and show increased risk), or may lead to irreversible changes causing visible symptoms of injury and changes in growth. Visible foliar injury may not be associated with growth reductions, but indicates that the tree has been forced to cope in a physiological or biochemical way with the pollutant. Chemical analysis of foliage is used to indicate the presence of pollutants, but this type of analysis is of little value for estimating plant responses to air pollution.

Because growth reductions occur in the absence of visible injury, research should be directed toward evaluating the extent of growth responses. The ease with which plant tissues develop visible symptoms is often used to classify a plant's sensitivity or resistance toward air pollutants. It is often overlooked that the foliage of trees, particularly of evergreens, has an anatomy that enables it to remain active for months or even years. It may endure climatically caused unfavorable situations without showing visible harm, although any stress may induce physiological or biochemical reactions. Therefore such biological indications and latent injury are of special importance in forestry. Evergreen species are particularly vulnerable to SO$_2$, and may show reactions long before (or even without) the appearance of visible symptoms. Under controlled conditions in the laboratory, physiological SO$_2$ effects can be detected very early with modern equipment. Such responses can be considered early warnings of a tree's decline.

Visible symptoms of injury are usually poor indicators of growth reductions: low SO$_2$ concentrations of long duration may not harm a tree visibly, but may impede its growth and finally cause its death. On the other hand, a short peak of SO$_2$ exposure may visibly "burn" a certain percentage of a tree's leaf area without affecting its growth.

When tree growth is used as a biological indication, one of the following is usually measured: (1) weight gain in foliage or other relevant plant parts, (2) foliage area, (3) shoot length, and (4) annual ring width. As a rule, only the first three are measured in the laboratory, because they are convenient when working with relatively small plants. Unfortunately, they do not necessarily have any direct relationship to the harvested product. In the laboratory, on the other hand, the fumigation regime can be controlled, and SO_2 concentration and duration usually are well-known. Consequently we get a clear-cut description of the effects of SO_2 on the particular growth parameter.

Annual ring width can be measured only on older trees. Ring measurement techniques have therefore been used very rarely under laboratory conditions, whereas they have been used in the field for a hundred years (Haselhoff and Lindau, 1903, pp. 105ff) and in many countries. When a complete cross section of the stem is not available, it is very convenient to use a borer that extracts an increment core on which the ring widths of past years can be measured. Care must be taken, however, to take the core horizontally, directed exactly toward the center of the tree, and well above the root collar. Because of reaction wood, small trees are usually not suitable for measurements of wood density.

Ring width gives a historical record that is relevant to the practical economic needs of today, because it shows wood production. This accounts for its wide application. For scientific purposes, however, it has the severe drawback that we usually do not know the fumigation regime in the forest that has led to this result, although we may know the emitting source. In addition, tree-ring width is affected by many other factors (Larcher, 1980) beyond just air pollution and may not be related to forest productivity of an entire forest stand.

Resistance to a Pollution Regime

In investigations of the resistance of a plant, there are two points of particular importance: (1) the relative resistance or sensitivity of the species and (2) the pollutant concentration or dose.

The relative resistance or sensitivity of a species is a matter of both practical importance and wide speculation. It is well known that there are great differences between species in sensitivity. The literature contains many attempts to list species with different susceptibilities to SO_2. Such lists are often derived from field observations, and must be regarded with caution. They may be of local value only, and usually neglect the fact that resistance is also connected to ecological conditions. In addition, local pollution history is usually unknown.

If, however, such a list is derived from screening tests based on a short fumigation, then pollutant concentration becomes important. Thus Garsed (1981) found ''a lack of relationship between the sensitivity of plants to acute injury at high SO_2 concentrations and the longer-term effects of low concentrations of SO_2 on growth.'' Later Garsed and Rutter (1982b) investigated the effect of 3.12 and 124.8 μmol m^{-3} of SO_2 on conifer populations. They obtained different orders of sensitivity after 35 days and after 67 days, and stated that the order obtained at 8,000 μg m^{-3} (125 μmol m^{-3}) was virtually the reverse of that at 200 μg m^{-3} (3.13 μmol m^{-3}). They concluded that the relative sensitivity depends almost entirely on the concentration and duration of exposure, and that ''short-term fumigation at high SO$_2$ concentrations cannot be used to predict responses to long-term exposure to SO$_2$ in the field.''

This leads to the second point, the question of pollutant concentration or dose. In nature the concentration of a pollutant usually changes constantly, particularly near point sources. In controlled fumigations, however, the concentration may remain constant, or may change occasionally or frequently. Constant-concentration fumigations have often been questioned as unnatural, although at low concentrations they may simulate an elevated background concentration, such as may occur in polluted areas with many pollution sources.

On the other hand, fluctuations may involve unnaturally high peak concentrations, as well as periods with additional pollutants (e.g., ozone) brought by the wind in unknown concentrations.

A change of concentration may allow detoxification of absorbed pollutants and regeneration of ''normal'' metabolic states, particularly when periods with pure air are included. Such changes were used by Garsed, Mueller, and Rutter (1982) in order to measure the effect of given peak concentrations. These authors fumigated pine seedlings (initially 3 years old) for 650 days with two peak concentrations (4.68 and 11.7 μmol m^{-3}) or with a constant lower concentration (0.036 ppm or 1.50 μmol m^{-3}). The peak concentrations lasted either 5 or 21 hours, and were applied at intervals of 1 or 22 days, respectively. The effect on growth, expressed as excess of dry-weight gain over that of the control during fumigation, is shown in Table 14.1, which indicates that the short but frequent peaks (SF), no matter which of the two concentrations was used, depressed growth about as much as the constant, continuous fumigation.

At either concentration, and at the same mean exposure, long occasional peaks (LO) were more detrimental to plant growth than either a constant fumigation or

TABLE 14.1
Effect of Different SO$_2$ Fumigations on Dry-Weight Gain (over Controls) of *Pinus sylvestris* Seedlings in 650 Days

Treatment	SO$_2$ concentration (μmol m^{-3})		Dry weight gain (g)
	Mean	Max.	
clean air	0.0	0.0	118.5
constant	1.46	1.46	101.9
SF[a]	1.46	4.36	101.8
SF[a]	1.46	10.9	100.9
LO[b]	1.46	4.36	92.8
LO[b]	1.46	10.9	91.2

SOURCE: Adapted from Garsed, Mueller, and Rutter (1982).
[a] SF: short, frequent peaks (5 hours with 1-d intervals).
[b] LO: long, occasional peaks (21 hours with 22-d intervals).

short, frequent peaks (SF). They depressed growth by an additional 10%. This seems to be of particular importance, because large tracts of forested land are exposed to variable SO_2 doses. Tree death in rural areas far from industry has recently been observed. This is thought to be caused by SO_2 transported from industrial areas over long distances by wind, and such transport is thought to be favored by high chimney stacks (Wentzel, 1982). In addition to long-distance transport of air pollutants, conditions of stagnation may have effects resembling those of the LO conditions of Table 14.1. More needs to be learned about the concentrations and duration of SO_2 in field sites, especially where forest decline is occurring.

Effects on Shoot Growth

Shoot growth in controlled fumigations can be assessed by means of seedlings, or by means of clonal material (cuttings, grafts, etc.). From the standpoint of experimental design, both approaches have advantages and disadvantages.

The use of seedlings. Large trees cannot be handled in a laboratory; so very young seedlings (shortly after germination) are sometimes used. Whereas older seedlings more closely resemble a grown tree, the very young seedlings have the following advantages (Constantinidou, Kozlowski, and Jensen, 1976): (1) they need little space, (2) they are highly susceptible, (3) they allow early detection of adverse effects, and (4) experimental results are easily related to natural regeneration.

Natural regeneration is quite important in forestry as an alternative to planting and for the maintenance of genetic diversity.

Growth effects in such small plants are usually measured in terms of increase in dry weight or development of foliage. Constantinidou and his co-workers found that seedling development as measured by dry-weight increase was inhibited even by short fumigations (15 to 120 minutes) of 0.5 to 4 ppm (21 to 166 μmol m^{-3}) of SO_2. Constantinidou and Kozlowski (1979a) detected a slowing down of leaf expansion and a reduction in the number of emerging leaves of elm seedlings after a fumigation of 6 hours at 2 ppm (83.2 μmol m^{-3}) of SO_2.

It was also shown that environmental conditions strongly influence seedling sensitivity; an increase in temperature after fumigation (Norby and Kozlowski, 1981a) or in humidity during fumigation (Norby and Kozlowski, 1982b) would increase sensitivity. The findings of Jones and Mansfield (1982) on grass confirm the observation that light may be an important factor in sensitivity (low light in winter increases sensitivity of grass).

The use of clonal material. The use of genetically uniform material limits variability; hence statistical evaluation may need fewer replicates for significance. On the other hand, we must keep in mind—especially when extrapolating results to a forest—that each tree has its own genetic makeup. Furthermore, we have to distinguish between cuttings and grafts, since the latter may have either a foreign rootstock or roots of their own. Only cuttings and grafts with their own roots are genetically homogeneous. Whereas cuttings can usually be made only from relatively young trees, grafts allow us to study SO_2 effects on old trees, with their different needle morphology. In either case, trees several feet high are normally used, and diameter increment is more often measured than dry weight. Nevertheless, increase in height may also be investigated, particularly for broadleaf cuttings, e.g., poplar (Biggs and Davis, 1981a).

CO_2 uptake varies widely in the course of a year, and Fig. 14.2 shows the

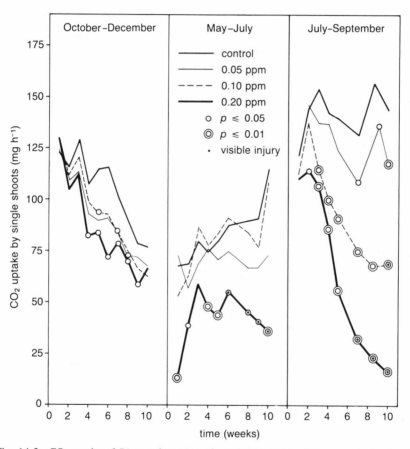

Fig. 14.2. CO_2 uptake of *Pinus sylvestris* grafts as influenced by SO_2 concentrations and seasons. Fumigations were continuous for 10-week periods. The grafts were 3 years old (the twigs were taken from a 30-year-old mother tree). Each point is the mean of five grafts. Care was taken to use grafts closely similar to each other to ensure comparability. They were about 60 cm long and contained 1- and 2-year-old needles. All differences from controls were checked for statistical significance by the u-test of Wilcoxon *et al.* Significance is denoted where given by this range test; 0.10 ppm of SO_2 is about 4.16 μmol m^{-3}.

effect of different continuous fumigations on CO_2 uptake of a Scots pine clone in three seasons. It shows distinctly that this physiological activity is significantly influenced long before symptoms of visible injury occur in perennial tissue. Since CO_2 uptake is the prerequisite for any organic growth, it is often measured in addition to other more obvious growth parameters.

Fig. 14.3 shows the correlation between relative CO_2 uptake and subsequent growth, expressed by comparing the number of cell rows in a radial file of the annual ring of *Picea abies*. Both parameters were related to their controls (control = 100%). With low SO_2 concentrations, it usually takes some time before CO_2 uptake is affected; therefore in the three spruce clones of Fig. 14.3, the CO_2 uptake of the fifth to seventh week of fumigation (June) was compared with the number of cell rows formed in the stem. Fig. 14.3 clearly illustrates an expected relationship between CO_2 uptake and number of cells, which determines ring width. Although in one clone the relative CO_2 uptake exceeded 100% (i.e., was greater than that of controls), the formation of cell rows was inhibited, because CO_2 is taken up for many processes, only some of which lead to wood formation.

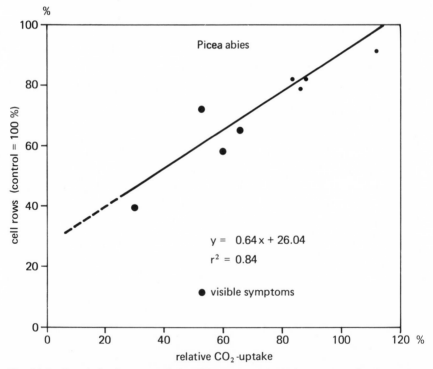

Fig. 14.3. Correlation between relative CO_2 uptake and relative number of cell rows in a radial file of *Picea abies*. Each value is the mean of five replicates. Three different clones were used. Reproduced from Keller (1980), p. 3.

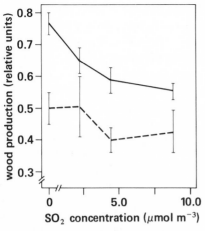

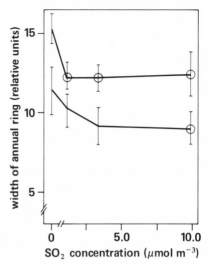

Fig. 14.4. Estimated relative wood production of a *Picea abies* clone after SO₂ fumigation in summer (solid line) and winter (dashed line). Continuous 10-week fumigations. Values are means of five replicates; vertical bars denote the standard error. Production is estimated from widths and densities of early wood and late wood as revealed by x-ray analysis of tree rings.

Fig. 14.5. Widths of the annual ring of two *Picea abies* clones fumigated continuously for 12 weeks during dormancy (October to end of December). Each value is the mean of nine replicates (± standard error). Circled values are statistically different from controls.

Visible symptoms of injury at the end of the seventh week appeared only where higher SO₂ concentrations were applied (4.16 μmol m⁻³ or more).

It is quite obvious that a decrease in ring width, especially when wood density is also affected, causes a decrease of wood production (fiber material), as shown for *P. abies* in Fig. 14.4. Ring width and wood production of the trees subjected to the ''winter'' fumigation (October-December) were calculated for the growing season after fumigation. The two parameters (ring width and wood production) investigated for this clone, however, showed no definite trend (only a slight tendency to decrease) when fumigation occurred during dormancy, but growth was appreciably depressed by fumigation during the growing season without the appearance of visible symptoms of injury.

Over large areas of countryside, higher SO₂ concentrations prevail during wintertime, however, than during the growing season. Winter is the dormant season, and vegetation is then supposedly less susceptible to SO₂. Therefore two other spruce clones were fumigated during three months of dormancy. Ring widths of the following growing season are illustrated in Fig. 14.5, which shows that the ring width of these clones was influenced even when the fumigation occurred during the dormant season. This effect is thought to be caused by a strong inhibition of CO_2 uptake during dormancy itself (Keller, 1983), which decreases the production of storage material needed for growth of early wood. Glerum and Balatinecz (1980), on the other hand, concluded in a recent study with *Pinus banksiana* that food reserves do not play a direct role in wood formation.

Effects on Root Growth

The periodicity of root growth in older forest trees and the fact that roots are hidden by the soil often make the investigation of root growth difficult. In herbs, however, it has been found that SO_2 depresses root growth as much as, or even more than, shoot growth (Dässler, 1976; Heck and Dunning, 1978). When root growth in fumigated (June 8 to November 18) seedlings of *Picea abies* was investigated in the subsequent year, these conifers exhibited a carryover effect. Root growth evidently reacted more strongly than shoot growth, just as in herbs: the fumigation with 0.1 ppm (4.16 μmol m^{-3}) of SO_2 did not cause any visible symptoms of injury to the shoot, but depressed subsequent root growth by 50% (Keller, 1979). Fig. 14.6 illustrates this aftereffect on root growth in two spruce provenances.

Similarly, Suwannapinunt and Kozlowski (1980) found in seedlings of two broadleaf species that the inhibition of dry-weight increment was a more sensitive indicator of susceptibility in roots than in shoots. Even a short SO_2 peak affected root dry weight of elm seedlings (Constantinidou and Kozlowski, 1979a) or of red pine seedlings (Norby and Kozlowski, 1981c). Likewise, Jensen (1981b) detected in fumigated poplar cuttings that available photosynthate was used more

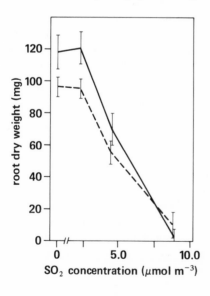

Fig. 14.6. The effect of SO_2 fumigations on root growth in the following year for *Picea abies* from Switzerland (solid line) and Czechoslovakia (dashed line). Values are the sum of the five longest roots of three seedlings each. Each value is the mean of five replicates ($\pm$ standard error as vertical bars).

for shoot growth than for root growth. On the other hand, the investigation by Garsed, Rutter, and Relton (1981) indicated that increment losses in Scots pine seedling roots were smaller than those in the shoots.

Quality Aspects

For a forester, growth is expressed not only quantitatively, by wood production, but also qualitatively. Thus it was found, after a wintertime fumigation of dormant beech (*Fagus sylvatica*), that many terminal buds failed to break in the following spring. This effect may lead to misshapen, bushy stems, whereas silvicultural practice aims at producing straight stems with wood of high quality (Keller, 1978a). In other beeches, it was observed that long-term fumigations not only depressed ring width, but also made the young trees more vulnerable to being bent by heavy snow (Keller and Beda-Puta, 1981). Such stems should be removed, with subsequent reductions in forest growth and possibly in ecosystem composition.

Although this overview deals mainly with quantitative growth manifestations,

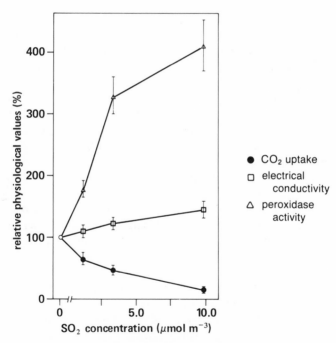

Fig. 14.7. The effect of 6-month continuous SO$_2$ fumigations during dormancy on CO$_2$ uptake, electrical conductivity, and peroxidase activity of needle diffusates in two clones of *Picea abies*. Values are relative to control values obtained from plants in SO$_2$-free air and are the means ($\pm$ standard error) of ten replicates.

we should bear in mind that growth is also related to the physiological status of the tree and that this status may be expressed in other ways. Fig. 14.7 shows three ecophysiological parameters in spruce needles affected by three continuous 6-month SO_2 fumigations during dormancy (October to April). CO_2 uptake, which, as we have seen, is related to wood production, was distinctly inhibited. At the same time, the activity of the detoxifying enzyme peroxidase was increased greatly. This enzyme is considered a valuable tool for screening sensitivity (Eckert and Houston, 1982). Shuwen and co-workers (1982) conclude that peroxidation of membrane lipids might be caused by oxidation of sulfite to sulfate. Sulfite is known to alter membrane function by attacking proteins, and Peiser, Lizada, and Yang (1982) point out that sulfite can cause an alteration of membranes by peroxidation. Since peroxidase activity may be connected to peroxidation, its increase might indicate a peroxidation of membrane lipids. Such a destruction would explain why membrane permeability, expressed as electrical conductivity, might be a measure of plant vitality related to growth.

Research Needs

1. *Experimental facilities and selection of experimental plants.* A prerequisite for research on SO_2 effects on trees is a fumigation system that allows for frequent checking of the air-pollution concentrations to which trees are exposed. Such fumigations should simulate ambient, naturally occurring air-pollution concentrations and environmental conditions as nearly as possible. The equipment must provide adequate change of air, little change in lighting, and allow a long-term application of low concentration of pollutants with occasional higher peaks. The ideal fumigation installation for trees should allow mixing of several pollutants and long exposure periods (months to years). Unfortunately, such a system has not yet been developed.

Scientists with easy access to experimental facilities with controlled environments, especially growth chambers with environmental controls, should use a greater variety of common tree species for studying interrelationships between SO_2 and environmental factors (soils, including mycorrhizae; water, light, temperature, etc.). We must also take care to use pots that are large enough to guarantee good rooting of trees.

In research with trees, progress has been hampered because genetically uniform (standardized) material has not been generally available. Different researchers use different breeds (even when working with the same species) and so their results cannot be compared. The simultaneous measurement of many different parameters is necessary, especially when other features in addition to growth are studied; so uniform stock needs to be made available for several species with a wide ecological range. It should include: *Pinus strobus*, *P. ponderosa*, *P. taeda*, or *P. echinata*; *Picea abies*; *Pseudotsuga menziesii*; *Juniperus virginiana*; *Acer saccharum*, *A. saccharinum*, or *A. rubrum*; *Fraxinus americana* or *F. pennsyl-*

vanica; *Ulmus americana*; *Prunus serotina*; *Populus* sp., because they are economically and ecologically important species.

2. *Biological indications of air pollution*. We need to know more about biological indications of air pollution and how they are related to tree growth and physiology. As examples, we should learn more about the effects of SO_2 on CO_2 uptake and subsequent metabolism to wood, pigment, and other cellular products. Such physiological processes are important for an early warning of increased susceptibility. In addition, we do not know enough about SO_2 effects on hormones that affect growth. Biological indications of this type may allow shorter experiments if they are sensitive to air-pollution exposures.

We have to bear in mind, however, that such biological indications, just like growth, are not necessarily specific for SO_2. They require a careful study of what environmental factors affect them. We must avoid increasing the risk of damage to trees by air pollution, because trees are already subject to many natural risks that may sap their vitality. We also do not know how SO_2 might increase their susceptibility to insects and pathogenic fungi, decrease their resistance to frost or drought, etc. These influences affect growth as well as competition and stand composition. We also need to consider the effect of SO_2 on tree crop quality; e.g., SO_2 is known to decrease vitamin-C content.

3. *Forest decline*. Forest decline observed in Europe and reported in Scandinavia and North America may be caused, at least in part, by SO_2. This pollutant may suppress growth, and reduce forest-tree vigor, without causing visible injury. In addition, SO_2 and its oxidation products contribute to rainfall acidity. Wet deposition of sulfur oxides has the potential to alter soil chemistry, litter decomposition rates, and leaf-surface morphology. Much more field research will be required to show how the wet and dry deposition of sulfur oxides can influence forest biology. This issue may emerge as one of the most important research topics for foresters during the next decade.

15

SO$_2$ Effects on the Growth of Native Plants

W. E. Westman, K. P. Preston, and L. B. Weeks

Given the relative abundance of information on SO$_2$ effects on cultivated plants, it seems germane to ask whether effects on native species differ in any qualitative way from those on cultivated species. We review here attributes of potential relevance to pollution sensitivity that differ in native species from those of crop plants as a class.

Firstly, owing to lack of breeding, there is greater genetic variability in a population of native plants, providing greater opportunity for rapid selection of resistant strains *in situ*. Secondly, whereas cultivated plants are grown in a managed environment, relatively free of environmental stressors, native plants are grown in the midst of growth stressors—drought, soil infertility, shading, herbivory, disease—that are likely to combine with air-pollution stress in additive or synergistic ways. Thirdly, whereas any growth inhibitions are likely to affect neighboring plants similarly in a monoculture, any growth inhibition in native species will immediately release resources for use by interspecific competitors, further aggravating the growth depression by enhancing the dominance of neighbors. Fourthly, whereas the cultivation of crop species typically involves harvest of only one plant organ, effects to nonharvested organs of native species and to the microorganisms of the ecosystem with which they interact become vital to the continuing health of the population, and to the functioning of the ecosystem. Fifthly, the reproduction of crop species occurs through artificial reseeding; in native species, by contrast, pollution effects on the reproductive cycle are critical to species survival. Sixthly, herbaceous (including graminoid) crop species typically are harvested within months of planting; the perennial nature of many native species, by contrast, poses new problems for survival under long-term stress. Finally, because there are only a few dozen major world crop species, whereas the number of native plant species exceeds 750,000, there is clearly a need to develop rapid-survey techniques for ascertaining the pollution sensitivity of native species. The detailed studies applied to single species are slow to yield information on the sensitivity of whole plant communities, and do not examine pollution stress in the ecosystem context. Nevertheless, such studies are extremely valuable in elucidating mechanisms of pollution sensitivity that may aid in the development of indices. They are also irreplaceable in testing hypotheses derived from, or applicable to, synecological studies. The literature on each of these points will be reviewed briefly.

Selection of Resistant Strains

Although variation in apparent SO$_2$ susceptibility within a species population has often been observed in the field (Tamm and Aronsson, 1972), experimental work to demonstrate the natural selection of strains resistant to SO$_2$ in the field is scant (see Karnofsky, this volume). Bell and Clough (1973) demonstrated by fumigation experiments that an SO$_2$-resistant strain of perennial ryegrass (*Lolium perenne*) had evolved at Helmshore, near Manchester, U.K., under exposure to coal smoke for an unknown period since the Industrial Revolution. Bell and Mudd (1976) further concluded that resistance was imparted by an unknown cellular tolerance mechanism, rather than by an avoidance mechanism (such as decreased stomatal conductance, stomatal closure, or decreased membrane permeability; however, see below). Horsman, Roberts, and Bradshaw (1978) showed that a resistant strain of *L. perenne* had also been selected for at Merseyside, U.K., after SO$_2$ exposure of unknown length. Recent work by Bell, Ayazloo, and Wilson (1982) suggests that resistance to acute injury in *L. perenne* and *Phleum pratense* may evolve in as little as 4 or 5 years; Taylor and Murdy (1975) have reported evolution of resistance in *Geranium carolinianum* within 31 years.

Our own recent work on Californian coastal sage scrub has revealed the evolution of an SO$_2$-resistant strain of brome grass, *Bromus rubens*, after exposure for 25 years to SO$_2$ from an oil-refinery/chemical complex in an otherwise clean-air area on the Nipomo mesa of the central coast of California. Preston (1980, 1983) studied the coastal sage shrubland, dominated by black sage (*Salvia mellifera*), growing on dunes downwind of the refinery, and documented both the extensive damage to black sage along the pollution gradient and the corresponding increase in herb growth, particularly of *Bromus rubens*, in areas released from black-sage domination. Preston (1983) found that the mean maximum daytime ground-level concentration of SO$_2$ downwind of the stacks was 0.09 ± 0.08 ppm (3.7 ± 3.3 μmol m^{-3}, ±95% confidence interval), based on atmospheric modeling; under extremely unfavorable atmospheric conditions, occurring 0.7% of the year, maximum ground-level concentrations may reach 0.33 ppm (13.73 μmol m^{-3}).

In subsequent fumigation experiments, we exposed *Bromus rubens* grown from seed collected from the Nipomo mesa population and from a clean-air site (Encinal Canyon, Santa Monica Mountains, Los Angeles County, California). Plants were exposed to SO$_2$ at 0.05, 0.2 and 0.5 ppm (2.1, 8.3, and 20.8 μmol m^{-3}) for 8 hours a day, 5 days a week, in open-topped, forced-draft fumigation chambers (two chambers per treatment). Seven seedlings were grown in each of six one-gallon pots per chamber. Pots were thermally insulated with an outer

TABLE 15.1

Stomatal Conductance and Visible Foliar Injury in Populations of *Bromus rubens* Collected from a Clean-Air Site and from a Site Exposed to SO_2 for 25 Years in Coastal California

| | *Bromus rubens* seedlings from: | |
Category	Clean-air population	SO_2-exposed population[a]
Stomatal conductance (mol m^{-2} s^{-1} ± 99% confidence interval):[b]		
Control treatment	0.220 ± 0.093	0.189 ± 0.116
0.5 ppm (20.8 μmol m^{-3}) SO_2	0.166 ± 0.118	0.045 ± 0.011
Blades with tip necrosis (percent):[c]		
Control treatment	0%	0%
0.05 ppm (2.1 μmol m^{-3}) SO_2	15	0
0.2 ppm (8.3 μmol m^{-3}) SO_2	20–25	5–10
0.5 ppm (20.8 μmol m^{-3}) SO_2	40	15

[a] Mean maximum daytime concentration at ground level was 0.09 ± 0.08 ppm (3.7 ± 3.3 μmol m^{-3}), but could, on about 2 or 3 days a year, reach 0.33 ppm (13.7 μmol m^{-3}).

[b] After one week of fumigation, 40 hours per week. Conductance was collected with Li-Cor Li-60 meter, with 1,800 ± 50 μE PAR light impinging on blade of each of 16 plants measured. Chamber temperatures were 28° to 29° C; relative humidity 44 to 47%.

[c] After five weeks of fumigation, 40 hours per week.

layer of humus, and the soil was regularly watered with 0.04 strength Hoagland's solution (except 0.01 strength phosphate). After one week, stomatal resistances were significantly lower in the grasses from the clean-air site than in those from the polluted site under high SO_2 treatment, but were not significantly different in the two populations in the absence of SO_2, i.e., in control chambers or in treatment chambers during nonfumigated periods (Table 15.1). After five weeks, substantially greater frequency of visible foliar injury was apparent in plants from the clean-air site than in plants from the polluted site (Table 15.1). The root systems of the plants from the clean-air site were also visibly stunted compared to those of plants from the polluted site after five weeks of fumigation (Fig. 15.1).

Resistance to SO_2 injury in *B. rubens* thus appears attributable at least in part to an avoidance mechanism (reduced stomatal conductance). Selection of the resistant strain appears to have occurred within 25 years. Although Bell and Mudd (1976) reported no differences in stomatal conductance between sensitive and resistant strains of *L. perenne*, their measurements of stomatal conductance were taken in clean air. Our measurements of stomatal conductance in *B. rubens* strains also failed to reveal differences in clean air; the differential conductances were only evident in the presence of SO_2. Thus stomatal conductance measurements in resistant strains of *L. perenne* should be repeated under SO_2 fumigation before conclusions about its mechanisms of resistance are drawn. Also, measurements of stomatal conductance in *B. rubens* should be repeated after a longer period of fumigation to find out if the stomatal response to SO_2 persists.

Fig. 15.1. Seedlings of *Bromus rubens* grown in open-top fumigation chambers under 0.50 ppm (20.8 μmol m^{-3}) of SO$_2$, 8 hours per day, 5 days per week for 5 weeks. Two individuals at left were grown from seeds collected in a clean-air area of coastal Ventura County, Calif. Two individuals at right were grown from seeds collected in an area of coastal San Luis Obispo County, Calif., exposed to the ambient concentration of SO$_2$ detailed in Table 15.1.

Interactions with Other Environmental Stressors

Soil and atmospheric moisture. Because injury to plants from SO_2 is directly correlated with degree of opening of stomates, factors that encourage gas exchange are generally considered to increase the susceptibility of plants to SO_2 damage. However, the relationships between stomatal movement, gas exchange, and pollution injury are complex. Adequate soil moisture is considered to increase SO_2 susceptibility to crop species by promoting stomatal opening (e.g. Markowski, Grezesialk, and Schramel, 1975; Schramel, 1975). However, one must distinguish moisture supplied via the roots from that entering the leaves. Studies on both crop plants (Majernik and Mansfield, 1970; McLaughlin and Taylor, 1981; *inter alia*) and tree species (Carlson, 1979; Norby and Kozlowski, 1982b; *inter alia*) have usually indicated greater SO_2 sensitivity under high than low relative humidity, although a relationship is not always found (Bressan, Wilson, and Gilner, 1978).

High relative humidity may promote SO_2 injury by different mechanisms in different species. In kidney beans (*Phaseolus vulgaris*), SO_2 damage increases under high relative humidity, apparently because of decreased internal leaf resistance (McLaughlin and Taylor, 1981). In seedlings of paper birch (*Betula papyrifera*), by contrast, the increased damage apparently results from increased stomatal conductance (Norby and Kozlowski, 1982b).

Further complicating the analysis of moisture effects on SO_2 injury is the fact that SO_2 can influence stomatal movement independent of moisture effects in some species. Thus stomatal closure under high SO_2 exposure has been reported for the native Californian chaparral species *Arctostaphylos densiflora*, *Ceanothus thyrsiflorus*, *C. maritimus*, *Heteromeles arbutifolia*, *Ribes sanguineum*, and *R. viburnifolium* (Winner and Mooney, 1980a,b; Winner, Koch, and Mooney, 1982), the coastal sage scrub species *Salvia mellifera* and *Diplacus aurantiacus* (Winner and Mooney, 1980a,b), the deciduous and evergreen oaks *Quercus kelloggii* and *Q. agrifolia* (Winner, Koch, and Mooney, 1982), several species of birch (Biggs and Davis, 1980; Norby and Kozlowski, 1981b), the Pacific tree, *Metrosideros collina* (Winner and Mooney, 1980d), and the C_4 herb, *Atriplex sabulosa* (Winner and Mooney, 1980c). Other native species, or the same species at lower concentrations, respond to SO_2 by opening their stomates; examples are *Atriplex triangularis* and *A. sabulosa* (Winner and Mooney, 1980c), some species of *Betula* (Biggs and Davis, 1980), and *Salvia mellifera* and *Ceanothus thyrsiflorus* (Winner, Koch, and Mooney, 1982). Still others apparently show no stomatal response to SO_2, e.g., *Dodonaea eriocarpa* (Winner and Mooney, 1980d). Crop species show similar variability: SO_2 causes stomates of *Vicia faba*, *Helianthus annuus*, and *Nicotiana tabacum* to open above a relative humidity of 40 or 50%, and to close below this threshold (Black and Unsworth, 1980; Majernik and Mansfield, 1970). *Phaseolus vulgaris* and *Raphanus sativus* opened their stomates under SO_2 exposure regardless of humidity (Black and Unsworth, 1980).

Recent attempts to generalize about the likely injury of SO_2 to native plants under various precipitation regimes (e.g., McLaughlin and Taylor, 1980) are therefore premature for several reasons. Humidity can act independently from soil moisture on stomatal movement, so that generalizations about "arid" or "mesic" climates should distinguish fog incidence and humidity levels (which can be quite high in areas of low precipitation) from actual precipitation and other factors affecting soil moisture. In addition, the variability in response of stomatal conductance to humidity, to SO_2, or to some combination, at different levels of exposure, further discourages generalizations at this stage.

It is tempting to conclude at least that native species whose stomates close during periods of water stress will generally be less susceptible to SO_2 damage than will well-irrigated crop species. Indeed, Zimmerman and Crocker (1934) found some wilted plants to be more SO_2-resistant. However, even such a generalization must be eyed with caution, since the conductances of many mesophyllous species are so much higher than those of sclerophyllous species that even under drought conditions the gas exchange rates of the former may equal or exceed those of the latter.

The shrubs occupying the drier portions of Mediterranean-climate regions are of interest here. Winner (1981) postulated that drought-deciduous, mesophyllous coastal sage scrub species would be more sensitive than evergreen, sclerophyllous chaparral species to SO_2 because of the higher levels of conductance in the former. Measurement of conductance in coastal sage scrub species is made more complex by the fact that *Salvia mellifera* and other California coastal sage species, as well as shrubs in analogous vegetation types in other Mediterranean-climate regions (e.g., Greek phrygana, Israeli batha, Chilean coastal matorral), produce two distinct leaf types. Such shrubs are seasonally dimorphic (Westman, 1981), producing a large early-season leaf (dolichoblast), which abscisses as the dry period progresses, and a smaller leaf (brachyblast), which initiates growth in the axil of a dolichoblast during the early to middle part of the growing season and remains on the branch until the onset of the new growing season.

Evidence from our fumigation-chamber studies (referred to earlier) indicates that, given adequate moisture and nutrition, the brachyblast leaves are more sensitive to SO_2 damage following prolonged exposure (ten weeks) than are the dolichoblasts (Table 15.2). This is so despite the natural occurrence and persistence of brachyblasts during the drier season, and their lower stomatal conductance when given adequate moisture and clean air (Table 15.3). The larger surface area for localized adsorption of SO_2 onto leaf surfaces of dolichoblast leaves may effectively reduce the SO_2 concentration entering stomates (cf. Winner, Koch, and Mooney, 1982). This mechanism seems more likely to create a differential effect if dolichoblast leaves also have a higher stomatal density, a feature not yet

TABLE 15.2
Growth of Large, Early-season (Dolichoblast) Leaves and Small, Mid- to Late-season (Brachyblast) Leaves in Seasonally Dimorphic Species Exposed to Ten Weeks of SO_2 Fumigation, as a Percent of Growth in Control (Filtered-air) Chambers, with Initiation of the Two Leaf Types Also Shown

Measured attribute	Treatment (ppm SO_2)[a]	Percent of control, *Salvia mellifera*[b]		
		Length	Width	Thickness
Growth:				
Dolichoblasts[c]	0.50	86 ± 4**	nsd[d]	52 ± 6**
	0.20	nsd	nsd	76 ± 4**
	0.05	nsd	nsd	80 ± 8**
Brachyblasts[c]	0.50	66 ± 8**	80 ± 10	49 ± 6**
	0.20	73 ± 6**	nsd	70 ± 7**
	0.05	nsd	nsd	76 ± 5**

		Percent of control[b]	
		Salvia mellifera	*Salvia leucophylla*
Initiation: Ratio of no. of brachyblasts to no. of dolichoblasts per branch[e]	0.50	30**	48*
	0.20	40**	nsd
	0.05	nsd	nsd

NOTE: Fumigation occurred 40 hours per week in open-top, forced-draft chambers, five plants per species per chamber, two chambers per treatment.

[a] In μmol m^{-3}, 0.50 ppm = 20.8, 0.20 ppm = 8.3, 0.05 ppm = 2.1

[b] Mean value ± 95% confidence interval.

[c] Leaf dimensions were measured on 45 leaves on a total of five plants per treatment.

[d] nsd, not significantly different from the control ($p < 0.01$).

[e] Leaf numbers were counted on three branches per plant, five plants per treatment.

TABLE 15.3
Stomatal Conductance of Four Coastal Sage Scrub Species Grown for Five Weeks in Filtered-air Fumigation Chambers

Species	Stomatal conductance[a] (mol m^{-2} s^{-1})	
	Dolichoblasts	Brachyblasts
Salvia mellifera	0.167 ± 0.042	0.043 ± 0.008
Salvia leucophylla	0.261 ± 0.082	0.042 ± 0.008
Artemisia californica		0.054 ± 0.011[b]
Encelia californica	0.159 ± 0.021	

NOTE: Conductance measured with Li-COr Li-60 diffusive resistance meter on lower surface of 30 leaves of each type or species, on one-year-old nursery stock in pots. Chamber temperatures 32° to 35° C; relative humidity, 34 to 37%;, PAR light 1,500 ± 50 μE; soil moisture, −0.2 bars.

[a] Mean ± 95% confidence intervals

[b] Because the distinction between dolichoblasts and brachyblasts in this species is tenuous (Westman, 1981), all leaves are described here as brachyblasts. The dolichoblast is considered to be a single, early-deciduous leaf at the base of the axillary cluster.

measured. The large area and protruding position of dolichoblast leaves, in contrast to the small, bunched, axillary brachyblasts, may also increase the boundary layer in the former. Under any given moisture conditions the two leaf types exhibit different SO_2 sensitivities (for reasons not yet known); therefore studies of SO_2 sensitivity should distinguish between leaf types when reporting results. The possibility that differential responses in leaf growth will not be reflected in growth of other plant parts also exists. Measurements of growth of all plant parts in these species under SO_2 treatment are in progress.

Temperature. Given the variation in response to SO_2 based on environmental moisture just described, it is not surprising that temperature, which affects moisture conditions and has other, independent effects, also elicits highly variable responses from species. This variation is nevertheless worth noting, since, until recently, some workers held to the generalization that high temperatures increase SO_2 injury. Norby and Kozlowski (1981b) review the recent literature and report on their own experiments with tree seedlings of three species (*Betula papyrifera*, *Fraxinus pennsylvanica*, and *Pinus resinosa*). Conductances were higher at 30°C than at 12°C in all species, but were lowered by SO_2 fumigation only in *Betula* and *Fraxinus*. Leaf sulfur concentrations were greater at 30°C in all species, but the effects on relative growth rates were various, and SO_2 absorption was not correlated with daytime leaf conductance. The last result may have been due to differences in boundary-layer conductances or in nighttime conductances, or to translocation of sulfur out of the leaf, according to the authors. Thus effects on growth were not a linear function of SO_2 uptake, and temperature effects on SO_2 inhibition of growth are correspondingly variable.

Nutrients. Relatively few studies deal with nutritional effects on SO_2 susceptibility in native species. Bjorkman (1970) reported that optimal soil-nutrient levels reduced SO_2 injury to Scotch pine (*Pinus sylvestris*). Fertilization of *Pinus strobus* in the field, however, did not reduce sensitivity to SO_2 and ozone (Dochinger and Seliskar, 1970). In a more controlled experiment, Cotrufo (1974) found clonal ramets of *P. strobus* were more susceptible to SO_2 with nitrogen additions, less susceptible with P additions, and not responsive to additions of K.

Mooney and Gulmon (1982) have reviewed evidence that leaf photosynthetic capacity is positively correlated with leaf nitrogen content. However, efforts to correlate substrate nitrogen levels with attributes of relevance to SO_2 susceptibility are frustrated by such processes as internal translocation of nitrogen, and non-substrate-related variations in foliar N concentration between species. Generalizations may prove sturdier when limited to particularly closely related vegetation types. Thus Winner, Koch, and Mooney (1982) note that drought-deciduous Californian coastal sage scrub species generally have higher conductances, photosynthetic rates, leaf nitrogen contents, and short-term SO_2 susceptibilities than evergreen chaparral species. How the short-term susceptibilities are translated

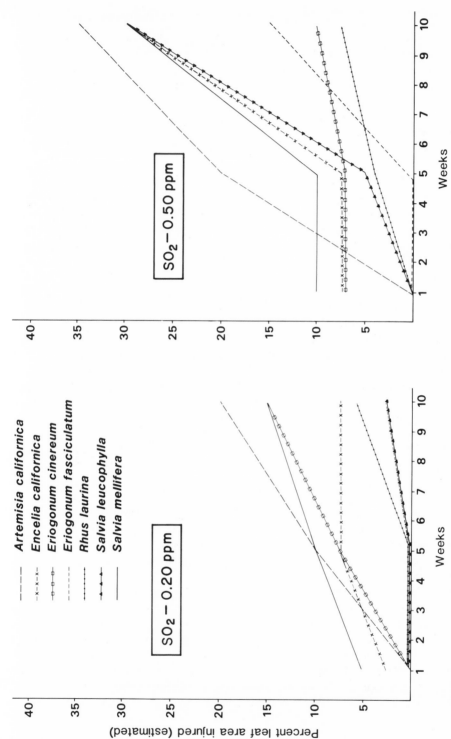

Fig. 15.2. Percent of leaf area injured in each of seven species of coastal sage scrub exposed for 10 weeks (8 hours per day, 5 days per week) to 0.20 ppm (8.3 μmol m⁻³) of SO₂ (*left*) and 0.50 ppm (20.8 μmol m⁻³) of SO₂ (*right*).

into growth differences over the life of the plants will depend on such additional factors as seasonal changes in foliar cover. The seasonal and annual leaf-area indices of the two vegetation types differ dramatically (see, e.g., Gray and Schlesinger, 1981; Westman, 1982).

Even within the coastal sage scrub, however, susceptibilities to chronic exposure (ten weeks) vary greatly, as measured by visual injury to foliage in our fumigation experiments (Fig. 15.2). Although foliar nutrient concentrations for the experimental plants are not yet available, foliar nitrogen concentrations for *Salvia leucophylla* and *Artemisia californica* collected at a site in the Santa Monica Mountains (mean ± coefficient of variation, 1.85 ± 31%, 1.88 ± 29%, respectively; Gray and Schlesinger, 1983) do not differ enough to account for the large differences in susceptibility shown in Fig. 15.2.

The effects of substrate fertility on plant susceptibility to SO_2 may be indirect. Thus plants growing on soils that are already acid or are poorly buffered are likely to become rapidly acidified under the influence of SO_2, which can reach the soil directly by gaseous absorption, precipitation, or dry deposition, or indirectly by excretion from plant roots, litterfall, and leachate. Acidification will change the levels of availability of soil nutrients, changing the fertility status and consequent growth of plants (see, e.g., Smith, 1981, Chap. 9).

Many other environmental variables, such as light intensity and quality (Dunning and Heck, 1973), wind (Brennan and Leone, 1968; Heagle, Heck, and Body, 1971), herbivory (Smith, 1981, Chaps. 12 and 15), and other air pollutants (Smith, 1981) affect the air-pollution susceptibility of plants. Although these will not be reviewed here, it seems fair to note that both the amount of stress to which species are exposed and the extent of adaptation to such exposure are likely to be greater for native species than for highly yield-bred crop species.

Competitive Relations of Species

Species-specific growth responses are extrapolated to ecosystem-level processes by other authors in this volume (see also, e.g., Miller and McBride, 1975), and will not be discussed in detail here. Nevertheless, it is relevant here to note that relative growth responses in a laboratory setting or a field monoculture may fail to reveal the observed relative growth of species in the field, because of competitive interactions between plants of native species.

Winner and Bewley (1978) noted effects of SO_2 from a natural-gas refinery in Alberta, Canada, on a white spruce (*Picea glauca*) stand downwind. They found that herbaceous angiosperms and mosses colonized areas exposed to light when overstory components were killed by SO_2. They classified species into resident species that flourished in the high SO_2 area, those relatively intolerant of SO_2, and SO_2-tolerant "weedy" or invader species. They noted that success of the resistant groups will vary both in time and in space. The classic study of Gordon and Gorham (1963) downwind of the Wawa iron-sintering plant also revealed a pollution-influenced vegetation gradient, in which increasers, decreasers, and invaders were arrayed by their individual tolerances along an axis of increasing distance from the SO_2 source.

Although species richness (e.g., Preston, 1983; Rosenberg, Hutnik, and Davis, 1979) and the Shannon-Weiner index of diversity (Rosenberg, Hutnik, and Davis, 1979) will sometimes increase as SO_2 concentrations decrease along the gradient, this will not always be the case (e.g., Winner and Bewley, 1978). The Shannon-Weiner index is a particularly inappropriate measure for use in pollution studies, since it confounds richness with equitability in a complex, nonlinear fashion (see, e.g., Peet, 1974). Further, all measures of diversity suffer from the fact that they are insensitive to the taxonomic identity and pollution sensitivity of species, so that an area in which the native vegetation has been destroyed and replaced by a mixture of pollution-tolerant "weedy" invaders can register an increase in both richness and equitability. Graphical techniques that separate richness from equitability as measures of diversity—e.g., dominance-diversity curves (Whittaker, 1965), log-normal curves (Westman, 1979)—and direct and indirect gradient analysis techniques (Whittaker, 1973), which preserve information on species identity, are more appropriate for synecological studies of pollution effects.

Whole-Plant Effects

Studies of pollution effects on crop species have tended to focus on harvested (or at least aboveground) plant parts. For native species, however, effects of pollution on all plant parts are relevant to the survival and growth of the species population. Studies of SO_2 effects on the growth of roots and rhizosphere microorganisms are few, but implications of initial results are important. Jensen and Kozlowski (1975) used labeled $^{35}SO_2$ to trace translocation of foliar-absorbed sulfur to the roots of five species of tree seedlings. The effects on belowground growth can be more dramatic than those on aboveground parts, and provide a useful index of SO_2 injury (see, e.g., Fig. 15.1).

In a finding of widespread significance, the U.S. Environmental Protection Agency (1976a) reported that concentrations of SO_2 in excess of 0.06 ppm (2.5 μmol m^{-3}) can significantly reduce nitrogen fixation by bacterial nodules in plant roots. Hällgren and Huss (1975) also reported SO_2-induced inhibition of nitrogen fixation in both free-living blue-green algae and the algal symbionts of two lichen species. The effects of SO_2 on mycorrhizal associations have scarcely been studied. Grzywacz (1971) reports that SO_2 and fluorine decreased the frequency of association of ectomycorrhizae with tree rootlets, and Sobotka (1964) found abnormal ectomycorrhizal development in SO_2-damaged spruce. The effects of SO_2 on endomycorrhizae have not been reported at all, despite their widespread occurrence in both crop and native species. McCool and Menge (1978) found that ozone reduced endomycorrhizal spore populations of citrus by 50%, and height of plants by 37%. We are currently studying effects of SO_2 and ozone on endomycorrhizae of ten coastal sage scrub species under controlled fumigation conditions.

Reproductive Effects

Following the example of studies with crop species, some workers have sought to rank native species' sensitivities to SO$_2$ based on short-term fumigations and subsequent visible foliar injury (e.g., O'Connor, Parbery, and Strauss, 1974; Hill et al., 1974). What is overlooked in such a procedure is that a plant whose foliage survives such treatment intact not only may be harboring latent or invisible injury, but may also suffer inhibition of reproduction because of damage to reproductive organs or seeds, or may change in susceptibility to damage during another part of its life cycle.

Several investigators have demonstrated interference of SO$_2$ with reproductive parts. Decreased pollen germination and tube elongation have been shown in both gymnosperms (Dopp, 1931; Houston and Dochinger, 1977; Karnofsky and Stairs, 1974) and angiosperms (Karnofsky and Stairs, 1974; Ma and Khan, 1976; Masuru, Syozo, and Saburo, 1976; Varshney and Varshney, 1981). Other reported effects in coniferous trees include reductions in cone length, in seed weight, in percent of filled seed, and in percent of seed germination (Houston and Dochinger, 1977; Mamajev and Shkarlet, 1970). Preston (1983) found reduction of the number of flower whorls per flowering spike in *Salvia mellifera* in the SO$_2$-influenced Nipomo site. We have since confirmed this result for *S. mellifera* and several other coastal sage scrub species by using fumigation chambers (Table 15.4).

TABLE 15.4
Effects of Ten Weeks of Fumigation with SO$_2$ on Flowering and Crown Sprouting in One-year-old Shrubs of Coastal Sage Scrub

Shrub	Treatment in ppm (μmol m^{-3}) of SO$_2$:		
	0.50 (20.8)	0.20 (8.3)	0.05 (2.1)
	NO. OF INFLORESCENCES AS PCT. OF CONTROL		
Flowering:			
Salvia mellifera	8.7	15	15
Artemisia californica	26	33	nsd[b]
Eriogonum fasciculatum	2.8	17	nsd
Encelia californica	34	32	nsd
	NO. OF BASAL SPROUTS AS PCT. OF CONTROL		
Crown sprouting:			
Salvia mellifera	180	nsd	nsd
Artemisia californica	325	nsd	nsd
Encelia californica	280	nsd	nsd

NOTE: Fumigation 8 hours per day, 5 days per week. Control chamber contained filtered air, forced draft; five plants per chamber, two chambers per treatment. All results reported are significant at $p = 0.05$.
[a] Number of flowering stems.
[b] nsd, not significantly different from the control.

Some species respond to the death of aboveground foliage by increasing the number of basal sprouts that arise and grow to maturity (Table 15.4). This is a particularly interesting response, since ecologists had tended previously to associate crown sprouting of Mediterranean-climate shrubs only with severe injury by fire or grazing. Malanson and Westman (unpublished manuscript) have recently shown, however, that coastal sage species will resprout, without external stress, as other stems on the plant mature and die. The effect of SO_2 is to accelerate this process.

Pollution Stress and Longevity of Plant Parts

Because the dominant vegetation of forests and shrublands is perennial, and because most of the vegetation in tropical, boreal, and Mediterranean climates (and much of the vegetation in temperate and arid climates) is evergreen, the issue of long-term stress from pollution is heightened in native species by comparison with herbaceous agricultural crops.

Resistance to SO_2 varies with leaf age. Linzon (1978) offers the generalization that middle-aged, fully grown leaves are the most sensitive in broadleaved species, but notes exceptions. Similarly, he suggests that mature, first-year needles of conifers are most susceptible to SO_2; yet Caput *et al.* (1978) report that one-month-old needles of *Pinus pinea* are more susceptible than needles more than a year old. The sensitivity of needles of pine appears related to their stomatal resistance (Caput *et al.*, 1978). One might expect, therefore, that sensitivity of leaves will decrease with age up to senescence, to the extent that secondary thickening, lignification, and cuticular thickening decrease stomatal conductance. There are insufficient data to test this hypothesis satisfactorily. The observation noted earlier, that Californian sclerophyllous shrub species generally exhibit lower stomatal conductances and lower SO_2 sensitivities than mesophyllous species (Winner and Mooney, 1980a), is only suggestive evidence for the hypothesis, since other resistance-imparting differences between the two sets of species have not been examined. Furthermore, fumigation and field studies with other floras (e.g., Gordon and Gorham, 1963; Rosenberg, Hutnik, and Davis, 1979; Davis and Gerhold, 1976; see Table 26.4 in Kozlowski, this volume) indicate that sclerophyllous softwood species may sometimes be more sensitive than mesophyllous hardwood species to moderate SO_2 exposures, and that the correlation between sclerophylly and SO_2 resistance in any event is not strong.

The increased duration of exposure of evergreen foliage to the atmosphere might be thought to counterbalance any resistance to SO_2 injury that its leaves possess. Linzon (1971) observed chronic damage to older needles of pine around Sudbury, resulting in sulfur accumulation and premature abscission. The findings of Keller (1978a, 1981b) on stem absorption of SO_2 and translocation of sulfur to leaves in deciduous species, however, opens even this generalization to question. Keller (1978a, 1981b) examined seven deciduous tree species, and found that in five of them SO$_2$ may be absorbed directly by leafless shoot tissue in winter and the sulfur translocated to leaves in spring. To the extent that this process induces unusually high initial levels of sulfur in young leaves (Keller, 1978a), it might make them more susceptible to accumulation of toxic levels of sulfur upon exposure to SO$_2$ in springtime. Terminal buds of deciduous species can actually be killed during the winter (Keller, 1978a). The sulfur translocated to the leaf is probably in the relatively nontoxic sulfate form, but its presence in large quantities may influence equilibria between other sulfur forms.

As noted by Winner, Koch, and Mooney (1982), loss of evergreen leaves is more costly in energy to the plant (Miller and Stoner, 1979), so that SO$_2$ stress may be more critical to evergreen species for this reason.

Indices of Pollution Sensitivity

Because so many native species have yet to be surveyed for SO$_2$ sensitivity, the development of readily measured indices of pollution sensitivity is a high-priority area for research. Winner, Koch, and Mooney (1982) have suggested that conductance values in SO$_2$-free air may be used to predict relative photosynthesis following fumigation, since conductance values were related to SO$_2$ absorption for ten California species ($p < .01$), and such absorption in turn was correlated with relative photosynthesis ($p < .05$). The correlation between pre-fumigation conductance and postfumigation photosynthesis will be less tight than either of its component regressions, however, since the error terms of the two regressions must be multiplied together. The California shrub data also seem to exhibit a direct correlation between leaf conductance and photosynthesis rate in the absence of SO$_2$ that may not hold up when a larger array of species is included. For example, Mooney and Gulmon (1982) generalize that leaves of greater longevity have lower photosynthetic rates; yet we noted in the preceding section that mesophyllous, deciduous leaves of high conductance and low longevity can be less sensitive to SO$_2$ than their evergreen sclerophyllous counterparts in non-Californian floras. Direct measurement of conductance values and photosynthetic rates in a larger array of SO$_2$-exposed mesophyllous and sclerophyllous species is needed to test these relationships more precisely.

Even among mesophyllous Californian coastal sage scrub species, we found substantially greater foliar injury (Fig. 15.2) and leaf drop (unpublished data) in *Artemisia californica* than in *Encelia californica*; yet the prefumigation conductances of *Artemisia* were significantly ($p < .05$) lower than those of *Encelia* (Table 15.3). Furthermore, the greater SO$_2$ sensitivity of brachyblast leaves in *Salvia mellifera*, despite their lower conductances, appears contrary to the proposed relationship between conductance and SO$_2$ injury. Beyond this, the relationship between leaf photosynthetic rates and observed plant growth in the field is complex and nonlinear. Despite these concerns, the use of conductance as an index deserves further testing, since additional data may yet elucidate trends.

Other physiological and morphological indicators deserve greater attention as well. Leaf size and shape, which can affect the configuration of the boundary layer, can alter pollutant absorption (Brennan and Leone, 1968). In addition, glands, hairs, resinous cuticles, leaf topography, and leaf size may play a role in adsorption and filtering of pollutants in the boundary layer (see, e.g., Elkiey and Ormrod, 1980b; Sharma, 1975; Sharma and Butler, 1975). Elkiey and Ormrod (1980b) found that petunia cultivars with larger and more numerous leaf trichomes were more resistant to SO_2. Sharma (1975) found higher trichome density on leaves of *Acer saccharum* growing in polluted vs. clean-air areas around Montreal. Trichomes also help retain rain and dew; Fowler and Unsworth (1974) note that surface wetness is important to dry deposition of SO_2 on leaves. Roughened cuticles may play a role in pollutant adsorption (Elkiey and Ormrod, 1980b). Bystrom *et al.* (1968) observed increased sensitivity in younger leaves of *Beta vulgaris*, which they attributed to lack of continuous cuticle in younger leaves. Cuticular growth (including differential growth on guard vs. adjacent epidermal cells; *viz.*, Black and Black, 1979a) may help explain some of the variation in SO_2 resistance with age of leaf that we discussed in the preceding section. It has been suggested that mosses and lichens are more susceptible to air pollutants because they lack a waxy cuticle. For further discussion of SO_2 effects on cryptogams, see Kozlowski (this volume).

Other leaf characteristics that hold promise as indicators, but which are somewhat less easily measured, include internal suberization and degree of surface hydration of the cell wall (Taylor 1978b; Glater, Solberg, and Scott, 1962), and biochemical pathways and associated chloroplast arrangement. Winner and Mooney (1980c) have suggested that C_4 plants may be more resistant to SO_2 in part because the chloroplast arrangement in bundle-sheath cells may enable the SO_2 to be dispersed or reduced on its way from stomates to vascular bundle. Preston (1983) suggested that a CAM plant on the Nipomo mesa had resisted SO_2 injury because nighttime wind patterns (when stomates are open in CAM plants) carried the stack plume away from the study area. CAM metabolism would similarly be relevant in sites where industrial emissions are reduced at night.

Areas for Future Research

In searching for morphological, anatomical, or physiological indicators of pollution sensitivity, one must be careful to test a series of alternative hypotheses simultaneously, rather than to examine correlations between a single characteristic and SO_2 sensitivity. Correlations gain strength as possible causal mechanisms to the extent that reasonable alternative hypotheses have been shown to have even lower predictive power. Multiple regressions, path models, or other multivariate statistical approaches are likely to provide greater predictive strength, and may aid in theoretical understanding, to the extent that they point to interactive elements. In this vein, much of the data on pollution sensitivity is already available for such a multivariate analysis. Investigators need only measure the more readily determined morphological and anatomical features of the tested species in order to amass a large data set for analysis. This would appear to be one of the more cost-efficient approaches to research on pollution sensitivity of native species currently available to researchers.

Summary

The genetic variability of native-species populations can result in selection of SO$_2$-resistant strains of annuals in as little as 25 years; an example is *Bromus rubens* in California, which has evolved strains that increase stomatal resistance upon exposure to SO$_2$. In general, stomatal closure may be induced by soil moisture, humidity, or SO$_2$, but there is much interspecific variability in response to each of these factors, so that earlier generalizations must be revised in light of new data. In particular, although conductance has seemed at times to be a good predictor of SO$_2$ susceptibility, important exceptions exist. In the leaf-dimorphic *Salvia mellifera*, for example, the small, dry-season leaves exhibit higher conductances in SO$_2$-free air than the large, early-season leaves under well-watered conditions, but exhibit lower levels of leaf initiation and growth after ten weeks of SO$_2$ fumigation. Earlier studies of SO$_2$ effects on *S. mellifera* and other seasonally dimorphic species did not report effects on the two leaf types separately. Species also show highly variable interspecific responses to temperature and substrate fertility as factors that affect SO$_2$ susceptibility.

In the community context, this interspecific variation registers as a gradient of responses to SO$_2$ stress based on the individual tolerances of indigenous species and of invading opportunists. Root systems can show even more dramatic growth inhibition from SO$_2$ exposure than aboveground parts, and SO$_2$ inhibition of nitrogen-fixing bacteria and ectomycorrhizae has been demonstrated. SO$_2$ can also injure native plant populations by interfering with a variety of aspects of reproduction, an effect generally missed by short-term fumigation experiments.

Deciduous species are not necessarily less sensitive to SO$_2$ damage than ever-green ones, both because of the greater susceptibility of leaves to SO$_2$ in some species, and because of absorption of SO$_2$ through bark during leafless periods. Although such indices of SO$_2$ sensitivity as leaf conductance, mesophylly, and leaf duration are proving to be imperfect predictors, further research into morphological, anatomical, and physiological features of leaves and other plant parts as indices of SO$_2$ susceptibility remains a promising and cost-efficient area for research into SO$_2$ effects on native species.

Acknowledgments

We thank C. R. Thompson, G. Kats (Statewide Air Pollution Research Center, University of California, Riverside), and field assistants from the Department of Geography, University of California, Los Angeles, for aid in the conduct of fumigation experiments described here. Unpublished data were collected with partial support from the National Science Foundation (DEB 76-81712) and the Academic Senate, U.C.L.A.

16

Effects of SO_2 and NO_x on Plant Growth

Mary E. Whitmore

SO_2 rarely occurs by itself in polluted atmospheres; other gaseous pollutants, notably oxides of nitrogen (NO_x) and ozone (O_3), usually accompany it. The mixture of NO_x and SO_2 is relevant for many field situations, because these pollutants are sometimes emitted from the same sources and often occur together. There have been several investigations into the effects that short exposures to SO_2 and nitrogen dioxide (NO_2) have on plants, especially in terms of visible leaf injury (Reinert, Heagle, and Heck, 1975). Our knowledge of effects on growth and yield is limited, but there is evidence that the growth of some species may be seriously inhibited by mixtures of SO_2 and NO_2, and that the effects may sometimes be more than additive (Mansfield and Freer-Smith, 1981). It appears, therefore, that the presence of NO_x may substantially increase the phytotoxicity of polluted air.

Before reviewing the combined effects of SO_2 and NO_x on plant growth, I think it useful to describe the concentrations of SO_2 and NO_x that crop plants are likely to experience, and to review very briefly what we know about the effects of NO_x by itself.

Concentrations of SO_2 and NO_x in Urban and Rural Areas

The range of SO_2 concentrations that may be experienced by plants in the field has been covered in preceding papers, and therefore my main objective here is to describe the concentrations of NO_x that occur. The major source of both SO_2 and NO_x is the burning of fossil fuels. These all contain sulfur, which is oxidized to SO_2 on combustion. In contrast NO_x is formed by the oxidation of nitrogen from the atmosphere; this occurs in the heat of the flame. The first to form is nitric oxide (NO), which is then slowly oxidized in the atmosphere to NO_2 (Spedding, 1974). The formation of NO is favored by high temperatures; since the internal-combustion engine operates at a high temperature, road traffic is a major emitter of this pollutant. To put NO_x emissions into perspective, Fowler and Cape (1982) quote annual emissions in North America and western Europe, together, of 30×10^9 kg NO_x (expressed as NO_2), compared with 50×10^9 kg SO_2.

The highest concentrations of NO_x are experienced close to sources. For example, beside a road in Genoa, Italy, Capannelli *et al.* (1977) measured peaks of up to 3.76 ppm (156 μmol m^{-3}) of NO and 0.28 ppm (11.6 μmol m^{-3}) of NO_2 at a height 2 m above the ground. Garnett (1979) surveyed NO_x concentrations at six locations in Sheffield, England, and confirmed that the highest NO_x levels were next to roads. At open sites the pollutants are usually rapidly dispersed; although 0.46 ppm (19.1 μmol m^{-3}) of NO_x was recorded 10 m from a Los Angeles freeway, the concentration fell to only 0.1 ppm (4.2 μmol m^{-3}) at a distance of 100 m (Rodes and Holland, 1981).

High concentrations may also occur near specific point sources. Harrison and McCartney (1979) surveyed NO_x concentrations at a rural site in the north west of England near an ammonium-nitrate fertilizer factory. Background concentrations of NO and NO_2 rarely exceeded 0.002 and 0.01 ppm (0.08 and 0.42 μmol m^{-3}), respectively, but when the plume from the factory passed overhead, the hourly mean of both pollutants often exceeded 0.1 ppm (4.2 μmol m^{-3}), with peaks of 0.5 ppm (20.8 μmol m^{-3}) of NO and 0.3 ppm (12.5 μmol m^{-3}) of NO_2 that lasted for 2 or 3 minutes.

The information about the concentrations of pollutants in rural areas is limited, and few investigators have made continuous measurements of SO_2, NO_x, and O_3 concentrations. However, two such studies were recently undertaken at rural sites in Britain, distant from obvious sources of pollutants (Martin and Barber, 1981; Nicholson *et al.*, 1980b). Concentrations of SO_2, NO, and NO_2 tended to be highest during the winter, at which season monthly means of between approximately 0.01 and 0.016 ppm (0.42 and 0.67 μmol m^{-3}) were recorded for each of these pollutants. Many agricultural areas in heavily populated countries are not, however, situated in such unpolluted areas, but are adjacent to towns. Fowler and Cape (1982) estimated that annual mean concentrations of both SO_2 and NO_2 exceeded 0.04 ppm (1.66 μmol m^{-3}) over approximately one million hectares of rural land in western Europe in the mid-1970's.

Toxicity of NO_x

The effects of NO_x on plants have been less extensively studied than those of SO_2, and much of the work has employed such high concentrations (more than 1 ppm or 44.6 μmol m^{-3}) that it could not be compared to any but the most polluted field situations (see review by Taylor *et al.*, 1975). There have, however, been some reports of the effects of lower concentrations of NO_x on plant growth. For example, Taylor and Eaton (1966) reported that concentrations of between 0.16 and 0.23 ppm (6.7 and 9.6 μmol m^{-3}) of NO_2 for 10 days inhibited dry-weight gain of *Lycopersicon esculentum*; and Spierings (1971) found that when this species was exposed to 0.25 ppm (10.4 μmol m^{-3}) of NO_2 throughout a 3-month growing period, both the number of fruit and the average fruit weight were reduced. Growth of *L. esculentum* may also be inhibited by NO; Capron and Mansfield (1977) found that plants exposed to 0.4 ppm (16.6 μmol m^{-3}) of

NO for 19 days weighed between 34 and 40% less than controls. Subsequently, Anderson and Mansfield (1979) observed not only that different cultivars showed great differences in sensitivity to NO, but also that the response could be modified by the nutrient status of the soil; growth of *L. esculentum* was stimulated by exposure to 0.4 ppm (16.6 μmol m^{-3}) of NO when soil nitrogen was limiting, but was inhibited when soil fertility was high. Concentrations of 0.1 to 0.5 ppm (4.2 to 20.8 μmol m^{-3}) of NO and NO$_2$ have also been found to inhibit rates of net photosynthesis in this species (Capron and Mansfield, 1976).

Studies have shown that NO$_x$ can be metabolized by plants. Yoneyama and Sasakawa (1979) exposed plants to ^{15}N-labeled NO$_2$ and found that it may be converted to nitrate, nitrite, and ammonia, and assimilated, through the glutamine synthetase / glutamate synthase pathway, into amino acids. In normal metabolism, nitrate is reduced to nitrite by nitrate reductase, and nitrite is converted to ammonia by nitrite reductase. Zeevaart (1974) observed that nitrate-reductase activity in *Pisum sativum* was stimulated by exposure to 4 or 12 ppm (166.3 or 498.8 μmol m^{-3}) of NO$_2$ when plants were grown on a nitrate-free medium. This parallels the well-established induction of this enzyme by nitrate (Beevers and Hageman, 1969). The precise fate of NO in the plant is not known, but it may dissolve in extracellular water to form nitrite. Exposure to NO has been found to stimulate nitrite-reductase activities (Wellburn *et al.*, 1976; Wellburn, Wilson, and Aldridge, 1980), which suggests that nitrogen from NO may also enter nitrogen pathways.

Comparatively little work has been specifically directed toward understanding the effects of NO$_x$ on plants, but "Much of the research work on mixtures . . . provides useful information on single-gas effects because direct comparisons are usually made with single-gas exposures" (Ormrod, 1982).

Toxicity of Mixtures of SO$_2$ and NO$_x$

Most investigators have used NO$_2$, rather than a mixture of NO and NO$_2$, when studying effects of mixtures of NO$_x$ and SO$_2$ on plants. Mansfield and Freer-Smith (1981) discuss this approximation in detail, and conclude that although NO and NO$_2$ may differ in toxicity because they differ in rate of uptake, their mechanisms of action at the cellular level may be considered to be similar.

Responses of plants to various pollutant combinations, including mixtures of SO$_2$ and NO$_2$, were reviewed by Reinert, Heagle, and Heck in 1975, at which time most of the investigations with SO$_2$ and NO$_2$ had involved assessment of visible leaf injury after only short periods of exposure to high concentrations. Only recently have effects on growth, after long-term exposures to lower concentrations, been reported.

Effects of short exposures. Investigations of the leaf injury arising from exposure to SO$_2$ and NO$_2$ have led to three important conclusions. First, concentrations that had little effect when plants were exposed to either SO$_2$ or NO$_2$ separately often induced extensive foliar injury when these gases were present together. For example, Tingey *et al.* (1971) observed leaf blemish on six crop species (*Nicotiana tabacum, Phaseolus vulgaris, Lycopersicon esculentum, Raphanus sativus, Avena sativa* and *Glycine max*) after the plants were exposed to mixtures of SO_2 and NO_2 at 0.05 to 0.25 ppm (2.1 to 10.4 μmol m^{-3}) for 4 hours; yet the threshold for damage by SO_2 alone was 0.5 ppm (20.8 μmol m^{-3}) and for NO_2 was 2 ppm (83 μmol m^{-3}). Although Bennett *et al.* (1975) found four of these species (*P. vulgaris, L. esculentum, R. sativus,* and *A. sativa*) to be less sensitive than Tingey *et al.* (1971) had found, they confirmed that a mixture of SO_2 plus NO_2 could induce injury at concentrations below the threshold limit for either gas by itself.

A second important conclusion drawn from these studies was that leaf injury caused by exposure to SO_2 plus NO_2 often resembled that caused by ozone (Tingey *et al.*, 1971; Reinert and Gray, 1981). Consequently, some of the leaf blemish previously attributed to ozone may have been caused by the presence of a mixture of SO_2 and NO_2.

The third noteworthy conclusion concerns the interaction between SO_2 and NO_2. Once concentrations are high enough that the individual pollutants could cause injury by themselves, responses to the combination of the two may be less than the sum of the individual responses but may also be greater than the sum. Matsushima (1971) exposed five crop species (*P. vulgaris, L. esculentum, A. sativa, Cucumis sativus,* and *Capsicum frutescens*) to very much higher concentrations—namely, 1.5 to 2.3 ppm (62.3 to 95.6 μmol m^{-3}) of SO_2 and 12 to 15 ppm (499 to 623 μmol m^{-3}) of NO_2—for periods of 40 to 70 minutes. Although this work has little practical application, it does illustrate that the interaction between SO_2 and NO_2 is variable. *L. esculentum* was injured by both SO_2 alone and NO_2 alone at these high concentrations. When *L. esculentum* plants were exposed to SO_2 plus NO_2, injury to the fifth leaf (youngest) was less than additive, that to the fourth leaf was additive, and that to the second and third leaves greater than additive.

Amundson and Weinstein (1981) observed that leaf blemish on three *G. max* cultivars was less than additive, because of increased leaf resistances in the combined gas exposure. Similarly, Ashenden (1979a) noted that, although 0.1 ppm (4.2 μmol m^{-3}) of SO_2 by itself or of NO_2 by itself caused short-term increases in the transpiration rate of *Phaseolus vulgaris*, transpiration decreased from the second day onward when they were supplied together. Although such a response may protect the plant somewhat, by reducing the flux of pollutants into the leaf, it also may inhibit the net photosynthetic rate.

White, Hill, and Bennett (1974) showed that two-hour exposures to 0.25 ppm (10.4 μmol m^{-3}) of only SO_2 or of only NO_2 had little effect on net photosynthesis of *Medicago sativa*, but the same concentration of the two combined caused a 9% inhibition. This study also showed that the interaction altered with concentration, for as the concentrations were raised to 0.5 to 0.55 ppm (20.8 to

22.9 μmol m^{-3}) of each gas, the interaction was lost and effects became additive. Furukawa and Totsuka (1979) observed a similar response in *Helianthus annuus*. Two-hour exposures to 1 ppm (41.6 μmol m^{-3}) of NO$_2$ alone or to 0.2 ppm (8.3 μmol m^{-3}) of SO$_2$ alone did not affect the rate of net photosynthesis, but exposure to them together induced an inhibition of approximately 15%.

Concentrations of SO$_2$ and NO$_2$ as high as those employed in these studies are uncommon in the field. However, Bull and Mansfield (1974) noted depressions in net photosynthesis of *Pisum sativum* at concentrations that may easily occur in some situations, namely, 0.05 to 0.25 ppm (2.1 to 10.4 μmol m^{-3}). A mixture of as little as 0.05 ppm (2.1 μmol m^{-3}) SO$_2$ and NO$_2$ inhibited net photosynthesis; although there was no evidence of interaction, the effects of the two pollutants were at least additive.

These studies have illustrated that NO$_2$, if present with SO$_2$, may lower the threshold concentrations required to induce leaf blemish or to inhibit net photosynthesis in several crop species. The response appears to vary among species and cultivars, and is related to concentration. It also seems likely that environmental factors and the physiological state of the plant will influence the response, but as yet these have not been investigated. These studies of visible injury do not allow us to predict the effects of SO$_2$ and NO$_2$ on plant growth, and in many cases the concentrations used would not occur widely or for extended periods in the field. However, studies of long-term exposure to lower levels of SO$_2$ and NO$_2$ have recently been conducted.

Effects of prolonged exposures. Ashenden and Mansfield (1978) exposed four common grasses (*Dactylis glomerata*, *Phleum pratense*, *Poa pratensis*, and *Lolium multiflorum*) to SO$_2$ and NO$_2$ for 20 weeks during a winter period. The exposure concentrations were 0.11 ppm (4.6 μmol m^{-3}) for 5 days each week, giving weekly means of 0.068 ppm (2.8 μmol m^{-3}). By the end of the exposure, growth of all four grasses was severely inhibited by the mixture of SO$_2$ plus NO$_2$; *L. multiflorum* weighed just 48% of controls, *Phleum pratense* only 14%. Responses to the single-gas exposures differed from species to species; whereas the effects of the mixture were additive in *Poa pratensis*, they were greater than additive in the other three. Consequently, Ashenden and Mansfield state, "If the toxicity . . . of this mixture of two pollutants had been estimated from their effects when applied separately, it would have been seriously underestimated for three out of (the) four species."

During this experiment, growth was measured every month, as reported by Ashenden (1979b) and Ashenden and Williams (1980). There was evidence that the responses altered as the exposure continued. For example, although SO$_2$ alone and NO$_2$ alone reduced dry weight of *L. multiflorum* at 84 days, these effects had disappeared by 140 days' exposure. In general, however, the magnitude of the inhibition caused by the combined gas exposure increased with the length of exposure.

Further investigations into the response of *P. pratensis* to SO$_2$ and NO$_2$ have been reported recently (Whitmore and Mansfield, 1983). Plants were grown in the outdoor fumigation system developed by Ashenden *et al.* (1982), and were exposed to concentrations of 0.1 ppm (4.2 μmol m^{-3}) for 5 days a week, the weekly means being 0.062 ppm (2.6 μmol m^{-3}). Seedlings of *Poa pratensis* were exposed from emergence in the autumn, and ten individuals were harvested on each of six occasions until the following May (Fig. 16.1). By this time total dry weight was only 55% of that of the controls for plants exposed to a single gas, and only 26% for plants exposed to both gases. Severe effects in the latter plants developed much earlier than did comparable effects in plants exposed to individual pollutants; the harvests in February and March revealed substantial growth reductions in plants exposed to both SO$_2$ and NO$_2$, but no significant effects on plants exposed to SO$_2$ or NO$_2$ alone (Fig. 16.1). Consequently at these times the interactions between SO$_2$ and NO$_2$ were statistically significant ($p <$

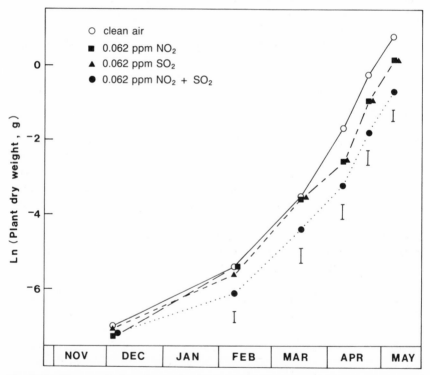

Fig. 16.1. The progression of growth of *Poa pratensis* cv. Monopoly from emergence, exposed to clean air or to air containing weekly mean concentrations of 0.062 ppm (2.6 μmol m^{-3}) of NO$_2$, of SO$_2$, or of NO$_2$ plus SO$_2$ during winter 1979–80. Values are means of ten replicates except at the first harvest where mean measurements only were made. Least significant differences ($p = 0.05$) are shown as bars. Reproduced from Whitmore and Mansfield (1983), p. 220.

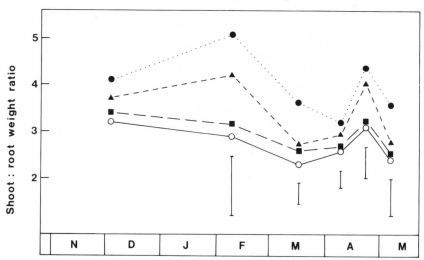

Fig. 16.2. Shoot : root weight ratios of *Poa pratensis* cv. Monopoly; same growing conditions and symbols as in Fig. 16.1. Values are means of ten replicates except at the first harvest where mean measurements only were made. LSDs ($p = 0.05$) are shown. Reproduced from Whitmore and Mansfield (1983), p. 221.

0.01). In the plants exposed to a single gas, there was an abrupt change between the absence of any effects on plant dry weight in March and the onset of marked growth reductions just 22 days later. Since the appearance of these large effects was accompanied by only a slight increase in the effect on plants exposed to both gases, the interaction was lost and effects then became additive.

Classic growth-analysis techniques were applied to the results to investigate the responses in more detail. The slower relative growth rates in plants subjected to the three pollution treatments were found to be caused by a lowering of their net assimilation rates. However, in plants subjected to SO_2 and to SO_2 plus NO_2, two responses were observed that increased the relative size of the leaf area and thus minimized the effect that the lower net assimilation rate had on growth. First, the partitioning of dry matter within the plants was altered in favor of the leaves at the expense of the roots, as was reflected by the increased shoot : root ratios in these treatments (Fig. 16.2). Even though NO_2 inhibited dry-weight gain as much as SO_2 alone, at no time did it significantly alter the partitioning of dry matter. The second response to exposures to SO_2 involved a reduction in leaf thickness (observed in February, April, and May). Both these responses effectively increased the relative size of the assimilatory surface and counteracted the reduced efficiency per unit leaf area.

There have been several reports that SO_2 reduces the roots of grasses more than the shoots (Ashenden and Mansfield, 1977; Bell, Rutter, and Relton, 1979; Ayazloo and Bell, 1981; Jones and Mansfield, 1982a), but specific leaf areas have rarely been quoted. Nevertheless, in one of the experiments of Bell, Rutter, and Relton (1979) and in those of Jones and Mansfield (1982a), SO_2 was reported to cause reductions in leaf thickness. The results of Whitmore and Mansfield (1983) show the same kinds of morphological responses to air that contains both SO_2 and NO_2.

A biochemical explanation for the severe growth effects that occur when grasses are exposed to a mixture of SO_2 and NO_2 was proposed by Wellburn *et al.* (1981). Four grasses (*Lolium perenne*, *Dactylis glomerata*, *Phleum pratense*, and *Poa pratensis*) were exposed to 0.068 ppm (2.8 μmol m^{-3}) of SO_2 and NO_2, both singly and in combination, as described by Ashenden and Mansfield (1978). Wellburn *et al.* found that after plants had been exposed for 20 weeks to NO_2 alone, their nitrite-reductase activities were approximately double those of the controls. Although SO_2 alone had little effect on *L. perenne* and *D. glomerata*, this treatment produced *lower* nitrite-reductase activities in *Phleum pratense* and *Poa pratensis*. When plants were exposed to both SO_2 and NO_2, the stimulation in activity of this enzyme was not seen; the presence of SO_2 prevented the additional activity normally induced by NO_2. Plants exposed to SO_2 and NO_2 together therefore could not detoxify the NO_2 as well, and were thus exposed to damage from both pollutants. Wellburn *et al.* also observed that SO_2, alone or combined with NO_2, reduced cyclic photophosphorylation, and as a result reduced the supply of ATP, which is essential for the detoxification of pollutant derivatives.

Whitmore and Mansfield (1983) investigated the effects of a longer exposure to SO_2 and NO_2 on the growth of *Poa pratensis*. Responses during the winter period were comparable to those previously described; interactive effects of SO_2

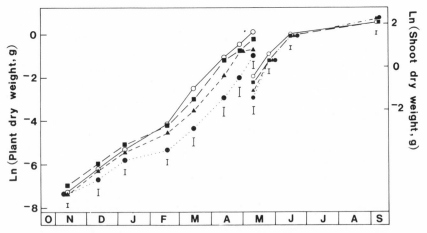

Fig. 16.3. The progression of growth of *Poa pratensis* cv. Monopoly; same growing conditions and symbols as in Fig. 16.1, except that these plants were grown during winter 1980–81. Up to early May 1981 total dry weights are shown; after early May only shoot dry weights are shown. Values are means of ten replicates. LSDs ($p = 0.05$) are shown. Reproduced from Whitmore and Mansfield (1983), p. 223.

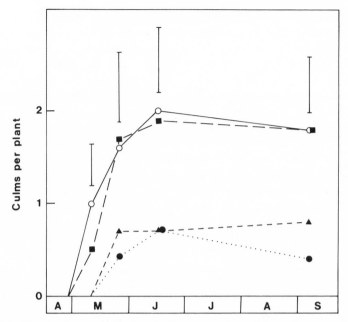

Fig. 16.4. The development of culms in *Poa pratensis* cv. Monopoly; same data as in Fig. 16.3, symbols as in Fig. 16.1. Values are means of ten replicates. LSDs ($p = 0.05$) are shown. Reproduced from Whitmore and Mansfield (1983), p. 225.

and NO$_2$ were observed at approximately the same time of year, before significant effects developed in the plants exposed to only a single gas (Fig. 16.3). After the effects of single gases were expressed, the net effect of the combined gases became additive. The harvest on May 11 was the last at which it was feasible to separate the roots from the shoots, and from this time onward in Fig. 16.3 the dry weights of only the shoots are shown. However, on May 11 the reductions in shoot dry weight were similar to those in plant dry weight, so we can identify trends that continue through the data even though a full growth analysis was not possible with the later harvests. There was a rapid recovery of the shoot dry weights between May and June, and at the end of the exposure in September the plants exposed to SO$_2$ or to SO$_2$ and NO$_2$ weighed slightly more than the controls ($p < 0.05$).

During this period flowering occurred. NO$_2$ slightly delayed the development of flowering stems, but SO$_2$, alone or combined with NO$_2$, not only delayed stem elongation significantly, but also severely reduced the number of flower heads (Fig. 16.4). The recovery of vegetative growth was therefore accompanied by a marked inhibition of reproductive growth in plants exposed to SO$_2$ or to SO$_2$ and NO$_2$. Since less of the shoot in these exposed plants was committed to reproductive growth, they had a greater capacity for leaf area development than the controls.

The effects of air pollutants on plant growth clearly differed at different times during this experiment, perhaps because of changes in environmental conditions, particularly in light and temperature, or perhaps because sensitivity to the pollutants depended partly on the physiological stage of development of the plant. In addition, there may have been some degree of acclimatization, for compensatory growth mechanisms were observed in the responses to SO_2. Moreover, the nature of the interaction between SO_2 and NO_2 also changed. The main effect of the interaction, noted in late winter, was to reduce growth sooner than in plants exposed to only single gases. Although the plants exposed to SO_2 and NO_2 weighed only 17% as much as the controls in March, they recovered during the summer, as did plants exposed to a single gas. It is likely, however, that the delayed development in early spring contributed to the inhibition of flowering observed in the summer.

This research has concentrated on grasses because of the importance of the productivity of grasslands in western Europe. Recently investigations have been extended to other species. Whitmore and Freer-Smith (1982) reported effects of SO_2 and NO_2 on six broadleaf tree species. Second-year cuttings of *Malus domestica*, *Betula pendula*, *B. pubescens*, *Populus nigra*, and *Alnus incana*, as well as seedlings of *Tilia cordata*, were exposed to SO_2 and NO_2, singly and in combination. Weekly mean concentrations of single and combined gases were 0.062 ppm (2.6 μmol m^{-3}), and plants were fumigated from before budbreak in March until mid-August. The range of responses to the pollutants is shown in Fig. 16.5, where mean dry weights are expressed as percentages of controls in histogram form. There were interactions in the effects of SO_2 and NO_2 in four species (*T. cordata*, *B. pendula*, *P. nigra*, and *B. pubescens*), which suffered greater than additive reductions in growth. Comparing the effects of a mixture of SO_2 + NO_2 on seedlings and on cuttings of *B. pendula* showed that a smaller response to exposure could be detected with clonal plant material, because there were fewer differences between individual plants than there were with tree seedlings.

Time-concentration-response relationships. O'Gara (1922; cited by Guderian, 1977) was the first to relate injury to the concentration of pollutant and the length of exposure, after studying SO_2-induced visible injury to alfalfa. Several other relationships have since been developed in studies describing plant responses to SO_2, O_3, or mixtures of the two (Heck, Dunning, and Hindawi, 1966; Macdowall and Cole, 1971; Temple, 1972; Jacobson and Colavito, 1976). These describe the acute injury caused by short exposures to high concentrations of pollutants and allow us to predict the injury that may result from a given exposure. Only rarely has this approach been used for the study of chronic pollutant injury. Recently Bell (1982) reviewed all the literature about chronic injury to *Lolium perenne* cv. S23 from exposure to SO_2, but was unable to detect a convincing relationship between dose (concentration × time) and reductions in shoot dry weight, probably because the experiments reviewed were conducted under diverse environ-

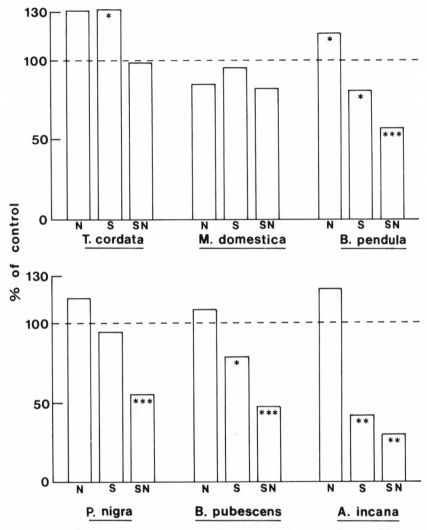

Fig. 16.5. Mean dry weights of six broadleaf tree species exposed to weekly mean concentrations of 0.062 ppm (2.6 μmol m^{-3}) of SO$_2$ (S), of NO$_2$ (N), or of both SO$_2$ and NO$_2$ (SN), expressed as percentage of dry weight of controls grown in clean air. Treatment means that are significantly different from control means are marked: *$p < 0.05$; **$p < 0.01$; ***$p < 0.001$. Reproduced from Whitmore and Freer-Smith (1982), p. 56.

mental and cultural conditions that alter plant response in different ways (Mansfield and Jones, this volume).

Recently an attempt was made to relate the dose of SO$_2$/NO$_2$ mixtures to effects on the growth of *Poa pratensis* (Whitmore, 1982). Experiments were conducted under closely controlled environmental conditions, and plants were exposed to mixtures of SO$_2$ and NO$_2$, in 1:1 ratio, from the time of emergence, for a period of 20, 34, or 38 days. Three exposure concentrations were used: 0.04 ppm (1.7 μmol m^{-3}) of each gas, a concentration that occurs over large areas of rural land in western Europe (Fowler and Cape, 1982); 0.07 ppm (2.9 μmol m^{-3}) of each gas; and 0.1 ppm (4.2 μmol m^{-3}) of each gas. Controls were grown for comparison in air containing approximately 0.007 ppm (0.3 μmol m^{-3}) of SO$_2$ and of NO$_2$. Plants were grown under 200 to 250 μmol m^{-2} s^{-1} PAR, 12-hour photoperiods, 22 to 25°C (day) and 16 to 19°C (night) temperatures, RH 50 to 60%.

The percentage reductions in plant dry weight of exposed plants compared

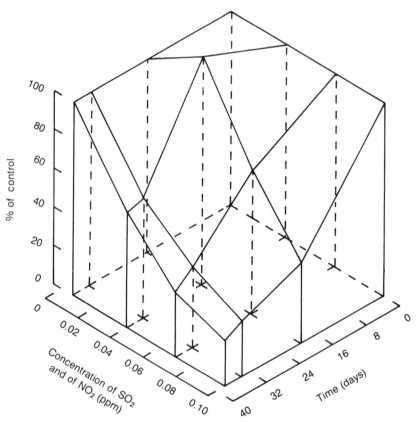

Fig. 16.6. Relationship between concentration of SO$_2$/NO$_2$ mixtures, duration of exposure, and percentage inhibition of dry weight gain in *Poa pratensis*. Plants were exposed to 0.007 ppm (0.29 μmol m^{-3}) of NO$_2$ and of SO$_2$ (control), 0.04 ppm (1.66 μmol m^{-3}) of NO$_2$ and of SO$_2$, 0.07 ppm (2.91 μmol m^{-3}) of NO$_2$ and of SO$_2$, and 0.1 ppm (4.16 μmol m^{-3}) of NO$_2$ and of SO$_2$ for periods of 20, 34, and 38 days, commencing at emergence.

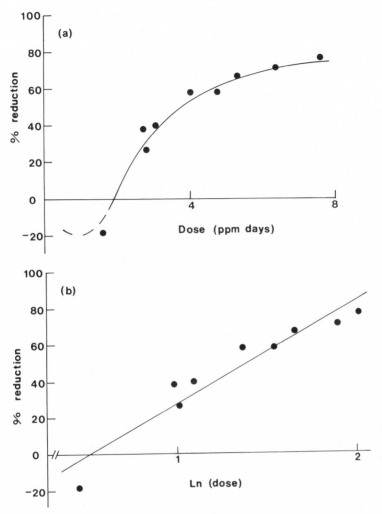

Fig. 16.7. Relationship between dose of SO$_2$/NO$_2$ mixture and response of *Poa pratensis* in terms of percentage inhibition in dry weight relative to dry weight of controls: (*a*) dose calculated as concentration of SO$_2$ plus that of NO$_2$ times the duration; (*b*) dose logarithmically transformed ($y = 55.88x - 29.15$; $p < 0.001$).

with controls are shown in an isometric projection in Fig. 16.6. It is noteworthy that exposure to only 0.04 ppm (1.66 μmol m^{-3}) of SO$_2$ and of NO$_2$ for 34 to 38 days severely inhibited dry-weight gain; these plants weighed only 60 to 62% as much as controls. In general, the response increased with both length of exposure and concentration. There was also some evidence of reciprocity between concentration and time, since equivalent doses (represented by the same symbol), composed of different concentrations and times, elicited similar responses. Although this hypothesis requires substantiating, it appears that, within the limitations of the experimental design, chronic injury from SO_2/NO_2 mixtures can be discussed in terms of a simple dose concept.

Dose, calculated as the concentration of SO_2 plus the concentration of NO_2 multiplied by the duration of exposure, is plotted against the percentage reduction in dry weight compared with dry weight of controls (Fig. 16.7,a). The lower end of the curve cannot be adequately characterized without further information on the effects of low doses. However, as dose increases, there is a sharp transition between no effect, or small stimulations, and the onset of severe growth reductions. When dose was logarithmically transformed, a linear relationship was revealed (Fig. 16.7,b). This relationship cannot be used to predict yield losses in the field, because response will be modified by the physiological state of the plant and by the multivariate microclimate. However, if we recognize that a clear dose-response relationship can be established under one precisely defined set of conditions, we may be able to discover how it is modified by other factors.

Suggestions for Future Research

It is clear that mixtures of SO_2 and NO_2 are often much more phytotoxic than SO_2 alone, and since SO_2 is usually accompanied by NO_x in polluted atmospheres, there is ample justification for studying their combined effects. Our knowledge of how mixtures of SO_2 and NO_x affect plant growth is limited to only a few species; so there is an urgent need to investigate responses of other species, especially major crops. It will be desirable to undertake comparative studies to identify species and cultivars that are particularly sensitive to these pollutants. However, comparisons based on results obtained at a single harvest may be of only limited value, because responses often change during an exposure of long duration.

Most investigators relate plant response to the concentration of pollutants in the surrounding atmosphere. However, since response will probably be more directly related to the amounts of pollutants entering the leaves, researchers might do well to measure fluxes of pollutants into the plant during exposures before examining how dose-response relationships may be modified by other factors.

This research needs to be extended to include other pollutants, especially ozone, but first we need to elucidate further the effects of NO_x. Generally NO_2 has been used as an approximation to mixtures of NO and NO_2 because the two gases act similarly in cells. However, a given concentration of NO_2 may be more (or less) toxic than the same concentration of a mixture of NO and NO_2 because of differences in rates of uptake. Since it is difficult to fumigate with NO without NO_2 being present, we cannot make a totally satisfactory comparison of their effects on plants. Nevertheless, it should be possible to compare the phytotoxicity of an SO_2/NO_2 mixture with that of a mixture of SO_2, NO, and NO_2.

Concentrations of SO$_2$ and NO$_x$ fluctuate continuously in the field; yet most experimenters employ steady exposure concentrations. Responses of plants to mixtures of SO$_2$ and NO$_x$ could be considerably modified by the pattern of these fluctuations, depending on whether peaks of these gases are sequential or simultaneous. Although more complex than research using steady exposure concentrations, investigations into the effects of fluctuating levels of SO$_2$ and NO$_x$ on plant growth are clearly required.

Acknowledgment

Financial support from the Natural Environment Research Council is gratefully acknowledged.

17

The Effects of SO$_2$ and O$_3$ on Plants

Robert Kohut

The need to investigate and understand the effects on vegetation produced by the interaction of SO$_2$ and ozone (O$_3$) arises from the known phytotoxic properties of the individual pollutants, their widespread distribution, and their frequent occurrence together or in sequence. Developing experiments to evaluate their interaction becomes problematic when we examine the individual pollutant exposure regimes and consider the possible combinations of these regimes. Sulfur dioxide is released from both isolated point sources and area sources. The exposure regimes produced by the two types of sources differ greatly in the maximum concentration produced as well as in the temporal and spatial distribution of SO$_2$. The type of source, the magnitude of SO$_2$ release, and the influences of local topography and climate make it extremely difficult to develop generalized SO$_2$ exposure regimes. Ozone, on the other hand, is ubiquitous in its distribution because it is a natural component of the atmosphere, where it generally is found in low concentrations. The presence of anthropogenic O$_3$ can result in short-term atmospheric concentrations that are ten or more times greater than the natural concentration. Because O$_3$ is generated by photochemical processes, anthropogenic O$_3$ concentrations generally follow a diurnal pattern of occurrence. Although the timing and magnitude of this pattern tend to be somewhat similar within a local area, there are significant differences in these characteristics between widely distributed sites. Weather patterns are also important, in that certain conditions result in long-distance transport of O$_3$.

Given the complexity of the exposure regime of an individual pollutant, researchers are confronted with an even wider range of exposure possibilities when they consider the pollutants in combination and their possible effects on plants. Such factors as the maximum concentrations of the individual pollutants, the frequency and duration of exposure, the use of sequential or simultaneous exposures, and the environmental conditions during exposure are all important. These factors may also play a role in determining whether the pollutants will interact when producing an effect on plants. Because of this complexity, research investigating the interaction of SO$_2$ and O$_3$ has moved slowly, and has generally employed simplified systems that provide only a preliminary understanding of the nature of the interactions, their significance in the field, and their modes of action.

This review addresses our understanding of the interaction of SO$_2$ and O$_3$ as it has been developed by laboratory and field studies. Although it is clear that exposures to these two pollutants are often accompanied by altered concentrations of NO, NO$_2$, and CO$_2$, studies of the interactions of these other pollutants as a third component are limited and will not be emphasized here. Although research may be undertaken for a variety of reasons, one common objective is to develop scientific knowledge that can be applied to the betterment of environmental quality. If this is, in fact, the case, three principal goals emerge (McCune, 1983): the diagnosis of pollutant-induced effects; the prevention of pollutant-induced effects by formulation of air-quality criteria and subsequent standards; and the ability to predict pollutant-induced effects. These goals should be kept in mind throughout this review. They represent problems of great practical significance that SO$_2$/O$_3$ interaction research should address.

Overview

Interest in the interactions of SO$_2$ and O$_3$ was spurred by the work of Menser and Heggestad (1966) with *Nicotiana tabacum* cvs. "Bel-W3" and "Bel-B" (tobacco). They exposed plants to subthreshold concentrations of the individual pollutants without producing foliar injury. However, when plants were exposed to the combined pollutants, each at a subthreshold concentration, foliar injury was produced. This response is a classic example of a pollutant interaction in which the combined pollutants produce a greater than additive effect, that is, an effect greater than the sum of the effects of the individual pollutants.

Since the publication of Menser and Heggestad (1966), many research projects to examine SO$_2$/O$_3$ interactions have been conducted. Most of these studies can be placed into one of two categories: (1) characterization studies, in which the objective is to examine the plant response, document the symptoms, and find out whether the effects were additive or not; or (2) sensitivity screenings, in which the responses of species or cultivars to combined pollutant exposures are compared. Only a few physiological and dose-response studies have been conducted with SO$_2$ and O$_3$ in combination.

Response characterization. Working with 11 plant species (*Medicago sativa* cv. "Vernal," alfalfa; *Allium cepa*, onion; *Glycine max* cv. "Scott," soybean; *Nicotiana tabacum* cvs. "Bel-B," "Bel-W3," and "White Gold"; *Phaseolus limensis* cv. "Thaxter," lima bean; *Brassica oleracea botrytis* cv. "Calabrese," broccoli; *Bromus inermis* cv. "Sac Smooth," Bromegrass; *Brassica oleracea capitata* cv. "All Season," cabbage; *Raphanus sativus* cv. "Cherry Belle," radish; *Spinacia oleracea* cf. "Northland," spinach; and *Lycopersicon esculentum* cv. "Roma VF," tomato) and several combinations of pollutant concentrations ranging from

0.10 to 1.00 ppm (4.16 to 41.6 μmol m^{-3}) of SO_2 and 0.05 to 0.10 ppm (2.08 to 4.16 μmol m^{-3}) of O_3, Tingey *et al.* (1973a) found significant differences in the degree of foliar injury produced by the combined pollutants. They concluded that the species, as well as the pollutant concentrations and ratios, played a major role in determining whether the pollutants would produce interactive effects. The symptoms produced by the combined pollutants differed from species to species, with both O_3-type and SO_2-type symptoms being found.

In similar studies conducted with cultivars, differences in the degree of injury produced by SO_2 and O_3 in combination were apparent on five cultivars of *Begonia × hiemalis*, Elatior begonia (Reinert and Nelson, 1980); eight cultivars of *Rhododendron* sp., azalea (Sanders and Reinert, 1982); 12 cultivars of *Poa pratensis*, Kentucky bluegrass (Elkiey and Ormrod, 1980a); and 33 cultivars of *Phaseolus vulgaris*, bean (Beckerson, Hofstra, and Wukasch, 1979). The response of *Pinus strobus* (eastern white pine) and *Populus tremuloides* (aspen) to SO_2 and O_3 singly and in combination has been evaluated using clonal stock. Houston (1974) found that *Pinus strobus* clones classified as sensitive and tolerant did not show consistent responses when exposed to the individual and combined pollutants. In addition, not all members of the sensitive clonal population had the same threshold concentration for injury. Five *Populus tremuloides* clones used by Karnosky (1976) also showed different thresholds for injury by the individual and combined pollutants both between and within clonal populations. At the injury threshold for the individual pollutants of either 0.35 ppm (14.56 μmol m^{-3}) of SO_2 or 0.05 ppm (2.08 μmol m^{-3}) of O_3 for 3 hours, 10 to 50% of the members of the sensitive clonal populations showed a response, but other clones remained uninjured. Exposure of the plants to the combined pollutants at their individual threshold concentrations increased the number of clones responding, the number of plants within a clone injured, and the percentage of leaves injured.

Many of these screening studies employed very little statistical analysis to evaluate the pollutant interactions. Most authors did, however, draw some conclusions about the interactions of the two pollutants. Additive or less than additive foliar injury responses to combinations of SO_2 and O_3 were reported for *Begonia × hiemalis* (Reinert and Nelson, 1980); *Rhododendron* sp. (Sanders and Reinert, 1982); *Petunia hybrida*, petunia (Elkiey and Ormrod, 1979a); *Phaseolus vulgaris*, white bean, and *Glycine max* (Hofstra and Ormrod, 1977); *Allium cepa, Phaseolus limensis, Bromus inermis, Spinacia oleracea, Brassica oleracea botrytis, Brassica oleracea capitata*, and *Lycopersicon esculentum* (Tingey *et al.*, 1973a); *Festuca arundinacea*, tall fescue (Flagler and Youngner, 1982); and 33 cultivars of *Phaseolus vulgaris* (Beckerson, Hofstra, and Wukasch, 1979). Greater than additive effects for the combined pollutants were found for *Raphanus sativus* and *Medicago sativa* (Tingey *et al.*, 1973a); *Pinus strobus* (Houston, 1974); *Populus tremuloides* (Karnosky, 1976); and several cultivars of *Nicotiana tabacum* (Menser and Hodges, 1970; Grosso *et al.*, 1971; Hodges, Menser, and Ogden, 1971; Menser, Hodges, and McKee, 1973).

These studies illustrate how much the responses of plants exposed to SO_2 and O_3 in combination can differ. The relative sensitivity of a species, cultivar, or clone to the individual pollutants is not a strong predictor of its response to the combined pollutants. Just as with SO_2 and O_3 singly, significant differences in sensitivity to the combined pollutants are found between species, cultivars, and clones. Even within clonal populations, thresholds for injury can differ greatly between individual plants.

Symptoms produced by the combined pollutants most often resemble those produced by O_3 alone (Elkiey and Ormrod, 1979a; Reinert and Nelson, 1980; Tingey et al., 1973a), although symptoms similar to those produced by SO_2 alone have been reported (Tingey et al., 1973a; Beckerson, Hofstra, and Wukasch, 1979). Sometimes the combined pollutants may produce a symptom that is unique and unlike those of either individual pollutant (Hofstra and Ormrod, 1977; Lewis and Brennan, 1978). Compared to the time required for the development of foliar injury produced by O_3 alone, symptom expression on plants exposed to SO_2 and O_3 in combination may be delayed several additional days (Hofstra and Ormrod, 1977; Beckerson, Hofstra, and Wukasch, 1979).

Growth effects. Research evaluating the effects of SO_2 and O_3 in combination on plant growth is limited, and has focused on *Raphanus sativus* and *Glycine max* as response models for the assessment of growth variables such as fresh and dry weights of total aboveground biomass, leaves, and roots and plant height. Working with *Glycine max*, Tingey et al. (1973b) found greater than additive effects by the combined pollutants in reducing top fresh weight and root fresh and dry weights of the cultivars "Hood" and "Dare." Exposures were for 40 hours a week for 3 weeks at 0.05 ppm (2.08 μmol m^{-3}) of each pollutant. When "Dare" was used in another study with an exposure regime of 4 hours of fumigation three times a week for 11 weeks at 0.25 ppm (10.4 μmol m^{-3}) of each pollutant, no indications of interactive reductions in shoot or root dry weights were found (Reinert and Weber, 1980). Differences in the exposure regime, and in other environmental variables, may be involved in the different growth responses of the soybean cultivar "Dare" in these two studies.

Studies of the growth of *Raphanus sativus* have generally indicated additive or less than additive effects for SO_2 and O_3 in combination (Tingey, Heck, and Reinert, 1971; Tingey and Reinert, 1975; Reinert and Gray, 1981; Reinert and Sanders, 1982). One study employing SO_2, O_3, and NO_2 statistically evaluated the interactions of SO_2 and O_3 across concentrations of NO_2, and found a greater than additive reduction in the fresh weight of *Raphanus sativus* foliage and roots (Reinert, Shriner, and Rawlings, 1982).

In two growth studies with *Populus deltoides* $\times$ *P. trichocarpa* (hybrid poplar), the combination of SO_2 and O_3 was found to have a less than additive effect in reducing foliar and shoot growth (Noble and Jensen, 1980a; Jensen, 1981b). Constantinidou and Kozlowski (1979a), evaluated the effects of the combined pollutants on the growth of *Ulmus americana* (American elm) seedlings. They assessed the response of new leaves, less than 1 cm, and young leaves, 1 to 10 cm, to acute exposure to 2.0 ppm (83.2 μmol m^{-3}) of SO_2 and 0.9 ppm (37.4 μmol m^{-3}) of O_3 singly for 5 hours, and combined at these concentrations for 5 hours with an additional hour of exposure to SO_2 alone. Young leaves showed reduced growth up to five weeks after exposure to the combined pollutants, but new leaves showed reduced growth for only one week. In both experiments the pollutants acted additively. No effect on stem dry weight was evident until the fifth week, when a greater than additive effect was detected. Evaluation of root dry weight showed an additive effect.

The evaluation of SO_2 and O_3 effects on plant growth is so fragmentary that few generalizations are possible. Most research has indicated that the pollutant effects are additive, but the limited number of species evaluated and the nature of the exposure regimes employed preclude drawing broad conclusions about growth effects. In addition, the growth studies were conducted under laboratory conditions, and may not accurately reflect plant response in the field.

The most clearly defined effect of SO_2 and O_3 on vegetation in the field is chlorotic dwarf of *Pinus strobus*. Intensive study of this disease failed to identify a biotic or abiotic pathogen responsible for the stunted, yellowed trees characteristic of the syndrome. Research conducted by Dochinger and Seliskar (1970) demonstrated that air pollution was responsible for the symptoms. Using charcoal-filtered air and controlled fumigations with SO_2 and O_3, they demonstrated that exposure to the pollutants induced the chlorotic dwarf symptoms and that the combined pollutants increased the extent and severity of injury. Chlorotic dwarf remains the most clearly documented example of an effect on vegetation in the field produced by the combined action of SO_2 and O_3.

Environmental variables. Since it has been clearly demonstrated that the response of plants to SO_2 and O_3 singly is influenced by environmental and exposure-regime variables, it is reasonable to conclude that plant response to the combined pollutants may also be affected by these variables. To date, however, the roles of environmental factors in mediating the interaction of SO_2 and O_3 have received very little attention. In an examination of the influence of temperature on plant response to combined pollutant exposures, Miller and Davis (1981) found that both the nature and the intensity of the foliar-injury symptoms were affected. When *Phaseolus vulgaris* cv. "Pinto" was exposed for 3 hours to 0.80 ppm (33.3 μmol m^{-3}) of SO_2 and 0.25 ppm (10.4 μmol m^{-3}) of O_3 in combination, O_3-type symptoms dominated at 32°C, SO_2-type symptoms at 15°C, and both were prevalent at 24°C. The individual pollutants produced the greatest amounts of injury at 15 and at 32°C, but maximum symptom production by the combined pollutants occurred at 15°C. The distribution of symptoms between the abaxial and adaxial leaf surfaces appeared to be a function of exposure temperature. On the adaxial surface, the combined pollutants produced less than additive foliar injury at 15 and 32°C and greater than additive injury at 24°C. At 15 and 24°C, the combined pollutants produced greater than additive injury on the abaxial and combined leaf surfaces, but less than additive injury at 32°C. If the response of *Phaseolus vulgaris* can be considered a model for a general effect, the temperature at which a plant is exposed to SO$_2$ and O$_3$ clearly influences the extent and type of foliar injury produced.

The interaction of leaf age and pollutant exposure concentration in influencing the response of *Pisum sativum* cv. "Alsweet" (pea) to SO$_2$ and O$_3$ has been examined (Olszyk and Tibbitts, 1982). Pollutant exposures designated high—1.25 ppm (52.0 μmol m^{-3}) of SO$_2$ and 0.13 ppm (5.41 μmol m^{-3}) of O$_3$ for 8 hours— and low—0.11 ppm (4.58 μmol m^{-3}) of SO$_2$ and 0.11 ppm (4.58 μmol m^{-3}) of O$_3$ for 4 hours—were used. The high concentration produced more injury on expanded than expanding leaves, but the low concentration produced similar degrees of injury on both classes of foliage. With both concentrations, however, the degree of foliar injury produced by the combined pollutants was less than additive.

Given the potential complex of pollutant concentrations, exposure, durations, and sequences of exposure that can arise when one is considering SO$_2$ and O$_3$ in combination, we could anticipate that the specific nature of the exposure regime would significantly influence plant response. Since the plant leaf surface represents a finite area within which a plant can express visible injury, the concentrations of SO$_2$ and O$_3$ and the magnitude of the plant's response to the individual pollutants can influence its ability to express enough injury to reveal a pollutant interaction. With *Glycine max* cvs. "Lee 60" and "Dare," less than additive, additive, and greater than additive responses to SO$_2$ and O$_3$ in combination were obtained, depending on the concentrations of the individual pollutants and the amount of injury they produced (Heagle and Johnston, 1979). Exposures were conducted using SO$_2$ concentrations of 0.50, 1.00, and 1.50 ppm (20.8, 41.6, and 62.4 μmol m^{-3}) and O$_3$ concentrations of 0.25, 0.50, and 1.0 ppm (10.4, 20.8, and 41.6 μmol m^{-3}) for .75, 1.5, and 3.0 hours. Greater than additive effects were produced when the individual pollutant concentrations were low and produced less than 45% foliar injury. When the degree of injury produced by pollutants was greater than 45%, the response to the combined pollutants was generally less than additive.

Although nearly all SO$_2$/O$_3$ interaction studies have employed simultaneous exposures to the two pollutants, the potentiation or predisposition effect of sequential exposures to the pollutants has received little attention. Pretreatment with SO$_2$ has been shown to alter the response of plants to a succeeding exposure to SO$_2$ and O$_3$ in combination (Hofstra and Beckerson, 1981). In an experiment that used plants exposed only to the combined pollutants for comparison, a two-day pretreatment with 0.15 ppm (6.24 μmol m^{-3}) of SO$_2$ prior to exposure to 0.15 ppm (6.24 μmol m^{-3}) of SO$_2$ and 0.15 ppm (6.24 μmol m^{-3}) of O$_3$ increased foliar injury on *Phaseolus vulgaris* cv. "Sanilac," decreased injury on *Lycopersicon esculentum* cv. "Tiny Tim" and *Cucumis sativus* cv. "National Pickling" (cucumber), and had no effect on *Glycine max* cv. "Harasoy" and *Raphanus sativus* cv. "Cherry Belle." Pretreatment delayed the onset of injury in *Raphanus sativus* and *Lycopersicon esculentum*, and advanced the onset of injury in *Phaseolus vulgaris*. When the pretreatment period was extended to three days, the responses to the combined pollutants were again altered. In comparison to injury to plants without an SO_2 pretreatment, foliar injury on pretreated *Phaseolus vulgaris* and *Raphanus sativus* was reduced, but injury on *Glycine max* and *Cucumis sativus* was unchanged. In an effort to explain the influence of the SO_2 pretreatment on subsequent response, Hofstra and Beckerson evaluated leaf diffusive resistance of *Phaseolus vulgaris* and *Cucumis sativus* during pretreatment and during the subsequent combined pollutant exposure. In both species, the SO_2 pretreatment tended to decrease diffusive resistance slightly. With *Phaseolus vulgaris*, the SO_2-pretreated plants had a much higher diffusive resistance after exposure to the combined pollutants for several days than did the plants without the pretreatment. In spite of this, foliar injury was greater in the pretreated plants. The pattern of diffusive resistance did not conform with or explain the pattern of injury development in *Phaseolus vulgaris*. Diffusive resistance was similar for both the pretreated and the non-pretreated *Cucumis sativus* plants after exposure to SO_2 and O_3 in combination.

Pollutant uptake and stomatal resistance. Stomates are known to play an important role in mediating the response of plants to air pollutants by affecting pollutant uptake. Evaluation of stomatal action in combined SO_2 and O_3 exposures could provide insight into mechanisms that control plant responses. Three *Petunia hybrida* cultivars differing in sensitivity to O_3 were used in a series of studies investigating the uptake of the combined pollutants (Elkiey and Ormrod, 1979a, 1980a,b, 1981a,c). The cultivars "Capri," "White Magic," and "White Cascade" were exposed to 0.80 ppm (33.3 μmol m^{-3}) of SO_2 and 0.40 ppm (16.6 μmol m^{-3}) of O_3 singly and in combination for 4 hours at a temperature of 25°C and at 51 and 90% relative humidity. Stomatal diffusive resistance was measured throughout the exposures. Although the three cultivars differed in their sensitivity to O_3, they all attained similar higher degrees of diffusive resistance when exposed at 51% relative humidity, but the rates at which their resistances increased were somewhat different. The O_3-sensitive cultivar "White Cascade" had the highest rate of increase in diffusive resistance, and the O_3-resistant cultivar "Capri" the lowest. Exposure to SO_2 did not produce any significant changes in the diffusive resistances of the cultivars. The combined pollutant treatment produced a more rapid and pronounced increase in diffusive resistance in all cultivars than did O_3. The resistance remained high throughout the exposure. In all treatments and cultivars, diffusive resistance quickly returned to normal after the exposures were terminated. At 90% relative humidity, there was little change in the diffusive resistance of any cultivar in any of the treatments. Pollutant absorption rates were also evaluated for each cultivar. Leaves of the least O$_3$-sensitive cultivar "Capri" absorbed less SO$_2$ and O$_3$ in single pollutant exposures than did the intermediately sensitive "White Magic" and the sensitive "White Cascade," which generally had the highest absorption rate. For all cultivars, there was less absorption of each of the gases from the pollutant mixture than of gases in the single-gas exposures. When the foliar sulfate concentration was evaluated as an index of pollutant uptake after exposure to SO$_2$ and O$_3$ together at 51% relative humidity, there was much less sulfate in the foliage of "White Magic" and "White Cascade" than after exposure to SO$_2$ alone. Sulfate concentrations in the foliage of "Capri" were similar for SO$_2$ alone and the SO$_2$/O$_3$ combination. Sulfate concentrations in tissue of all three cultivars exposed to SO$_2$ alone were similar. When the same set of exposures was conducted at 91% relative humidity, sulfate accumulation from the SO$_2$ treatment was substantially increased in all cultivars, and all showed lower sulfate concentrations after the combined pollutant treatment. Evaluation of the adsorption rates of the three cultivars indicated that the least sensitive ("Capri") adsorbed the most SO$_2$ and O$_3$, but the most sensitive ("White Cascade") adsorbed the least. Adsorption rates of the individual pollutants by each cultivar were similar whether the pollutants were applied singly or in combination.

This series of experiments with *Petunia hybrida* provides evidence for differences among cultivars in stomatal resistance to O$_3$ alone early in the exposure, followed by similar degrees of resistance as exposure progressed. The rates of increase in stomatal resistance were not directly related to the degree of O$_3$ resistance of the cultivar. The decreased uptake during the combined pollutant exposure appears to have been a result of the rapid increase in stomatal resistance. Sulfate concentrations in the foliage of plants exposed at 51% relative humidity indicate that stomatal closure reduced SO$_2$ uptake. However, at 90% relative humidity there was little change in stomatal resistance in any treatment, but reduced sulfate concentrations still occurred in the plants exposed to the combined pollutants.

Nine cultivars of *Poa pratensis* that differ in their sensitivity to SO$_2$ and O$_3$ were used to evaluate uptake rates of the gases when applied singly and in combination (Elkiey and Ormrod, 1981b). Leaves of the less-sensitive cultivars generally absorbed less of the single gases than did those of the more-sensitive cultivars. Uptake of each gas from the mixture was less than in the single-gas exposures. Adsorption of the gases by the cultivars differed, and was not generally related to cultivar sensitivity. The adsorption rates for SO$_2$ by the cultivars were similar in the single and mixed pollutant exposures, but the rates for O$_3$ were generally higher in the single-gas exposure than in the combined exposure.

The pollutant-uptake studies have focused on evaluating a plant's ability to remove a specific pollutant from a mixture, in contrast to its ability to remove the pollutant when it occurs alone. Unfortunately, none of the studies assessed foliar injury or growth to allow the changes in uptake rates to be related to effects produced by the combined pollutant exposures.

An evaluation of leaf diffusive resistance and associated degrees of foliar injury in *Raphanus sativus* cv. "Champion," *Cucumis sativus* cv. "National Pickling," and *Glycine max* cv. "Harosoy" was conducted by Beckerson and Hofstra (1979a). They found that the levels of foliar injury produced by the combined pollutants, 0.15 ppm (6.24 μmol m^{-3}) of SO$_2$ and of O$_3$ for 6 hours a day for 5 days, could not be explained in terms of leaf diffusive resistance. In all three species, SO$_2$ reduced diffusive resistance, O$_3$ increased diffusive resistance, and the combined pollutants increased diffusive resistance much more than O$_3$ alone. In contrast, foliar injury produced by the combined pollutants was greater than additive for *Raphanus sativus* and *Cucumis sativus* and less than additive for *Glycine max*. A delay observed in symptom expression in *Glycine max* after the combined pollutant exposure may have been associated with the increased stomatal resistance. However, the early appearance and increased severity of injury in *Raphanus sativus* and *Cucumis sativus* after exposure to SO$_2$ and O$_3$ are inconsistent with the increased diffusive resistance in these species.

In a similar study using *Phaseolus vulgaris* cv. "Sanilac" (Beckerson and Hofstra, 1979b), 0.15 ppm (6.24 μmol m^{-3}) of SO$_2$ produced a slight decrease in diffusive resistance, 0.15 ppm (6.24 μmol m^{-3}) of O$_3$ an increase, and the combined pollutants an increase greater than that of O$_3$. Injury in the combined pollutant treatment was substantially less than that produced by O$_3$, but the decrease was of a magnitude not directly relatable to the differences in diffuse resistance between the two treatments.

In the studies evaluating leaf diffusive resistance in the single and combined pollutant exposures and the concomitant foliar injury, changes in resistance during the combined pollutant exposures show no relationship to the amount of injury produced. Differences in injury must be related to other physiological or biochemical mechanisms that alter the effects of the pollutant mixture.

Photosynthesis. The effect of exposure to combinations of SO$_2$ and O$_3$ on photosynthesis has been examined in a few studies. *Vicia faba* (field bean) was exposed to 0.04 (1.66 μmol m^{-3}) of SO$_2$ and to a range of O$_3$ concentrations from 0.04 to 0.20 ppm (1.66 to 8.32 μmol m^{-3}) for 4 hours, and the effects of the individual and combined pollutants on photosynthesis were then evaluated (Ormrod, Black, and Unsworth, 1981; Black, Ormrod, and Unsworth, 1982). As the O$_3$ treatment concentration increased above 0.04 ppm (1.66 μmol m^{-3}), the photosynthetic rate decreased. When SO$_2$ was added to the O$_3$, photosynthesis was decreased at all O$_3$ concentrations, including those below 0.04 ppm. However, the rate at which photosynthesis decreased with increasing O$_3$ concentrations was greater for O$_3$ alone than for the combined pollutants. For this reason, the combined pollutants dominated in reducing photosynthesis at the lower O$_3$ concentrations, but at approximately 0.15 ppm (6.24 μmol m^{-3}) O$_3$ the reduction due to O$_3$ alone began to exceed that produced by the combined pollutants. Photosynthetic reductions produced by O$_3$ alone at higher concentrations were adequate to mask the presence and influence of SO$_2$. The importance of pollutant concentrations and their ratios in determining the combined pollutant response is clear in this study.

Interacting with the pollutant-induced effects on photosynthesis are the influences of other environmental variables. The roles of light intensity and relative humidity in conditioning the foliar injury and photosynthetic responses of *Acer saccharum* (sugar maple) and *Fraxinus americana* (white ash) to combinations of SO$_2$ and O$_3$ have been explored (Carlson, 1979). Three levels of light intensity, low, medium, and high, and two levels of relative humidity, 22 to 43% and 55 to 92%, were used. A greater than additive reduction in photosynthesis was produced in both species by the combined pollutants at 0.50 ppm (20.8 μmol m^{-3}) each after two days of exposure. After one week of exposure the greater than additive reduction persisted only in *Fraxinus americana*. Although the combined pollutants decreased photosynthesis under all light intensities, the greatest reductions occurred at the high and low intensities. Photosynthetic rates for plants exposed to the combined pollutants at the higher relative humidity were slightly lower than for those exposed at the lower relative humidity. Evaluation of the foliar injury produced by the combined pollutants at different combinations of light and humidity indicated that injury was produced in all exposures at high light intensity independent of the percentage of relative humidity. Neither species was injured by exposures at low light and low relative humidity. Fumigation at high humidity and low to intermediate light intensity caused foliar injury in some, but not all, exposures. Comparison of foliar injury and photosynthesis indicated that the photosynthetic rate for visibly injured foliage was not always lower than that for uninjured foliage. Fumigation at low humidity and low light, though not inducing visible injury, resulted in significant reductions in photosynthesis. The use of foliar injury as an indirect indicator of reduced photosynthesis may not be reliable.

The limited research examining the combined effects of SO$_2$ and O$_3$ on photosynthesis indicates that the pollutants can act in a greater than additive fashion in reducing photosynthetic rates, although the exposure bounds over which this effect is produced may be limited. In addition, environmental factors during fumigation can modify the effects of the combined pollutants on photosynthesis. Careful consideration needs to be directed toward environmental variables and pollutant concentrations in experiments assessing the effects of combinations of SO$_2$ and O$_3$ on photosynthesis.

Dose-response relationship. Dose-response studies employing several concentrations of SO$_2$ and O$_3$ and evaluating their individual and combined effects on plants can provide valuable insight into the ranges over which single pollutant and combined pollutant effects dominate the interactive effects, as well as define their nature. In a laboratory dose-response study with *Nicotiana tabacum* cv. "Bel-W3" and *Phaseolus vulgaris* cvs. "Tempo" and "Red Kidney," plants were exposed to 0.10, 0.20, 0.30, 0.40, and 0.80 ppm (4.16, 8.32, 12.49, 16.64, and 33.28 μmol m^{-3}) of O_3 with and without 0.40 ppm (16.64 μmol m^{-3}) of SO_2, or to 0.40, 0.60, 0.80, 1.20, and 1.10 ppm (16.64, 24.96, 33.28, 49.92, and 45.76 μmol m^{-3}) of SO_2 with or without 0.70 ppm (29.12 μmol m^{-3}) of O_3 for 4 hours (Jacobson and Colavito, 1976). Evaluation of the effects of SO_2 on the response of plants to O_3 indicated that foliar injury in *Phaseolus vulgaris* was reduced, but that in *Nicotiana tabacum* was increased. The effects of O_3 on the response of plants to SO_2 were similar. When dose-response relationships were evaluated using probit analysis, it was found that SO_2 raised the ED_{50} for O_3 on *Phaseolus vulgaris* by 6 to 21%, and lowered the ED_{50} on *Nicotiana tabacum* by 77%. Similar analyses for the effects of O_3 on SO_2 indicated the ED_{50} on *Phaseolus vulgaris* was raised 25 to 45%, and the ED_{50} on *Nicotiana tabacum* lowered by 25%.

The nature and magnitudes of these interactions could not have been predicted from the sensitivities of the species to the individual pollutants, since preliminary exposures had indicated that all three were much more sensitive to O_3 than to SO_2. Probit analysis allowed a clear interpretation of the influence of one pollutant on the response of the plants to the other.

The response of *Glycine max* cv. "Dare" to long-term exposures to SO_2 and O_3 has been evaluated under field conditions. In one study (Heagle, Body, and Neely, 1974), closed-top field chambers were used to expose plants to charcoal-filtered air, or to charcoal-filtered air to which 0.10 ppm (4.16 μmol m^{-3}) of SO_2 and/or of O_3 were added. Exposures were conducted for 133 days during the growing season. The addition of SO_2 had no significant effect on the plant response to O_3. Foliar injury was somewhat higher and yield slightly lower in the combined pollutant treatments, but neither was significantly different from that produced by O_3 alone.

In contrast to the study just described, which employed only single concentrations of SO_2 and O_3 to evaluate the interaction, a follow-up field study with *Glycine max* cv. "Davis" used multiple levels of each pollutant (Heagle *et al.*, 1982). Charcoal-filtered air and nonfiltered air with O_3 additions were used to produce 7-hour average treatment concentrations of 0.025, 0.055, 0.069, 0.086, 0.106, and 0.125 ppm (1.040, 2.288, 2.870, 3.577, 4.409, and 5.120 μmol m^{-3}). Four concentrations of SO_2 were employed: 0.0, 0.026, 0.085, and 0.367 ppm (1.082, 3.536, and 15.266 μmol m^{-3}) in 4-hour fumigations. Exposures were conducted in open-top chambers from June to October. As the O_3 treatment concentration increased, yield was reduced. Additions of SO_2 generally resulted in greater yield reductions at all O_3 concentrations, although there were some exceptions at the lower SO_2 concentrations. The authors concluded that, although chronic exposure to O_3 resulted in yield reductions, the response was not affected by SO_2 at concentrations known to exist regionally in soybean production areas.

A similar field study conducted with *Glycine max* cvs. "Essex" and "Williams" employed 0.014, 0.039, 0.032, and 0.057 ppm (0.582, 1.622, 1.331, and 2.371 μmol m^{-3}) of O$_3$ and 0.005, 0.030, 0.052, 0.110, 0.223, and 0.435 ppm (0.208, 1.248, 2.163, 4.576, 9.276, and 18.095 μmol m^{-3}) of SO$_2$ (Heggestad, Bennett, and Douglass, 1982). Exposures were conducted in open-top chambers on 64 days during the growing season. Analysis of the harvest data indicated no significant interaction of the two pollutants in reducing yield.

Sulfur dioxide was added to charcoal-filtered air and nonfiltered air to find out whether the yield of *Lycopersicon esculentum* cv. "Jet Star" would be affected by the interaction (Heggestad, Bennett, and Lee, 1981). Four concentrations of SO$_2$, 0.06, 0.12, 0.24, and 0.48 ppm (2.50, 4.99, 9.98, and 19.97 μmol m^{-3}), were added to the nonfiltered air with 0.056 ppm (2.329 μmol m^{-3}) of O$_3$, and two concentrations of SO$_2$, 0.12 and 0.48 ppm (4.99 and 19.97 μmol m^{-3}), were added to filtered air with 0.015 ppm (0.624 μmol m^{-3}) of O$_3$. Exposures were conducted throughout the growing season. Evaluation of the weekly harvest data indicated that the pollutant effects were additive. A similar study was conducted with *Phaseolus vulgaris*, using 0.06, 0.12, and 0.30 ppm (2.50, 4.99, and 12.48 μmol m^{-3}) of SO$_2$ added to nonfiltered air, and 0.30 ppm (12.48 μmol m^{-3}) of SO$_2$ added to filtered air (Heggestad and Bennett, 1981). Addition of SO$_2$ to nonfiltered air produced a threefold greater reduction in yield than did the addition in charcoal-filtered air.

Summary. This review of the research assessing SO$_2$/O$_3$ interactions indicates that the two pollutants can produce additive, less than additive, or greater than additive effects in the production of foliar injury, the reduction of plant biomass, and the reduction of crop yield. Most of the greater than additive effects have been observed in laboratory studies employing single or short-term exposures. Dose-response research conducted in the field has generally indicated that the pollutants do not interact and that their effects are additive. Very little is known about the roles of environmental variables in regulating the pollutant interaction, although it has been shown that temperature, light intensity, and relative humidity can be important.

Interaction studies have, for the most part, used pollutant-exposure regimes unlike those found in the ambient environment. Little attention has been paid to the relevance of the pollutant concentration and its variation, the exposure duration, and the exposure frequency in establishing the experimental treatment regimes. Many experimental exposure regimes have used SO$_2$ concentrations that are characteristic of maxima associated with point sources, and exposure durations and frequencies that are more characteristic of area sources. Many O$_3$ exposure regimes have used concentrations and exposure durations considerably greater than those found under ambient conditions. Responses to the combined pollutants that are produced with these unrealistic exposure regimes may have little if any relevance to effects found under field conditions. This concern is supported by the demonstrated importance of pollutant concentrations and ratios in determining the nature of the response in the combined pollutant exposures.

Although the foliar injury symptoms produced by the combined pollutants can be similar to those produced by either SO_2 or O_3, the symptoms usually most resemble those caused by O_3; certain rare symptoms were unlike those commonly associated with either pollutant. It appears that the type of symptom produced may be a function of the relative concentrations of the pollutants and of the relative sensitivity of the exposed plant to each pollutant.

Dose-response studies have been few in number and have employed only a few plant species. The field studies have not been replicated over time, and thus do not permit evaluation of differences in response or of the influence of environmental variables. Few laboratory dose-response studies have been conducted to evaluate the effects of the combined pollutants on plant growth and physiology, and to assess influences of environmental variables on the pollutant interactions.

The physiological mechanisms by which SO_2 and O_3 produce interactive effects are not understood. Studies of stomatal resistances and uptake rates during individual and combined pollutant exposures fail to explain the interactive impacts that are observed. Very few biochemical investigations of SO_2/O_3 interactions have been conducted.

Collectively, SO_2/O_3 interaction studies have used few plant species. Most studies have been conducted with *Raphanus sativus*, *Glycine max*, *Phaseolus vulgaris*, and *Petunia hybrida*. Other crop, ornamental, grass, and tree species have been used only in isolated studies.

Future Research Needs

The delineation of future research needs can be approached by an evaluation of how well our present understanding of SO_2/O_3 interactions satisfies the three research goals concerning the betterment of environmental quality that were presented at the beginning of this chapter. For the first goal, diagnosing pollutant-induced effects, exposure to SO_2 and O_3 in combination does not produce a unique foliar injury symptom that can be used in field diagnosis. Injury produced by the two pollutants acting in concert will most likely be diagnosed as having been produced by one pollutant acting independently. A histological study of the impacts of the individual and combined pollutants on *Pinus ponderosa* (ponderosa pine) provided some indication that the causal agent could be identified by the pattern of cellular injury (Evans and Miller, 1975). Histological studies have not been conducted with other species or with plants exposed to lower concentrations of the pollutants under field conditions.

Extant research provides little information useful for developing air-quality standards in order to prevent pollutant effects. Other than indicating that SO_2 and O_3 can sometimes interact in a greater than additive fashion, research data are inadequate to allow these interactions to be considered by those concerned with establishing air-quality standards for the individual pollutants. Greater understanding of the roles of absolute and relative pollutant concentrations, the influences of environmental variables, and the range of response differences within and between plant species must be attained before the consideration of interactive effects can be incorporated into the standard-setting process in a meaningful and supportable manner. This is particularly true in light of the limited field dose-response studies that have been conducted and that, for the most part, show the pollutants to act in an additive fashion.

The last goal, developing the ability to predict combined pollutant effects, remains unfulfilled. Attempts to answer practical questions about potential vegetation injury from interactions of SO_2 and O_3 in the field under specific exposure regimes and on plant species of interest quickly confront a fragmented database that is clearly inadequate for the task. We cannot predict combined pollutant effects, for several reasons. In addition to the constraint posed by the small number of species used in interaction research, it is unclear whether any of the species that have received more in-depth evaluation can serve as a response model for a broader sample of plant species. Experimental exposure regimes have been adopted without due consideration of whether they relate to ambient exposure regimes; so the research results produced by these exposure regimes often cannot be used to answer questions about field effects. The lack of a generalized response model is a major hindrance to the assessment of potential field effects. The comprehensive databases from field and laboratory research needed to develop either empirical or mechanistic response models are not currently available.

In general, our understanding of SO_2 and O_3 interactions falls far short of what would be needed in order to make meaningful decisions about establishing air quality standards and about their significance in producing field effects. Consideration of several specific factors in future research will improve our understanding of SO_2/O_3 interactions, and extend our ability to apply this understanding to practical issues and questions.

Evaluation of SO_2 and O_3 interactions could be most effectively advanced by adopting a dose-response approach to research. Use of multiple concentrations of each pollutant, and using them singly and in combination, allows response surfaces to be developed, and additive and interactive response regions to be delineated. Studies of this type would be much more useful for understanding interactions and for predicting their occurrence than studies employing only one or two concentrations of each pollutant.

Additional attention needs to be directed toward field research in the assessment of pollutant interactions. These studies should take an integrated approach to the identification and quantification of plant responses to the combined pollutants, as well as explore the mechanisms that produce the effects. An in-depth assessment of plant response should be conducted to facilitate evaluation of the pollutant treatment effects at several response levels. Physiological studies evaluating photosynthesis and stomatal conductance can provide information about pollutant uptake and subsequent effects on carbon fixation. Sequential harvests of plants throughout the growing season allow biomass accumulation and partitioning to be evaluated for each treatment. Ultimately the plants are harvested, and the effects of the treatments on yield evaluated. This systematic approach to assessment of effects allows plant response to be evaluated as a series of integrated processes culminating in the alteration of a growth variable, usually yield or biomass.

Much more effort should be made to insure that the experimental exposure regimes employed for SO_2 and O_3 resemble ambient conditions. The regime components of concentration, duration, and frequency should be carefully considered. In general, the O_3 exposure regime should reflect the dynamic, diurnal cyclic pattern of exposure that is characteristic of ambient O_3 concentrations under anthropogenic influences. In selecting an SO_2 exposure regime, an initial decision of whether a point or area source is to be simulated must be made. Depending on the type of SO_2 source selected, significantly different exposure regimes will be implemented. If a point source is selected, the treatment regime will be composed of exposures that are generally of higher concentration, shorter duration, and less frequent occurrence than if an area source is being simulated. In addition to the individual pollutant regimes, the issues of sequential exposure and co-occurrence should be considered. If SO_2 fumigations tend to occur during early morning inversion breakup and prior to the period of significant photochemical generation of O_3, sequential exposures may be appropriate. If the SO_2 fumigation is the result of a power-plant plume reaching the Earth's surface, O_3 scavenging by nitric oxide in the plume would lead to depressed O_3 concentrations during the SO_2 exposure. The dynamic interactions of the pollutant regimes in the ambient environment have not been considered in the development of experimental treatments in the past. Assessment of the individual pollutant regimes, and of their dynamic temporal and chemical interaction, is essential to the development of SO_2 and O_3 interaction research programs that can be used to assess plant response in the field (Noggle and Jones, 1981).

Laboratory research has a critical role to play in understanding the effects of the combined pollutants on plants. This work is essential in the development of simulation or mechanistic models of the effects of SO_2/O_3 exposures on plants. When feasible, laboratory research should employ a dose-response approach in the assessment of pollutant impacts. Among the areas in which important research contributions could be made are the identification of pollutant-sensitive crop-growth stages, elucidation of the roles of environmental variables in mediating plant response to the combined pollutants, and description of the physiological basis for the pollutant interactions. Although each of these research areas constitutes an element in the development of an empirical system of effect assessment, collectively they would be a major advance toward the development of a mechanistic model of plant response.

Improved methods of statistical analysis should be adopted in all aspects of interaction research. Much of the foliar response, growth, and screening research that has been conducted is bereft of statistical analyses of any nature.

Sometimes analytical techniques inappropriate for the data have been used. Some very useful statistical approaches using analysis of variance and the partitioning of the sums of squares into single-degree-of-freedom tests for main and interactive effects have recently been employed (Reinert and Gray, 1981; Reinert, Shriner, and Rawlings, 1982). More complete and rigorous analysis of data not only is necessary for an accurate interpretation of effects, but also provides the type of mathematical output essential to model development.

A persistent problem in air-pollution research is our inability to detect small, subtle plant responses against the larger background of natural plant variability. One approach to help resolve the problem is to use covariate measurements to account for significant within-treatment differences in plant growth (Ormrod, Tingey, and Gumpertz, 1983). In a test of the power of such an approach, *Lactuca sativa* cv. "Grand Rapids" and *Raphanus sativus* cv. "Cherry Belle" were exposed to 0.80 ppm (33.28 μmol m^{-3}) of SO$_2$ and 0.40 ppm (16.64 μmol m^{-3}) of O$_3$ singly and in combination for 6 hours. Nondestructive measurements made prior to exposure for analysis of covariance were planar leaf area as a covariate for leaf area, fresh weight, and dry weight at harvest; plastochron index as a covariate for plastochron index at harvest; and hypocotyl diameter as a covariate for *Raphanus* hypocotyl weight at harvest. Use of the covariates reduced the variability of the response measurements, and increased the precision of the statistical tests. With *Lactuca sativa*, analysis of variance indicated that there were no pollutant effects on leaf area or leaf fresh weight. When an analysis of covariance was conducted, a less than additive interaction of SO$_2$ and O$_3$ on both leaf area and leaf fresh weight was detected. Analysis of covariance was not as useful with *Raphanus sativus*, probably because there were much greater differences between individual plants. With careful selection of covariate factors and the collection of nondestructive measurements prior to treatment, analysis of covariance appears to be useful for reducing experimental variability and improving the detection of subtle treatment effects.

The forms of the potential interactions of two air pollutants and how they can change over a range of pollutant concentrations have made it difficult to describe dose-response relationships in interaction studies. The use of response surfaces may be adequate when the interaction is simple, but as the interaction becomes increasingly complex, the response surface becomes difficult to depict and interpret. An alternative approach, using contour plots composed of sections taken through the response surface, clearly illustrates the nature of the pollutant interactions (Ormrod *et al.*, 1983). The contour-plot format makes it easier to identify pollutant interactions and to understand how the interactions change as pollutant concentrations change. This method of depicting plant response should find wide application in interaction studies that employ a dose-response experimental design.

The use of an alternative to the full factorial design for pollutant interaction studies has been evaluated by Ormrod *et al.* (1983). Alternative designs would be advantageous, since they would reduce the amount of experimental equipment required for dose-response interaction studies. Use of a central composite rotatable design minimizes the number of treatments required to cover the desired range of concentrations of two air pollutants, and allows evaluation of potential interactions over the entire range. The central rotatable design as used in interaction studies by Ormrod and associates shows considerable promise.

The development of future research programs evaluating SO_2/O_3 interactions should give clear consideration to the eventual use of the data in field-effect assessment and in establishing ambient air-quality standards. Development of extensive criteria bases for SO_2 and O_3 individually has documented their potential effects on vegetation and has led to measures to control their atmospheric concentrations. The limited information available on the interactions of SO_2 and O_3 suggests that the responses of plants to the combined pollutants may be important for effect assessment and for developing air-quality standards, but finding out how important will require a much deeper understanding of the interactions of SO_2 and O_3. The added complexity introduced into a vegetation-effects study by the use of two pollutants has deterred the development of comprehensive, indepth research programs. However, the establishment of programs that incorporate the research considerations on exposure regimes, experimental designs, response evaluation, and statistical analysis dealt with here is a challenge that must be addressed if the significance of SO_2 and O_3 interactions is to be understood.

18

Plant Response to SO_2 and CO_2

Roger W. Carlson and F. A. Bazzaz

It is well documented that global CO_2 concentrations have been rising since the beginning of the industrial revolution. The average concentration of atmospheric CO_2 is estimated to have been 290 ppm (12.1 mmol m^{-3})) for 1860 compared with the present concentration of 335 ppm (13.9 mmol m^{-3}) (data by C. D. Keeling and coworkers, cited in U.S. Council on Environmental Quality, 1981). The increase has been caused by the accelerated rate of fossil fuel consumption and perhaps by the rapid destruction of forests (Adams, Montovan, and Sundell, 1977; Bolin, 1977; Woodwell and Houghton, 1977; Woodwell, 1978; Woodwell et al., 1978; Stuvier, 1978; Olson, Pfruderer, and Chan, 1978; C. S. Wong, 1978, 1979). The latter, however, may be balanced by reforestation (Revelle and Munk, 1977) and by the production of charcoal (Seiler and Crutzen, 1980). Industrial processes such as kiln operation and cement processing also add CO_2 to the atmosphere. It is predicted that CO_2 concentration will have doubled by around the middle of the twenty-first century (Bacastow and Keeling, 1973; Baes et al., 1977). Continuous measurements taken since 1958 in remote areas (e.g., Mauna Loa, Hawaii, and the South Pole) have revealed a continuous increase in CO_2 concentrations, with a superimposed annual cycle (Machta, 1973; Ekdahl and Keeling, 1973). The northern hemisphere is subject to seasonal variations of from 4 to 6 ppm (0.17 to 0.25 mmol m^{-3}) in Hawaii and of 14 ppm (0.58 mmol m^{-3}) in Point Barrow, Alaska (Machta, 1979). This seasonal variation has been attributed to seasonal cycles of vegetation activity and CO_2 release. The seasonal variation is less pronounced in the southern hemisphere because of the smaller land mass relative to oceans and lower rate of industrial activity.

The continued increase of CO_2 concentration in the atmosphere, and the inability of the oceans to absorb and the forests to fix all released CO_2, have raised concern about the ecological effects caused directly by changes in plant processes and indirectly by CO_2-induced climatic change (Woodwell and Pecan, 1973; Woodwell, 1978; Woodwell et al., 1978; Olson, Pfruderer, and Chan, 1978; Siegenthaler and Oeschger, 1978; Stuiver, 1978).

Increased atmospheric concentrations of CO_2 and of SO_2 frequently occur simultaneously, since both are products of fossil-fuel combustion. High concentrations of CO_2 can partially or wholly offset pollutant-induced reductions in photosynthesis (Mansfield and Majernik, 1970; Coyne and Bingham, 1977;

Carlson, 1983a). It is therefore important to understand the response of plants to SO_2 in the presence of both "normal" and above-normal atmospheric concentrations of CO_2. Processes associated with both gaseous transfer and plant response are involved. In this paper we will examine the aerodynamic processes that control gaseous mass transfer to plant canopies, examine plant uptake of CO_2 and SO_2, and consider the effects of SO_2 on plant physiology, growth, and yield in the presence of increased concentrations of CO_2.

Uptake and Release of Gaseous Substances

Early air-pollution studies suggested that there are threshold concentrations for the effects of SO_2 on plants. These results may be more a function of experimental techniques than of plant biology, since the rate of air movement in some of the exposure systems may have been very low. As we will see, the rate of air movement is very important in controlling pollutant uptake by plant leaves, and it is pollutant uptake, not external concentration, that determines plant response.

It is useful to describe the transfer of gaseous substances in terms of an analogy with electrical resistance (Bennett and Hill, 1973a; Jarvis, 1981; Unsworth, 1981; Hosker and Lindberg, 1982). In this analogy, gaseous flux across an interface is proportional to the concentration difference between two points and directly proportional to the conductance associated with the interface. Flux (J) is expressed as mass per unit time per unit area, concentration (C) in mass per unit volume, and conductance (g) in distance per unit time. Conductance may also be referred to as "deposition velocity" (v_d); this term is used in describing the rate of deposition to a surface, and does not apply to flux across an interface.

Absorption of gaseous substances by plant canopies is influenced by environmental parameters that may exhibit large diurnal and spatial fluctuation, as well as by surrounding sources of air contamination. Of primary concern are fluctuations in factors that affect the rate at which mixing occurs in the layer of air surrounding the plant canopy. Mixing in this layer is controlled by the rate of air movement, which is a function of free stream wind velocity above the canopy, the density of the plant canopy, and thermal boundaries due to the heat balance of the canopy (Hosker and Lindberg, 1982). The fumigation regime can vary markedly, depending on whether a site is adjacent to a single pollutant source or is surrounded by many point sources. With a single source, plants may be exposed to short-term peak concentrations, but with many point sources, plants may be exposed continuously to elevated concentrations of the pollutant. The effects of these two fumigation regimes on plants could be very different. For example, Dreisinger and McGovern (1970) found vegetation damage around the Sudbury smelter to be correlated more with peak concentration episodes than with time-averaged pollutant concentration.

A canopy of vegetation contains many sources and sinks for CO_2, only sinks for SO_2, and usually only sources for water vapor. At night, vegetation surfaces could become a sink for water vapor because of radiative cooling and subsequent dew formation. The soil can be an important source of CO_2 for a plant canopy as well as an important sink for SO_2 and other air pollutants. Consequently, a concentration profile develops in the canopy for each gaseous component in the air (Bennett and Hill, 1973a). The profile is likely to be quite different for each gas because of differing rates of absorption and release. Diurnal variation in the shape of the profile is also to be expected (Lemon, Stewart, and Shawcroft, 1971). This is especially true for CO_2, since net CO_2 flux changes direction, going into leaves during the day, out of leaves at night.

Each gaseous component must be considered individually, since conductance values may be different for each layer within the canopy. Some representative estimates of transfer conductances for gaseous exchange are listed in Table 18.1, and a schematic diagram showing the location of each conductance (identified as the inverse of conductance, i.e., resistance, r) is presented in Fig. 18.1. Aerodynamic conductance (g_a), defined as the conductance to mass transfer from the bulk air above the canopy to the top of the canopy, is estimated from eddy correlation or flux gradient analyses (see Unsworth, 1977, 1981; Hosker and Lindberg, 1982). The more direct method, eddy correlation, requires the use of fast-response measurement instruments. In flux gradient analysis, an analogy with molecular diffusion is used, and the change in concentration with time at different heights is measured. The eddy-correlation method appears to give acceptable

TABLE 18.1
Maximum Values of Conductance for the Transfer of Gaseous Substances

Type of conductance	Range of maximum value $(mol\ m^{-2}\ s^{-1})$[a]	Symbol
Aerodynamic	0.45–4.5	g_a
Canopy	2–10	g_p
Leaf boundary layer	0.2–4.5	g_b
Leaf		
external (dry)	0.01	g_e
external (wet)	2	
cuticle	0.01	g_c
stomatal		g_s
Herbaceous crops	0.2–0.9	
Trees	0.03–0.5	
Xerophytes	0.03–0.09	
Mesophyll		
Surface		g_{ms}
Soluble gases	0.5–5.0	
Poorly soluble gases	0.05–0.5	
Influx	0.5	g_{mi}
Efflux	?	g_{me}
Ground surface	0.2–2.0	g_{gs}

SOURCES: Unsworth (1981); Hosker and Lindberg (1982).
[a] $1\ mol\ m^{-2}\ s^{-1} = 2.2\ cm\ s^{-1}$.

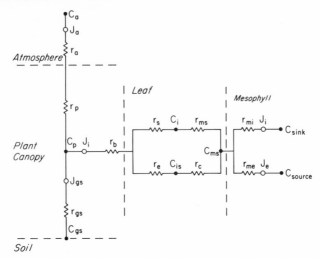

Fig. 18.1. A generalized resistance analog for gaseous exchange in a plant canopy (adapted from Black and Unsworth, 1979b, and Unsworth, 1981). Symbols represent concentration (C), net flux (J), and resistance (r) at different locations: (a) above the canopy, (p) within the canopy, (gs) ground surface, (l) leaf, (b) leaf boundary layer, (s) stomata, (i) substomatal cavity, (ms) mesophyll surface, (e) external leaf surface, (ls) leaf surface, (c) cuticle plus epidermis, (mi) influx through the mesophyll, and (me) efflux through the mesophyll. Conductance (g), the term used in the text, is equal to the reciprocal of resistance.

results for short, relatively "smooth" canopies, but tends to lose some accuracy when applied to tall, "rough" vegetation. The inaccuracy increases where there is strong surface heating and increased atmospheric instability (see Unsworth, 1981). Aerodynamic resistance is rarely limiting under natural conditions during daylight hours. For example, Fowler and Unsworth (1979) found g_a to be 2.5 to 5.6 mol m^{-2}s^{-1} (1 mol m^{-2}s^{-1} = 2.22 cm s^{-1}) at wind speeds of 1–5 m s^{-1} for a wheat crop. However, at night there may be very little air movement, and g_a could become quite small.

Canopy conductance (g_p) is the conductance associated with the transfer of material from the top of the canopy to the boundary layer of surfaces within the canopy. It is a function of the density of vegetative elements and the rate of air movement within the canopy. Although it is difficult to distinguish between aerodynamic, canopy, and boundary-layer conductance for short grass communities, a unique term for canopy conductance is desirable for taller plant communities.

Leaf boundary-layer conductance (g_b) can be estimated either by heat-balance analysis (Raschke, 1960; Gates, 1962), by wind-tunnel formulations of heat and gas transfer to leaves (Chamberlain, 1974; Gates, 1965), or by measurement of the rate of water-vapor evaporation from leaf models saturated with water (Gaastra, 1959; Sesták, Cătský, and Jarvis, 1971). Each technique has its own particular advantages in different situations. The heat-balance analysis is useful in the field, where it is possible to measure leaf temperature, radiation, and wind speed, but not possible to directly measure transpiration. Calculation of g_b by measuring evaporation from a leaf model is particularly well-suited to small-chamber mass-balance experiments.

Leaf conductance is estimated by measuring the rate at which water vapor evaporates from leaves, either with a chamber that encloses a leaf or a small whole plant, or with a porometer that samples a part of the leaf (Sesták, Cătský, and Jarvis, 1971; Black and Unsworth, 1979c; Jarvis, 1981). Leaf conductance is made up of two components in parallel, i.e., stomatal conductance (g_s) and an external leaf-surface conductance (g_e). External leaf-surface conductance for undamaged dry leaves is usually very low for both CO_2 and water vapor. In such cases, stomatal conductance is equal to total leaf conductance. However, surface wetness will increase the uptake of soluble gases by leaves and also increase evaporative water loss (Garland and Branson, 1977). The presence of surface wetness, from aerially applied water droplets, dew formation, or moisture leaking from damaged cells, could, therefore, lead to an overestimate of stomatal conductance if stomatal conductance is set equal to total leaf conductance. Water vapor exiting through stomata originates from internal leaf surfaces, which are considered to be saturated with water. Therefore the internal concentration of water vapor is taken to be equal to saturation at the temperature of the leaf.

The passage of CO_2 from the stomatal cavity to the sites of CO_2 fixation involves two additional terms. These are the conductances for absorption by the mesophyll surface (g_{ms}) and influx through the mesophyll (g_{mi}). These two conductances are difficult to separate experimentally, and are therefore usually combined into one term called mesophyll or residual conductance. However, it is important to realize that since r_{ms} is a function of solubility, its value could be quite different for different gases (Table 18.1). Residual conductance to CO_2 is usually large for healthy plant tissue, but may decrease significantly for injured plants (Barton, McLaughlin, and McConathy, 1980; Carlson, 1983a,b). Mesophyll surface conductance to SO_2 has been found to be large relative to the conductance associated with other factors that control the uptake of SO_2 (Black and Unsworth, 1979b; Carlson, 1983a).

SO_2 uptake by leaves is more complex than CO_2 uptake, since the leaf surface may absorb large amounts of SO_2. The transfer of SO_2 to plant surfaces through the boundary layer and into leaves has been analyzed by Unsworth and co-workers in terms of three parallel pathways to a point where the concentration of SO_2 is zero. The pathways include (1) passage through the stomata and internal air spaces to the water layer adhering to internal cell walls, (2) passage through the cuticle to internal cells, and (3) absorption by the leaf surface. The third pathway is not specifically identified in Fig. 18.1, since SO_2 absorption by the leaf surface is part of the second pathway. The series and parallel portions of the conductance analog for SO_2 uptake were solved to give a linear expression with terms for flux,

g_b, g_s, g_e, and g_{ms} (Unsworth, 1981). SO_2 flux was measured directly, and g_b and g_s for SO_2 were calculated from measurements of water-vapor efflux (Unsworth, Biscoe, and Black, 1976). Estimates were made for g_{ms} until a straight-line graph was obtained for the relationship between $(1/g_t - 1/g_b)^{-1}$ and $(1/g_s + 1/g_{ms})^{-1}$, where g_t (total leaf conductance to SO_2) = SO_2 flux divided by concentration. The intercept of such a plot is equal to the sum of SO_2 conductance through the cuticle. Black and Unsworth (1979b) used this technique to partition SO_2 uptake by *Vicia faba*. Cuticular conductance to SO_2 was found to be 0.013 mol m^{-2}s^{-1}, but conductance varied from 0.012 mol m^{-2}s^{-1} for closed stomata to 0.23 mol m^{-2}s^{-1} for fully open stomata. The fact that a slope of near unity was obtained when $1/g_m$ was assumed to be zero indirectly confirms that the internal concentration of SO_2 was close to zero.

The pathway of SO_2 into the leaf, though opposite in direction, is believed to be very similar to that of water vapor leaving the leaf. For example, $^{35}SO_2$ uptake and stomatal conductance were measured on adjacent shoots of *Pinus sylvestris* fumigated with SO_2 (Garland and Branson, 1977). A linear correlation between SO_2 uptake and water-vapor efflux was obtained with a slope very close to the ratio of diffusivities between SO_2 and water vapor. These data also indicate that $1/g_m$ for SO_2 was negligible.

Studies with a variety of other species have also shown a close correlation between stomatal conductance to water vapor and the rate of SO_2 uptake. This has prompted the use of stomatal conductance as an indicator to predict SO_2 resistance of plants (Winner, 1981; Winner, Koch, and Mooney, 1981). Using stomatal conductance as measured in "clean" air as an indicator for potential SO_2 uptake will be most successful for species in which the stomata are not highly sensitive to the presence of SO_2. Extrapolating from SO_2 absorption into leaf mesophyll tissues to the effect of SO_2 on metabolic processes is likely to be successful only for closely related genotypes and ecotypes, since different species have different ways in which to deal with toxic compounds. Environmental variation will also alter plant response to a given amount of absorbed SO_2.

Large amounts of SO_2 are adsorbed on the surfaces of leaves. The ratio of adsorbed to absorbed SO_2 has been found to vary from 0.2 to 1.2 for different varieties of *Petunia* (Elkiey and Ormrod, 1980b) and from 0.3 to 1.1 for different varieties of *Poa pratensis* (Elkiey and Ormrod, 1981b). Ratios greater than 1.0 indicate that cuticular conductance was higher than stomatal conductance to SO_2 uptake for some of the cultivars, perhaps partly because the measurements were made under very low light (less than 15% of full sunlight) and the stomata may have been, therefore, partially closed. It would be instructive to examine the differences between these cultivars at higher light intensity using the conductance analysis scheme outlined above.

Resistance analogs are firmly established for the analysis of the exchange of momentum, heat, water vapor, and CO_2 in plant canopies (Unsworth, 1981). The experience gained from this extensive body of literature is very useful for the study of air-pollutant flux. Aerodynamic and flux-gradient methods are useful for calculations dealing with large-scale phenomena, and small-chamber and porometry techniques are useful for calculating uptake rates and transfer functions for small portions of the canopy. Thoughtful experimental design and careful measurements are required to bridge the present large gap between the two levels of research. The papers by Jarvis (1981) and Unsworth (1981) should be consulted for guidance in this endeavor.

Response to SO$_2$

Stomata. Stomata are not only important ports of entry for gaseous substances, but are also important points of attack for injurious air pollutants. Increased stomatal opening can occur at low SO$_2$ concentrations (Black and Black, 1979a), with closure being the dominant stomatal response at high SO$_2$ (Carlson, 1983a). Increased stomatal conductance would both allow the entry of additional SO$_2$, other air pollutants, and CO$_2$, and increase the rate of water-vapor loss. Entry of additional CO$_2$ could be beneficial to the plant if photosynthetic processes could reduce it. However, an increase in the rate of transpiration could be detrimental, especially for water-stressed plants (Mansfield, 1973).

SO$_2$ seems to induce stomatal opening because it has different effects on different cell types within the leaf. Subsidiary cells, those surrounding the stomata, appear to be more sensitive to SO$_2$ than the guard cells. In fact, the guard cells may be more resistant to SO$_2$ than any other cells within the leaf. Squire and Mansfield (1977) found that turgor for all epidermal cells (with the exception of guard cells) was very sensitive to low *p*H. Lowering the turgor of epidermal cells could cause stomata to open. The alternative hypothesis of increased guard-cell turgor in response to SO$_2$ is much less likely, since SO$_2$ inhibits metabolism and is likely to reduce the supply of metabolic energy available for active transport and other processes associated with stomatal opening.

Evidence that stomatal opening in response to SO$_2$ is due to the loss of turgor in cells adjacent to the guard cells comes from work with *Vicia faba* (Black and Black, 1979a,b). SO$_2$ entering the leaf is likely to take the shortest pathway to an absorbing surface. Guard cells have thickened cell walls, and may, like the epidermis, be protected from SO$_2$ injury by an impermeable cuticular layer (Black and Unsworth, 1980; Black, 1982). Therefore the most likely SO$_2$ absorption sites are the internal walls of epidermal cells immediately adjacent to the guard cells (Fig. 18.2). At low SO$_2$ concentrations, small numbers of epidermal cells become weakened and lose turgor, thereby increasing stomatal conductance, but recovery to normal conductance is possible. As the concentration of SO$_2$ is increased, larger numbers of cells are injured; the stomata remain open, but recovery to control conductance does not occur. At higher SO$_2$ concentrations (more than 7.5 μmol m^{-3}), guard cells were injured and stomata closed.

Stomatal opening in response to SO$_2$ is sensitive to vapor-pressure deficit

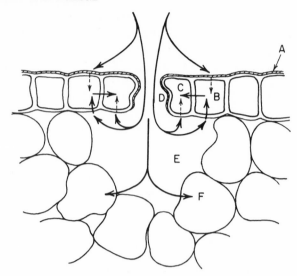

Fig. 18.2. Cross section of a substomatal cavity of a leaf, indicating major (solid line) and minor (broken line) pathways for pollutant uptake through the cuticle (*A*), into epidermal cells (*B*), into guard cells (*C*), through the stomata (*D*), into the substomatal cavity (*E*), and into mesophyll cells (*F*). From Black (1982).

(vpd) in *Vicia faba* and certain other species (Black and Unsworth, 1980). Upon continued exposure at low SO_2 concentrations, exposure to high SO_2 concentrations, or with increasing vpd, stomatal conductance in vpd-sensitive species declines to levels below that of control plants. Black and Unsworth (1980) suggested that this change in conductance is due to water loss from the guard cells. Apparently only a small loss of water from guard cells is required to close stomata if the surrounding cells are sufficiently dehydrated. Species in which stomatal conductance does not decline with increasing vpd do not exhibit a similar reduction in conductance in response to SO_2. Stomatal closure that occurs in response to high concentrations of SO_2 may also be caused indirectly by an increase in CO_2 concentration within the leaf because of a reduction in the rate of photosynthesis.

Several studies suggest that stomatal conductance may return to prefumigation levels within hours to days after the end of fumigation (Black, 1982). Recovery is difficult to demonstrate unequivocally, since stomata respond to many environmental factors, all of which must be maintained within close tolerances before, during, and after fumigation. It is not known if stomatal function, as opposed to stomatal response, can also recover from SO_2 injury.

Photosynthesis and respiration. SO_2 at concentrations of less than 0.1 ppm (4.2 μmol m^{-3}) have been found to reduce photosynthesis in sensitive species (Black, 1982). Photosynthesis may increase at low SO_2 or during initial exposure to higher concentrations of SO_2 (Black and Unsworth, 1979a; Winner and Mooney, 1980c). This stimulation is most likely due to SO_2-induced stomatal opening. Reduction in photosynthesis depends on the concentration of SO_2, and is usually reversible within 3 to 24 hours after fumigation at exposure concentrations that are not high enough to cause visible symptoms. The wide range in sensitivity to SO_2 is due both to differences in SO_2 uptake capacity between species and to differences in the ability of plants to metabolize absorbed SO_2.

Stimulation, inhibition, and no effect of SO_2 on respiration have been reported (see Black, 1982). Black suggested that differences in experimental protocol in the various studies may have been largely responsible for the differences in the results. Measurement of respiration for plants fumigated in the dark is likely to show that respiration is insensitive to SO_2, as was found by Sij and Swanson (1974), because SO_2 uptake in the dark may be very low due to closed stomata. Respiratory rates have been found to increase for *Vicia faba* fumigated in the light (Black and Unsworth, 1979a). Further experiments are required to examine the effects of SO_2 on respiratory processes of intact leaves. An examination of both dark respiration and photorespiration is necessary, and measurements of sufficient duration must be made in order to describe the dynamics of recovery.

The effects of SO_2 on photosynthesis vary in response to changes in environmental conditions. Environmental variation can influence stomata and thereby have a direct effect on the rate of air-pollution uptake. Environmental factors, such as temperature, soil-moisture availability, and plant nutrition, could also influence respiratory rates and metabolic functions that are associated with the detoxification of absorbed pollutants. Photosynthetic reduction in response to SO_2 has been found to be larger at high light intensity than at low light (Black and Unsworth, 1979a; Winner and Mooney, 1980c). Carlson (1979) found photosynthetic reduction to be larger at low and at high light intensity than at an intermediate light intensity for tree species exposed to SO_2 plus O_3. This response may be due to the fact that at low irradiance the energy available for repair processes is limited, but at high irradiance the production of reactive oxygen species is increased. Thus there seems to be an optimum light intensity at which SO_2-caused reductions in photosynthesis are at a minimum for many plant species, except possibly for those adapted to high light environments.

Stomatal conductance decreases in response to increased concentrations of CO_2. This suggests that increases in CO_2 may help protect plants from air pollutants if the stomata respond rapidly (Mansfield, 1973). Field data that support this contention have been obtained for the effects of O_3 on photosynthesis (Coyne and Bingham, 1977).

Other data suggest that increased CO_2 not only reduces the effect of SO_2 on photosynthesis by inducing stomatal closure, but may also reduce photosynthetic inhibition via nonstomatal processes. Fig. 18.3 demonstrates SO_2-induced nonstomatal reduction of photosynthesis for *Vicia faba* (Black, 1982), and Table 18.2 also shows this for *Glycine max* (Carlson, 1983a). At comparable rates of SO_2 uptake, the observed reduction in photosynthesis was larger for SO_2-fumi-

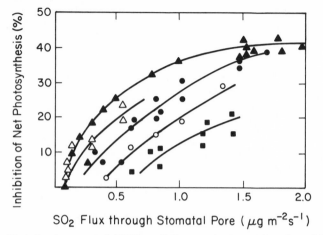

Fig. 18.3. Variation in percent inhibition of photosynthesis as a function of SO_2 uptake into the stomata for *Vicia faba* (from Black, 1982). CO_2 concentrations were manipulated to change stomatal conductance and included (left to right) 13.7, 17.9, 20.0, 24.1, and 28.3 mmol m^{-3} (1 μg m^{-2}s^{-1} = 0.016 μmol m^{-2}s^{-1}).

TABLE 18.2
Mesophyll Conductance to CO_2 for *Glycine max* Fumigated for Two Hours
with Various Concentrations of SO_2

SO_2 concentration (μmol m^{-3})	CO_2 concentration (μmol m^{-3})			
	12.5	18.7	25	50
0	0.51	0.43	0.39	0.39[a]
9	95	92	100	89[b]
20	76	78	101	98
32	24	56	103	93
47	18	41	78	87

SOURCE: Carlson (1983a).
[a] Mesophyll conductance for nonfumigated plants (mol m^{-2} s^{-1}).
[b] Mesophyll conductance for fumigated plants (percent of control).

gated plants at 330 ppm (13.7 mmol m^{-3}) of CO_2 than it was for plants fumigated at 500 to 1200 ppm (20.8 to 49.9 mmol m^{-3}) of CO_2. Carlson found residual conductance to CO_2 uptake to be less sensitive to SO_2 as the concentration of CO_2 was increased, indicating that a large portion of the plants' photosynthetic capacity remained intact at high CO_2, but was severely reduced at 330 ppm (13.7 mmol m^{-3}). Higher photosynthetic rates at higher concentrations of CO_2 may have supplied the plant with more reduced carbon, either as carbohydrate or as additional carbon skeletons, to combine with or detoxify absorbed SO_2.

Plant growth. Several recent reviews are available for a more comprehensive analysis of the effects of SO_2 on plant growth than will be presented here (IERE,

1981; Bell, 1982; Godzik and Krupa, 1982). Here we will merely summarize some of the conclusions found by these authors, and extend their discussion by means of more recent work.

The IERE study cites more than 200 references, and comes to the following assessment about the development of SO_2-induced visible symptoms in plants. Short-term experiments showed that threshold concentrations causing 5% coverage of leaf area with visible necrosis occurred at 22 to 44 μmol m^{-3} for 1 hour, 14 to 27 μmol m^{-3} for 3 hours, and 8.5 to 11 μmol m^{-3} for 6- to 8-hour exposures to SO_2. Resistant species were found to have threshold levels for the development of visible symptoms at three times these concentrations. For recurrent short-term fumigation studies, IERE found a threshold concentration for yield reduction of 8.5 μmol m^{-3}, and for long-term continuous exposures a threshold concentration of 3.4 μmol m^{-3}. The conclusion from IERE about long-term fumigation of tree species is that growth of conifer species and yield of fruit trees can be reduced by fumigation with 4.4 μmol m^{-3} of SO_2 for several weeks and by 2.2 μmol m^{-3} of SO_2 for fumigations spanning a 3-year period. The threshold concentration was lower for multiyear fumigations because of the cumulative effect of reduced growth rate on yield. Bell (1982) concluded that it is still not possible to establish reliable threshold concentrations for long-term exposure of grasses to SO_2. Although some studies have shown reduced growth and premature senescence at 1.7 μmol m^{-3}, there is much inconsistency in the results of various experiments with SO_2 concentrations up to 12 μmol m^{-3}.

There are many possible reasons for these and other inconsistencies. Some have been discussed for SO_2 uptake (aerodynamic, canopy, boundary layer, and stomatal resistance) and photosynthesis (environmental variation). Environmental variation is also an important factor in the effect of SO_2 on growth. For example, sensitivity of oats to SO_2 was found to increase as temperature and relative humidity increased. Foliar injury was also highly correlated with reduced growth (Heck and Dunning, 1978). Studies on *Lolium perenne* (Cowling and Lockyer, 1978; Bell, Rutter, and Relton, 1979) and *Phleum pratense* (Davies, 1980a; Jones and Mansfield, 1982a) indicated that slow-growing plants were more sensitive to SO_2 than fast-growing plants. The environmental conditions used in these studies to induce slow plant growth were either low light intensity, low temperature, reduced soil nitrogen, or the winter growing season in England.

Significant reductions in yield can occur in the absence of visible symptoms. Extended exposure with SO_2 added to filtered air reduced yield in several crop species when the concentration exceeded 3.0 μmol m^{-3} (Table 18.3). Reductions at the lower concentrations listed in this table were not accompanied by visible foliar necrosis, but were sometimes associated with premature senescence. Although reductions in yield have occasionally been measured for SO_2 concentrations below 3.0 μmol m^{-3}, the results of fumigations with low concentrations have been inconclusive (Bell, Rutter, and Relton, 1979; Bell, 1982). Recurrent short-term exposure (4 to 5-hr exposures on each of 24 to 28 days

TABLE 18.3
SO$_2$ Concentrations and Duration of Exposure Producing Significant Yield Reductions
after Continuous Long-term Exposure to SO$_2$

Species	SO$_2$ concentration (μmol m^{-3})	Duration (d)	Vis. symp.	Source
Lolium perenne	6.3	77	Y	Lockyer *et al.* (1976); Cowling *et al.* (1981)
Medicago sativa	4.5	22	Y	Thomas and Hill (1937b)
	4.1	40	Y	
Lolium perenne	4.5	28	Y	Ashenden and Mansfield (1977)
Lolium perenne	2.9	140	Y	*Ibid.*
Dactylis glomerata			Y	
Poa pratensis			Y	
Helianthus annuus	4.1	35	Y	Shimizu, Furukawa, and
	2.0	35	N	Totsuka (1980)
Lolium perenne	0.7 to	63 to	N	Bell and Clough (1973); Bell, Rutter, and Relton (1979)
	6.6	194	Y	Ayazloo, Bell, and Garsed (1980)

TABLE 18.4
Yield of Soybeans Subjected to Varying Concentrations of SO$_2$
in an Open-Air Fumigation System

Plot	Mean SO$_2$ concentration (μmol m^{-3})	Yield reduction (percent)[a]
1977		
Low	4.8	12.3 (4.3)
Medium	12.5	20.5 (3.1)
High	32.8	45.3 (3.7)
1978		
Low 1	4.0	6.4 (2.1)
Low 2	4.2	5.2 (2.5)
Medium 1	8.0	12.2 (3.1)
Medium 2	10.6	19.2 (2.0)
High	15.0	15.9 (3.0)

SOURCE: Sprugel *et al.* (1978, 1980); Miller *et al.* (1978).
[a] Values given in parentheses are standard errors of the mean.

during a 2-month period) reduced yields of *Glycine max* by 15 to 20% at SO$_2$ concentrations ranging from 4.2 to 14 μmol m^{-3} (Table 18.4). Visible injury occurred on some plants when the mean SO$_2$ concentration exceeded 15 μmol m^{-3}.

Some studies have indicated that loss in yield may be significantly affected by the timing of exposure to SO$_2$ (Godzik and Krupa, 1982). For example, larger reductions in yield occurred in wheat when plants were fumigated during the flowering stage than at other times during the season. On the other hand, final weight for radish was low for plants fumigated either at a young age (10% of control) or when approximately half-grown (40% of control). Plants that show extreme sensitivity to SO_2 at a particular stage of development may not have time to recover from the stress by the end of the growing season.

Root growth may be reduced upon exposure to SO_2 before any changes in shoot growth become apparent (Bell, 1982; Jones and Mansfield, 1982b). Studies using growth-analysis techniques have found SO_2-induced increases in specific leaf area and leaf-area ratio, but relative growth rate and root:shoot ratios have declined (Bell, Rutter, and Relton, 1979; Ayazloo, Bell, and Garsed, 1980). The study by Bell, Rutter, and Relton (1979) with *Lolium perenne* found a larger reduction in relative growth rate for both young and old plants than for plants of an intermediate age. Premature leaf senescence, observed in some of the experiments, was not the principal factor accounting for reduced seasonal growth.

Reduced root growth at the expense of shoot growth and other changes in plant structural relationships suggest that translocation is altered by SO_2. Teh and Swanson (1977) found translocation in *Phaseolus vulgaris* to be more sensitive than photosynthesis to SO_2. Subsequent studies by Noyes (1980) with *P. vulgaris* showed that reductions in translocation of 39, 44, and 69% occurred upon exposure to 4.5, 45, and 135 μmol m^{-3} of SO_2 for a few hours, but photosynthesis was reduced by 0, 13, and 73%, respectively. Labeled $^{14}CO_2$ was fixed by leaves, and accumulated near or in minor veins, suggesting that phloem loading processes or axial transport in sieve tubes was inhibited by SO_2. Labeled carbon was diverted to other tillers on the plant at the expense of translocation to roots in *Phleum pratense* fumigated with SO_2 (Jones and Mansfield, 1982b). It is not known how SO_2 can induce a change in the pattern of carbon translocation. Perhaps certain energy-requiring repair processes are stimulated by exposure to SO_2, and energy is thereby diverted that would normally be available for transport processes.

The sulfur absorbed by plant leaves may be beneficial in geographical areas where the level of soil sulfur is low (IERE, 1981). The replacement of ammonium sulfate by ammonium nitrate in nitrogen fertilizers may lead to increased sulfur deficiency in agricultural soils. Crop plants take up between 10 and 50 kg S ha^{-1} yr^{-1} (Metson, 1973; Terman, 1978). Since some sulfur is available from the soil, an aerial application of this amount of sulfur as SO_2 would be more than sufficient to meet the needs of most crop plants. Assuming a typical average deposition velocity over cropland of 0.36 mol m^{-2}s^{-1}, an aerial concentration of 10 to 50 μg m^{-3} (0.16 to 0.78 μmol m^{-3}) would be sufficient to supply the sulfur requirement for most crops in the absence of any other sulfur input (IERE, 1981). Addition of sulfur fertilizer to crops in regions where total deposition exceeds 20 kg S ha^{-1} yr^{-1} has not resulted in increased yield.

Response to CO_2

Many metabolic and growth processes of plants are influenced by CO_2 concentration (Strain and Armentano, 1980). The effects of CO_2 differ markedly for different species and plant processes. There is much evidence that increased CO_2 concentration increases the growth and photosynthetic rate of many species (Baker, 1965; Bishop and Whittingham, 1968; Egli, Pendleton, and Peters, 1970; Wittwer, 1980; Akita and Moss, 1972, 1973; Akita and Tanaka, 1973; Regehr, Bazzaz, and Boggess, 1975; Green and Wright, 1974; Ho, 1977; Strain, 1978; Reynolds, Cunningham, and Syvertsen, 1979; Carlson and Bazzaz, 1980). The rate of photosynthesis is usually found to increase more with increased CO_2 for C_3 plants than for C_4 plants, perhaps because of a difference in the bicarbonate affinity of the initial CO_2 fixation enzymes of these two plant groups (Ziegler, 1972, 1973). Transpiration declines with increasing CO_2 concentration for both plant types, leading to large increases in water-use efficiency.

These generalizations are derived for the most part from studies with agricultural species (Moss, Musgrave, and Lemon, 1961; van Bavel, 1974; Tinus, 1974; Sionit, Helmers, and Strain, 1980). Until recently, very little information has been available for plants of unmanaged ecosystems (Strain and Armentano, 1980). We therefore began a study in which growth and photosynthesis were measured for a number of species grown under CO_2 concentrations of 13.7, 25,

TABLE 18.5
Change in Total Weight and Change in Photosynthetic Rate for Plants Grown
at Different Concentrations of CO_2
(percent of value for plants grown at 13.7 mmol m^{-3} of CO_2)

Species	Weight change for CO_2 concentration (in mmol m^{-3}) of:		Photosynthetic change for CO_2 concentration (in mmol m^{-3}) of:	
	25	50	25	50
C_3 plants				
Datura stramonium (1)	74%	96%	74%	115%
Datura stramonium (2)	67	83	60	107
Chenopodium album	57	79	76	140
Polygonum pensylvanicum	51	64	48	100
Abutilon theophrasti	44	75	38	65
Ambrosia artemisiifolia	10	24	68	112
Acer saccharinum	61	89	32	63
Populus deltoides	65	74	29	20
Platanus occidentalis	13	30	33	33
Glycine max	58	75	47	120
Helianthus annuus	20	38	40	55
C_4 plants				
Setaria faberii	13%	37%	42%	106%
Setaria lutescens	40	20	70	45
Amaranthus retroflexus (1)	41	21	36	59
Amaranthus retroflexus (2)	27	33	29	48
Zea mays	24	−7	21	10

SOURCE: Carlson and Bazzaz (1980, 1982).
NOTE: Each value is the mean for a minimum of six plants.

TABLE 18.6
Percent Increase (D_2) in Water-Use Efficiency of Plants per Doubling of the
CO_2 Concentration between 13.7 and 50 mmol m^{-3}

Species	Common name	D_2
Ambrosia artemisiifolia	Common ragweed	128%
Abutilon theophrasti	Velvet leaf	87
Datura stramonium	Jimson weed	84
Amaranthus retroflexus	Pigweed	76
Platanus occidentalis	Sycamore	71
Populus deltoides	Cottonwood	56
Acer saccharinum	Silver maple	38
Helianthus annuus	Sunflower	55
Zea mays	Maize	54
Glycine max	Soybean	48

SOURCE: Carlson and Bazzaz (1980).
NOTE: Water-use efficiency was calculated for each plant (six plants for each species) by dividing the product of transpiration rate times total leaf area into total plant weight. This value was then divided by the mean value of water-use efficiency for each species obtained at 13.7 mmol m^{-3} CO_2. The resulting normalized values of water-use efficiency were fit by least-squares regression analysis to a linear form of $y = a x^b$, where y = water-use efficiency, x = CO_2 concentration, and a and b are constants. The exponent b was used to calculate the percent of change in water use (D_2) that occurred with a doubling of the CO_2 concentration by solving $D_2 = (2^b - 1)/100$.

and 50 mmol m^{-3}. We found that the growth of C_3 species is stimulated more by increased CO_2 than was the growth of C_4 species (Table 18.5). However, response to CO_2 was quite variable. At 25 mmol m^{-3} of CO_2, the stimulation in growth for C_3 species varied from 10 to 74% of the weight of plants grown at 13.7 mmol m^{-3} of CO_2. The variation observed for C_4 species (13 to 41%) was also quite large. Stimulation of growth by increased CO_2 for three of the ten C_3 species was actually less than that observed for some of the C_4 species. A similar variability was found for the rate of photosynthesis. In general, transpiration decreased with increasing CO_2 for all species. The amount of biomass per unit of transpired water increased by an average of 52% for crop species, 76 to 128% for weed species, and 38 to 71% for tree species with a doubling of the CO_2 concentration within the range of 13.7 to 50 mmol m^{-3} (Table 18.6). The large differences between species in water-use efficiency and growth response indicate a large potential for change in the competitive relationships between species in response to an increase in the concentration of CO_2 in the atmosphere.

Response to SO_2 plus CO_2

We have recently begun an investigation of the combined influences of SO_2 and CO_2 on plant growth, biomass allocation, and plant water use. Six annual weed species (three C_3 and three C_4 plants) were exposed for up to 4 weeks to 0.0 or 10 μmol m^{-3} of SO_2 at CO_2 concentrations of 13.7, 25, or 50 mmol m^{-3}. The plants were grown in specially designed growth cabinets with glass sides and

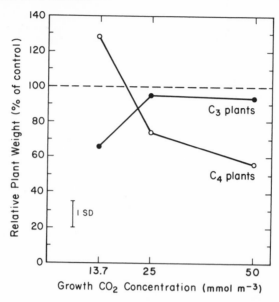

Fig. 18.4. Average growth of C_3 and C_4 plants fumigated with 11 mmol m^{-3} of SO$_2$ expressed as a percent of the growth of nonfumigated control plants at 13.7, 25, and 50 mmol m^{-3} of CO$_2$. Each point is the mean of 24 plants, i.e., eight plants for each species. From Carlson and Bazzaz (1982).

tops, and were exposed to ambient summertime insolation, supplemented with illumination from high-intensity metal halide lamps to provide irradiances in excess of 800 μE m^{-2} s^{-1}. High light intensity is necessary to allow for the full expression of CO$_2$ on plant growth. SO$_2$ at 10 μmol m^{-3} was administered for 8 hours each day for 5 days during both the second and the fourth week of CO$_2$ treatment. The amount of water transpired by each plant was measured during the second 5-day period of fumigation with SO$_2$, and the plants were harvested within 3 to 4 days after the end of the second week of fumigation with SO$_2$.

The biomass of SO$_2$-fumigated C_3 plants was reduced to an average of 65% at 13.7 mmol m^{-3} of CO$_2$ but was not significantly different from nonfumigated plants at the higher CO$_2$ concentrations (Fig. 18.4). Conversely, biomass of fumigated C_4 plants at 13.7 mmol m^{-3} averaged 27% greater than nonfumigated plants. Growth of fumigated C_4 plants decreased sharply as CO$_2$ concentrations increased to 25 and 50 mmol m^{-3}. Thus even though C_4 plants may be more resistant to SO$_2$ in an environment of 13.7 mmol m^{-3} of CO$_2$, it appears that these plants will be at a disadvantage compared to C_3 plants if global concentrations of both CO$_2$ and SO$_2$ increase.

Leaf area increased with increasing CO$_2$ for all nonfumigated plants, but was reduced by exposure to SO$_2$ (Table 18.7). Fumigation with SO$_2$ caused a larger reduction in the leaf area of C_3 species at 13.7 mmol m^{-3} of CO$_2$ than at the higher

CO_2 concentrations. Conversely, leaf area of C_4 plants was reduced more by fumigation with SO_2 at the higher concentrations of CO_2 than at 13.7 mmol m^{-3}.

The ratio of photosynthetic to nonphotosynthetic material within a plant (leaf-area ratio) declined with increasing CO_2 for C_3 species, did not change with CO_2 concentrations for C_4 species, and was not appreciably different between fumigation treatments (Table 18.7). The shoot:root ratio of plants not fumigated with SO_2 was nearly constant for all SO_2 concentration for the C_3 plants, decreased with increasing CO_2 for *Amaranthus*, and increased slightly with increasing CO_2 for both *Setaria* species. In general, for plants fumigated with SO_2, the trend was for the shoot:root ratio in C_3 plants to decline, but that of C_4 plants increased slightly with CO_2 concentration.

Differences in the sensitivity of photosynthesis to SO_2 have been found between C_3 and C_4 species of *Atriplex* (Winner and Mooney, 1980c). Increased photosynthetic susceptibility of C_3 plants to SO_2 was found to be correlated with higher stomatal conductance and higher sensitivity of the photosynthetic mechanism to SO_2. Leaves with high stomatal conductance will absorb larger amounts of SO_2 and therefore be effectively exposed to higher amounts of the pollutant than leaves with low stomatal conductance. The data for whole plant water uptake at 13.7 mmol m^{-3} of CO_2 in our C_3-C_4 study (Carlson and Bazzaz, 1982) suggest that SO_2 uptake may have been higher for two of the C_3 species (*Chenopodium* and *Polygonum*) than it was for the C_4 species. This correlates with the greater reduction in growth observed for C_3 species at 13.7 mmol m^{-3} of CO_2. Whole-plant water uptake for fumigated *Datura* was about the same as the higher values observed for C_4 species. This agrees with the growth data for *Datura*, which indicates that SO_2 had less effect on growth at 13.7 mmol m^{-3} of CO_2 than was observed for the other C_3 species.

PEP carboxylase, the initial CO_2-fixation enzyme of C_4 species, has a higher bicarbonate affinity than RuBP carboxylase, the initial enzyme of C_3 species (Ziegler, 1972, 1973). Ziegler suggested that sulfite competitively binds to both enzymes, but PEP carboxylase is better able to discriminate between bicarbonate and sulfite, and therefore shows better sulfite exclusion than RuBP carboxylase. This may be a cause of the difference in susceptibility to SO_2 between C_3 and C_4 species (Ziegler, 1972, 1973, 1975; Winner and Mooney, 1980c). At high CO_2 concentrations, C_3 plants are less sensitive to SO_2 than the C_4 plants, perhaps because of differences in internal CO_2 concentrations between C_3 and C_4 plants caused by differences in stomatal response. Three different types of stomatal behavior in response to increased CO_2 have been identified (Louwerse, 1980). These are (1) stomata that stay open, resulting in internal CO_2 concentrations that remain near to that of the air surrounding the leaf, (2) stomata that tend to keep the internal CO_2 concentration constant and therefore independent of the external CO_2 concentration, and (3) stomata that maintain a constant ratio between external and internal CO_2 concentrations. Many C_4 species maintain internal CO_2 concentrations nearly constant with increasing CO_2, but C_3 plants tend

TABLE 18.7
Response of Early Successional Annual Species to Fumigation with SO₂ at Different Concentrations of CO_2

Plant; and CO₂ concentration (mmol m⁻³)	Plant response; and SO₂ concentration (μmol m⁻³)									
	Shoot:root ratio		Leaf area (dm²/plant)		Leaf area ratio		Leaf water use efficiency (mg CO₂/g H₂O)		Whole plant water uptake during fumigation (ml/day)	
	0	10	0	10	0	10	0	10	0	10
C₃ SPECIES:										
C. album										
13.7	6.8	8.2	13.0	9.8	1.9	2.1	4.6	5.2	10.6	9.4
25	6.4	6.2	15.9	14.7	1.5	1.4	11.0	12.8	6.7	7.3
50	6.4	6.0	16.2	15.3	1.3	1.4	23.8	18.9	7.6	7.7
D. stramonium										
13.7	6.6	7.8	9.6	7.8	3.5	3.9	4.9	4.4	8.7	6.6
25	5.4	6.5	14.0	13.9	2.9	3.1	10.1	13.1	4.6	4.6
50	6.5	5.6	14.9	14.2	2.9	2.9	17.8	26.2	3.5	4.4
P. pensylvanicum										
13.5	6.4	6.5	24.3	19.5	1.7	2.3	6.1	4.8	13.3	8.8
25	6.5	6.2	27.9	27.1	1.3	1.3	11.6	8.7	10.2	11.2
50	6.8	5.7	29.2	27.5	1.1	1.3	18.1	14.9	8.8	11.5
C₄ SPECIES:										
A. retroflexus										
13.7	9.9	8.7	2.8	4.9	1.0	1.3	10.8	11.1	6.7	6.8
25	8.4	9.7	3.5	2.9	1.0	1.2	24.7	23.7	6.6	6.9
50	7.0	10.6	3.9	3.1	1.0	1.4	53.2	38.2	5.9	5.4
S. faberii										
13.7	9.0	8.5	10.4	10.3	2.0	1.9	12.9	11.8	4.0	5.2
25	11.0	10.4	10.5	11.4	2.1	2.0	23.8	25.1	3.0	3.4
50	11.3	10.8	15.0	9.4	2.1	2.1	36.2	39.7	3.0	3.7
S. lutescens										
13.7	2.8	4.0	5.2	6.5	2.5	1.8	12.6	12.3	3.0	4.4
25	4.6	5.6	7.6	3.0	1.9	1.8	24.8	23.3	3.2	3.6
50	4.8	4.5	6.4	3.3	2.0	2.1	39.3	37.1	3.0	3.8

SOURCE: Carlson and Bazzaz (1982).

to exhibit a response similar to either 1 or 3, in which internal CO_2 increases in proportion to the ambient CO_2 concentration (Akita and Moss, 1972; Goudriaan and van Laar, 1978; Louwerse, 1980). If this difference in response holds for plants being fumigated with SO_2, the photosynthetic rate of C_3 plants could increase with increased CO_2 and compensate for rate reductions caused by SO_2. The C_4 plants would not be able to compensate in a similar fashion and would thus exhibit lower photosynthetic rates. We are presently making measurements of gas exchange and stomatal conductance during fumigation at increased CO_2 concentrations in an attempt to resolve this question.

19

Growth/Environment Interactions in SO$_2$ Responses of Grasses

T. A. Mansfield and Teresa Jones

The productivity of grasslands is a major component of the agricultural economy of the United Kingdom and of other densely populated countries of western Europe. Pastures and meadows are often located in or near urban and industrial areas, where there are elevated levels of SO$_2$ in the atmosphere. Serious attention has therefore been paid to the influence of SO$_2$ on the growth and physiology of grasses, but extensive research has not yet enabled us to predict the effect of a known amount of pollution on crop yield. Progress has been made, however, insofar as we are beginning to understand some of the factors that affect the responses of grasses to SO$_2$, and thus we are learning how to explain discrepancies between the results of different experiments. Pursuit of these factors should enable us to develop models that will predict yield reductions.

Some of the differences in the magnitude of responses to SO$_2$ that have been reported can be ascribed to idiosyncrasies of fumigation equipment. These differences can be classified according to whether they originate mainly in the artificial conditions of fumigations, or reflect the sorts of differences that might be found between environmental conditions in the field. We will be concerned particularly with some factors that need to be considered in order to predict responses in the field, but we will start by looking briefly at errors that originate in the design of fumigation equipment.

Characteristics of Fumigation Chambers

The main feature of exposure systems that has led to errors in estimating effects of SO$_2$ is too little air movement across the leaves. In slow-moving air the resistance to molecular diffusion in the thin boundary layer around each leaf can become very large, and may be the principal factor determining how fast a pollutant diffuses from the bulk of the air in a chamber into sensitive sites in the cells. The types of experiment most open to criticism are those in which plants have been put into a single chamber that has a small inlet through which a pollutant/air mixture can be passed at a known rate. If the air movement within the chamber depends solely on the rate at which gas passes through the inlet, there will almost certainly be a high boundary-layer resistance around the leaves. When such a chamber is used to measure the threshold concentration at which SO$_2$ damage occurs, a completely false impression of SO$_2$ tolerance can result. This was demonstrated by Ashenden and Mansfield (1977) in experiments with perennial ryegrass (*Lolium perenne* L.) in wind tunnels. Plants were grown for 4 weeks in a relatively high concentration of SO$_2$ (0.110 ppm, or approximately 4.57 μmol m^{-3}), but when the wind speed was 17 cm s^{-1}, there was no detectable growth inhibition. However, at a wind speed of 42 cm s^{-1} there were appreciable effects of the pollutant on the main growth parameters (e.g., leaf area, root and shoot dry weights), and the average growth reduction in two experiments was 43%. The differences in response were attributed to changes in the boundary-layer resistances, which for water vapor were 797 and 89 s m^{-1} at the lower and higher wind speeds, respectively. The average stomatal resistances for water vapor were 440 s m^{-1} in light and 1,570 s m^{-1} in darkness; so it is clear that at the lower wind speed the gaseous flux into the leaf in the light would not be dominated by changes in stomatal aperture as it is in the field.

Even the lower wind speed in these experiments (17 cm s^{-1}) represents more air movement than is found in most fumigation chambers that are not specially equipped to achieve adequate throughflow or internal stirring. It is clear that a false impression of the tolerance of *L. perenne* to SO$_2$ was given by the experiment performed at this wind speed, and we must conclude that the same will be true for some published data. Unfortunately, it is difficult to decide which particular experiments fall into this category, because in many papers the fumigation apparatus used is not described in enough detail. We have to recommend, therefore, that where doubt exists about the adequacy of techniques used in the past, the effect of SO$_2$ on a particular species should be reinvestigated. Previous estimates of SO$_2$ tolerance or of threshold concentrations that produce injury should not be accepted uncritically.

Separate problems are raised by the methods researchers have used for measuring the concentrations to which plants in chambers are exposed. A critical examination of the procedure adopted in one experimental system that has generated many published data on effects of SO$_2$ on grasses showed that concentrations were much less than stated (Unsworth and Mansfield, 1980). The problem here was that the exposure concentration was defined in terms of that at the inlet, but the uptake into the plants and adsorption onto chamber walls were overlooked. Koziol (1980) showed that in his experiments on soybean, in which the uptake of SO$_2$ into leaves may be as high as 3 mg m^{-2} h^{-1}, the supply concentration would have overestimated the exposure concentration by a factor of ten. His chambers had a volume of 0.48 m^3, contained ten soybean plants, and were provided with one air change every 12 min.

We conclude that fumigations with SO$_2$ must be conducted with enough air movement across the leaves to reduce boundary-layer resistance to a value typical for plants in the open air. The exposure concentration quoted must be that inside the plant chamber, which may be much less than that at the inlet. Disregard of the importance of boundary-layer resistance, and of the depletion of SO_2 within a chamber, will lead to serious underestimates of the sensitivity of plants to SO_2.

Environmental Factors

The problems discussed above, which relate to the plant's environment in fumigation chambers, result from factors that are largely artificial. Boundary-layer resistances can be very high in nearly still air, which will occur only rarely in a natural environment. Other differences between experimental fumigations, however, may arise from differences between factors within the ranges actually encountered in the outside environment; so variations in the results of experimental fumigations that derive from such differences are important, because they represent possible responses of plants in the outside environment. Our own recent research has been concerned with the way in which growth rate, which is easily varied in growth chambers by altering factors such as light intensity and temperature, can modify the responses of grasses to SO_2. During the course of these studies we have also identified certain growth-compensation mechanisms that need to be taken into account when the overall effect of SO_2 is assessed.

Growth rate and pollution response. Bell, Rutter, and Relton (1979) fumigated *Lolium perenne* for periods of up to 194 days with SO_2 concentrations ranging from 0.015 to 0.148 ppm (0.62 to 6.15 μmol m^{-3}) in both summer and winter conditions in the U.K. In one experiment, 0.62 μmol m^{-3} for 173 days in winter produced a 68% reduction in shoot dry weight, but this response was not found in some other experiments in winter, even with higher concentrations. The collected data from nine different fumigations conducted in different outdoor conditions showed great inconsistency; the authors attributed the difficulty of obtaining a clear dose-response relationship to the interaction of environmental factors with effects of SO_2. These data pointed strongly to the need for careful investigation of the pollution responses of grasses under different environments in controlled conditions.

The main subject of our own studies of this question has been timothy grass (*Phleum pratense* L.), whose response to SO_2 was shown to vary in preliminary studies. Experiments were conducted in wind tunnels (Horsman and Wellburn, 1977; Jones and Mansfield, 1982a) in different light intensities and temperatures, and the effects of a fairly high concentration of SO_2 (0.120 ppm or 4.99 μmol m^{-3}) were observed. The light intensities (photon fluxes) were 100 or 400 μmol m^{-2} s^{-1} of photosynthetically active radiation (PAR), and the temperatures were 19°C (night) / 30°C (day) or 12°C (night) / 26°C (day). Individual plants in the different treatments were ranked according to their final size, and equivalent plants in the rank order from the control and SO_2 treatments were then paired. In Fig. 19.1, each point represents two such plants, and the dry weights of SO_2-

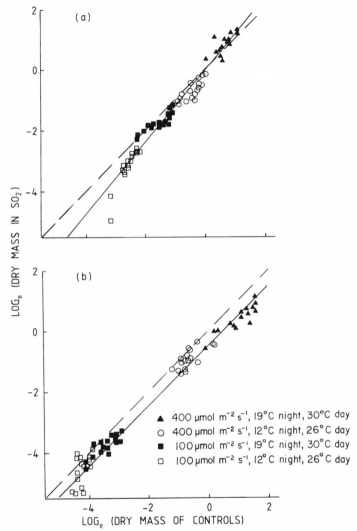

Fig. 19.1. Relationship between dry mass of *Phleum pratense* grown in 0.120 ppm (4.99 μmol m^{-3}) of SO$_2$ and that of clean air controls: (*a*) shoot weight; (*b*) root weight. The broken line is the "no-effect" line ($y = x$), and the solid line is a fitted regression line. Growth conditions are shown in the key. Redrawn from Jones and Mansfield (1982a), p. 63.

treated and control plants are plotted against each other. If there were no influence of the pollutant on growth, the points would tend to lie about the dotted "no effect" line ($y = x$); if there were an effect of similar magnitude in all treatments, the relationship would be described by a line parallel to the "no effect" line. In fact, the effect of SO$_2$ on shoot growth (Fig. 19.1a) depended on treatment, and the continuous line fitted to the points differs significantly from the broken line. When growth was slowed in low light and/or low temperature, there was a greater effect of SO_2 on shoot dry weight. However, root growth was always affected in the same way by SO_2, no matter what treatment was employed (Fig. 19.1b).

The results of this experiment are not definitive, because the SO_2 concentrations, duration of fumigation (44 days), and environmental conditions were chosen arbitrarily. A different choice of conditions would probably produce a different picture of how the SO_2 response of *P. pratense* is controlled by environmental factors. The value of the data in Fig. 19.1 is that they suggest the existence of certain relationships, and these, if confirmed by subsequent studies, might enable us to predict responses to SO_2 under different conditions in the field.

Detailed analyses of growth. Further understanding of how *P. pratense* responds to SO_2 has emerged from experiments in which growth parameters have been studied in more detail. Seven-day-old seedlings were fumigated with 0.120 ppm

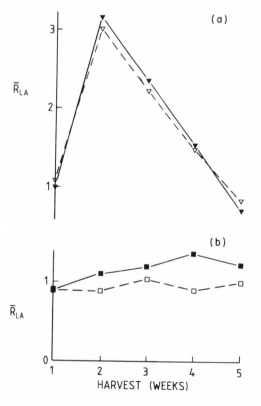

Fig. 19.2. Mean relative growth rates of the leaf area ($\bar{R}_{LA}$) of *Phleum pratense* grown in 0.120 ppm (4.99 μmol m^{-3}) of SO_2 (broken line) or in clean air (solid line): (*a*) plants grown in high photon flux and 16-h photoperiods; (*b*) plants grown in low photon flux and 12-h photoperiods. Redrawn from Davies (1980a), p. 483.

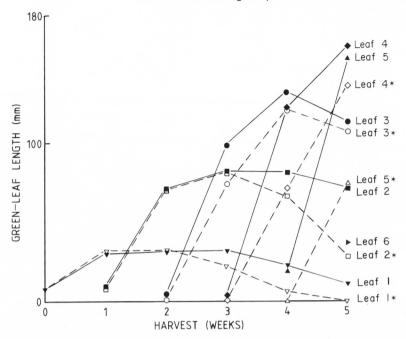

Fig. 19.3. Changes in leaf length with time for plants of *Phleum pratense* grown under low photon flux and 12-h photoperiods in clean air (solid line) or 0.120 ppm (4.99 μmol m^{-3}) of SO₂ (broken line). The asterisks denote the leaves of polluted plants. See also Fig. 19.2b. Redrawn from Davies (1980a), p. 484.

(4.99 μmol m^{-3}) of SO₂ for 35 days in two different light regimes, namely, 125 μmol m^{-2} s^{-1} PAR in 12-h photoperiods, and 480 μmol m^{-2} s^{-1} PAR in 16-h photoperiods. The temperatures were identical for the two treatments (20°C day, 16°C night), and relative humidity was 60%. Leaf dimensions were measured each week, and relative growth of leaf area ($\bar{R}_{LA}$) was calculated in terms of the weekly increment, $\bar{R}_{LA}$ per week = log$_e$ (area at week $x + 1$) $-$ log$_e$ (area at week x). There was a large difference between the two light regimes, both in the magnitude of $\bar{R}_{LA}$ and in the effect of SO₂ (Fig. 19.2). The SO₂ caused a reduction in $\bar{R}_{LA}$ in the slow-growing plants that was statistically significant ($p <$ 0.05) starting between the first and second harvests, but $\bar{R}_{LA}$ in the more rapidly growing plants was little affected by SO₂. The influence of SO₂ on leaf development for the plants in low-light/short photoperiods is clearly seen in the detailed record of the development of individual leaves on the main shoot (Fig. 19.3). There was little difference in leaf length until after the second week of fumigation, by which time the first and second leaves had reached almost their maximum rate of expansion. During the following weeks the growth of these and subsequent leaves declined, and senescence was greatly accelerated by the presence

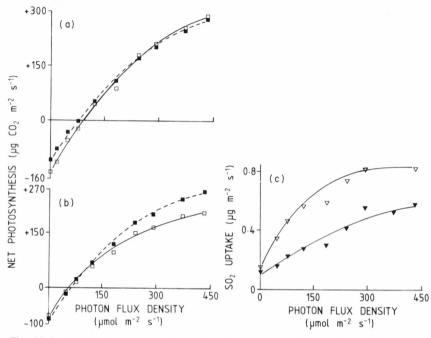

Fig. 19.4. Effects of increasing photon-flux density on net CO_2 exchange of plants of *Phleum pratense* raised in 15-h photoperiods with light of (*a*) 360 μmol m^{-2} s^{-1}, or (*b*) 135 μmol m^{-2} s^{-1}. In *a* and *b* the solid lines are for plants exposed to 0.110 ppm (4.57 μmol m^{-3}) of SO_2 during growth, the broken lines for the clean air controls. Part *c* shows rates of SO_2 uptake in the plants grown in 0.110 ppm (4.57 μmol m^{-3}) of SO_2; the upper curve is for the data in *a*, the lower curve for the data in *b*.

of SO_2. By the last harvest the sixth leaf had not appeared in the SO_2-treated plants, although its development was well-advanced in many of the controls.

The data in Figs. 19.2 and 19.3 provide a useful picture of the inhibition of leaf growth caused by SO_2, but they do not contain all the information needed to understand events taking place within the plant. Studies of the photosynthetic characteristics of the leaves, and of the rates of uptake of SO_2, have also revealed differences between plants grown under low light intensities and those grown under high light intensities. Seedlings of *P. pratense* were grown in 15-h photoperiods in PAR of 135 μmol m^{-2} s^{-1} or 360 μmol m^{-2} s^{-1} in temperatures of 16°C (night) and 21°C (day), and exposed to 0.110 ppm (4.57 μmol m^{-3}) of SO_2 for 4 to 6 weeks. Net photosynthesis and net sulfur flux were measured in plants of similar total leaf area (i.e., the plants from the low light-intensity treatment were approximately 12 days older). There were important differences between the plants grown in the two photoenvironments (Fig. 19.4). There were no significant differences in net photosynthesis between the controls and the SO$_2$-fumigated plants grown in the high-light environment, throughout the range of light intensities, but below the light-compensation point a stimulation in dark respiration in the polluted plants was evident. In contrast, the plants raised in the low-light environment showed differences only above the light-compensation point, and no stimulation of dark respiration was seen in the polluted plants. The rates of SO$_2$ uptake per unit leaf area were on average about 50% higher for the plants grown in the higher light intensity; so the smaller effect of SO$_2$ on net photosynthesis in these plants could not be attributed to lower doses.

The data for the plants grown in a low-intensity light (Fig. 19.4b) have features like those obtained for *Vicia faba* by Black and Unsworth (1979a). Photosynthesis was little affected by SO$_2$ at low light intensities, but as intensity increased the inhibition became progressively greater. In *V. faba*, photosynthesis was little affected by SO$_2$ when it was light-limited, but was strongly inhibited when it was light-saturated. This behavior was consistent with the view that SO$_2$ competes with CO$_2$ for binding sites on ribulose bisphosphate carboxylase (RuBPC). Our data for the CO$_2$ exchange in high-intensity light (Fig. 19.4a) show clear evidence that dark respiration is stimulated in polluted plants, as was also observed by Black and Unsworth. Perhaps higher mitochondrial activity is necessary for repair or detoxification mechanisms, but these can operate only when given sufficient respiratory substrate. The resistance of photosynthesis to SO$_2$ when there is substrate for additional respiration in high-intensity plants (Fig. 19.4b), and the inhibitory effect of SO$_2$ with no detectable effect on respiration in low-intensity plants (Fig. 19.4a), support this interpretation. A difference between these data and those of Black and Unsworth is the duration of the SO$_2$ treatment prior to the measurements. Our plants grown in the high-intensity light and designated as SO$_2$-treated in Fig. 19.4a had been exposed to 0.110 ppm (4.57 μmol m^{-3}) of SO$_2$ for at least 26 days prior to the measurements of photosynthesis, and those from the low intensity were pretreated for a minimum of 35 days. Our plants thus had sufficient time to carry out any metabolic adjustments needed for repair or detoxification, but the plants given the shorter fumigations performed by Black and Unsworth may not have had enough time. Consequently, differences between their observations and our own may result from this difference in experimental procedure as well as from differences between the species studied.

A detailed analysis of growth parameters during the exposure of *P. pratense* to 0.120 ppm (4.99 μmol m^{-3}) of SO$_2$ has indicated the existence of other mechanisms that would further complicate our assessment of the effects of SO$_2$ on plant productivity. These observations were made on seedlings that were fumigated for 40 days beginning 10 days after sowing. The conditions used—light intensity 220 μmol m^{-2} s^{-1}, photoperiod 15 h, temperature 17°C (night) and 23°C (day), RH 47 to 57%—led to relatively slow growth, and it was anticipated that photo-

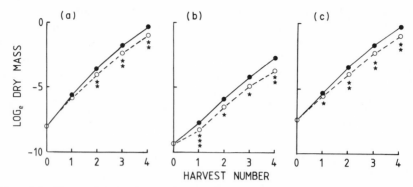

Fig. 19.5. Changes in dry mass of *Phleum pratense* with time (harvests at 10-day intervals) in clean air (solid line) or in 0.120 ppm (4.99 μmol m⁻³) of SO₂ (broken line): (*a*) shoots; (*b*) roots; (*c*) whole plant. Significance of SO₂ effect at each harvest: *$p < 0.05$; **$p < 0.01$; ***$p < 0.001$. Redrawn from Jones and Mansfield (1982a), p. 65.

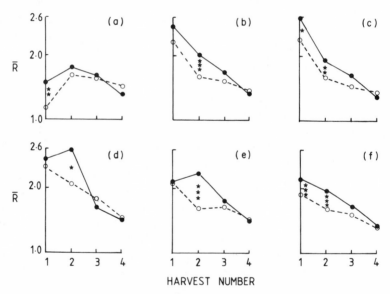

Fig. 19.6. Mean relative growth rates ($\bar{R}$) of *Phleum pratense* exposed to clean air (solid line) or to 0.120 ppm (4.99 μmol m⁻³) of SO₂ (broken line): (*a*) root dry mass; (*b*) shoot dry mass; (*c*) leaf dry mass; (*d*) leaf area; (*e*) stem dry mass; (*f*) total dry mass. Statistical significance as in Fig. 19.5. Redrawn from Jones and Mansfield (1982a), p. 66.

synthesis would be inhibited by SO_2 throughout the period of the observations (cf. Fig. 19.4b). Ten randomly selected plants were removed at intervals of 10 days for destructive growth analysis. Fig. 19.5 shows the increases in shoot, root, and total plant weight with time.

The root weight of the SO_2-treated plants was significantly less than that of the controls after 10 days, but at this stage an effect on shoot growth was barely detectable. The effects of SO_2 on the shoot were, however, very marked (41% reduction in dry weight) by the second harvest, and at the end of the experiment there were 62% and 51% reductions in dry weights of roots and shoots, respectively. It proved valuable to plot the mean relative growth rates ($\bar{R}$) for each plant part (Fig. 19.6). Large differences in $\bar{R}$ between the control and SO_2 treatments became apparent in the first 20 days, but by 40 days there was a recovery in every case. This restoration of $\bar{R}$ occurred in spite of the large difference in total plant weight at the end of the experiment. The gains in dry weight per unit leaf area (net assimilation rate, NAR) were slightly reduced by SO_2, and effects were of similar magnitude throughout the experiment (Fig. 19.7a). This response to SO_2 is just what would be predicted from the measurements of CO_2 exchange of plants grown in low-intensity light (Fig. 19.4b). The changes that proved to be of the greatest interest were in leaf-area ratio (LAR) and specific leaf area (SLA). There was no evidence that SO_2 affected these parameters during the first three harvest intervals, but both showed a highly significant increase after 40 days (Fig. 19.7b,c). LAR is the proportion of assimilatory area to the plant's dry weight, and SLA is the leaf area divided by leaf dry weight. Both LAR and SLA

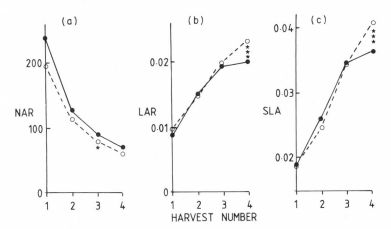

Fig. 19.7. Mean growth parameters for *Phleum pratense* exposed to clean air (solid line) or to 0.120 ppm (4.99 μmol m^{-3}) of SO_2 (broken line): (*a*) net assimilation rate (NAR); (*b*) leaf area ratio (LAR); (*c*) specific leaf area (SLA). Statistical significance as in Fig. 19.5. Redrawn from Jones and Mansfield (1982a), p. 67.

are parameters that display plasticity as a plant adjusts to different environmental conditions, and similar mechanisms may operate following an inhibition of NAR by SO_2.

Interpretation of Responses of *P. pratense* to SO_2

These observations have allowed us to formulate the following interpretation of the behavior of *P. pratense* in 0.110 to 0.120 ppm SO_2 (4.57 to 4.99 μmol m^{-3}) in low- and high-intensity-light environments.

High-intensity light. The effect of SO_2 on net photosynthesis is apparently countered by repair mechanisms that depend on increased mitochondrial activity. The rate of photosynthesis is high enough to provide the necessary respiratory substrate. Net photosynthesis is therefore maintained in polluted plants (Fig. 19.4a), but only at the cost of increasing respiratory activity, which must reduce the amount of carbohydrate available for use elsewhere in the plant. The observed reduction of root growth (Fig. 19.1b) may be the consequence. The net effect can be a maintenance of the dry weight of the plant as a whole, but with an increase in the shoot:root ratio (Jones and Mansfield, 1982b).

Low-intensity light. Under low light, not enough carbohydrate is available to support additional respiration; so repair processes do not take place, and net photosynthesis continues to be inhibited by SO_2 (Figs. 19.4b, 19.7a). Initially, relative growth rates are reduced in all parts of the plant, but these recover progressively with time (Fig. 19.6). The recovery can be attributed to increases in LAR and SLA; i.e., reduced assimilatory ability per unit leaf area is countered by an increase in the total leaf area. The plant thus overcomes the effects of SO_2, but only at the cost of a relatively higher investment of dry matter in leaf production. Just as in plants grown in high-intensity light, leaf production seems to take place at the expense of the roots, whose growth is inhibited at an early stage (Fig. 19.6). The effect of SO_2 on shoot:root ratio is similar in plants grown in low- and high-intensity-light environments (Fig. 19.1b).

Our conclusions, that plants seem to have growth-compensatory mechanisms that enable them to overcome some of the effects of air pollutants, find support in other published studies. Oshima, Bennett, and Braegelmann (1978), and Oshima *et al.* (1979), found evidence of changes in assimilate partitioning in parsley and cotton exposed to ozone pollution. Root dry weight in parsley was decreased by 43% in 0.20 ppm (8.3 μmol m^{-3}) of O_3, but there was little effect on the leaves. Relative growth rate was reduced initially but eventually recovered. We can also compare the effects of some foliar pathogens, for example, the powdery mildews and rusts. In wheat and barley infected with *Erysiphe graminis*, growth of the

mental procedures for measuring genetic differences in growth responses caused by SO_2, and discuss research needs in this area.

Growth Responses and Evolutionary Implications

Differences in experiments. Although the literature is fairly consistent about relative vulnerabilities of plants to foliar injury caused by SO_2, there is much confusion about relative sensitivities of plants to SO_2 as manifested by growth responses. The main reasons for this confusion are as follows: (1) measuring growth responses is more complex than simply observing foliar injury; (2) the extreme heterogeneity of experimental regimes used to measure SO_2-induced growth responses makes it difficult to compare the results from experiments in different locations; and (3) growth responses have not been measured in any uniform manner.

Most foliar sensitivity studies have been conducted with acute fumigations, in which short-term, high-concentration fumigations are used to cause visible injury. These types of studies are generally inadequate for measuring growth responses to SO_2. Such responses can be studied only with low-level, long-term fumigations. Thus, these growth-response studies are more complex in their space, equipment, and labor needs.

Diverse experimental methods for examining SO_2 effects on plant growth have been used. Some experiments have been conducted in growth chambers, others in greenhouses, field chambers, or field plots. The dosages of SO_2 have differed greatly, as has the timing of SO_2 exposure (i.e., the number of hours per day and days per week, the total number of days, and the season when SO_2 fumigation occurred). The genetic makeup and the ages of plants used have also differed between experiments. For example, in perennial ryegrass studies, some authors have used tillers, and others have used seedlings. Similarly, for trees, where little is known about juvenile-mature correlations in pollutant responses, studies have been done on young seedlings (Constantinidou and Kozlowski, 1979a; Constantinidou, Kozlowski, and Jensen, 1976; Farrar, Relton, and Rutter, 1977; Jensen, 1981a; Jensen and Dochinger, 1979; Marshall and Furnier, 1981; Roberts, 1975; Suwannapinunt and Kozlowski, 1980), saplings (Garsed, Rutter, and Relton, 1981), and older trees (Legge *et al.*, 1977; Stone and Skelly, 1974; Thor and Gall, 1978). The foliar age of plant material has also differed greatly from study to study. Other environmental factors shown to cause differences in plant growth responses to SO_2 include soil fertility (Ayazloo, Bell, and Garsed, 1980; Eaton, Olmstead, and Taylor, 1971; Leone and Brennan, 1972b; Milchunas *et al.*, 1981) and wind speed (Ashenden and Mansfield, 1977).

The variables measured in growth-response studies have differed from study to study. Height; diameter; radial increment; fresh and dry weights of green leaves, dead leaves, shoots, flowers, fruits, and roots; crop yields; flower, fruit, and leaf size and number; the numbers of tillers; and root and leaf elongation have all been used in attempts to measure SO$_2$ effects on plant growth. Relative growth rates have also been used to compare SO$_2$ effects on growth (Farrar, Relton, and Rutter, 1977; Garsed, Rutter, and Relton, 1981; Jensen, 1981a,b; Marshall and Furnier, 1981).

Given the difficulties outlined above, it is very important to use only comparable data in identifying the effects of genetic factors. The rest of this section therefore examines differences in plant growth responses within and between species as measured by researchers who compared multiple genotypes in their experiments.

Acute exposures. The use of acute exposures to measure pollution effects is advantageous in that shorter-duration experiments can be run; so more plants can be tested in a given time period than could be done with chronic fumigations. For some species, there seems to be a good correlation between responses to chronic SO$_2$ exposures and those to acute exposures. For example, Dochinger and Jensen (1975) found the relative growth responses of two hybrid poplar (*Populus deltoides* Bartr. × *P. trichocarpa* Torr. and Gray) clones were correlated with the degrees of foliar injury caused by SO$_2$, in both chronic and acute fumigations. Clone 211 was consistently more sensitive than clone 207 to both chronic and acute exposures. Other researchers, however, have reported a lack of correlation between acute foliar injury and chronic growth reductions (Ayazloo and Bell, 1981; Garsed and Rutter, 1982a; Horsman, Roberts, and Bradshaw, 1979).

Dochinger *et al.* (1972) found significant differences in growth responses of four hybrid poplar clones exposed to 5 ppm (208 μmol m^{-3}) of SO$_2$ for 3 to 12 hours. Posthumus (1978) reported substantial differences in dry-weight production of several plant species following exposure to either 0.6 or 0.8 ppm (25 or 33.3 μmol m^{-3}) of SO$_2$ for continuous exposure during 14 days. Jensen and Dochinger (1979) found differences in growth responses of *Acer saccharinum* L., *Liriodendron tulipifera* L., and *Populus deltoides* Bartr. exposed to 4 ppm (166.4 μmol m^{-3}) of SO$_2$ for 5 to 8 hours. In general, growth reduction for these three species was inversely correlated with an increase in foliar injury. Species sensitivity rankings from the acute fumigations were the same as those for relative growth responses to a chronic SO$_2$ exposure of 0.25 ppm (10.4 μmol m^{-3}) of SO$_2$ for 8 hours per day, 5 days per week, for 12 weeks (Jensen and Dochinger, 1979).

Significant reductions in height growth and leaf dry weight were found for *Robinia pseudoacacia* L. but not for *Ulmus americana* L. following exposure to 2 ppm (83.2 μmol m^{-3}) of SO$_2$ for 4 hours (Suwannapinunt and Kozlowski, 1980). Thompson, Kats, and Lennox (1980) found differences in growth responses to 2.0 ppm (83.2 μmol m^{-3}) of SO$_2$ for 25 hours per week for 8 to 32 weeks in five perennials and five annuals native to the Mojave Desert. Reductions in linear growth and dry weight were found in the perennials *Larrea divaricata* Cav., *Chilopsis linearis* Cav., and *Ambrosia dumosa* (Gray) Payne, but not in *Atriplex canescens* (Pursh) Nutt. or *Encelia farinosa* Gray ex Torr. All five annuals tested were severely damaged by the same SO₂ exposure. Mortality was noted in all plants of *Phacelia crenulata* Torr., *Plantago insularis* Eastw., *Erodium cicutarium* (L.) L'Her., and *Chaenactis carphoclinia* Gray.

Reinert and Nelson (1980) found *Begonia × hiemalis* Fotsch 'Fantasy' to be the only Elatior begonia cultivar of five tested to be sensitive to 0.5 ppm (20.8 μmol m^{-3}) of SO₂ for four exposures of 4 hours each, 6 days apart. 'Fantasy' showed a decreased flower production without showing visible injury.

Chronic exposures. Studies with perennial ryegrass (*Lolium perenne* L.) fumigated at low SO₂ concentrations show how difficult it is to detect subtle, chronic growth effects. Bell and Clough (1973) demonstrated an apparent genetic difference in the growth response of *Lolium perenne* L. to low SO₂ concentrations in chamber fumigations. The dry weight of shoots of cultivar S23 was depressed by 52% in comparison to controls, but that of the tolerant Helmshore plants was not reduced. Cowling, Jones, and Lockyer (1973) reported a stimulatory effect of low SO₂ concentrations on the S23 cultivar of *L. perenne*, sparking a debate that has run for the past 10 years on whether low SO₂ concentrations increase or decrease *L. perenne* growth. In subsequent studies, Ashenden and Mansfield (1977), Ayazloo and Bell (1981), Ayazloo, Bell, and Garsed (1980), Bell, Rutter, and Relton (1979), and Crittenden and Read (1978b) have shown reductions in growth of S23 in SO₂-polluted air, but Cowling and Koziol (1978), Cowling and Lockyer (1978), and Lockyer, Cowling, and Jones (1976) found no effect on growth of the same cultivar following SO₂ exposure. Bell, Rutter, and Relton (1979) suggested that some of the differences between these studies could have resulted because fumigations were done on plants of different ages. Younger plants of S23 appear to be more sensitive to SO₂ than older plants. Other possible contributing factors are different environmental conditions, such as temperature, wind speed, and light conditions at the various test sites, and the presence of other air pollutants that may be interacting with SO₂ at some sites. The season of the year when the plants are exposed to SO₂ is also important for relative sensitivities.

Horsman, Roberts, and Bradshaw (1978) and Horsman *et al.* (1979) found wide differences in growth responses of *L. perenne* clones exposed to 0.25 ppm (20.8 μmol m^{-3}) of SO₂ for 8 weeks. They found tolerant clones with little reduction in yield, very sensitive clones that had more than 60% reduction in yield, and many gradations of sensitivity in between. No correlation was found between visible injury due to acute fumigation (0.98 ppm, or 40.8 μmol m^{-3}, of SO₂ for 2 weeks) and the reduction in yield due to chronic SO₂ exposure.

Dactylis glomerata L. 'S37' was found to be slightly more sensitive to SO₂ than *Lolium perenne* 'S23' (Ashenden, 1978), and *Poa pratensis* L. 'Monopoly' was found to be intermediate in sensitivity between *L. perenne* 'S23' and *D. glomerata* 'S37' (Ashenden, 1979b). Ashenden and Williams (1980) found

Phleum pratense L. 'Eskimo' to be extremely sensitive to low-concentration SO_2 exposures. They noted a 60% reduction in dry weights of plants exposed to 0.068 ppm (2.8 μmol m^{-3}) for 140 days as compared to control plants. *Lolium multiflorum* Lam. 'Milamo' was less sensitive to SO_2, and the only dry-weight fraction significantly different from control plants was that of the green leaves, which showed a 28% reduction. Crittenden and Read (1979) found both *L. multiflorum* 'S22' and *D. glomerata* 'S143' to be sensitive to very low SO_2 concentrations. Reductions in shoot dry weight of both species were in the range of 30 to 40% after 8 to 10 weeks in 0.019 to 0.034 ppm (0.79 to 1.4 μmol m^{-3}) of SO_2. There was evidence, however, that late-season fumigations of *L. multiflorum* had much less effect on yield than fumigations of young plants.

Ayazloo and Bell (1981) compared growth rates of wild populations and selected cultivars of four English pasture grasses exposed to low concentrations of SO_2. *Dactylis glomerata* 'S37,' *Festuca rubra* L. 'Engina,' and *Phleum bertolonii* all had whole-plant dry-weight reductions of about 30%, and *Lolium perenne* 'S23' had about a 20% reduction. For wild collected plants, *Holcus lanatus* L. was reduced in total dry weight by more than 50%, as compared to 37% for *Dactylis glomerata*, 31% for *Festuca rubra*, and 9% for *Lolium perenne*. The authors presented evidence for evolution of SO_2 tolerance in all four species tested. Again, no correlation was found between acute and chronic injury either within or between species.

Tingey and Reinert (1975) reported a 22% reduction in dry weight of *Nicotiana tobacum* L. 'Bel W-3,' but no significant effect was seen in *N. tobacum* 'Burley-21' following 0.05 ppm (2.08 μmol m^{-3}) of SO_2 for 8 hours a day, 5 days a week, for 4 weeks. The foliage and most dry weights of *Medicago sativa* L. 'Vernal' were reduced 26 and 49%, respectively. In a similar study, Reinert and Sanders (1982) found that nine exposures to 0.3 ppm (12.5 μmol m^{-3}) SO_2 over a three-week period did not affect the shoot or root dry weights of *Raphanus sativus* L. 'Cherry Belle' or *Tagetes patula* L. 'King Tut,' but a reduction in *T. patula* 'King Tut' flower dry weight was detected.

Laurence (1979) compared the growth rates of four *Zea mays* L. and seven *Triticum aestivum* L. cultivars exposed to 0.1 to 0.6 ppm (4.16 to 25.0 μmol m^{-3}) of SO_2 for up to 100 hours. The maize cultivars (*Zea mays* 'EX114,' 'RX94,' 'RX57,' and 'ATC75') all showed no growth reductions caused by SO_2. The exposure to 0.6 ppm (25.0 μmol m^{-3}) of SO_2 for 100 hours caused significant decreases in dry weight of all *Triticum aestivum* wheat cultivars except 'Prelude,' the most SO_2-tolerant wheat cultivar.

Eaton, Olmstead, and Taylor (1971) showed that 0.2 ppm (8.3 μmol m^{-3}) for 3 days increased the dry weight of *Gossypium hirsutum* L. 'SJ-1' by 7%, but decreased that of *Lycopersicon esculentum* Mill. 'Pearson's Improved' by 8%.

Roberts (1976) and Houston (1974) have both shown differences in the needle growth of tolerant and sensitive clones of *Pinus strobus* L. following SO_2 exposure. Tolerant clones had significantly more needle growth in both studies than did sensitive clones. Significant height-growth reduction was found in one family of *Pinus taeda* L., but not in another family, following exposure to 0.14 ppm (5.8 μmol m^{-3}) of SO_2 for 6 hours each day for 28 consecutive days in 1976 and again in 1977 (Kress, Skelly, and Hinkelmann, 1982a).

Garsed and Rutter (1982a) reported variation in dry-matter production for *Picea abies* (L.) Karst, *P. sitchensis* (Bong.) Carr, *Pinus contorta* Dougl., and *P. sylvestris* L., exposed to ambient SO_2 in the United Kingdom. They found that the order of sensitivity of these species as measured for chronic exposures virtually reversed for acute fumigations.

Jensen (1981a) found that exposure to 0.2 ppm (8.3 μmol m^{-3}) of SO_2 for 12 hours a day, 7 days a week, for 6 weeks reduced the relative growth rates of one-year-old *Populus deltoides* and *Liriodendron tulipifera* seedlings by 30% and 50%, respectively. *Fraxinus americana* L. seedlings were described as being tolerant, since they were not affected by the same SO_2 exposure.

Roberts (1975) found that a relatively high SO_2 environment stimulated growth of *Betula papyrifera* Marsh., but decreased the growth of *Quercus palustris* Muenchh. Roberts speculated that the growth differences resulted primarily because SO_2 had a fertilizer effect on the *B. papyrifera*, which is generally considered to be a SO_2-sensitive species (Davis and Wilhour, 1976).

Whitmore and Freer-Smith (1982) reported many differences in the growth of *Alnus incana* L., *Betula pendula* Roth., *B. pubescens* Ehrh., *Malus domestica* Borkh., *Populus nigra* L., and *Tilia cordata* Mill. plants exposed to 0.062 ppm (2.6 μmol m^{-3}) of SO_2 for 150 days. A stimulation of growth occurred for *Tilia cordata*, no significant differences from controls were noted for *Malus domestica* and *Populus nigra*, and growth reductions were found for *Betula pendula*, *B. pubescens*, and *Alnus incana*.

Significant reductions of stem dry weight were reported for three of eight hybrid azalea cultivars exposed to six 3-hour fumigations of 0.25 ppm (10.4 μmol m^{-3}) of SO_2 during a 4-week period (Sanders and Reinert, 1982). The three sensitive cultivars were 'Mrs. G.G. Gerbing,' 'Glacier,' and 'Red Luann.' Tolerant cultivars included 'Hershey Red,' 'Mme. Pericat,' 'Pink Gumpo,' 'Red Wing,' and 'Snow.'

Combinations of pollutants. Comparisons of the relative growth effects of SO_2 on plants are complicated by the fact that SO_2 often occurs in nature with one or more additional pollutants. The effects of SO_2 in combination with other pollutants on plant growth can be less than additive or antagonistic (Jensen, 1981a; Noble and Jensen, 1980a,b; Reinert and Sanders, 1982; Tingey and Reinert, 1975; Tingey, Heck, and Reinert, 1971), additive (Ashenden and Mansfield, 1978; Mandl *et al.*, 1980; Reinert and Weber, 1980; Thompson, Kats, and Lennox, 1980; Tingey and Reinert, 1975), or more than additive or synergistic

(Ashenden, 1979b; Ashenden and Mansfield, 1978; Ashenden and Williams, 1980; Kress, Skelly, and Hinkelmann, 1982b; Mooi, 1981; Reinert and Gray, 1981; Tingey *et al.*, 1973b).

Pollutant combinations with SO_2 can induce different types of responses, even within the same plants. Tingey, Heck, and Reinert (1971) exposed *Raphanus sativus* seedlings for 5 weeks to 0.05 ppm (2.08 μmol m^{-3}) of SO_2 and of O_3 for 8 hours per day, 5 days a week, and found two different interaction effects in the same experiment. The inhibition of top growth was additive, but the inhibition of root growth was significantly less than additive. Similarly, Constantinidou and Kozlowski (1979a) found that acute exposure to an SO_2 and ozone (O_3) caused an additive effect on leaf expansion and leaf initiation, but a synergistic effect on stem dry weight of *Ulmus americana*.

Differences between species in plant growth responses to SO_2 in combination with other pollutants were shown by Reinert and Sanders (1982). No significant interactions between 0.3 ppm (12.5 μmol m^{-3}) of NO_2, of SO_2, and/or of O_3 for nine exposures of 3 hours each during a 3-week period were detected in measurements of *Raphanus sativus* 'Cherry Belle' foliage and root dry weights. However, interactions among the three pollutants occurred in shoots and roots of *Tagetes patula* 'King Tut.' Sulfur dioxide alone reduced *T. patula* shoot and root dry weight, but this effect was reversed in the presence of O_3 or NO_2. Ashenden and Mansfield (1978) found synergistic effects caused by 0.068 ppm (7.8 μmol m^{-3}) of SO_2 and of NO_2 on total dry weights of *Dactylis glomerata, Lolium multiflorum*, and *Phleum pratense*, but additive effects on *Poa pratensis*.

Differences in height growth of *Pinus strobus* sensitive and tolerant trees exposed periodically in nature to nitrogen oxides (NO_x) and SO_2 were reported by Skelly, Moore, and Stone (1972). Later field studies of radial increments, however, showed similar growth reductions by sensitive and tolerant trees (Stone and Skelly, 1974). Skelly, Moore, and Stone (1972) found forest-grown trees of *Pinus echinata* Mill., *P. taeda*, and *P. virginiana* Mill. to be less susceptible than *P. strobus* to the two pollutants. Stone and Skelly (1974) reported that forest-grown *Liriodendron tulipifera* trees differed greatly in their growth response to this same NO_x and SO_2 problem.

Differences in the responses of seedlings from two *Pinus taeda* full-sib families and two *Platanus occidentalis* L. half-sib families exposed to combinations of O_3, SO_2, and NO_2 were demonstrated in *Pinus taeda* (Kress, Skelly, and Hinkelmann, 1982a,b). Kress (1978a,b) showed significant differences in the growth responses of young seedlings from two tree species following exposure to 0.05 ppm (2.08 μmol m^{-3}) of O_3, 0.1 ppm (4.16 μmol m^{-3}) of NO_2, and 0.14 ppm (5.8 μmol m^{-3}) of SO_2. *Pinus taeda* seedlings showed a 26% growth reduction, and seedlings from ozone-sensitive and ozone-tolerant families of *Platanus occidentalis* were reduced by 45 and 34%, respectively.

Krause and Kaiser (1977) demonstrated different species' responses to combinations of SO_2 and heavy metal dust. The total yield of *Raphanus sativus radi-*

cula L. was significantly reduced when SO$_2$ plus heavy metal dust were combined, but not when the single pollutants were applied. In contrast, the dust alone caused significant reductions that were not amplified by SO$_2$ in *Lactuca sativa* L., *Raphanus sativus oleifera* L., and *Setaria italica* L.

Evolution of SO$_2$ tolerance. The fact that there has been natural selection for SO$_2$ tolerance has been documented for grasses in the United Kingdom (Bell and Mudd, 1976; Bell, Ayazloo, and Wilson, 1982; Horsman, Roberts, and Bradshaw, 1979). Populations of *Lolium perenne*, *Festuca rubra*, *Holcus lanatus*, *Dactylis glomerata*, and *Phleum bertolonii* DC. collected from polluted sites were shown to have more tolerance to chronic SO$_2$ injury than cultivars or populations of the same species from relatively clean-air sites (Bell, Ayazloo, and Wilson, 1982). Screening of turfgrass cultivars in experimental plots at a polluted urban site revealed selection for tolerance to acute SO$_2$ injury within 4 to 5 years (Bell, Ayazloo, and Wilson, 1982).

Evidence of selection for SO$_2$ tolerance in *Geranium carolinianum* L. plants from a polluted site has been shown by Taylor and Murdy (1975) and in *Pinus strobus* by Thor and Gall (1978). The phenomenon in *P. strobus* can be at least partly explained by the much higher mortality rate among the air-pollution-sensitive individuals in the forest stand. Karnosky (1981b) showed that the natural mortality in young *P. strobus* stands was almost five times as great for the air-pollution-sensitive trees as it was for the tolerant trees. Most of this mortality resulted because the slower-growing sensitive trees could not compete with the neighboring, faster-growing tolerant trees around them.

The subtle effects of SO$_2$ stress on plant populations in terms of genotypic losses or gene-frequency changes are not well-understood. However, excellent reviews of the implications of air pollution as a selective force have been done by Reinert, Heggestad, and Heck (1982), Scholz (1981), and Sinclair (1969).

Assessment of Experimental Techniques

As described in this article, the experimental techniques used to study genetic differences in growth effects in response to SO$_2$, alone or in combination with other pollutants, have differed greatly. I cannot here examine comprehensively the many different experimental techniques, but will instead attempt to discuss some important concepts to be considered in developing future experiments to study genetic differences in growth responses caused by SO$_2$.

Although there have been a few exceptions, such as the studies by Dochinger and Jensen (1975) and Jensen and Dochinger (1979), it is becoming increasingly evident that the response of plants correlates much less with acute exposures than with chronic exposures (Ayazloo and Bell, 1981; Garsed and Rutter, 1982a; Horsman *et al.*, 1979). Thus, a first consideration in testing the relative differences in SO$_2$-induced growth responses should be to test the plants under realistic conditions, which generally consist of chronic exposure to low concentrations of SO_2.

The findings with *Lolium perenne* 'S23,' that plant age, season of the year, and soil fertility all affect growth responses caused by SO_2, suggest that care must be taken to fumigate the plants under conditions as near as possible to those in nature. This is particularly important for plant ages of forest trees and for chamber conditions. It would be highly desirable to compare growth effects on older trees than have generally been used. One way to do this is to expose plants being tested in outplanted field plots, as was done with trees by Roberts (1975, 1976) and with ryegrass by Horsman, Roberts, and Bradshaw (1979). An experimental system that has gained popularity in the United States and that approximates field conditions better than most chamber designs is the open-top chamber system. This system has been used to study differences in SO_2-induced growth effects by Mandl *et al.* (1980) and by Thompson, Kats, and Lennox (1980). The advantages of this system are that the plants are grown in soil in the field, and environmental conditions, such as light intensity and relative humidity, are very close to those in the field.

Numerous indoor chamber types have been used in testing genetic differences in SO_2-induced growth effects. I will not attempt to evaluate the relative merits of these various chamber designs, but again emphasize that chamber conditions need to mimic field conditions closely in order for the study results to be meaningful.

The correct pollutant exposure—number of hours per day, number of days per week—is still an open question for debate. Many European studies have used continuous SO_2 exposures (Ayazloo and Bell, 1981; Bell, Rutter, and Relton, 1979; Cowling and Koziol, 1978; Crittenden and Read, 1978b; Farrar, Relton,

TABLE 20.1
Examples of Intraspecific Variation in Growth Responses to SO_2

Species	Plant materials	Source
Begonia hiemalis	Cultivars	Reinert and Nelson (1980)
Dactylis glomerata	Cultivars, clones	Ayazloo and Bell (1981)
Festuca rubra	Cultivars, clones	*Ibid.*
Holcus lanatus	Clones	*Ibid.*
Lolium perenne	Cultivars	Bell and Clough (1973); Horsman, Roberts, and Bradshaw (1978, 1979)
Nicotiana tabacum	Cultivars	Tingey and Reinert (1975)
Phleum bertolonii	Cultivars, clones	Ayazloo and Bell (1981)
Pinus strobus	Clones	Houston (1974); Roberts (1976)
Pinus sylvestris	1/2-sib families, provenances	Bialobok, Karolewski, and Oleksyn (1980); Huttunen and Törmälehto (1982)
Pinus taeda	Full-sib families	Kress (1978a); Kress, Skelly, and Hinkelmann (1982a)
Triticum aestivum	Cultivars	Laurence (1979)

and Rutter, 1977; Horsman *et al.*, 1979), but those in the United States have generally been for 8 to 12 hours per day, 3 to 7 days a week (Jensen, 1981a; Jensen and Dochinger, 1979; Kress, 1978a,b; Noble and Jensen, 1980a,b; Reinert and Weber, 1980; Thompson, Kats, and Lennox, 1980; Tingey, Heck, and Reinert, 1971; Tingey *et al.*, 1973b). Again, it would seem that fumigations should closely mimic actual SO_2 concentrations to be useful. These pollutant fumigations in nature will differ considerably from location to location.

In the few studies where multiple genotypes of a given species have been tested for growth responses induced by SO_2, many differences have been found (Table 20.1). In order to develop species rankings for susceptibility to SO_2, multiple genotypes should be tested, especially for tree species, which generally exhibit a wider range of differences than carefully bred agricultural or horticultural crops.

Directions for Future Research

The study of differences in growth responses caused by SO_2 has shown that some differences between and within species appear to be genetically controlled. However, there are many open research questions remaining, even in species such as *Lolium perenne*, which has probably been more intensely studied in terms of SO_2 effects on plant growth than any other species. With *L. perenne* and with other important agricultural crops, additional research is needed to discover mechanisms of tolerance, so that SO_2 tolerance can be bred into cultivars for use in polluted areas (Ayazloo and Bell, 1981). The studies with *L. perenne* have revealed evolutionary trends in SO_2 tolerances, but additional work is needed to calculate rates of evolution of SO_2 tolerance (Horsman, Roberts, and Bradshaw, 1979). Changes in the genetic structure of forest-tree populations caused by different growth responses to air-pollution exposure need to be examined (Scholz, 1981), especially in terms of loss of genetic variability.

Studies are needed to learn genotype-environment interactions for such factors as the effects of age, wind speed, soil fertility, and season of the year on plant susceptibility to SO_2. These studies are needed for grasses in the United Kingdom, but also for many other plant species, especially commercially important tree species such as *Pinus strobus*, *P. sylvestris*, and *Picea abies*.

There is little published information on the relationship between the responses of individual plants to acute SO_2 exposures and their responses to chronic exposures (Ayazloo and Bell, 1981). The genetic variation in growth responses to SO_2 in combination with other pollutants is not well-understood, and additional studies are needed to measure tolerances to combinations of multiple pollutants at low concentrations (Kress, Skelly, and Hinkelmann, 1982a).

Genetic variation in growth responses to SO_2 exposure has been studied for only a few commercially important plant species and very few genotypes. Thus, additional work is needed to test the extent of genetic variation in SO_2-induced growth responses in many plant species. Studies of heritability are also needed to verify that the variation seen is heritable in nature.

Comparisons of genotypic responses to continuous SO_2 exposure, as is generally found in European studies, and to daytime-only fumigations, as has generally occurred in American studies, would be useful, to find out if the results of American and European studies can be compared for species' responses to SO_2.

The comparison of results from studies of the same genotypes exposed in growth chambers, greenhouse chambers, field chambers, and field plots would be valuable for finding out what method gives results most comparable to those found in natural conditions. Studies of relative growth rates around large SO_2 sources, such as the studies by Linzon (1971), Skelly, Moore, and Stone (1972), and Stone and Skelly (1974), would seem to be ideal for comparing species' sensitivities.

21

Modeling Effects of SO₂ on the Productivity and Growth of Plants

J. R. Kercher and D. A. King

Much experimental effort has been expended over many years in investigating the effects of air pollutants in general, and sulfur dioxide in particular, on vegetation. There has been much less work on interpreting this mass of experimental data with a comprehensive, mathematical scheme that can integrate the various observed phenomena. Sulfur dioxide may affect the growth of individual plants directly, and it may alter community composition by shifting the competitive relationships of plants (West, McLaughlin, and Shugart, 1980; Kercher and Axelrod, 1981; Shugart and McLaughlin, this volume). In this paper, we wish to describe the relatively few models that have been constructed for predicting or analyzing effects of sulfur dioxide on the productivity and growth of individual plants. (We will not deal with effects on entire populations or communities of plants.) Our goal is to define growth models, describe the types of such models, and review the work to date on the various types. We will assess the current approaches being used and their utility for various problems, and suggest what we believe to be the outstanding problems in modeling the effects of sulfur dioxide on plant growth. We will also suggest the benefits that the field of plant-effects research can derive from working with such models by discussing the power and usefulness of plant-growth models. Finally, we will suggest future directions that should be pursued in modeling plant effects.

There are several issues in developing a model of the effects of sulfur dioxide on plant growth. One must decide what type of model to use, and choose the appropriate scale of resolution for the model. The scope of the model or of the phenomena to be covered by the model must also be decided. Once the phenomena are chosen, one must decide how much detail to use in modeling these phenomena, by considering the uses for which the model is being developed. That is, the goal of the project will set certain conditions that the model must satisfy. The role of the model within the project will be defined by certain questions that the model must answer. Once the objectives are articulated, the questions of scale of resolution, scope of phenomena, detail of study, and type of approach are readily answered. By considering the capabilities of the existing techniques, one can then decide which approach is most feasible given the demands of the project.

Review of Modeling Approaches

A model of the effects of sulfur dioxide on plant growth is a mathematical transformation that uses some measure of sulfur-dioxide concentration to calculate a resulting effect on plant growth. The measure of sulfur-dioxide concentration may be a very simple statistic, such as average atmospheric concentration during the growing season, or a very complex measure, such as an annual time series of concentrations taken at very short time intervals (e.g., subhourly). The resulting effect may be the change in some indicator of plant productivity, such as the yield of the economic portion of the plant, a growth increment, the total growth of the plant, or a change in plant allocation. Which measure of pollution should be used is a central question in pollution research, and is intimately connected with what biological phenomena one considers important and what mathematical transformation one chooses. The level of detail is also a crucial decision, as outlined above.

Types of models. For convenience we will divide existing models of plant-growth effects into two types: (1) *descriptive* and (2) *process*. The descriptive type is usually characterized by one relatively simple equation, and usually relates yield to an aggregated statistic of sulfur-dioxide concentration. The descriptive model may include other environmental factors, such as temperature or water stress, and may include interactions of the pollutant with these factors. The descriptive models are sometimes called "empirical" or "statistical" models, and are usually cast in the form of a regressive equation. They usually operate at one level of resolution. The criterion for selecting a particular model is that it should provide the best statistical fit to the data. Such models are weak outside the data base on which they were constructed.

Process models are models in which the processes of the system that control the variables of interest are specifically considered. Usually many phenomena are considered in a process model, with one or more processes used to explain each phenomenon. That is, a process model usually consists of submodels, each of which models a subprocess of the system. For plants, the internal physiology of the plant is modeled to some degree. The degree of detail used in describing the internal physiology varies from model to model. Process models can also be characterized as *mechanistic*, *physiological*, or *phenomenological*. Phenomenological models, for example, are constructed to reproduce the observed phenomena, but do not explicitly make use of biochemical or physiological mechanisms. Process models are often simulators; they simulate the dynamics of the system over time. Process models bridge two levels of resolution; they explain the results of the overall simulation or process by using constituent subprocesses.

Since process models are more complicated than descriptive models, the criteria for selecting them are more complicated. Usually the constituent pieces (subprocesses) of the model are chosen by a combination of hypothesis and theory, which are then fit to data, and the subprocesses are then integrated together, again on the basis of hypothesis and theory. The models are then compared to experiments at the top level of resolution; for us, these might be field experiments on growth and yield. These models are usually much stronger than descriptive models when extrapolated beyond the data base from which they were constructed. In fact, such extrapolation is often carried out in order to find out where the models break down, and what additional processes or phenomena therefore need to be considered.

Models of pollutant effect on plant growth are useful in two capacities, as research tools and as assessment tools. As a research tool they provide a convenient way for us to analyze data and compare experiments done at different locations, with different crops, with different fumigation regimes, with different environmental characteristics, or by different researchers. Process models allow the integration and synthesis of the different processes that bear on the final result. Since subprocesses are often studied by different researchers, models allow the integration of information within a project and across project boundaries. To a lesser extent, descriptive models also facilitate integration. The discipline of developing a model, which relates all the processes under consideration, forces a coherent structure to emerge and exposes any areas for which information or data is missing. Construction of the process model greatly facilitates the generation of hypotheses about the functional relationships of plant processes. Developing descriptive models also tends to generate new hypotheses that need to be tested. However, because of the inherent differences between the two approaches, the hypotheses tend to be more detailed for process models than for descriptive models.

A developed model is then useful as a predictive tool that forecasts the result of a pollutant exposure on vegetation growth. The model can be used to assess future scenarios for pollutant impacts, predict the results of field experiments, estimate current impacts under existing ambient conditions, or estimate the costs or benefits that would result from a marginal change in current emission rates. Such assessment could be made on a local, regional, or national basis, depending on resources and data bases available. We will review examples of such assessments later.

Descriptive models. Oshima *et al.* (1976) introduced the notion that one could find a direct relationship between a pollutant dose and a resulting reduction in plant yield. They presented regressions of yield reductions of alfalfa as a linear function of dose of ozone (ppm-hours). Larsen and Heck (1976) quantified the relationship between injury, concentration, and duration of exposure (a concept that was introduced by O'Gara, 1922), and found that the probit of the injury was well fitted by the log of the product of the duration and the concentration raised to a power. This relationship suggests a threshold at low doses and difficulty in achieving total injury at high SO_2 concentrations delivered in short time periods (less than 8 hours). Stevens and Hazleton (1976) used O'Gara's relationship directly in a calculation that estimated effects on leaf area of many SO_2 episodes during a growing season. They predicted yield from leaf area. Sprugel *et al.* (1978) developed an equation for yield of soybeans as a quadratic function of SO_2 dose expressed in ppm-hours. Armentano (1980) reviewed the literature for available studies from which one could construct dose-response functions of crop loss due to air pollution, particularly by sulfur dioxide. Armentano (1980) used the log of the dose (in ppm-hours) as the independent variable in the regression, and assessed the current losses in the Ohio River basin based on these equations. Heagle and Heck (1980) regressed yield reductions of corn, wheat, soybeans, and spinach against seasonal 7-h/day mean ozone concentrations. They suggested a national crop-loss estimate based on these dose-response functions. Heck *et al.* (1981) have reported many dose-response functions for several crops. These functions are for both sulfur dioxide and ozone, and use both average ppm and ppm-hours as the measure of dose.

Benson *et al.* (1982) developed a descriptive model of the effects of sulfur dioxide on yield that was used to evaluate impacts of air pollution in Minnesota. Their model is based on an approach that estimates the effects on final yield that a disease incident at different times during the season will have (James *et al.*, 1972). The approach breaks the growing season up into weekly intervals and develops weekly coefficients of loss. This approach requires input of daily SO_2 concentrations for each of the weekly intervals in order to accurately calculate the weekly loss coefficients; so the model is highly time-dependent. The authors refer to the calculation as a "macrosimulation."

Descriptive models have also been used to quantify the effects of pollution on tree growth. Kercher and Axelrod (1981) used data of Lathe and McCallum (1939) and Katz (1939) to develop a dose-response function of ponderosa pine to sulfur dioxide. This regression also included other environmental variables, and it was found that the best predictors of incremental tree growth were dose (ppm), growing degree-days, and annual precipitation. We cite this study as an example in which environmental variables in addition to pollution were quantified and used as additional independent variables.

There are two application areas in which descriptive models are especially useful because of their relative simplicity. First, they can generate first-order estimates of pollutant effects easily and quickly. These models require a minimum of analytical effort to construct and fit to data. They quickly establish magnitudes of effects; and they are easily used with a minimum of data. Second, they can be used as submodels in large models of whole communities, which consist of individual plants (Shugart and McLaughlin, this volume; Kercher and Axelrod, 1981). In such models, physiological submodels usually go into details on a much smaller spatial and temporal scale than appropriate for interactions between plants or for the resulting community dynamics (Goldstein and Mankin, 1972; Goodall, 1976). The mismatching of scales of resolution leads to practical calculational problems when the models are run on computers, e.g., stiff differential equations. These difficulties can be avoided by using simple empirical relationships to describe internal plant physiology, and descriptive models are very suitable for this purpose.

Process models. Kickert and Benenati (1983) have reviewed the literature on simulation models of chemical stress on agricultural plants. Gaud (1974) developed a model of effects of sulfur dioxide on desert grasses. A very simple growth algorithm was used in which available water controlled plant reproduction and growth. However, there was no modeled interaction between water usage and sulfur-dioxide effects. Sulfur dioxide affected assimilation independent of considerations of water stress. Morgan (1975) developed a model of effects of sulfur dioxide on alfalfa. He did not explicitly consider photosynthesis or allocation; rather, he used a simple plant-growth algorithm, and water relations were not included in the model. Morgan did have a detailed model of sulfur dynamics, in which sulfite converts to sulfate and is accumulated. Injury and ultimately biomass reduction were calculated from sulfite and accumulated sulfate levels. Morgan found that damage was a function of the time-history of exposure. He suggested that damage cannot be predicted "simply from total integrated exposure." Kercher (1977) developed a general model of plant growth in which there was a detailed photosynthesis model and an allocation submodel. The effects of pollutants were introduced into the model by way of their effect on photosynthesis, which was modeled as an empirical dose-response function. Kercher (1978) developed a process model of pollutant effects on photosynthesis, and applied it to H$_2$S effects on sugar beets. This model was used to predict effects of gaseous effluents from proposed geothermal development in the Imperial Valley, California (Kercher 1982). Luxmoore (1980) developed a productivity model of an oak-hickory forest. This model contains a photosynthesis submodel that supplies photosynthate to an allocation submodel. The effects of sulfur dioxide on photosynthesis were modeled as a simple dose-response function. Heasley, Lauenroth, and Dodd (1981) developed an ecosystem model that simulates a grassland. Photosynthesis and allocation were modeled, as were the sulfur dynamics of sulfite conversion to sulfate. Sulfur diffuses into the leaf and can either stimulate or inhibit enzymatic processes of photosynthesis. Kirchner, Heasley, and Lauenroth (1981) modeled the effects of sulfur dioxide on lichens. This model also contained a photosynthesis and allocation submodel. The effects of sulfur dioxide were introduced as an effect on photosynthesis based on an empirical data set. Kercher, King, and Bingham (1982) have developed a model of the pollutant effects on soybeans by coupling a general photosynthesis and growth model, BACROS (Wit *et al.*, 1978), with a specific soybean growth and allocation model, CROP (Wilkerson *et al.*, 1983). This model was developed for estimating the effects of ozone. Once again the pollutant effect was introduced as an effect on photosynthesis, which is translated into an effect on growth by using the existing allocation algorithm of the nominal, unfumigated system. King, Kercher, and Bingham (1982) used the soybean model to make a preliminary prediction of soybean losses in Indiana and Illinois.

Typically, process models span two levels of research. The submodels deal with physiological and metabolic processes. The results of the model are generated in terms of the whole plant. In fact, the results are in terms of growth and yield, and thus relate directly to season-long growing experiments. Process models allow us to hypothesize the relationships between variables that represent the plant components that take part in the processes. A model allows us to examine the logical and mathematical consequences of the hypotheses. Models facilitate the testing of hypotheses by producing results that are directly comparable to experiment. Furthermore, a good plant-growth model integrates the processes of photosynthesis, water uptake and transpiration, allocation, and nutrient dynamics. The plant processes are followed over time, and critical phenological stages can be included in the model, to allow the time-dependent variation in plant sensitivity to air pollution to be made explicit.

Process Models: Examples of Present Content

In process plant-growth models, photosynthesis and allocation processes are typically taken as the core of the model, because photosynthesis is the plant's energy supply, and because photosynthetic products are the building blocks for much of the plant's structure. A complete description of plant growth includes all other processes that interact with the photosynthesis and allocation processes, e.g., transpiration, nutrient incorporation and utilization, or pollutant uptake. The general approach taken in process models is to consider the plant as a box (Fig. 21.1) whose internal physiology is acted on by the environment (Kercher, 1980). Pollutants may act on the stomata, and thereby affect the growth of the plant indirectly. Once inside the plant, the pollutants directly affect growth by their action on plant metabolism. Uptake is controlled by the stomata. This conceptualization suggests that the internal concentration of the pollutant is the critical independent variable that determines plant response. The external concentration should be used to calculate the internal concentration.

To date, process models have emphasized results in three areas: (1) calculation of internal dose; (2) effect of pollution on photosynthesis; and (3) converting photosynthesis changes into growth and yield changes. The calculation of dose is a central problem in predicting plant response. In fact, a wide range of expressions for dose have been considered in descriptive models. However, descriptive models calculate an external dose, whereas process models can calculate the internal concentration, which, as we said, is what actually affects plant response.

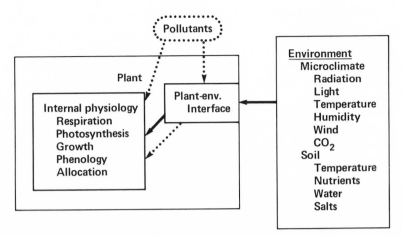

Fig. 21.1. Conceptualization of plant response to environmental conditions and pollutant concentration. Pollutants affect internal physiology both directly and indirectly, through their action on the stomata.

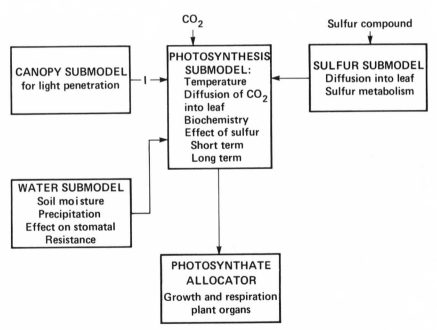

Fig. 21.2. Schematic structure of GROW1, showing submodules.

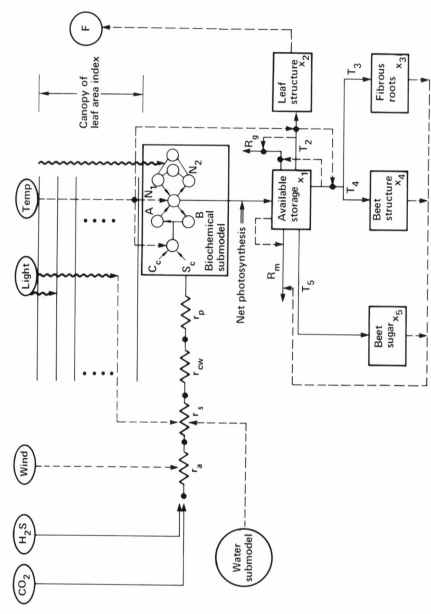

Fig. 21.3. Detailed diagram of framework of GROW1 applied to sugar beets, showing canopy, photosynthesis, and allocation submodels.

Process models estimate internal dose from external concentrations by calculating diffusion into the leaf through the various resistances: canopy, leaf-boundary layer, stomatal, cell wall, etc. For example, King, Kercher, and Bingham (1983) calculated the total uptake rate for ozone into leaves of soybean using BACROS. The calculation used external O_3 concentration and calculated leaf resistances based on plant water status, etc. Kercher (1978, 1982) used calculated internal sulfite as the independent variable of pollutant effect. We will use this model, GROW1, applied to sugar beets as an example of all three areas mentioned above.

GROW1 is conceived with five main submodels (Fig. 21.2). The canopy submodel calculates light penetration into the canopy, and the geometry of sun and leaf angles. The water submodel simulates soil moisture and its effect on stomatal resistance. The sulfur submodel calculates internal concentration of sulfite in leaves. The photosynthesis submodel calculates photosynthesis and effects based on both short- and long-term fumigation with sulfur. Photosynthate is passed from the photosynthesis submodel to the allocation submodel. In Fig. 21.3, we show a more detailed view of GROW1 with internal structure. CO_2 and H_2S or SO_2 diffuse into the leaf, encountering the illustrated resistances. A biochemical portion of the photosynthesis submodel generates photosynthate from CO_2, using light that has penetrated into the canopy. Photosynthate is deposited into a storage compartment, from which it is allocated to leaf structure, beet sugar, beet structure, or fibrous roots. Note that there is feedback in the model. Photosynthate that is converted to leaf structure increases the canopy leaf area, F, which may allow more production of photosynthate if all available light is not already being used.

Uptake. Consider the processes of sulfur uptake and metabolism. An apparent threshold for deleterious effects of sulfur dioxide has long been observed. Can a process model account for this threshold? The two processes that were modeled (Kercher, 1978) were diffusion of SO_2 into the leaf (via the resistances discussed above) and sulfur metabolism, using simple enzyme kinetics expressed in a Michaelis-Menten form. If the uptake rates in the equations for the two processes are set equal to each other, then we can calculate either the uptake rate or the internal sulfite concentration. The latter turns out to be the solution to a quadratic equation, and the threshold appears as seen in Fig. 21.4.

Photosynthesis. The photosynthesis submodel is similar to a group of photosynthesis models (e.g., Chartier, 1970; Hall and Bjorkman, 1975; Murphy, Sinclair, and Knoerr, 1974; Stewart, 1970) in which a light reaction, driven by sunlight, produces NADPH, which makes energy available to a dark reaction that assimilates CO_2. Sulfite was included in the photosynthesis model in two ways. First, sulfite can enter into the biochemical-enzymatic cycles directly, and interact either positively or negatively. For example, Ziegler (1975) suggests that sulfite competes for the same sites as CO_2 on ribulose diphosphate carboxy-

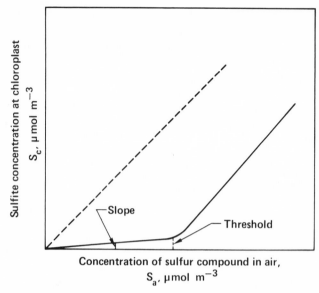

Fig. 21.4. Internal concentration of sulfite as a function of external surfur-dioxide concentration for fixed stomatal resistance.

lase. Ziegler also suggests that sulfite enhances the light reaction. One can see that these effects would manifest themselves on subhourly time-scales. Second, sulfite or its hypothesized proxy agents (e.g., superoxide) could alter the photosynthetic process by changing the concentrations of the enzymes, substrates, or sites for the reactions to be carried out. In the type of model used by Kercher (1978), the biochemical portion of the model leads to an equation of the form

$$P_g = F_1(\text{total enzymes}) \times F_2(\text{total substrates})$$
$$\times F_3(\text{functional chloroplast concentration}) \times F_4(T, C_c, S_c, I), \quad (1)$$

where P_g is gross photosynthesis; F_1, F_2, and F_3 are functions that increase with increasing arguments; F_4 contains relationships concerned with the biochemistry of the light and dark reactions; T is temperature; C_c and S_c are the concentrations at the chloroplast of carbon dioxide and sulfite, respectively; and I is light. We loosely call the product of the functions F_1, F_2, and F_3 productive capacity. For light levels that saturate F_4, and for CO_2 levels and temperature that maximize F_4, it is clear that F_1, F_2, and F_3 determine maximum photosynthetic rates. Since F_1, F_2, and F_3 multiply F_4, they also determine the slopes of the light curve for low light (quantum efficiency) and the slopes of the photosynthetic response to CO_2. F_1, F_2, and F_3 are usually proportional to their arguments. If a quantity, say, total enzymes, were reduced by sulfite or another agent, then its rate of return to its nominal level after sulfite was removed would depend on the rate at which the plant normally turned over that enzyme. We characterize F_3 as long-term; F_1

and F_2 as intermediate to short-term; and F_4 as short-term. Consider one component F_i, $i = 1, 2$, or 3. Kercher (1978) suggested that one could write an equation for change in F_i, which we generalize to

$$\frac{dF_i}{dt} = B(\text{available reserves, } S_c, \text{ nutrients}) - L(d_1, S_c, F_i), \tag{2}$$

where B is a term for generation or repair of this portion of the productive capacity, and L is a term for turnover and loss. Sulfur fertilization would occur in the B term. One would anticipate that B should rise linearly at low values of nutrients when they are limiting, and saturate at high values when they are no longer limiting. In fact, Kercher (1978) hypothesized such a function for CO_2 and sulfur in the form of a Michaelis-Menten expression. In the negative term (L), d_1 stands for the normal turnover rate. This negative term for loss could take many forms, but to first order we might expect that the turnover rate in absence of S_c would be proportional to the amount of F_i in place, $d_1 F_i$. With S_c present, to first order, losses in F_i should be equal to the probability of interaction between sulfur and F_i, which is given approximately by the product of S_c and F_i. In Fig. 21.5, we

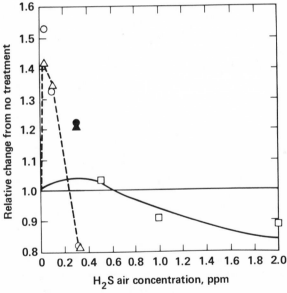

Fig. 21.5. Comparisons of models and data for relative photosynthesis and growth for sugar beets fumigated with H$_2$S. Photosynthesis data taken after 4-h fumigation. Growth data from season-length fumigation. Open squares are relative photosynthesis data from J. Shinn (unpublished). Circles are relative growth data from Thompson and Kats (1978). Solid line is photosynthesis model result. Triangles are results of growth model (GROW1). Open circles and triangles are for ambient CO_2 conditions. Solid circle and triangle are for ambient CO_2 with 50 ppm CO_2 added. The dashed line connects the growth-model calculations.

compare the results of the photosynthesis model with the H_2S data used to adjust it. The solid curved line is essentially determined by F_4. This curve is the relative change in photosynthesis at high light resulting from a few hours' fumigation of H_2S in field-grown sugar beets. The dotted line and triangles are for the full growth-model calculations of relative change in total plant biomass as a function of applied H_2S. The changes in relative growth are a function of the product of F_1, F_2, F_3, and F_4. That is, effects on long- and short-residence-time systems are exhibited by the triangles, whereas only results from short-residence-time systems and biochemical kinetics are manifested in the photosynthesis data.

In this approach, the model was used as a tool to reproduce results of growth and yield experiments. A more satisfying approach would be to monitor photosynthesis during the growing season, and compare it directly to the long-term model. Note that these data suggest that complicated responses are possible, and one must be prepared to examine many different concentrations of sulfur pollution. Eq. 1 also suggests that many experiments with different exposure lengths would be useful in fixing model parameters.

Allocation. The allocation problem in plant response is a third area in which process models can make a contribution. The contribution here lies in part in formulating allocation mechanisms, and in part in the analyses that can be made with existing phenomenological models. For example, there are some observations that root and storage organs undergo proportionally less growth when the plant is under pollutant stress (Bennett and Oshima, 1976). Allocation mechanisms should be introduced into process models to simulate these effects. But current phenomenological process models can also be used to examine such an allocation effect. Consider the example of GROW1. The allocation part of the model in Fig. 21.3 (Kercher, 1977) was patterned after TEEM (Shugart *et al.*, 1974). In the application to sugar beets, extensive use was made of the formulation of SUBGRO (Fick, Loomis, and Williams, 1975). As net photosynthate is deposited in the available storage compartment at rate P_n, it is removed to plant compartments at rates T_2 through T_5. The reserve compartment pays the maintenance respiration cost R_m proportional to biomass. Growth is defined as the transfer T_2 through T_4; growth respiration, R_g, proportional to the sum of these transfers, is also paid by the reserve compartment. In future work we might hypothesize that there is a cost for metabolism of SO_2, R_p, associated with elevated SO_2 concentrations, and a repair cost, R_r, associated with SO_2 injury. The equation for the rate of change of the reserve compartment is then

$$\frac{d(\text{reserve})}{dt} = P_n - (\Sigma T_i) - R_m - R_g - R_r - R_p. \tag{3}$$

Each of the other plant-part compartments with content C_i has change equations such as

$$\frac{dC_i}{dt} = T_i - \text{loss}. \tag{4}$$

Here T_i is the input, and loss is senescent death or any losses that the reserve compartment cannot pay. It should be clear that the basic biology is contained in the T_i functions, which Fick, Loomis, and Williams (1975) take to be of the form

$$T_i = f_i(C_i) g_i(\text{reserve}). \qquad (5)$$

The function f is dependent on the sink for the transfer C_i. The function g is dependent on the availability of reserves. The function f is sometimes referred to as the "sink strength" and g as the "source strength." For sugar beets, GROW1 used functions for g as given in Fig. 21.6, where g is expressed as a relative growth rate. The threshold for delivery of carbohydrate to leaves is the lowest of all thresholds. This implies that leaves have the highest priority for photosynthate, and beet has the lowest. The general properties of the curves, i.e., relative positions of thresholds, were arrived at by Fick, Loomis, and Williams, based on the logic of the model, in which these organs had to grow with these priorities in order to survive. For example, Fick (1971) noted that if the thresholds for leaves and fibrous roots were reversed, the plant eventually died. When the simulation was tried with GROW1, the same result was obtained. If the reserve compartment is decreased (starting from any value), then the transfer to beet decreases relative to transfer to fibrous roots and leaves, and the transfer to fibrous roots decreases relative to transfer to leaves. Thus the model, developed before its application to effects of pollution, would predict a change in allocation caused by a reduction in available reserve. This change would favor leaves over beets. Two comments should be made. First, the model is not based on mechanisms; rather, it was developed solely to agree with an observed phe-

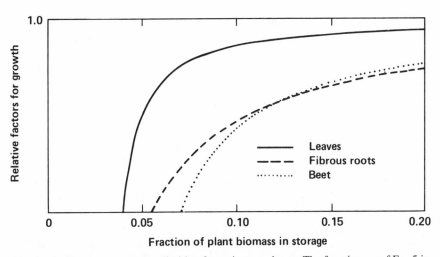

Fig. 21.6. Storage content as a limiting factor in sugar beets. The functions g_i of Eq. 5 in the text are plotted for leaves, roots, and beets. This is the relative growth rate as a function of available carbohydrate reserve (compartment 1 in Fig. 21.3), which is expressed as the fraction of total biomass.

nomenon. Second, it is not known whether or not the same allocation equations apply to a plant under fumigation stress. That is, does the form of the functions *f* and *g* change under pollution?

Summary of discussion of process models. Work to date on process models has shown that it is both feasible and practical to model the interactions of pollutants with growth. To date, special value of process modeling has been found in (1) estimating pollutant uptake and internal dose, (2) estimating the resulting effect on photosynthesis, and (3) estimating the effect of air pollutants on total plant growth. Furthermore, process models have been used to make predictions covering some environmental conditions of both pollutant concentrations and climatic regime that were outside the domain of the original data set. Predictions will usually need to use dose-response data that are outside the experimental extremes of the original data set.

Directions for Future Research

A discussion of directions for future research in modeling the effects of SO_2 on growth cannot be isolated from discussion of all future research on plant effects, because modeling is simply a quantification of what we already know or believe to be true. With that in mind, we can point out several areas in which our modeling needs to be more fully developed, and other areas in which we need completely new modeling initiatives. We will deal with process models, simply because they represent the most complete description of the plant. However, many of the following remarks can be applied to the more limited descriptive approach.

More work needs to be done on the interaction of water stress with pollutant effects. It is known that plant-water relations have great bearing on SO_2 pollutant effects (McCune and Weidensaul, 1978), because water-stress affects pollutant uptake through stomata. Existing models do address this effect, but current efforts should be improved because of the importance of water stress on pollutant uptake. Also, we must consider the effect that water stress has on the plant's physiology. As water stress alters internal physiology, the subsequent response of the plant to SO_2 may be altered. To correctly assess water-stress/pollutant interactions, the physiological mechanism of SO_2 damage to photosynthesis must be modeled correctly. Different hypothesized SO_2 damage mechanisms can be fitted to the same data base, but may provide differing predictions of the interaction between water stress and SO_2.

Future work should consider the possibility that pollutants may alter water-use efficiency. Crop productivity is frequently more limited by water than by pollutants. If SO_2 changes the rate at which plants extract water from the soil, then the water stress encountered by plants will vary with pollutant exposure. A proper accounting for the interaction of water stress and pollutant effects requires a process model that follows soil water as well as plant growth.

We can suggest several questions about investigations of mechanisms of productivity reduction. Research should be formulated to examine the magnitudes of direct loss in photosynthetic or productive capacity, including the change over time in maximum photosynthesis rates, and recovery rates of photosynthesis when stress is removed. The research and modeling should be designed to separate short-term effects from long-term. In particular, experiments should be undertaken to test the validity of the approach outlined in Eqs. 1 and 2, or to test future improvements. The experiments should be designed to determine parameters for photosynthetic capacity models. Researchers need to ask whether the energy costs of repair and of SO$_2$ metabolism should also be included in models. These hypotheses must be tested, and quantified if they turn out to be reasonable.

There is also substantial work to be done on allocation. Current modeling of allocation is phenomenological rather than mechanistic, because we have little theoretical understanding of allocation mechanisms; so one area of future research will be to develop mechanistic models of allocation. After this, however, one might ask a series of questions. A fundamental question is whether the effects of pollution are translated into growth effects correctly by an existing allocation scheme that works correctly in the absence of pollution. That is, are pollutant effects on allocation due simply to either reduced available photosynthate or increased repair or metabolism costs? There are several associated subissues. For example, if an existing allocation scheme cannot explain changes in allocation patterns, are there other mechanisms brought into play that might, e.g., accelerated senescence? What effect does altered allocation have on storage organs? Are they mined for repair costs? If we have an allocation scheme like that described in Eqs. 3 through 5, does pollutant damage cause the functions of f_i and g_i to change? In particular, do the thresholds or initial slopes of the curves in Fig. 21.6 change?

Future modeling of effects of SO$_2$ needs to look beyond models in which the stated variables are solely compartments of plant-part biomass or reduced carbon content. Specifically, we must simulate the nutrient pools in the plant, beginning with sulfur and nitrogen. We must explicitly consider the conversion of sulfite to sulfate and its subsequent use in the economy of the plant. We should incorporate the effects of the nitrogen contents of plant parts on allocation of photosynthate. This has been done for many crop models (e.g., Wilkerson *et al.*, 1983), and investigations of pollutant effects on plants should do likewise. We may then investigate in detail the relevant questions of the effects of nitrogen or sulfur availability on plant response and the interaction of nitrogen with sulfur. To do this will require information that relates the nutrient content of plant parts to patterns of accumulation of plant biomass. Some of our research-oriented models should include details of nitrogen and sulfur biochemical metabolism.

Finally, we must include one other consideration in our models. Plants rarely encounter only one gas, SO$_2$, in their lifetime. There are interactions between gases because of the ways which plant metabolism is affected by the gases individually. Clearly, if one pollutant damages the mechanism by which the plant metabolizes a second pollutant, addition of the second pollutant can introduce profound effects. To include the interaction between gases accurately, we should be working toward including the more intricate details of plant-cell biochemistry in some of our models. Other models should address these effects empirically.

To facilitate the exploration of the preceding suggestions, we suggest that, whenever circumstances allow, modeling efforts should be coupled to experimental programs. In practice, the researchers carrying out the experiments should be instrumental in suggesting or hypothesizing processes and interpreting model results. Experimental researchers should be active in analysis of the data to be included in the model as parameters. Modeling researchers should be active in suggesting experiments to be done or data sets that need to be gathered. Modeling researchers should take an active role in planning and interpreting experiments. These activities should be in addition to, not instead of, the experimenters' traditional activity in conducting experiments, and the modelers' current task of building computer-simulation codes.

Acknowledgments

Work for this paper was carried out under the auspices of the U.S. Department of Energy by the Lawrence Livermore National Laboratory under contract number W-7405-Eng-48. We would like to thank Dr. F. W. T. Penning de Vries for a discussion on turnover times of cellular components.

Part 4
SO$_2$ Effects on Plant Communities

22

Pathways for Sulfur
from the Atmosphere to Plants and Soil

*M. H. Unsworth, D. V. Crawford, S. K. Gregson,
and S. M. Rowlatt*

There have been many reviews of the sulfur cycle, both global (Robinson and Robbins, 1968; Granat, Rodhe, and Hallberg, 1976) and national (National Research Council of Canada, 1977; National Academy of Sciences, 1978). These reviews have usually considered emission, transformation, and deposition on an annual basis, and there has been little detailed discussion of pathways to vegetation and soils. This paper is intended to identify the various pathways by which SO_2 in the atmosphere is transferred to plants and to the soil, and especially to identify paths by which SO_2 can influence plant productivity directly and indirectly. We exclude some indirect actions, such as effects on insect pests and disease.

The time-scale of a year, typical of most reviews of the sulfur cycle, is too long for our purpose. Deposition conditions vary day by day, and plant growth varies seasonally; and it is essential with SO_2, which can have both nutritional and damaging action (Cowling and Koziol, 1982), to assess relationships between rates of supply and demand in the growing season. We begin by summarizing pathways for SO_2 between the atmosphere, plants, and soil. We then discuss techniques for quantifying rates of transfer along pathways, using examples from our own studies of the sulfur cycle of wheat. Finally we suggest areas for further study.

This book is concerned specifically with SO_2, and we therefore confine ourselves to pathways that can be traced at some stage to SO_2 in the atmosphere. Examples of direct cycles of reduced sulfur gases (e.g., H_2S) between the atmosphere and the surface are therefore omitted, but we emphasize strongly that it is an oversimplification to consider *only* direct plant responses to SO_2 in evaluating effects of SO_2 on plant productivity. Transformations of SO_2 in the atmosphere and at plant and soil surfaces may lead to other chemical forms of sulfur, including acidic precipitation, which may influence plant productivity; we believe that it is necessary to develop research to study responses to these simultaneous and interrelated pathways.

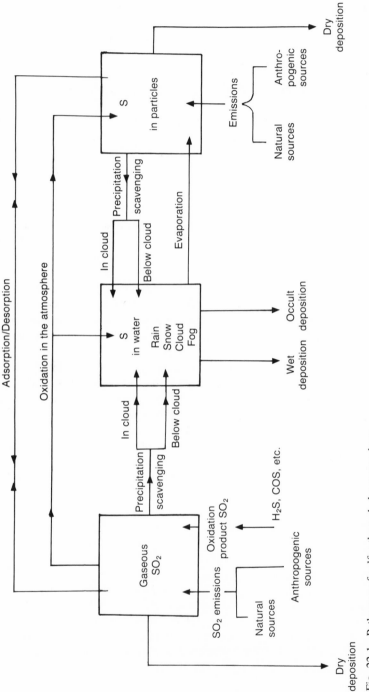

Fig. 22.1. Pathways of sulfur through the atmosphere.

Pathways for SO_2 Through the Atmosphere

Gaseous SO_2 reaches the atmosphere directly as natural emissions from vol-canoes, etc., and as a combustion product of fossil fuels (Fig. 22.1). Natural SO_2 emission is probably small on a global and annual scale (Granat, Rodhe, and Hallberg, 1976); most of the SO_2 in the atmosphere from natural sources prob-ably results from oxidation of H_2S, CS_2, and COS emitted from soils (Adams *et al.*, 1980), plants (Winner *et al.*, 1981), and the oceans (Bandy and Marculis, 1980). Anthropogenic emissions are probably the dominant SO_2 source in the atmosphere of Europe and the industrialized areas of North America.

Sulfur dioxide is removed from the atmosphere by several mechanisms. Dry deposition, meaning turbulent transfer in the air of the gas to surfaces (in contrast to transfer in rain, etc.), operates whether the surface is wet or dry. Rates of dry deposition depend on SO_2 concentration, atmospheric mixing, and surface af-finity for SO_2. In wet deposition, SO_2, which is a soluble and reactive gas, is scavenged by water drops, ice, and snow, either as they fall to earth as precipita-tion or within clouds and fog. The scavenging rate for SO_2 is not particularly rapid, but depends on other chemicals present in liquid or gaseous form that act as catalysts or oxidizing agents (Garland, 1978).

An important removal pathway arises from chemical reactions that oxidize SO_2 in the atmosphere and lead to the formation of sulfuric acid and particulate sulfur. There is rapid scavenging of these products in clouds and by precipitation; hence much of the sulfur reaching the ground as wet deposition existed in the atmosphere at some stage as SO_2. Wet deposition may be measured regularly in networks of rain gauges (Fowler *et al.*, 1982), but upland vegetation also collects sulfur efficiently in drops of wind-driven cloud, which are not captured by rain gauges; sulfur input by this "occult deposition" may be significant for upland forests (Dollard, Unsworth, and Harvey, 1983).

Aerosol particles suspended in the atmosphere may have evolved directly from gas-phase reactions involving SO_2, or they may result from the evaporation of cloud drops, and contain sulfur scavenged as described in the preceding para-graph. Particles may also provide surfaces on which SO_2 can be adsorbed. Spe-cial cases are particles emitted from industrial smoke stacks or volcanoes, where high concentrations of SO_2 can lead to adsorption followed by desorption in the atmosphere.

Particles are removed from the atmosphere by dry deposition at rates that de-pend on size and concentration, and by precipitation scavenging, which is espe-cially effective for removing particles that act as condensation nuclei in clouds (Garland, 1978).

Rates of transfer and transformation of sulfur from gas-phase SO_2 to liquid drops and particles are relatively slow; close to SO_2 sources, dry deposition of SO_2 is therefore often the most efficient pathway for removal from the atmo-sphere, provided that the SO_2 is well-mixed throughout the atmosphere. At remote sites downwind of SO_2 sources, much of the SO_2 will have been oxidized to sulfate in aerosol and cloud drops; so wet and "occult" deposition are the most effective transfer mechanisms in rainy regions. In arid areas far from emission sources, dry deposition of sulfate aerosol is the only significant removal mechanism, and there is much uncertainty about the rates at which this occurs to vegetation (Hicks *et al.*, 1982; Chamberlain, 1975).

Pathways for Sulfur Through Plants

Fig. 22.2 illustrates pathways for sulfur, originating as SO_2 in the atmosphere, through vegetation. A distinction is made between sulfur on the surfaces of leaves, stems, etc., and sulfur that is located and transported internally.

Sulfur reaches plants by dry and wet deposition (including the occult pathway) and by root uptake. Dry deposition of particles is to leaf surfaces, and depends on wind speed, particle size, leaf size, and leaf-surface structure (Chamberlain and Little, 1981). Dry deposition of SO_2 may be to leaf surfaces or internally, after entering through stomata. Rates of deposition to the surface depend on con-

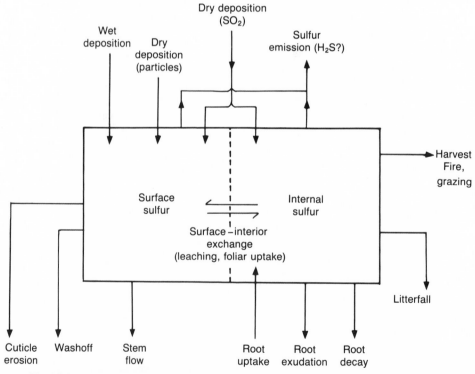

Fig. 22.2. Pathways of sulfur through plants.

centration, and are limited by boundary-layer resistance (a function of wind speed and leaf size) and surface affinity for SO_2, often expressed as a surface resistance (Unsworth, 1981). Internal uptake of SO_2 is further restricted by the stomatal resistance and by the ability of internal surfaces to absorb SO_2. This last factor can be described by an internal resistance, but its magnitude is not yet well-defined (Black and Unsworth, 1979a; Winner and Mooney, 1980b).

Sulfur entering through stomata can be assimilated and translocated (Garsed and Read, 1977a); the fate of SO_2 dry-deposited on surfaces has not been studied closely, but a small amount is likely to diffuse through the cuticle to the internal sulfur pool.

There have been several recent reports of light-dependent emission of reduced sulfur compounds, probably H_2S, from plants either exposed to low concentrations of SO_2 (Hällgren et al., 1982) or supplied with sulfate solution at the roots (Wilson, Bressan, and Filner, 1978; Winner et al., 1981). It is not clear whether this emission arises from the interior of leaves, as a result of the reduction of sulfate as part of the synthesis of amino acids, or whether it comes from microflora on the leaf surface; in principle both seem possible. Different rates of sulfur loss in this manner have been reported; they are often about 10% of typical rates of sulfur uptake by dry deposition or root action. It is still not clear whether emissions of reduced sulfur compounds commonly occur over wide areas. However, almost certainly the recent reports of H_2S emissions following exposure at low SO_2 concentrations represent a different mechanism than previous descriptions of H_2S emissions from plants that had been exposed to high concentrations of SO_2 causing gross destruction of tissue.

Precipitation moderates sulfur exchange in several ways.

1. It may wash off sulfur that had been previously dry-deposited on leaves. This process is most important at the beginning of a rain shower, and may depend on the chemical and physical properties of the rain as well as on leaf-surface structure.

2. Leaves wet with rain or dew may act as a good sink for dry deposition of SO_2 (Fowler and Unsworth, 1979). The sink strength depends on the pH of leaf surface, and decreases if the water becomes too acid (Brimblecombe, 1978).

3. Precipitation promotes chemical transfer in solution between the water film and the leaf interior: there may be foliar uptake of sulfur or leaching of internal sulfur, depending on species, plant nutrition, and maturity of leaf (Tukey, 1970). Processes (2) and (3) are most effective if the foliage is wet for long periods; so frequency of rain and rate of canopy drying are important for this part of the sulfur cycle.

It is often useful to distinguish between the two pathways whereby sulfur is removed from plants by precipitation. "Throughfall" contains material removed by water falling from leaves directly to the ground, whereas "stem flow" involves removal by water running down the stem or trunk to the ground. The spatial distributions of throughfall and stem flow on soil below a canopy are thus quite different, and the relative amounts of sulfur in each component depend on plant structure and on variations in surface-internal exchange over a plant.

When there is no precipitation, sulfur is removed from plants in root exudates or when leaf litter falls to the ground; both these represent local recycling. More distant removal is associated with harvesting of agricultural or tree crops, burning, and grazing by animals and insect pests. Cuticle erosion may also occur.

Root uptake is normally considered the most important pathway of sulfur into plants, but there is evidence that net uptake of sulfur from the atmosphere increases to compensate when soil sulfur is deficient (Cowling and Koziol, 1982). Pathways that take sulfur *out* of roots appear to exist, but are not well-quantified (Raybould, Unsworth, and Gregory, 1977). There is evidence from experiments in solution culture (Faller, 1972) and in soil (Thomas and Hill, 1937b) that exchange of S between roots and the surroundings occurs continuously but becomes a net loss when plants become senescent. This path is difficult to distinguish from sulfur losses as roots die and decay.

Pathways of Sulfur Through Soils

A large proportion of soil sulfur occurs within organic compounds (Bardsley and Lancaster, 1960; Scott and Anderson, 1976); the remainder normally exists as inorganic sulfate, except under anoxic conditions, where sulfides are formed. Sulfate is either available and mobile within the soil solution, sorbed to soil constituents, or precipitated as insoluble forms. Fig. 22.3 shows that there are exchanges between soluble, sorbed, and insoluble forms. An important factor influencing sulfur availability is pH. Below pH 6.0, sulfate becomes strongly adsorbed onto surfaces, especially those coated with hydrous aluminium and iron oxides (Harward and Reisenauer, 1966).

Organic sulfur compounds have been loosely categorized into three groups:

1. Sulfate esters bonded to carbon with C-O-S or C-N-S bonds: HI-reducible (Freney, 1961).

2. Carbon-bonded sulfur (C-S) as in amino acids, proteins, and humic acids (Freney, Melville, and Williams, 1971).

3. Heterocyclic sulfur (Lowe, 1969).

It is thought that HI-reducible S is in a transition state between inorganic sulfate and the more stable C-bonded form (Freney, Melville, and Williams, 1971; McLaren and Swift, 1977). Since sulfate is the form of sulfur available to plants, the balance between assimilation of inorganic sulfur and mineralization of organic sulfur is a primary factor that determines the sulfur-supplying capacity of the soil. The rate of mineralization appears to increase in the presence of plants (Cowling and Jones, 1970), and it increases with increasing temperature (Tabatabai and Al-Khafaji, 1980). However, there is evidence that sulfur mineralization occurs by a different metabolic process from nitrogen mineralization (Swift and Posner, 1972; Tabatabai and Al-Khafaji, 1980).

Sulfur from the atmosphere may reach the soil directly by dry and wet deposi-

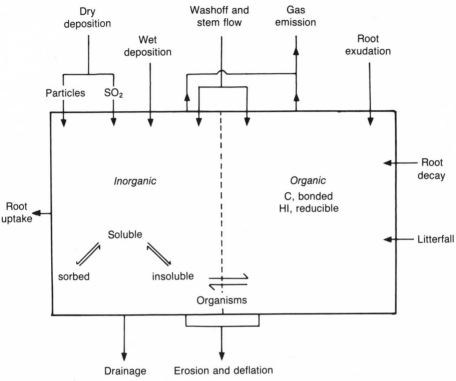

Fig. 22.3. Pathways of sulfur through soils.

tion. Rates of dry deposition are influenced by soil type and moisture content, decreasing as soils become more acid and/or drier (Chamberlain, 1980). Below dense canopies, rates of dry deposition are low for all soils, because of the restriction on turbulent transfer imposed by the canopy (Fowler and Unsworth, 1979). Deposition from the atmosphere adds predominantly to the inorganic sulfur pool (Fig. 22.3).

Sulfur also reaches soils below vegetation canopies in washoff and stem flow, probably as both inorganic and organic forms. Further sources of organic sulfur from plants are leaf litter, root decay, and root exudation.

Sulfur is lost from soils by root uptake of soluble sulfate, by water erosion and deflation by wind of surface material, by leaching of soluble sulfate to the water table and subsequently to watercourses, and by emission of organic and inorganic gases to the atmosphere. Plants are known to absorb some of the gases emitted (Cope and Spedding, 1982); so emissions from soils below canopies may be absorbed by the vegetation, much as CO_2 from soil is absorbed in the daytime by crops (Biscoe, Scott, and Monteith, 1975). If this is wholly or partially true, the recycled sulfur would not represent a loss to the soil-plant system.

Experimental Techniques

The cycle of sulfur in a wheat crop—a case study. For several years we have investigated pathways of sulfur in agricultural crops. As an illustration of some useful experimental approaches, and to identify areas for further research, we summarize here the sulfur cycle for the final 100 days before harvest of a winter-sown wheat crop, covering the period beginning when the field crop was about 7 cm high and had eight leaves.

Measurements were made in an area of about 0.25 ha in a commercial crop of winter-sown wheat (*Triticum aestivum* c.v. Maris Huntsman), growing at Sutton Bonington in a sandy loam soil.

Rainfall was sampled at nine locations, about a meter above the crop canopy, using glass funnels draining into polypropylene bottles. Throughfall was collected at eight locations below the canopy, using Perspex (Plexiglass) gutters (500×30 mm) placed perpendicular to the rows. The gutters and funnels tended to contaminate the sample, because sulfur dry-deposited to them during dry spells could be washed into the collection vessels during rain. To minimize this error, we washed the funnels and gutters with distilled water at least every 48 hours (preferably daily) during dry periods. Estimates of error caused by dry deposition were made by analyzing the funnel and gutter washings for sulfur.

Sulfate was measured in rain and throughfall samples by means of a modification of the thoronol method used by Healy and Atkins (1975); interfering cations were removed by an ion-exchange resin (Dowex 50W-X8). Total sulfur in throughfall was measured after digestion using Cunningham's technique (see below). Volume of stemflow R_s could not be measured, but was assumed to account for the difference between rainfall R above and throughfall R_t below the canopy, ignoring the small amount of water lost by evaporation of intercepted rain (Raybould, Unsworth, and Gregory, 1977); i.e.,

$$R_s = R - R_t,$$

where each term refers to unit ground area. The total quantity of sulfur lost from the crop by leaching in rainfall was estimated as the difference between the total quantity in throughfall plus stemflow and the quantity in rain; i.e.,

$$\text{leaching} = R_s S_s + R_t S_t - R S_r,$$

where S_s, S_t, and S_r are the measured sulfur concentrations in each component.

The sulfur content of the wheat was measured in 1-m sections of wheat rows, collected weekly. The plants were separated into leaves (laminae), stems (including sheaths), ears, and tillers. The dried plant material was ground, and subsamples were digested by means of Cunningham's (1962) digestion technique. Sulfate in the digests was measured by precipitation with a known amount of barium, the excess barium being measured by atomic absorption spectrophotometry.

Litterfall below the plants was collected in filter funnels (150-mm diam.) sunk into the ground so that the rims were flush with the soil surface. The funnels were fitted with nylon gauze at the neck to trap particles. The debris was collected weekly, and analyzed by the same technique used for the harvested plants.

Daily mean atmospheric SO_2 concentrations were measured by drawing air through hydrogen peroxide solution and measuring the sulfate formed by the method of Healy and Atkins (1975). Particulate atmospheric sulfur was measured by filtration of known volumes of air through filter paper (Whatman 41). The sulfate was then extracted from the filter paper with distilled water and measured by means of the same automatic technique employed for the rainfall samples.

Plant-available soil sulfur was measured by extracting a sample of soil with ammonium acetate solution (Bardsley and Lancaster, 1965). The uptake of sulfate by the plants was investigated by injecting ^{35}S as sulfate into the soil at the field site.

The dry deposition of SO_2 to the wheat crop was estimated from the measured atmospheric SO_2 concentration and a deposition velocity, according to the relationship $F = V_g\chi$, where F is the flux of SO_2 to the crop, V_g the deposition velocity, and χ the atmospheric SO_2 concentration. Fowler and Unsworth (1979) analyzed a detailed series of measurements of V_g for a wheat crop at Sutton Bonington. For the actively growing crop, a daytime value of $V_g = 8$ mm s^{-1} was appropriate; at night, with stomata shut, $V_g = 4$ mm s^{-1}. As the crop senesced and the stomata were less active, the cuticle was the major sink for SO_2, and V_g for the dry senescent crop was about 3.5 mm s^{-1}, irrespective of time of day.

Using the values of V_g from Fowler and Unsworth (1979), we estimate the total flux of sulfur as SO_2 to the crop we have studied as 6.8 kg ha^{-1} during the 100-day experimental period, divided into 5.0 kg ha^{-1} when the crop was growing and 1.8 kg ha^{-1} as it senesced. Dry deposition of SO_2 to the soil below the crop was assumed to be negligibly small. The estimated deposition of particulate sulfur to the crop was about 0.3 kg ha^{-1}, a small term in the sulfur budget because the atmospheric concentration was only about 3 μg m^{-3} (0.047 μmol m^{-3}) and V_g was taken as 1 mm s^{-1}. About 4 kg of S ha^{-1} were added to the soil in rainfall during the experiment, and ammonium-nitrate fertilizer, applied to the crop as part of the normal cultural treatment, added less than 0.5 kg of S ha^{-1} to the soil.

Fig. 22.4 shows that the sulfur content of the crop reached a maximum at about a week after anthesis, and then decreased until harvest. These data are expressed as uptake rates in Fig. 22.5.

Analyses of sulfur in throughfall, and estimates of the quantity of stemflow, indicated that the amount of sulfur leached from the crop by rain accounted for only about 30% of that lost by the crop after anthesis. The remainder was probably lost from the roots (Gregory, Crawford, and McGowan, 1979). Measurements of soil-moisture deficit at the experimental site (McGowan, pers. comm.) indicate that there was little drainage of water from the soil and consequently an insignificant loss of sulfur via this pathway. At harvest, about 10 kg of S ha^{-1}

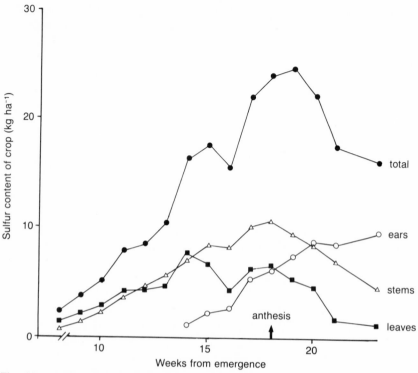

Fig. 22.4. Sulfur content of leaves, stems, and ears of winter-sown wheat at Sutton Bonington, 1977.

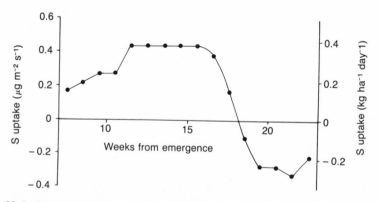

Fig. 22.5. Rate of change of total sulfur content of winter-sown wheat at Sutton Bonington, 1977, calculated from a smoothed curve based on Fig. 23.4.

were removed as grain and straw; stubble burning released about 4 kg of S ha⁻¹
to the atmosphere and 2 kg ha⁻¹ to the soil. Fig. 22.6 summarizes sulfur concentrations in atmosphere, crop, and soil during the study, and shows the magnitudes of transfer by each pathway.

Sulfur from the atmosphere may have supplied a substantial proportion of the
sulfur harvested in this wheat crop, but Figs. 22.4 and 22.5 illustrate some important aspects that are often ignored (e.g., I.E.R.E., 1981). Some of the dry-
deposited sulfur is on the surface of leaves, and it is not clear whether this fraction, which is perhaps 50% of the total dry deposition (Fowler, 1978), makes any
contribution to sulfur nutrition, either by entering leaves through cuticles or by
entering roots after reaching the soil through washoff. The period of maximum
sulfur demand by the crop was during weeks 11 to 16 (Fig. 22.5), when about

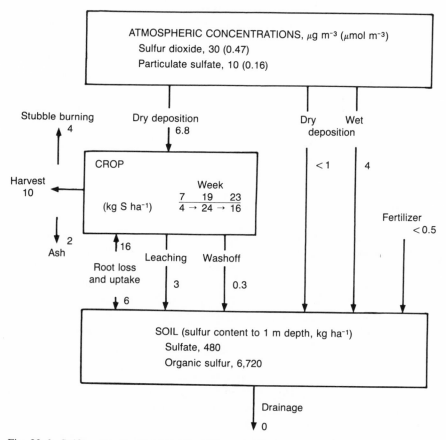

Fig. 22.6. Sulfur exchange between atmosphere, crop, and soil for winter-sown wheat at
Sutton Bonington during the final 100 days before harvest, 1977. Arrows represent transfers of sulfur in kg ha⁻¹ per 100 days.

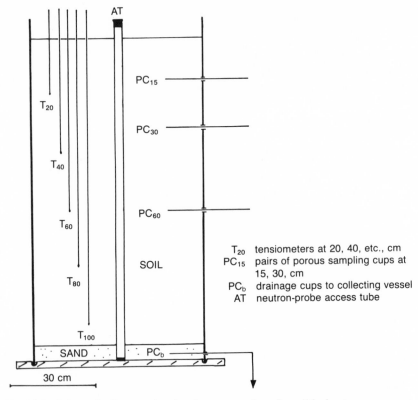

Fig. 22.7. Schematic diagram showing the construction of a soil lysimeter.

0.4 μg of S m^{-2}s^{-1} (0.3 kg of S ha^{-1} day^{-1}) were required. Dry deposition at our site was unlikely to supply sulfur at a rate larger than about 0.2 μg of S m^{-2}s^{-1} (30 μg m^{-3} × 0.008 m s^{-1}), and only part of that sulfur would have been available for assimilation. Consequently, soil sulfur is an important feature of the crop sulfur cycle, even in a relatively polluted area such as ours. Knowledge of sources of soil sulfur is therefore important.

Much sulfur reaches agricultural soils at Sutton Bonington in winter by dry deposition of SO$_2$ and wet deposition in rain and snow. To evaluate the contribution of this source to sulfur nutrition of crops the following spring, we have recently established monolith lysimeters, using undisturbed soil cores collected by the method of Belford (1979). Fig. 22.7 illustrates the construction.

The monolith cylinders are 1.2 m high and 0.60 m in diameter, constructed of heavy-duty industrial PVC. The basal layer (10 cm) of each core was replaced with coarse sand, and two porous cups were inserted in the sand for drainage; the base was sealed with a PVC plate. To allow removal of soil solution, porous cups were inserted horizontally at 15, 30, and 60 cm below the soil surface. Access tubes for a neutron moisture probe were installed down the axis of each cylinder, and tensiometers were fixed at five depths. Twelve lysimeters have been installed in access pits at Sutton Bonington, so that their soil surfaces are level with the adjacent field. Tension is maintained on the basal porous cups to simulate natural drainage. The lysimeters are being used for three types of experiments:

1. To measure rates of input and output of sulfur during the year.

2. To study movement of ^{35}S-sulfate tracer in cropped and uncropped systems.

3. To investigate relationships between ^{35}S distribution and root activity.

Techniques such as these offer controlled situations in the outdoor environment for assessing the contribution of atmospheric sulfur to plant productivity via soil pathways. We see this as an important step in any complete evaluation of the effects of SO_2 on plant productivity.

Directions for Future Research

Many pathways for sulfur have not been investigated thoroughly, in either agricultural or natural systems. New techniques for measuring dry deposition in the

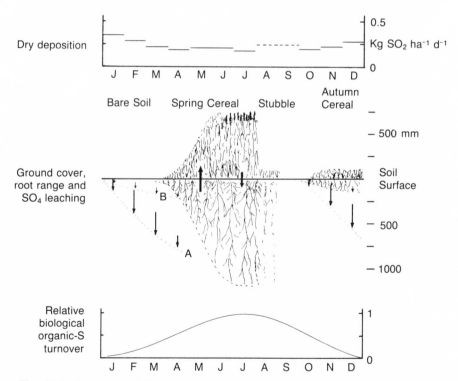

Fig. 22.8. Seasonal variation of some parameters in the atmosphere-plant-soil sulfur cycle. Dry deposition is dependent on SO_2 concentration, soil/crop surface; sulfate leaching on effective rainfall, void ratio, sulfate sorption on soil (A, freely mobile; B, adsorbed); biological turnover on organic matter present, temperature, and microbial activity.

field are needed to define the primary direct path of SO_2 into plants. To generalize from such research techniques, it is necessary to know the SO_2 concentrations in the air of regions at risk; monitoring networks for SO_2 have often concentrated on urban or remote sites, omitting many areas where plant productivity is important.

To understand the relationship between SO_2 exposure and plant response, we need to learn more about the fate of sulfur that enters stomata or is deposited on surfaces. In particular, the significance of sulfur-gas emission from plants needs to be clarified.

The ability of soils to store sulfur supplied by the atmosphere and to release it for plant growth at times of maximum sulfur demand is not well-understood, but seems crucial to any assessment of the contribution of SO_2 to plant nutrition. Some of the seasonal variations that must be considered are indicated in Fig. 22.8. Dry deposition is largest in winter, when SO_2 concentrations are high. Sulfate leaching in soil water, of both freely mobile and absorbed sulfur, moves sulfur rapidly from the surface before spring crops develop. Rooting depth in relation to leaching therefore needs study. The turnover of organic sulfur reaches a peak in summer, but its relation to temperature and microbiological activity in the outdoor environment needs further study.

In this review we have attempted to show that the influence of SO_2 on plant productivity is not merely a question of direct physiological and biochemical responses to SO_2 gas. The complex pathways and fates of SO_2 in the atmosphere-plant-soil continuum make it impossible in any realistic situation to separate responses to SO_2 from those to sulfur derived from SO_2, supplied in rain, and stored in the soil. It is this interrelationship between all aspects of the sulfur cycle originating from SO_2 that makes the subject of this meeting such a challenge, but it is a challenge that must be met before the economic impact of sulfur dioxide on plant productivity can be assessed completely.

Acknowledgments

This work was supported by the U.K. Natural Environment Research Council, Allied Breweries, and the Central Electricity Generating Board.

23

Deposition of SO$_2$ onto Plant Canopies

David Fowler

In considering deposition on canopies of vegetation, the emphasis is on events occurring in the field. This paper will therefore concentrate on field observations, but will also show where the lines of laboratory and field studies overlap, and whether they yield consistent results. This last point is important, because the two lines of research have been directed toward different objectives. Laboratory studies have generally been linked to direct effects of absorbed SO$_2$ and other gaseous pollutants, whereas the field measurements of deposition have largely been concerned with the fate of industrial emissions of SO$_2$. More simply, laboratory studies have concentrated on gas effects, whereas field measurements of deposition have concentrated on long-range transport and the acid-rain problem. The treatment here is restricted entirely to details of the deposition process, showing (where possible) how rates of deposition may be estimated for various crops and conditions from information currently available. Some aspects of the process are poorly understood, and the uncertainties will be discussed.

The Deposition Process

Consider an atmosphere containing SO$_2$ over a field crop. Molecules of SO$_2$ in the atmosphere to be absorbed by the crop must first be transported through the turbulent boundary layer and through the laminar boundary layer very close to elements of the surface. On arrival at the "surface," the molecules may be sorbed onto any of a range of chemically distinct sites. To describe the overall process, it is convenient to separate processes in the atmosphere from those at the surface. The atmospheric components of SO$_2$ deposition are very similar to those for a range of other entities (H$_2$O, CO$_2$, heat, and momentum) for which a good understanding has been obtained from research during the last 30 years (Thom, 1976). For surface processes, however, the chemical properties of SO$_2$ are quite unlike those of other gases. We can therefore draw on a large body of information to deduce appropriate rates of transfer in the atmosphere above a range of crops, but for the surface processes we must rely on work specific to SO$_2$.

Resistances and deposition velocities. In order to quantify the influence of atmospheric and/or surface processes on deposition rates, a resistance analog is commonly used. The resistance to transfer is analogous to electrical resistance in Ohm's law, with the potential difference here being the difference between SO_2 concentration in the free atmosphere and that at the site of uptake. This form of analysis has been widely used in recent years to describe the exchange of gases between vegetation and the atmosphere (Monteith, 1973; Thom, 1976), and specifically for air pollutants (Unsworth, 1981; Fowler, 1980). The symbols introduced for the resistance analogy and used throughout this paper are listed in Table 23.1. If the concentration of SO_2 at the site of uptake is assumed to be zero, the air concentration at some reference level (χ_s) then defines the potential difference, and the total resistance to transfer r_t is obtained from measurements of the air concentration and the vertical flux (F_{SO_2}):

$$F_{SO_2} = \chi_s/r_t. \tag{1}$$

The reciprocal of total resistance has the dimensions of a velocity, and is identical to the velocity of deposition (Chamberlain, 1976). The total resistance to transfer may be subdivided into atmospheric and surface components (Fig. 23.1) to show which processes exert most control in the overall mechanism.

TABLE 23.1
Symbols Used for Atmospheric and Surface Processes in the Deposition of SO_2 on Crops

Symbol	Representing	Units
F_{SO_2}	Vertical flux density of SO_2	$\mu g\ m^{-2}\ s^{-1}$
χ_s	Air concentration of SO_2	$\mu\ mol\ m^{-3}$ ($\mu g\ m^{-3}$)
$r_{t(z)}$	Total resistance to SO_2 transfer from a defined height above the crop (z) to the height within the crop at which $\chi_s = 0$, where z is omitted, 2 m is assumed.	$s\ m^{-1}$
r_{am}	Atmospheric resistance for momentum transfer	$s\ m^{-1}$
r_{bg}	Boundary-layer resistance for mass transfer	$s\ m^{-1}$
r_c	Bulk canopy resistance for SO_2 deposition	$s\ m^{-1}$
r_{wv}	Bulk stomatal resistance for water-vapor loss	$s\ m^{-1}$
r_{c1}	Stomatal component of canopy resistance for SO_2 deposition	$s\ m^{-1}$
r_{c2}	Component of canopy resistance for SO_2 uptake on external surfaces of plants	$s\ m^{-1}$
r_{c3}	Component of canopy resistance for uptake by soil	$s\ m^{-1}$
h	Crop height	m
Z_0	Roughness length ($\approx 0.13\ h$)	m
R_n	Net radiation flux density	$W\ m^{-2}$
C	Flux density of heat loss by convection	$W\ m^{-2}$
λE	Flux density of latent heat loss	$W\ m^{-2}$
v_g	Deposition velocity for SO_2	$mm\ s^{-1}$
U	Wind velocity in horizontal direction	$m\ s^{-1}$

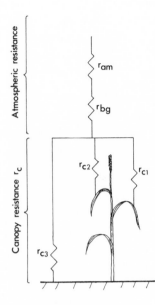

Fig. 23.1. Resistances to SO_2 deposition on wheat: r_{am}, atmospheric resistance for momentum transfer; r_{bg}, extra boundary-layer resistance for mass transfer; r_{c1}, r_{c2}, r_{c3} are the stomatal, cuticular, and soil components of canopy resistance, respectively.

Total resistance r_t may therefore be obtained from measurements of the vertical flux (F_s) and the concentration (χ_s) of SO_2. At suitable sites the components of atmospheric resistance r_{am} and r_{bg} (see Table 23.1) may be calculated from vertical gradients of air temperature and wind velocity. If the total resistance to transfer (r_t) is found to be equal to $r_{am} + r_{bg}$, then the surface is a "perfect sink." In such conditions rates of deposition depend solely on atmospheric processes. In practice, however, for vegetation $r_{am} + r_{bg}$ seldom accounts for more than 20% of r_t; so the bulk of the transfer resistance is contributed by surface processes. The residual resistance is known as the bulk canopy resistance r_c and is also known as a surface resistance, though the latter term is usually applied to nonvegetative surfaces (soil, water, and snow). For vegetation, the canopy resistance may be subdivided into a stomatal component (r_{c1}), a nonstomatal component (r_{c2}), and a soil uptake term (r_{c3}), as in Fig. 23.1.

Atmospheric resistances. In the free atmosphere, turbulent or eddy diffusion is the mechanism of gas transfer (rates of molecular diffusion are several orders of magnitude too small to be important). The rate of eddy diffusion above crop canopies is determined by the aerodynamic roughness of the crop, wind velocity, and vertical temperature gradients in the air over the surface: the rougher the surface and/or the higher the wind velocity, the greater the rate of eddy diffusion. Vertical gradients of air temperature that decrease with distance from the surface augment turbulence (unstable conditions), whereas air temperatures that increase with distance from the surface (stable conditions) suppress turbulence. For suitably flat and uniform crop surfaces, relationships between rates of turbulent diffusion and vertical wind and temperature gradients have been established (Thom, 1976). These may be used to deduce the value of the atmospheric resistance components from measurements of wind and temperature profiles. The resistance r_{am} is the atmospheric resistance for momentum, which is transferred to the crop by frictional drag and by bluff-body forces acting on individual leaves and stems. For the transfer of H_2O, SO_2, CO_2, or heat, there is no equivalent of bluff-body forces. Atmospheric resistances for gas and heat transport are therefore larger and an additional resistance for gas transport (r_{bg}) is introduced (Chamberlain, 1966) to allow appropriate atmospheric resistances for gas transport to be obtained from wind profiles over vegetation. In plant-effects studies, chambers are widely used for exposure of vegetation to pollutant gases. In such chambers the atmospheric resistance to gas transfer is commonly calculated by measuring the water loss from wetted replicas of the foliage for a known difference in water-vapor concentration between the leaf replica and the air. In such measurements the resistance obtained is equivalent to $r_{am} + r_{bg}$.

Atmospheric resistance ($r_{am} + r_{bg}$) decreases with wind speed and with crop height, and appropriate values may be calculated from wind speed at a defined height above the crop and from crop height, by means of the wind-profile equations (Monteith, 1973) or may be interpolated from a nomogram of the form shown in Fig. 23.2.

As a crop grows, atmospheric resistance becomes progressively smaller, and the maximum rates of deposition (for a surface that behaves like a perfect sink, $r_c = 0$) increase. For most agricultural crops, the resistance is more than 200 s m^{-1} for a crop height of 5 cm or less and for typical wind speeds of 2 m s^{-1} above the crop surface, but decreases to about 40 s m^{-1} when the crop is 1 m tall. Potential rates of deposition increase from 5 mm s^{-1} to 25 mm s^{-1} during this period.

Table 23.2 shows values of atmospheric resistance for a range of crops at two wind speeds. For this table the atmosphere has been assumed to be neutral (i.e., that temperature decreases with height at the adiabatic lapse rate of 0.01 °C m^{-1}). In unstable conditions, temperature decreases with height faster than 0.01 °C m^{-1}, and atmospheric resistance is decreased by the additional mixing caused by buoyancy of the air close to the ground. In stable conditions air temperature increases with height above ground, which reduces the vertical exchange of air and increases atmospheric resistance. The effects of stability on the exchange of gases between vegetation and the atmosphere are discussed in detail by Webb (1965), but stability corrections generally modify atmospheric resistances by only 10 to 20% of neutral values.

Surface processes. The importance of atmospheric transfer to rates of deposition is ultimately governed by the magnitude of canopy resistance (r_c). If r_c accounts for 70% or more of the total resistance (r_t), then rates of deposition will not be influenced strongly by changes in wind speed or in atmospheric stability. Measured rates of SO_2 deposition onto grass and cereal crops generally yield total resistances between 10^2 and 10^3 s m^{-1}, of which only 10 to 20% is generally

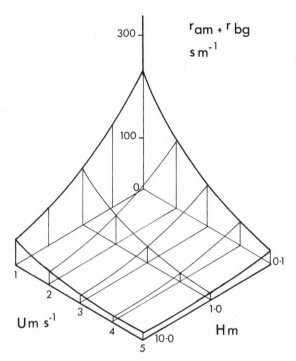

Fig. 23.2. Variation in atmospheric resistance ($r_{am} + r_{bg}$) with wind speed (U_m) and vegetation height (H_m).

TABLE 23.2
Typical Values of Atmospheric Resistance ($r_{am} + r_{bg}$) for Gas Transfer to Crop Canopies

Crop	Crop height in m	Roughness length in m	($r_{am} + r_{bg}$) in s m^{-1} at a wind speed of:[a]	
			1 m s^{-1}	4 m s^{-1}
Grass	0.1	0.013	235	58
Grass	0.4	0.052	133	33
Potatoes	0.6	0.12	90	22
Sugar beet	0.4	0.08	110	28
Cereals	1.0	0.13	73	18
Cotton	1.27	0.13	61	15
Maize	2.3	0.15	47	12
Coniferous forest	10	1.0	26	9

NOTE: Neutral stability is assumed for the atmosphere above the crop.
[a] Reference height for atmospheric resistance and windspeed 2 m or crop height, whichever is greater.

due to atmospheric transfer. It follows that surface processes generally control deposition, and that we must understand the properties of canopy resistance in order to predict rates of deposition.

There is abundant information from laboratory (Spedding, 1969; Black, 1982) measurements that the substomatal cavity represents an important site of uptake for SO_2 molecules. From measurements of SO_2 deposition onto a rapidly growing wheat crop, Fowler and Unsworth (1979) showed that daytime rates of deposition were due mainly to stomatal uptake. In these conditions 70% of the total flux was absorbed within stomata, the remaining 30% representing the combined uptake by external plant surfaces and beneath the crop. Cuvette measurements by Garland and Branson (1977) and field measurements by Fowler and Cape (1983) of SO_2 deposition have confirmed the importance of stomatal uptake for Scots pine canopies. Considering the simplified representation of SO_2 deposition in Fig. 23.1, atmospheric components of the resistance network may be estimated as has been described. Since stomata are responsible for most of the SO_2 deposited in dry conditions, we must be able to deduce appropriate values for the stomatal component of the transfer resistance in order to estimate rates of SO_2 deposition onto crop canopies.

At its simplest, the stomatal uptake of SO_2 may be regarded as proportional to water loss through the same apertures. In theory, well-established micrometeorological methods for estimating evapotranspiration (Thom, 1976) could be used to deduce appropriate rates of stomatal uptake of SO_2. However, in practice this would require that the substomatal cavity be a perfect sink for SO_2 (or at least would require its properties for uptake to be constant), and that the presence of SO_2 not influence the stomatal mechanism. Neither of these conditions have been shown to be valid. Laboratory studies have shown that stomata respond to ambient SO_2 concentrations, and that the stomatal conductance varies in direction and magnitude depending on the species and environmental conditions (Mansfield and Majernik, 1970; Black, 1982; Winner and Mooney, 1980b).

If concentrations of SO_2 commonly found in the field influence rates of exchange of SO_2, H_2O, and CO_2 between crops and the atmosphere, then predictions based on the micrometeorological literature for evapotranspiration would fail. The laboratory measurements cover a wide range of conditions and species, but generally test a response to large SO_2 concentrations (more than 1.6 μmol SO_2 m^{-3} or 100 μg SO_2 m^{-3}). The effect of SO_2 concentrations typical of polluted agricultural districts has not been widely tested. The variability of air concentrations of SO_2 (and other pollutants) present an additional complication, so that in the longer term we need to know the dynamic response by stomata in a field crop to changes in air concentration. Field measurements have generally lacked the precision necessary to detect small changes in canopy resistance as a response to changes in SO_2 concentration, but there is no field evidence to date that the rate of SO_2 deposition onto crop canopies depends on ambient SO_2 concentration. Independent of stomatal effects, which have been studied exclusively in the laboratory, the validity of assumptions about the mechanism of sorption within the substomatal cavity has been the subject of both laboratory and field investigation. Hällgren *et al.* (1982), in a series of experiments enclosing shoots of Scots pine in a small chamber in the field at SO$_2$ concentrations between 40 and 250 μg m^{-3} (0.63 and 3.91 μmol m^{-3}), showed that the rate of stomatal uptake was much smaller than would be expected from the conductivity of the stomatal pore to SO$_2$ transport. There appears, therefore, to be a mesophyll resistance, equivalent to that for CO$_2$ transport, but probably due to different processes of sorption, transport, and chemical transformation within the plant. A light-dependent release of H$_2$S from the plant was also observed. For measurements of net sulfur exchange between crops and the atmosphere, H$_2$S emission would generate an apparent mesophyll resistance if the instruments were equally sensitive to SO$_2$ and H$_2$S. The measurements of Hällgren *et al.* are field-based, but still involve the use of enclosure methods.

For the same crops (Scots pine), Fowler and Cape (1983) made measurements of sulfur fluxes using a micrometeorological method. The measurements showed that canopy resistance generally accounted for more than 90% of the total resistance to transfer, and provided further evidence that stomata (when open) are the main site of uptake in the crop. By measuring heat and water-vapor fluxes concurrently with the SO$_2$ fluxes, they could calculate the stomatal component of canopy resistance to SO$_2$ uptake and the canopy resistance for water loss (Fig. 23.3 and Table 23.3).

Flux measurements were made by the eddy-correlation technique (Wesley *et al.*, 1978), in which turbulent fluctuations in the vertical component of wind velocity (w'), and air concentrations of SO$_2$ (x') are correlated. The majority of the signal from the sulfur detector is contributed by the ambient SO$_2$ gas concentration, but the turbulent fluctuations cause variations of the order of 2 to 5% in this signal. A small upward flux of H$_2$S from the forest would appear in the turbulent fluctuation term, and would therefore be difficult to detect as a component of the ambient concentration of sulfur-containing gases. During these measurements no attempt was made to exclude H$_2$S from the eddy-correlation gas analyzer, and no evidence of H$_2$S was obtained from separate measurements of the air concentration of all sulfur-containing gases with and without an H$_2$S scrubber. It is not possible therefore to deduce the relative importance of chemical (or biological) restrictions in the uptake mechanism and a possible upward flux of H$_2$S. Clearly more field measurements are required.

The bulk of the SO$_2$ deposition shown in Fig. 23.3 took place during daylight hours; more detailed information on the exchange of net radiation (R_n), convective heat (C), and latent heat (λE) above the forest canopy for March 25, 1982, is provided in Table 23.2. The bulk canopy resistances for water-vapor loss (r_{wv}) and stomatal SO$_2$ uptake (r_{c1}) differ on average during the 6-h daylight period by a factor of 5. In the absence of a mesophyll resistance, r_{c1} would be approximately $2r_{wv}$ because of the differences in molecular diffusivity of SO$_2$ and H$_2$O. The

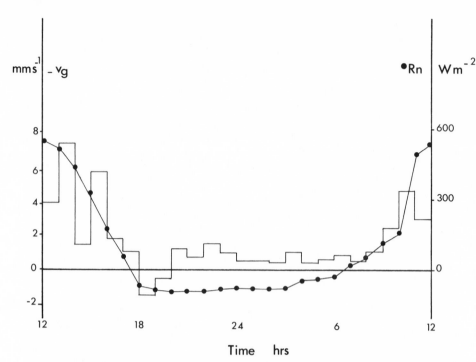

Fig. 23.3. The diurnal variation in deposition velocity (v_g) and net radiation (R_n) above a Scots pine canopy, for March 25–26, 1982.

TABLE 23.3
Hourly Average Heat Fluxes and the Bulk Stomatal Component of Canopy Resistance
for SO_2 Deposition (r_{c1}) and H_2O Loss (r_{wv}) for 3/25/82, Showing the Presence
of a Mesophyll Resistance to SO_2 Uptake

Time (GMT)	R_n (W m^{-2})	C (W m^{-2})	λE (W m^{-2})	v_g (mm s^{-1})	r_{wv}[a] (s m^{-1})	r_{c1}[a,b] (s m^{-1})
1230–1330	−580	323	207	6.98	78	151
1330–1430	−440	193	247	3.21	56	356
1430–1530	−380	150	230	4.72	52	232
1530–1630	−270	141	129	1.83	100	714
1630–1730	−140	73	67	1.52	170	909
1730–1830	0	0	0	0.40	∞	∞

NOTE: All symbols are defined in Table 23.1.

[a] r_a taken as 10 s m^{-1} (U at 5 m above zero plane ~3 m s^{-1}).

[b] r_{c1} (the stomatal component of canopy resistance to SO_2 transfer) = $(r_c^{-1} - r_{c2}^{-1})^{-1}$, r_{c2} taken as 2,500 s m^{-1}, value for dry surface with closed stomata during 1730–1830 GMT. Average ratio r_{c1}/r_{wv} = 5.03, value for no mesophyll resistance 2.0; thus on average mesophyll resistance = ~50% of r_t.

TABLE 23.4
The Stomatal Component of Canopy Resistance for SO$_2$ Uptake r_{c1} for Various Crops

	r_{c1} in s m^{-1}		
Crop	Daytime range[a]	Typical value[b]	Source
Grass	50–300	100	Ripley and Redman (1976)
Potatoes	60–200	90	Brown (1976)
Sugar beet	40–200	70	Brown (1976)
Wheat/barley	30–300	120	Monteith (1973)
Cotton	40–300	80	Stanhill (1976)
Maize	100–500	200	Uchijima (1976)
Coniferous forest	100–500	150	Jarvis, James, and Landsberg (1976)

NOTE: Mesophyll resistance is assumed negligible. From canopy resistances to water-vapor loss (r_{c1}SO$_2$ $\simeq 2r_{c1}$H$_2$O).
[a] The lower end of range represents plants in optimum growing conditions; upper end of range represents conditions of stress, especially water, or end-of-day values.
[b] These generally represent midday values from the literature cited.

mesophyll resistance is therefore approximately 50% of r_c, and since the atmospheric resistance is small above forests (less than 20 s m^{-1}), the mesophyll resistance is approximately 50% of r_l. The possibility of a small upward H$_2$S flux remains as an uncertainty in these measurements.

The complications introduced by recent observations of mesophyll resistance and the possibility of significant H$_2$S loss from crop canopies in polluted areas indicate likely topics for development of research on this subject. It would be surprising, given the number of sulfur-containing gases that may have biogenic origins—SO$_2$, H$_2$S, OCS, (CH$_3$)$_2$S, (CH$_3$)$_2$S$_2$, CS$_2$—if our concept of a simple undirectional flux of SO$_2$ from the atmosphere into the substomatal cavities of leaves proves satisfactory for all vegetation, all the time. However, current evidence shows that SO$_2$ uptake by stomata dominates this exchange for cereal crops and for pine forests in dry conditions. It is therefore reasonable to use literature values of canopy resistances for widely grown crops to deduce minimum values for the stomatal component of canopy resistance for SO$_2$ deposition. These values (Table 23.4) may then be used to estimate probable deposition rates of SO$_2$ onto different crops in the field.

The site and mechanism of SO$_2$ uptake within the substomatal cavity remain obscure. The traditional view of uptake by intercellular fluids has yet to be convincingly demonstrated, and details of appropriate reaction-rate coefficients for oxidation of SO$_2$ inside leaves have yet to be established.

Nonstomatal uptake. Uptake by surfaces other than stomata has been measured in the laboratory by Spedding (1969) and in the field by Fowler and Unsworth (1979). The rates of SO$_2$ deposition onto these surfaces are small relative to typical rates of deposition onto dry surfaces when stomata are open, and are typically 0.5 to 2.0 mm s^{-1}. These small values represent large canopy resistances, of 500 to 2,000 s m^{-1}, so that when stomata are closed, rates of deposition are small and insensitive to atmospheric resistance. Laboratory measurements show that cuticular surfaces of plants absorb SO_2. Garland and Branson (1977) have shown that the dead foliage and bark in a pine canopy also absorb SO_2. To further complicate the process, sorption onto vegetation or soil below the main crop canopy may contribute to nonstomatal uptake. Uptake by soil beneath a wheat canopy was shown by Fowler and Unsworth (1979) to account for up to 10% of the net SO_2 flux. The affinity of SO_2 for leaf surfaces (i.e., excluding stomata) of the same crop changed with time, from $r_{c2} = 250$ s m^{-1} during the vegetative phase of growth, to more than 1,000 s m^{-1} during senescence. Immediately prior to harvest, rates of deposition were too small to be detected.

In the absence of measurements for other crops, values of 200 to 300 s m^{-1} for r_{c2} (Fig. 23.1) for the vegetative stage, 500 s m^{-1} during flowering, and 1,000 s m^{-1} during senescence seem appropriate for grass and cereal crops.

Wet surfaces. Water on the surface of vegetation may arise from condensation processes (dew) or from rain or wind-driven cloud droplets. The extent of field data is very limited, but the main features seem clear.

Dew. The condensation of pure water onto surfaces of vegetation presents SO_2 molecules with an additional site for uptake. Uptake by a water layer proceeds rapidly until the S(IV) in solution and the SO_2 in air approach equilibrium. In practice, oxidation in the water layer removes S(IV) species, but hydrogen ions from the solution and oxidation of SO_2 increase the acidity of the liquid film and eventually limit the solution of SO_2. Details of the solution chemistry of SO_2 in laboratory conditions have demonstrated these features (Terraglio and Manganelli, 1967), but in the field, leaf surfaces are contaminated by many ions that may be washed from the foliage (Miller and Miller, 1980) and that probably influence chemistry in the liquid film. Measurements of SO_2 deposition onto a dew-wetted wheat crop show a variation in deposition rate with time that is consistent with the preceding argument (Fig. 23.4).

The period of rapid uptake showed the surface to be behaving as a perfect sink while dew was condensing onto the crop surface (from 1900 hrs on 7/24/73 to 0100 hrs on 7/25/73), but as rates of dewfall decreased after the appearance of 8/8 cloud cover (0200 hrs), the rate of SO_2 uptake also decreased, long before the dew evaporated. Clearly the surface was no longer behaving as a perfect sink. Acidification of the dew during oxidation of the S(IV) to S(VI) and a decrease in SO_2 solubility (Terraglio and Manganelli, 1967) provide a chemical explanation for these observations. In conditions favorable for dew formation, atmospheric resistance is usually large (100 s m^{-1} or more), so that expected maximum deposition velocities would be ~ 10 mm s^{-1} for agricultural crops in these conditions. Chemical limitations to the uptake process, in addition to the large atmospheric

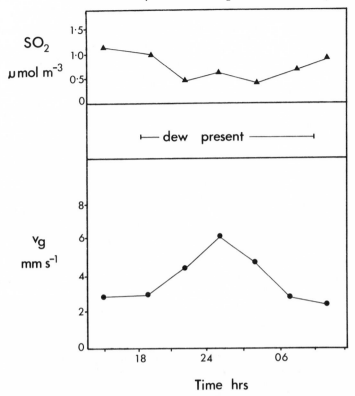

Fig. 23.4. The diurnal variation in deposition velocity (v_g) and SO_2 concentration 1 m above a wheat canopy for July 24–25, 1973. The presence of dew on the crop is also shown.

resistances during the presence of dew, reduce the likely average values for deposition rate onto dew-wetted crops. It is unlikely, therefore, that the average deposition velocity appropriate for SO_2 deposition on short crops in these conditions exceeds 5 mm s^{-1}.

Rain. A rain-wetted canopy of vegetation may differ chemically from dew-wetted surfaces. Except in conditions where a pronounced vertical gradient in SO_2 concentration exists (e.g., over a city or close to an SO_2 source), rain is usually in equilibrium with ambient SO_2 concentrations by the time it arrives at the surface. The layer of water on vegetation would be expected to absorb SO_2 in response to removal of S(IV) species from solution if acidity is not limiting.

Some measurements of SO_2 deposition onto a rain-wetted Scots pine canopy showed interesting features (Fig. 23.5). In general, rates of deposition were quite small, 2 to 3 mm s^{-1}. Since it was dark and surfaces were fully wetted, the

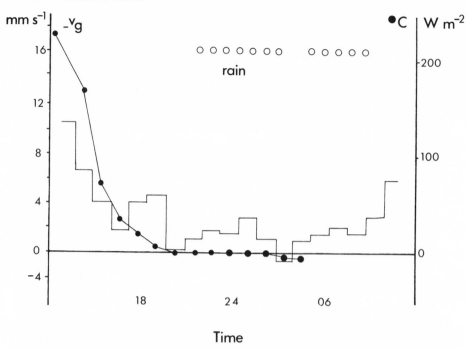

Fig. 23.5. Deposition velocity (v_g) and convective heat loss (C) above a canopy of Scots pine for April 21–22, 1982. The presence and duration of rain is also indicated.

water layer could be assumed to be the only sink; values of canopy resistance for uptake by this water layer were 300 to 500 s m^{-1}. Another occasion with rain in similar conditions also showed small rates of deposition. Since atmospheric resistances during these measurements were generally small (less than 20 s m^{-1}), at least 90% of the resistance was due to the surface processes. The wet surface therefore was far from a perfect sink, but too few measurements of the ions present within the layer of water were made to show which step was limiting. In the absence of more measurements of SO$_2$ deposition onto rain-wetted vegetation, the values reported here provide a basis for estimating the importance of SO$_2$ deposition onto wet surfaces as a component of annual or regional sulfur budgets.

Conclusions

The deposition process outlined here has been greatly simplified to show the mechanisms and quantify each component. This simplification necessarily contains uncertainties, and further complexity will probably need to be introduced to obtain a satisfactory description of processes occurring in the field. The emphasis on field measurements shows many areas where more data are required, in particular to examine predictions based on laboratory measurements.

To bring together the sections of this paper, atmospheric resistances for a particular crop from Table 23.2 may be combined with stomatal component r_{c1} from Table 23.3 and a "typical" resistance to deposition onto surfaces (other than stomata) of 1,000 s m^{-1} for forests, and 300 s m^{-1} for other surfaces, to yield a total resistance to transfer and hence deposition velocity.

Table 23.5 shows typical values of the resistance components for six different crops for dry and rain-wetted conditions. The uncertainties introduced by the presence of a mesophyll resistance to SO$_2$ uptake cannot be satisfactorily incorporated in this table, and remains one of the areas requiring more field data.

A feature of the values in Table 23.4 is the gradual increase in the relative importance of canopy resistance with crop height. For crops taller than 1 m, can-

TABLE 23.5
Rates of SO$_2$ Deposition onto Crops

Crop; and time	Surface	Atmospheric resistance[a] (s m^{-1})	Stomatal uptake[b] (s m^{-1})	Uptake by other surfaces (s m^{-1})	r_t (s m^{-1})	Deposition velocity (mm s^{-1})
Grass						
Day	Dry	100	100	300	175	5.7
Night	Dry	150	∞	300	450	2.2
Night/day	Wet (rain)	100	∞	200	300	3.3
Sugarbeet						
Day	Dry	50	70	300	107	9.3
Night	Dry	75	∞	300	375	2.6
Night/day	Wet (rain)	50	∞	200	250	4.0
Wheat/barley						
Day	Dry	40	120	300	126	8.0
Day/night	Senescent	40	∞	1,000	1,040	1.0
Night	Dry	60	∞	300	360	2.8
Day/night	Wet (rain)	40	∞	200	240	4.2
Cotton						
Day	Dry	30	80	300	993	10.7
Night	Dry	60	∞	300	360	2.8
Day/night	Wet	30	∞	200	230	4.3
Maize						
Day	Dry	20	200	300	140	7.1
Night	Dry	40	∞	300	340	2.9
Day/night	Wet (rain)	20	∞	200	220	4.5
Coniferous forest						
Day	Dry	10	150	1,000	140	7.1
Night	Dry	20	∞	1,000	1,020	1.0
Day/night	Wet (rain)	10	∞	200	210	4.8

[a] Values from Table 23.2.
[b] Values from Table 23.4. Stomatal uptake during rain will be partly governed by wettability of the surface; no measurements are available to allow separation of r_{c1} and r_{c2}.

opy resistances account for most of the transfer resistance, and for forests atmospheric resistances are generally so small that rates of deposition are controlled almost exclusively by the affinity of the surface.

Acknowledgment

The author gratefully acknowledges the financial support of the U.K. Department of the Environment for research included in this paper.

24

Plant Response to SO_2:
An Ecosystem Perspective

Robert A. Goldstein, Carl W. Chen,
and Steven A. Gherini

The organizers of the workshop from which this chapter derives asked us to place SO_2-plant interactions into the context of ecosystem element cycling. In preparing to frame our response, we posed several questions. How does SO_2 uptake by plant canopies relate to the movement of sulfur throughout the ecosystem? How does the amount of SO_2 absorbed through stomata relate to the total atmospheric inputs of sulfur into an ecosystem and to the growth requirements of a plant community for sulfur? How useful is an ecosystem element-cycling construct for analyzing the effect of SO_2 on plant productivity? These questions will be examined for forest ecosystems in the eastern U.S., Sweden, and West Germany, and especially for three forested watersheds in the Adirondack Park region of New York State.

Hierarchical Structure of Ecological Systems

Conceptually, ecological systems can be structured hierarchically by level of organization. As one proceeds up the hierarchy, what was initially a system by itself becomes an element or subsystem of an organizationally more complex system. Hence, one can start with an individual leaf, and analyze its interactions with its environment. For the leaf, the rest of the plant and the atmosphere are part of an environment that exists beyond the boundary of the leaf. Properties of the environment that affect the leaf may be defined (e.g., atmospheric humidity or plant nutrient status), but the structure and dynamics of the environment (e.g., atmospheric thermal fluxes or allocation of labile carbohydrates to different plant organs) are generally not. As we move up to higher levels of organization, the leaf system can be expanded to an individual plant, to a plant plus the immediate atmosphere and soil surrounding it (soil-plant-atmosphere system), to a community of plants, and to an entire ecosystem. Different ecological processes are associated with each level of organization; e.g., photosynthesis with the leaf, carbohydrate allocation with the whole plant, transpiration with the soil-plant-atmosphere system, succession with the community, and landscape element cycling with the ecosystem.

At each higher level of organization, the boundary of the system is enlarged to encompass what were previously elements of the environment. For an expanded system, the elements that are added are given structural and dynamic definition, whereas the elements that remain from the previous system, by necessity, have their structural and dynamic resolution decreased. Hence the model of community response of Shugart and McLaughlin (this volume) will not explicitly contain the detailed definition of plant physiology found in the process models of plant growth discussed by Kercher and King (this volume).

Whether a specific ecological system is appropriate as a framework for analysis of an SO_2 effect depends on the specific effect one is concerned with. Historically, analysis of effects of SO_2 (and gaseous pollutants in general) on plants focused on the leaf. For many years, the major variable for quantifying plant injury by air pollution was visible leaf injury. This early emphasis was followed by the adoption of a concept of SO_2 as being absorbed into the plant through stomata and reacting in the interior of the plant to produce some effect. (Such a concept is pictured in Figs. 21.1 and 21.3 in the chapter by Kercher and King.) The effect may involve short-term changes in biochemical or physiological processes (e.g., carbon fixation), annual changes in quantity or quality of yield, or longer-term changes in community structure. A whole array of ecological factors, many of which are time-dependent, affect plant response to SO_2. These include atmospheric (e.g., temperature, insolation, moisture), soil (e.g., moisture, nutrients), plant (e.g., species, cultivar, nutrient status, growth rate, developmental stage, preconditioning), and other factors (e.g., disease, pests, competition).

The movement of sulfur within ecosystems is largely related to the movement

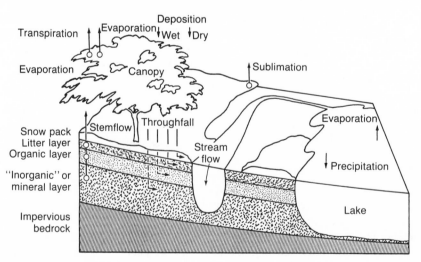

Fig. 24.1. Flowpaths for water through a forested ecosystem. Adapted from Chen *et al.* (1982).

of water. Fig. 24.1 illustrates the flowpaths for water through a forested eco-system containing streams, bogs, and a lake. Water is the major vehicle for carrying most elements throughout the ecosystem; hence the flowpaths of water establish a network for the ecosystem cycling of elements. Within this ecosystem framework, vegetation is just one of several major elements, and is not neces-sarily the most important element. This contrasts with the soil-plant-atmosphere system, where vegetation is the central element that binds the system together.

Relation Between Acidic Deposition and Plant Uptake of SO_2

During the last decade, much attention has been focused on the phenomenon of acidic deposition. If precipitation were simply distilled water in equilibrium with atmospheric CO_2, then its pH would be approximately 5.6. Throughout the United States, in Canada, and in many parts of Europe, precipitation whose pH is appreciably lower because of the presence of sulfuric and nitric acids, has been observed (U.S. Department of Agriculture Forest Service, 1976). The term acidic precipitation has been applied to this phenomenon. Wetfall and dryfall can also introduce acidity into an ecosystem by depositing substances (e.g., NH_4^+, NO_x, SO_2) that can be readily converted into strong acids by nitrification and oxidation. The term acidic deposition encompasses both acidic precipitation and acidic dryfall.

Various ecological effects have been hypothesized to result from acidic deposi-tion. These include acidification of surface waters, loss of fish populations, de-creased productivity of natural and agricultural vegetation, and decreased soil fertility. Acidic deposition is an ecosystem-level perturbation. It directly per-turbs the cycling of sulfur and nitrogen (which are macronutrients) and acidity. Many ecosystem components and processes can potentially be affected, since water carries acidity (or alkalinity), sulfate, and nitrate throughout the ecosystem. Many ecosystem processes naturally produce or consume acidity. Any analysis of acidic deposition effects should include consideration of these processes.

Conceptually, acidic deposition perturbs an ecological system differently from SO_2. Acidic deposition is an ecosystem-level perturbant that will probably affect vegetation, whether beneficially or adversely, indirectly, because of some altera-tion of soil conditions, e.g., increased availability of nitrogen and sulfur, in-creased leaching of essential base cations from the root zone, or elevated con-centration of Al in the soil water in the root zone. In contrast, the effect of SO_2 on vegetation is primarily related to direct uptake by foliage. Another conceptual difference between the two pollutants is that acidic deposition tends to be a regional-scale phenomenon. The acidity of precipitation tends to be fairly con-stant throughout large regions. The greatest acidity often occurs in rural areas distant from large point sources of SO_2. SO_2, on the other hand, only occurs in high concentrations, with a distribution that is highly nonuniform both in space and in time, in the vicinity of point sources.

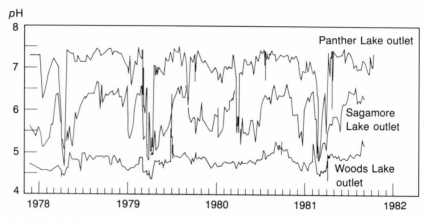

Fig. 24.2. Outlet air equilibrated pH of Woods, Panther, and Sagamore Lakes throughout the ILWAS study. Adapted from Goldstein *et al.* (1984).

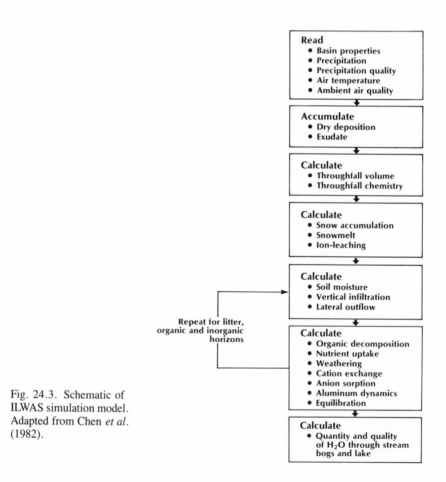

Read
- Basin properties
- Precipitation
- Precipitation quality
- Air temperature
- Ambient air quality

Accumulate
- Dry deposition
- Exudate

Calculate
- Throughfall volume
- Throughfall chemistry

Calculate
- Snow accumulation
- Snowmelt
- Ion-leaching

Calculate
- Soil moisture
- Vertical infiltration
- Lateral outflow

Repeat for litter, organic and inorganic horizons

Calculate
- Organic decomposition
- Nutrient uptake
- Weathering
- Cation exchange
- Anion sorption
- Aluminum dynamics
- Equilibration

Calculate
- Quantity and quality of H_2O through stream bogs and lake

Fig. 24.3. Schematic of ILWAS simulation model. Adapted from Chen *et al.* (1982).

Integrated Lake-Watershed Acidification Study (ILWAS)

The Integrated Lake-Watershed Acidification Study (ILWAS) is an intensive, integrated, five-year ecosystem analysis of three forested watersheds (Goldstein *et al.*, 1980). Its major objective is to develop a general, mechanistic, quantitative relationship between the deposition of atmospheric acids and the acidity of surface waters. The three watersheds are located within 30 kilometers of each other in the Adirondack Park region of New York State. Although each watershed receives similar atmospheric inputs, each has a lake with a different *p*H and with different acidic dynamics (Fig. 24.2). The bedrock of the area is granitic,

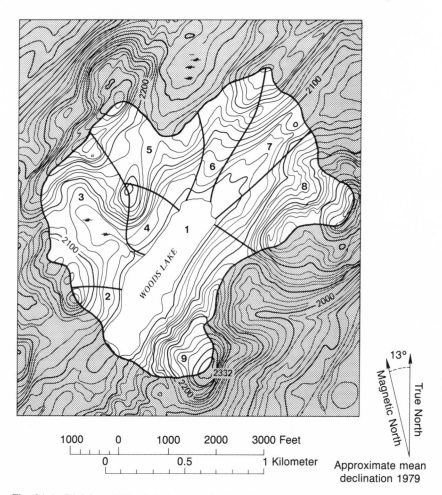

Fig. 24.4. Division of Woods Lake watershed into homogeneous subwatersheds for purposes of simulation. Adapted from Chen *et al.* (1982).

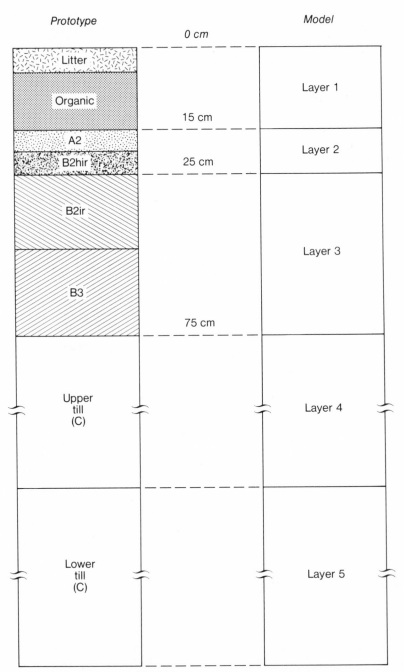

Fig. 24.5. Characteristic division of soil into homogeneous vertical layers for purposes of simulation.

and the plant communities are coniferous and deciduous. The coniferous communities tend to be dominated by spruce (*Picea rubens*) and fir (*Abies balsamea*), and the deciduous communities by beech (*Fagus grandifolia*), birch (*Betula* spp.), and maple (*Acer* spp.). The upper soil layers are strongly acidic. Detailed information about the geology, hydrology, soils, vegetation, and water chemistry of the watersheds is given in Electric Power Research Institute reports (1981 and 1983).

Based on a conceptual model (Chen, Gherini, and Goldstein, 1979), the watersheds have been divided into a cascade of compartments: atmosphere, canopy, snowpack, soil system, bogs, streams, and lake. Data have been collected for each compartment to evaluate its acid-producing and acid-consuming processes, and to verify a mathematical model that simulates the quantity and quality of water moving through the ecosystem (Fig. 24.3). Data collection started in the fall of 1977, and was completed at the end of 1981. Data records for each compartment do not necessarily extend for the total period of data collection.

Within the study, the role of the model is to organize the measurement of lake-watershed acidification processes into an integrated theoretical framework. To accomplish this, the model simulates each of the acid-producing and acid-consuming processes, calculates the quantity and quality of water in each component of the lake-watershed system, and integrates the results. To apply the model, a watershed is divided into horizontally homogeneous subareas (Fig. 24.4). The connectivity of water flow among the subareas is defined, and each subarea is stratified into a series of homogeneous vertical compartments, as shown in Fig. 24.5 (Chen *et al.*, 1982).

Since the model is based on fundamental biogeochemical processes, it should be applicable to geographical areas other than the Adirondack region where the data set has been collected, and may therefore be a vehicle for extrapolating to other geographical areas the understanding of surface-water acidification derived from ILWAS.

Ecosystem Sulfur Cycle

Since the ILWAS model calculates alkalinity as the difference between the sums of base cations and acid anions (Stumm and Morgan, 1981), the study and model trace the cycling, production, and consumption of all major base cations and acid anions, including sulfate (Fig. 24.6). Sulfur enters the ecosystem primarily as sulfate and SO_2 from the atmosphere in wetfall and dryfall. SO_2 adsorbed to the canopy or absorbed by the soil may be oxidized to sulfate. SO_2 absorbed by the canopy may be converted to biomass. Sulfate can enter the soil solution as a result of throughfall, organic matter decomposition, mineral weathering, and oxidation. Processes that remove sulfate from the soil solution are plant uptake, sulfate reduction under anoxic conditions, and lateral flow and drainage to streams and lakes. Sulfate adsorption sites in the mineral soil (Johnson, 1980) may act as either a sink or a source of sulfate in the soil solution.

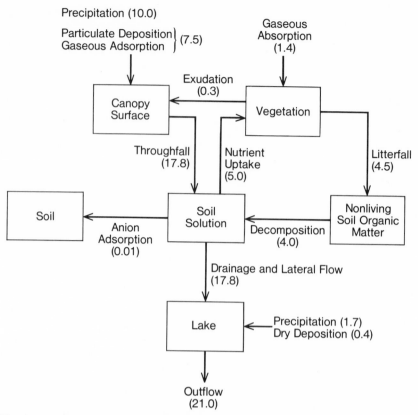

Fig. 24.6. Schematic of ecosystem sulfur cycle, displaying annual fluxes (Kg ha^{-1}) characteristic of Panther Lake watershed. Fluxes are based on simulation of Panther system with ILWAS model. All fluxes (including lake outlet) are normalized with respect to terrestrial area of watershed.

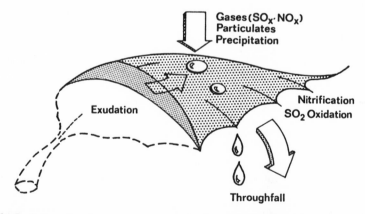

Fig. 24.7. Schematic of canopy processes simulated by ILWAS model. Adapted from Chen *et al.* (1983).

In the ILWAS watersheds, mineral weathering does not appear to be a significant source, nor anoxic reduction a significant sink, for sulfate.

Role of SO₂ in the ILWAS system. The fluxes of sulfur entering an ecosystem from the atmosphere can be simulated using the canopy module of the ILWAS model. The formulation of this module and its calibration are described in detail by Chen *et al.* (1983). Canopy processes simulated are: (1) interception and evaporation of precipitation, (2) dry deposition of particulates, (3) dry deposition of SO_2 and NO_x, (4) SO_2 and NO_x absorption by foliage, (5) leaf exudation, (6) SO_2 and NO_x oxidation, (7) ammonium nitrification, (8) aqueous-phase equilibration of pH, alkalinity, and total inorganic carbon, and (9) wash-off. A schematic representation of the processes as they occur on an individual leaf is shown in Fig. 24.7.

Simulated annual budgets for atmospheric sulfur inputs and throughfall for a coniferous forest stand, a deciduous forest stand, and an open area are given in Table 24.1. The simulation is for the ILWAS site.

The annual sulfur transfer from the atmosphere to the coniferous system is almost twice that of the transfer to the deciduous system (Table 24.1), because the two canopies have different collection capabilities for particulates and gases. The coniferous canopy has greater surface area, and retains its foliage throughout the entire year. Maximum leaf-area index of the coniferous canopy is estimated to be 10, but that of the deciduous canopy is 5 (Chen *et al.*, 1983).

Although annual throughfall flux under the deciduous canopy is richer in sulfate and nitrate than precipitation, annual throughfall flux of alkalinity is greater than that of precipitation (Chen *et al.*, 1983) because of exudation of basic substances from the foliage. Annual throughfall flux of acidity (negative alkalinity) observed from May 1, 1980, through April 30, 1981, under deciduous canopy was approximately two-thirds of that observed in precipitation (540 eq ha^{-1} versus 830 eq ha^{-1}). Within the coniferous canopy, exudation is not sufficient to counterbalance the addition of acidity from dry deposition and to produce a

TABLE 24.1
Simulated Annual Sulfur Input Budget
(kg S / ha)

Flux	Coniferous stand	Deciduous stand	Open area
Precipitation	11.0	11.0	11.0
Dry deposition			
Particulate	12.0	3.5	1.6
Total gaseous[a]	8.3	2.1	0.5
Total deposition	31.3	16.6	13.1
Canopy SO₂ absorption	−4.5	−1.0	
Canopy exudation	0.02	0.02	
Simulated throughfall	26.8	15.6	13.1

[a] Sum of absorbed and adsorbed fluxes.

net neutralization of throughfall relative to precipitation. The annual acidity throughfall flux observed is 1,420 eq ha^{-1}. It is essential to note that ratios between different annual fluxes (e.g., throughfall S versus precipitation S) are functions of ecosystem properties, precipitation quality and quantity, and atmospheric quality; hence identical systems experiencing different air-quality situations can exhibit very different ratios between fluxes. The ILWAS model provides a general framework for calculating fluxes in terms of the fundamental factors that determine them.

Because the ILWAS site is remote from any large point source of atmospheric sulfur pollution, would be considered to have good air quality (in terms of gaseous and particulate sulfur), and has high precipitation with high sulfate concentrations, one might expect dry deposition to account for a very minute fraction of atmospheric sulfur input. However, the simulated annual input by dry deposition is approximately two-thirds of the total atmospheric input for coniferous stands and one-third for deciduous stands. Gaseous deposition accounts for 27% and 13% of total atmospheric input for coniferous and deciduous stands, respectively. Gaseous absorption represents approximately 14% and 6%.

How does the amount of SO$_2$ absorbed by the forest canopy relate to the amount of sulfur used by the vegetation? Whittaker *et al.* (1979) estimated the annual accumulation of sulfur by trees in net woody growth to be 1.2 kg ha^{-1} for a watershed in Hubbard Brook Experimental Forest in New Hampshire. The species composition, biomass, and net production of the three communities of the Hubbard Brook Forest and the ILWAS site are similar. Aboveground tree biomass at Hubbard Brook ranges from 102,000 to 162,000 kg ha^{-1}, and annual aboveground net production from 7,700 to 11,200 kg ha^{-1} (Whittaker *et al.*, 1979). The ranges in biomass and productivity correspond to changes in elevation. Both biomass and productivity decrease as elevation increases. Cronan and DesMeules (in preparation) estimate aboveground tree biomass on the three ILWAS watersheds to range from 143,000 kg ha^{-1} on Woods Lake watershed to 199,000 kg ha^{-1} on Sagamore Lake watershed, and aboveground net productivity from 6,400 kg ha^{-1} on Panther Lake watershed to 8,800 kg ha^{-1} on Sagamore Lake watershed. Differences between the three ILWAS watersheds are related to differences in species and size distributions but not to elevation. Given the similarity between the Hubbard Brook and ILWAS forests, it is reasonable to assume that their net woody growths are similar.

Comparing annual sulfur deposition fluxes (Table 24.1) to the sulfur accumulated in net annual woody growth ($\sim$1.2 kg ha^{-1}), we observe that the simulated annual input of S absorbed as SO$_2$ by the canopy is comparable to the amount needed to satisfy net woody production. Total simulated SO$_2$ deposition exceeds net growth demand for S, and total sulfur deposition exceeds it by an order of magnitude.

Let us consider the amount of sulfur required to satisfy annual net tree primary production. Net primary production includes, in addition to net growth, the production of living organic material (leaves, roots, branches, fruits, flowers, budscales, stems) that is removed from the vegetation during the year by processes such as litterfall, insect consumption, and root death. Whittaker *et al.* (1979) calculate net primary production for the tree stratum of Hubbard Brook to use 8.6 kg S ha^{-1}. Looking at the simulations of S deposition (Table 24.1), we see that SO_2 absorption by the foliage could account for from 10 to 50% of this requirement. Total S deposition would be sufficient to meet the entire requirement; hence if the organic and mineral layers of soil were to be depleted of stored sulfur, all the sulfur needed for net primary production could potentially be supplied by atmospheric deposition.

If we wished, we could continue at this point to use the results of ILWAS to examine sulfur movement in the soil profile and sulfur export in lake outflow in order to trace redistribution of sulfur in the system and find out whether the system on the whole is losing or accumulating sulfur. But let us instead return to one of the major questions we raised at the beginning of this chapter. How much insight is ecosystem analysis providing about the effects of SO_2 on plant productivity? To what degree is SO_2 absorbed by the canopy used to satisfy the vegetation's requirements for sulfur? Is S taken up as SO_2 through the foliage used more efficiently than S taken up as sulfate through the roots? Sulfur in the environment seems to be abundant relative to the vegetation's requirements. How does the vegetation regulate its overall nutrient status? Can the nutrient status of the vegetation control its S uptake through either roots or foliage? How do plants handle excess sulfur? How does nutrient status of the vegetation affect its vulnerability to injury by SO_2 uptake?

Using the ILWAS database and model, we can analyze the dynamics of the ecosystem cycling of all major cations and anions, but the critical questions about plant response to SO_2 appear not to involve the ecosystem cycling of nutrients as much as the plant's internal cycling of nutrients. We need to understand how the uptake of nutrients through the root system is controlled, and how the plant uses nutrients taken in through both roots and foliage to satisfy its metabolic and growth requirements. The focal points of the conceptual framework needed to carry out this analysis are the vegetation and its interfaces with the atmosphere and soil. We need to analyze the mechanisms that control nutrient movement across the boundaries of the vegetation and the internal cycling of the nutrients. Soil-plant-atmospheric models appear more useful at this time than ecosystem element-cycling models for such analyses.

Comparison of sulfur budgets. Sulfur input budgets for several temperate forest sites are compared in Table 24.2. In addition to the ILWAS site and Hubbard Brook, there are the Lake Gårdsjön site in southwestern Sweden (Hultberg, Grennfelt, and Olsson, 1982), the Solling site in West Germany (Meiwes and Khanna, 1981) and Walker Branch Watershed in eastern Tennessee (Johnson *et al.*, 1982; Shriner and Henderson, 1978).

TABLE 24.2
Comparison of Annual Sulfur Input Budgets for Several Temperate Forest Sites

Flux[a]	ILWAS coniferous	ILWAS deciduous	(1) Gårdsjön	(2) Hubbard Brook
Wetfall	11.0	11.0	11.4	12.0
Particulate	12.0	3.5	5.0–11.3	0.3
Gaseous	8.3	2.1	8.4	6.5
Net deposition	26.8	15.6	24.8–31.1	18.8
Bulk precipitation	15.9	15.9		12.7
Throughfall	27.0	19.0	27.1	21.0
Rainfall[b]	122.0	122.0	130.0	132.0

Flux[a]	(3) Walker Branch I	(4) Walker Branch II	(5) Solling coniferous	(5) Solling deciduous
Wetfall	14.0	14.9		
Particulate	3.0–6.0	3.2[c]		
Gaseous	8.0			
Net deposition	25.0–28.0	18.1		
Bulk precipitation	26.0	18.1[d]	23.0	23.0
Throughfall	28.0	44.2	87.7	52.3
Rainfall[b]	146.0	146.0	98.5	98.5

SOURCES: 1, Hultberg, Grennfelt, and Olsson (1982). 2, Eaton, Likens, and Bormann (1980) and Likens *et al.* (1977). 3, Johnson *et al.* (1982). 4, Shriner and Henderson (1978). 5, Ulrich (1980).
[a] Kg of S ha^{-1}.
[b] Cm of H_2O. Characteristic values do not necessarily correspond to those of the time period for which the budget was calculated.
[c] Collection of dry bucket of wet/dry collector; includes gaseous and particulate deposition.
[d] Measured by summing buckets of wet/dry collector.

The ILWAS budgets are for a deciduous and a coniferous stand rather than a given watershed. The Gårdsjön budget is for a 2.6-ha subcatchment that is predominantly old mesic mixed coniferous dominated by Norway Spruce (*Picea abies*). Hubbard Brook is predominately deciduous, mixed northern hardwood (*Fagus grandifolia*, *Betula* spp., *Acer* spp.). The Solling budgets are for a coniferous and a deciduous stand (*Picea abies* and *Fagus sylvatica*). The Walker Branch I budget is for a deciduous chestnut-oak stand (*Quercus prinus*). Walker Branch II is for the entire watershed and thus averages deciduous (predominately *Quercus* spp., *Carya* spp., *Liriodendron tulipifera*) and coniferous stands (predominately *Pinus echinata*).

Annual precipitation (Table 24.2) is often a characteristic or multiyear average value, and is not necessarily identical to the precipitation during the time for which the budget was calculated. All the sites with the exception of Solling have low atmospheric gaseous and particulate sulfur concentrations that vary with time. Typical measured values for SO_2 are zero to 8 μg S m^{-3} (0.13 μmol m^{-3}) and for particulate sulfate zero to 10 μg S m^{-3} (0.16 μmol m^{-3}) (Altwicker and Johannes, 1983; Eaton, Likens, and Bormann, 1980; Hultberg, Grennfelt, and

Olsson, 1982; Lindberg *et al.*, 1979; Johnson *et al.*, 1982). We are not aware of atmospheric gaseous or particulate S measurements for Solling. Meiwes and Khanna (1981) describe the site as being 100 km away from a major oil-consuming industry. The high enrichment of S in throughfall relative to bulk precipitation indicates much higher atmospheric loadings of S than at the other sites.

In Table 24.2, wetfall, bulk precipitation, and throughfall were measured directly, and dryfall (with one exception) was calculated either by multiplying atmospheric concentrations by average deposition velocities or by differencing other fluxes. The one exception is Walker Branch II, where dryfall is measured using the dry bucket of a wet/dry collector (Henderson, personal communication).

Shriner and Henderson (1978) point out that, "Portrayals of landscape element cycles have not yet reached the level of sophistication whereby statistical error terms can be assigned to distribution and transfer functions." Whether or not realistic error or uncertainty terms ("statistical" or not) can be estimated, estimation is rarely attempted. Lack of error estimates makes comparisons of different budgets semiquantitative at best. When only a single number with no uncertainty is reported, there is a great danger of thinking that number to be exact. The reader is strongly cautioned against doing this. Sources of uncertainty include sample collection, chemical analysis, numerical analysis, biological variation, and landscape heterogeneity.

Accurate estimation is very difficult for dry deposition. There are major technical difficulties with all measurement techniques (Hicks, Wesely, and Durham, 1982). Mayer and Ulrich (1980) have suggested that for many elements and systems, throughfall flux, which is relatively easy to measure, may be a good estimator of total atmospheric input. For those budgets in Table 24.2 where dryfall was calculated "independently" of the other fluxes (i.e., ILWAS, Gårdsjön, and Walker Branch I), there appears to be reasonable agreement between net deposition (total deposition minus canopy absorption) and throughfall. Note that only the ILWAS budget included a separately calculated flux for absorption of SO_2 by the canopy; hence for all the other budgets net is the same as total (wet plus dry). The adequacy of using throughfall as an estimator of total deposition should be judged not solely in terms of the absolute error that may result from assuming canopy exudation, leaching, and uptake to be zero, but also in terms of how this error compares to those of other techniques used to measure dry deposition.

Bulk precipitation tends to underestimate total atmospheric input, because the canopy is much better at capturing gases and particulates than the bulk collector is. This is especially dramatic for the coniferous vegetation and Solling deciduous sites. In the past, many studies have estimated net ecosystem accumulation (or depletion) of a specific element by assuming bulk precipitation to be a good estimate of total atmospheric input. Such an assumption can lead to a very large underestimation of accumulation or overestimation of depletion.

The two Walker Branch budgets are dramatically different. Johnson *et al.* (1982) give some possible explanations for the differences in throughfall, but do not address the differences in bulk collection. It is puzzling that the bulk collection of II is 30% less than that of I, whereas the throughfall of II is 58% more than the throughfall of I. Rather than speculate about the reasons for this, we wish to emphasize the value of attempting to rigorously estimate and report error terms associated with landscape element fluxes.

Just as with the ILWAS budgets, SO_2 deposition for the "clean air" sites appears comparable to, and total sulfur deposition an order of magnitude greater than, the amount of sulfur needed by the net growth of trees. From the perspective of atmospheric inputs, the trees at all sites appear to have a superabundance of sulfur available. These observations hold true even though estimates of net growth may be higher for some of the other sites than the 1.2 kg of S ha^{-1} estimated for Hubbard Brook. Shriner and Henderson (1978) estimated 2.3 kg of S ha^{-1} for Walker Branch watershed. Johnson, Turner, and Kelly (1982) compared net aboveground growth for temperate forest sites of diverse locations, species, and ages. Among the sites compared were Hubbard Brook, Walker Branch, and Solling. Walker Branch placed at the upper end of the distribution of net annual aboveground growth among the compared sites. Given that the climate at Gårdsjön is more severe than that at Walker Branch, it is reasonable to assume that net growth at Walker Branch is also greater. Of course, the critical questions—which ecosystem analysis does not lend itself to—are: To what degree is SO_2 absorbed by the foliage used by the vegetation to satisfy growth needs? To what degree, if any, may the SO_2 produce detrimental effects? How is the internal sulfur balance of the vegetation regulated in the presence of the apparent superabundance of available sulfur?

Summary and Directions for Future Research

In summary we conclude that:

1. Sulfur budgets for several temperate forest sites are consistent with the suggestion of Mayer and Ulrich (1980) that, under appropriate conditions, throughfall flux of sulfur is a good estimator of total atmospheric input (excluding gaseous absorption) of sulfur to the ecosystem.

2. Dry deposition as measured by bulk collection in the open can grossly underestimate total deposition to forested watersheds. In mass balances, where bulk deposition is used to estimate inputs of elements to a basin, basin losses (net outputs) may be overestimated.

3. In some forested areas where low SO_2 concentrations exist, it appears that gaseous uptake could provide adequate sulfur to meet net annual growth needs.

4. Ecosystem analysis of element cycles helps quantify the environment in which SO_2-plant interactions occur. Soil-plant-atmosphere models are likely to be more useful for evaluating direct plant effects.

Ecosystem analysis is an appropriate framework in which to study the effects of perturbations of ecosystem element fluxes. However, the absorption of SO_2 by vegetation may not be so much a perturbation of the ecosystem cycling of sulfur as of the plant's internal cycling of sulfur compounds. To understand how SO_2 can affect the plant requires detailed study of the plant's material cycling, which includes internal cycling, and exchange of elements with the atmosphere and soil. Studies of this nature should be supported by conceptual and mathematical mechanistic process models of the soil-plant-atmosphere system, where the internal cycling of materials by the plant is described explicitly. This does not mean that the ecosystem should be ignored, but it should serve as a background to frame the detailed study of the soil-plant-atmosphere system. Studies of this nature require an integrated interdisciplinary approach involving micrometeorologists, plant and soil scientists, and mathematical modelers.

In order to make rigorous quantitative intersite comparisons of element cycling or multiyear comparisons for a single site, greater effort needs to be placed on estimating error or uncertainty terms. There is no simple, straightforward way to estimate and treat uncertainties, and much research is needed on developing appropriate methods in this area that Goldstein and Ricci (1981) have called "Ecological Risk Uncertainty Analysis."

For ecosystem analysis, it is important in the field to study simultaneously the different compartments of the system as was done in ILWAS. When different compartments are studied at different times, differing environmental conditions make integration exceedingly difficult. Furthermore, it is important to conduct an ecosystem analysis with a mechanistic ecosystem process perspective. Without such a foundation, it is not possible to predict how system behavior will be changed by differences in atmospheric input or by system perturbations, or to understand interannual and intersite differences in system behavior.

25

SO$_2$ Effects on Agricultural Systems: A Regional Outlook

Walter W. Heck and Allen S. Heagle

This chapter addresses aspects of the agricultural system that should be understood and included in an assessment of SO$_2$ effects on plant productivity. Other chapters have covered additional background information that is essential to the regional assessment of SO$_2$ effects on agricultural systems. Although we have highlighted some of the informational needs, we have not included, or have only given examples of, specific information needed for regional assessments.

Historically, SO$_2$ problems were associated with releases from point sources that affected vegetation growing near them (Thomas, 1951). Such problems still occur from small point sources, old smelters, and power plants with low stacks. However, with the increased use of tall stacks for power plants and smelters, SO$_2$ and other emissions are being transported regionally and even globally (Environmental Protection Agency, 1982a). Thus, it is critical that SO$_2$ be considered a regional problem and that effects be assessed regionally. The contribution of SO$_2$ toward the formation of acidic precipitation further complicates the regional assessment of SO$_2$ effects, but acidic precipitation is not addressed here (Linthurst and Altshuller, 1983).

With the regional dispersal of SO$_2$, there is greater potential for SO$_2$ to occur with other pollutants (e.g., ozone and nitrogen dioxide). There is evidence of interactive effects of these mixtures on plant growth and productivity. Although adequate information is not available to assess these interactive effects on a regional basis, they must eventually be included in a regional assessment of SO$_2$ effects on agricultural production.

This chapter will emphasize three areas of critical importance in developing regional assessments of the impact of SO$_2$ on agricultural production and will discuss research needs. Specifically, we will deal with: (1) concepts of the air pollution system, with emphasis on monitoring and air-quality data analysis needed for assessment activities, (2) examples of research protocols with selected field data necessary for regional effects assessment, (3) an overview of how to do regional assessments and expected shortcomings, and (4) future research directions.

The Air Pollution System

An understanding of the air pollution system (Heck, 1982 and 1983) permits one to visualize pollutant sources, emissions, transport, transformations, and deposition interacting to determine the pollutant dose that can affect plants (Fig. 25.1). Major factors of relevance to the biologist will be discussed here, with an emphasis on monitoring needs.

Sources and emissions. An understanding of sources and emissions is important in undertaking a regional assessment of SO$_2$ effects, especially when a predictive component is part of the assessment. Such an understanding is also important for a reasonable interpretation of the regional assessment.

Although we are primarily interested in anthropogenic emissions, we must understand how natural emissions relate to anthropogenic emissions. Volcanoes and fumerols, and oxidation of H$_2$S and other reduced sulfur compounds, are examples of natural SO$_2$ sources (U.S. Environmental Protection Agency, 1982a). Although some background SO$_2$ is present, and might make a sizeable contribution to the global sulfur budget, it has a relatively short half-life in the atmosphere and may be below the detection limit of SO$_2$ analyzers. Thus, for purposes of discussion, we will ignore possible background concentrations of SO$_2$.

Anthropogenic sources of SO$_2$ are primarily stationary point sources where S-containing fuels are burned (i.e., power plants) or where sulfide ores are smelted; SO$_2$ is directly produced in the combustion/smelting process. Maps showing the distribution of power plants and county-level emissions of SO$_2$ for the U.S.A. are found in McLaughlin and Taylor (this volume). These maps show that problems associated with SO$_2$ are not uniformly spaced over the country. Sulfur-dioxide emissions in the U.S.A. (U.S. Environmental Protection Agency, 1978b and 1980a) were about 19.5 $\times$ 10^6 metric tons yr^{-1} in 1940, and came from many small emissions from many sources. Although total emissions were about the same in 1960 (21.4 $\times$ 10^6 metric tons), there were fewer sources, with larger emissions per source. By 1970, total emissions of SO$_2$ were 29.8 $\times$ 10^6 metric tons, mostly from large power plants (many with tall stacks). Emissions decreased slightly during the 1970's (27.0 $\times$ 10^6 metric tons in 1978), because of the installation of SO$_2$ scrubbers and the use of relatively low-S gas and oil as alternative fuels. Many old and most new sources now use tall stacks for dispersion over greater distances. The use of scrubbers and low-S fuels has led to predictions of lower emission rates in the 1980's (Linthurst and Altshuller, 1983), followed by a slight increase, from new power plants, to the year 2000 (26.6 $\times$ 10^6 metric tons).

The projections assume a set rate of replacement of old power plants and the continued use of low-S fuels (i.e., gas, oil). In actuality, some power plants have switched back to coal, some are using a mixture of high-S and low-S coals to meet emission rates, and many older plants are not being phased out as the

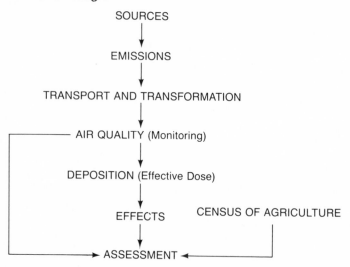

Fig. 25.1. The air pollution system (selected components) from source through effects assessment. Adapted from Heck (1983).

projection assumed. These factors and the continued interest in coal as our major fuel source suggest that SO_2 emission rates predicted for the year 2000 are probably too low.

Transport and transformation. Transport and transformation processes should be understood in regional assessments, if monitoring data is to be interpreted in relation to emissions and sources. Although SO_2 is not as ubiquitous as ozone or the nitrogen oxides, because it comes from stationary rather than mobile sources, it is a regional problem.

Meteorology and atmospheric chemistry have elucidated some of the relationships between long-distance transport of pollutants in air masses and the SO_2 concentrations (Linthurst and Altshuller, 1983). The increase in tall stacks and large coal-burning power plants suggests that SO_2 is more prevalent in air masses that cover large areas, especially in the northeastern U.S., than in the past, when low stacks were prevalent. There are two primary reasons for understanding the rate and form of atmospheric SO_2 transformations: (1) to find out how far SO_2 can be transported, and (2) to find out how much SO_2 is transformed during transport into sulfite and sulfate, which are components of acid precipitation. Transformation rates are higher in the eastern U.S. and in summer; maximum transformation rates are estimated at $25 \pm 10\%$ for SO_2 in power-plant plumes during a 24-hour summer day (Linthurst and Altshuller, 1983).

We know less about the occurrence of SO_2 in the western half of the U.S., except in the vicinity of multiple SO_2 sources. Many of the western power plants are fairly well isolated, burn low-S coal, and probably do not pose a major problem. However, the dry and less polluted atmosphere in the West could be conducive to transport of SO$_2$ over longer distances because of low transformation rates.

Monitoring. Knowledge of SO$_2$ concentrations is fundamental to assessing the effects of SO$_2$ on agricultural systems. An accurate measure of atmosphere concentrations of SO$_2$, usually measured at ground level (from 1 to 5 m), is essential to establish the relationship between the atmospheric concentration of SO$_2$ and the flux of SO$_2$ to the plant. The atmospheric SO$_2$ concentration is often called the exposure concentration, and it, with the exposure duration, defines the exposure dose. The exposure dose can be considered the air-quality unit that links the atmospheric scientist, the plant scientist, and the air-pollution-control official. It is a measure of air quality that lends itself to the concept of a standard, can be used to measure SO$_2$ flux (uptake) to plants, and is an essential component of any regional assessment effort.

Since knowing the exposure dose is pivotal to understanding the air pollution system and to assessing effects, it is surprising that monitoring of ambient SO$_2$ concentrations is still inadequate. Monitoring for SO$_2$ is done mostly around urban areas (to protect human health) or close to point sources to help relate SO$_2$ concentration to effects on vegetation close to point sources. Very little monitoring is done in rural areas away from point sources; so the agricultural air-quality database needed to assess SO$_2$ effects on agricultural systems is inadequate. In addition, methods of handling and interpreting monitoring data for assessing the effects of SO$_2$ on agricultural production are just being developed. The available SO$_2$ database is primarily made up of hourly averages that can be averaged and interpreted in many ways. Currently, the general averaging techniques use 24-hour, monthly, and yearly means; the vegetation standard is a 3-hour average. These averaging techniques are easy to understand and use, but have limited applicability for long-term agricultural assessment. Averaging techniques to relate plant effects to SO$_2$ doses (concentration and exposure duration) are being developed and tested by biologists (Oshima, this volume).

A regional assessment of SO$_2$ effects on agricultural production requires some form of seasonal SO$_2$ averaging. Short-term averaging times are inadequate unless the correlation with seasonal values is known. To interpret potential SO$_2$ problems on vegetation, we must understand the concentration-time-frequency relationships for different situations (e.g., urban-industrial-multiple source, point source with low and high stacks, and topographical differences in location of the source). Some plant scientists have suggested use of a concentration-time summation-dose value (e.g., ppm-hours), as a substitute for an averaging technique. This method is arbitrary, does not consider concentration-time-frequency relationships, is impossible to interpret in biological terms, and is impractical for the

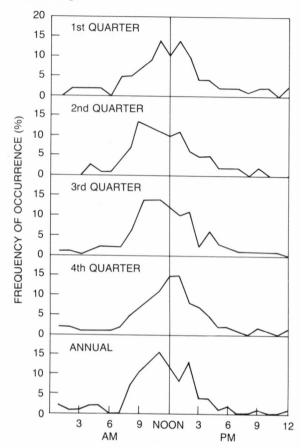

Fig. 25.2. Distribution of the typical hour of maximum SO_2 among 153 monitoring locations, 1975–78. Redrawn from Frank, Blagun, and Slater (1981).

control official. However, for seasonal effects in locations with similar sources and environments, good correlations of effects and summation-dose values have been shown (McLaughlin and Taylor, this volume).

It may be impossible to develop a single biologically meaningful averaging technique that will apply to all source situations. However, innovative efforts could lead to averaging techniques, acceptable to biologists, atmospheric scientists, and control officials, that could be used in assessment efforts.

A technique that was used to develop an exposure design (Heagle *et al.*, 1983) made use of quarterly averaging data (Frank, Blagun, and Slater, 1981) from 153 highly affected monitoring sites in urban areas during 1975–78. The data were summarized to show the diurnal distribution of the maximum SO_2 hourly averages (Fig. 25.2). These curves suggested that the maximum values occurred between 0800 and 1400 hours across all quarters. When the average annual and quarterly diurnal patterns were developed for individual sites, the maximum values usually occurred during these same hours across years and quarters (Fig. 25.3). Possibly in some locations averages across these midday time periods would provide a good correlation with plant response and be usable in assessment methods.

Plant scientists must have representative ambient exposure doses for various situations in order to better plan exposure regimes for dose-response experiments that can then be used in assessment activities. For example, if one assumes that the maximum values cause the greatest effect on plants, then, for dose-response studies, a set period of experimental exposures performed during 0800 to 1400 hours would be fairly representative of a real-world situation. If the experimental design was such that a uniform concentration was not prescribed, but was to depend on meteorological conditions (i.e., wind speed) and not be the same each day, then such an exposure regime may be a fair simulation of a "real world" situation for a regional source. This experimental approach is probably realistic for simulating regional problems, but would not be representative of point-source exposures, for which SO₂ concentrations would be much more variable. One attempt to simulate point-source SO₂ releases was done by Male (1982), who used so many different types of SO₂ exposure regimes that general conclusions about the results of these various exposure regimes on plants could not be drawn.

One cannot develop an exposure simulation for SO₂ that models all point sources, because each point source has a unique meteorology and topography; exposure dynamics would be different from day to day, month to month, and year to year, even for the same source. Study of the effects of the exposure dy-

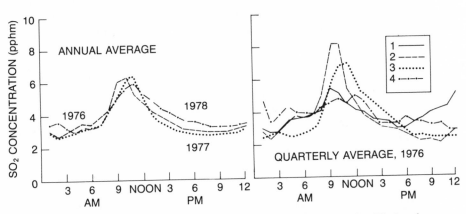

Fig. 25.3. Annual and quarterly diurnal patterns for the Glassport SO₂ site, Pittsburgh, Pennsylvania, 1976–78 (Saroad site-393640001G01). Redrawn from Frank, Blagun, and Slater (1981).

namics of a given source on crop production would have to be done in the near vicinity of the source, using the SO_2 from the source for several years. It would be difficult, but possible, to develop a reasonable dose-response function under these conditions. For simulation of experimental designs for sites representing the area near SO_2 point sources, it seems reasonable to use a "worst case" exposure regime and develop a dose-response design using some multiple of this "worst case." Data from such experimental designs is a necessary component of assessment technology.

There are known interactions between SO_2, O_3, and NO_2 for at least some biological systems; so data analysis efforts should be directed toward understanding the relationships of these gases in ambient air. Preliminary information from the Sulfate Regional Experiment (SURE) (Mueller *et al.*, 1980) showed that SO_2 tended to peak at midday (on O_3 days above 0.05 ppm), NO_2 in midmorning, and O_3 in midafternoon. These types of analyses will eventually be necessary to adequately assess regional effects of SO_2 on agricultural production.

Deposition and plant uptake. Deposition is the mechanism that removes pollutants from the atmosphere. This process consists of both wet and dry deposition to surfaces. Wet deposition of SO_2 is less important to plant systems than dry deposition. Dry deposition is by gaseous sorption processes of the leaf. Sorption processes include SO_2 adsorption on plant surfaces, and absorption of SO_2 through stomata or cuticle. Rates of absorption through stomates are related to flow resistances associated with wind speed and internal plant factors.

An understanding of deposition and deposition rates is essential to an understanding of SO_2 effects on plants. Sulfur-dioxide uptake by the plant involves the physical-biological link of the plant leaf. It is at this point that the plant scientist attempts to quantify the amount of pollutant received by the plant. Heck (1982) has discussed several terms that can be used to bridge the gap between the atmospheric and plant scientists. These terms are exposure dose (the monitored concentration of SO_2 at the plant during some finite time frame), uptake dose (the amount of SO_2 absorbed by the plant), and effective dose (the amount of absorbed SO_2 greater than the amount that can be tolerated by the plant). The effective dose, then, is that portion of the uptake dose that induces leaf pathology and/or altered physiology (the stress-causing dose). For regional assessment of the effects of SO_2 on agricultural production, the exposure dose is the most important concept.

Effects of SO_2 on Crop Yield

The primary goal of studying the effects of SO_2 on crops for regional assessment is to discover possible effects on crop yield or productivity. Much research done over the years has permitted some indirect assessment of effects, but the methodology has involved greenhouse research, closed field chambers, or indirect assessment based on percent of leaf injury. During the 1970's, investigators developed several experimental facilities to study the effects of SO$_2$ on crops when the crops were grown in the field and under exposure conditions as realistic as possible. Use of these facilities permits the collection of crop-yield data that is useful for regional assessment. Results from open-top chambers are discussed in this section.

Early field studies on productivity. Early work made use of relationships between leaf injury and yield without special regard for SO$_2$ concentration (Thomas, 1951). Many of these experiments were carried out in closed field chambers. Field surveys of foliar injury were used to estimate yield reductions. A second field technique was to measure yield differences with distance from a source, assuming a decrease in SO$_2$ with distance (Guderian and Stratmann, 1968). These studies were often done in natural systems (Linzon, 1971). In both of these studies, SO$_2$ was extensively monitored.

One of the best studies was the Biersdorf, Germany, study (Guderian and Stratmann, 1968) near a major SO$_2$ source. Plants were grown in large pots; so growth and nutritional variables could be controlled. Air-quality data was presented in several ways, but concentrations under 0.10 ppm (4.2 μmol m^{-3}) of SO$_2$ were considered below the response threshold. This assumption is probably acceptable for point-source assessments.

All of these studies were for point sources where concentrations were relatively high and usually caused acute injury. These led many to believe that acute injury was the only visible symptom of SO$_2$ effects on crops. During the 1960's and early 1970's, effects of chronic exposures (simulating SO$_2$ from multiple sources) showed that chlorosis and early leaf senescence could also occur. These studies led to some reappraisal of early research and a realization that innovative field-experimental techniques were needed.

Approaches to field experiments: perils and pitfalls. An acceptable field system to develop relationships between dose and plant response is essential for the regional assessment of the effects of SO$_2$ on agricultural production. Such a system must allow exposure to several prescribed SO$_2$ concentrations under the same environmental conditions, without greatly changing the field environment.

The first and most extensively studied innovation was the open-top chamber, which is 3 m in diam and about 2.5 m in height (Heagle, Body, and Heck, 1973). Open-top chambers, with an SO$_2$ dispensing and monitoring system, provide relatively uniform and controllable SO$_2$ concentrations throughout the chamber. Air can be filtered and enters the chamber at about 2,100 cfm. The primary concern expressed for the open-top chamber is the slight change in environmental conditions induced by the chamber. The wind pattern and slight reduction in light intensity are probably the major differences (Heagle *et al.*, 1979). Chamber effects on plant growth and yield are routinely measured by comparing chamber results with those plots in ambient air at the same pollutant concentration. Chamber effects found for a few species have been related to frost injury (fall and spring studies); some plants grew slightly taller because of light exposures, and some differences resulted from differences in insect population within and outside the chambers. Most species tested have not been significantly affected by the chambers.

A second technique that has considerable appeal is the zonal air pollution system (ZAPS), which releases SO_2 from holes in pipes into a field at plant canopy height. The original ZAPS (Lee and Lewis, 1978) used release pipes on all four sides of an experimental field. Variations have used a single line (or double parallel lines), where SO_2 is released only under certain wind conditions (Miller *et al.*, 1980). The single-line ZAPS is used primarily to simulate a point source without possible chamber influences on plant sensitivity. However, small changes in wind velocity cause large changes in SO_2 concentration, and thus make it difficult to control or describe the SO_2 dose. In addition, there is no way to provide an experimental control for ubiquitous pollutants such as O_3. Another ZAPS variation uses standpipes throughout a field, so that the same concentration regime can be maintained within a prescribed section of the field (Unsworth, personal communication). The last variant does not permit a dose/response design within a single field, whereas the single-line release does. As with all ZAPS variations, this system does not allow for a control when other pollutants are present in ambient air.

Another field technique uses plastic ducts, perforated at various intervals, to provide pollutant concentration gradients along sections of crop row (Shinn, Clegg, and Stuart, 1977). Linear gradient systems have been used in dose-response experiments for single pollutants or for pollutant mixtures (Reich, Amundson, and Lassoie, 1982). The control and description of pollutant doses in linear gradient systems is probably adequate when used in closed-canopy situations, but not when plants are small. An experimental control treatment for ambient pollutants has not been shown.

Plastic ducts between plant rows, aspirated with charcoal filtered air that escapes through holes in the ducts, have been used to exclude ambient pollutants from the area of plant canopies. The "exclusion" system is still being tested and its effectiveness has not been described (H. C. Jones, personal communication).

Some workers have suggested that dose-response relationships can be measured by comparing yields of important crops grown in the open in areas with different ambient pollutant levels. The basic approach is an epidemiological one. The protocols would include careful measurement of pollutant dose and all factors known to affect plant growth. Through multiple-regression techniques, the pollutant effects would be separated from effects of other factors. This method could work where SO_2 doses are variable within an area of uniform climate and soil (e.g., near a point source). However, the distances required to obtain differences in SO$_2$ doses for sources with tall stacks are too great to insure uniform climate and soil. Not enough is known about environmental effects on plant yield and on plant response to pollutants to make this "epidemiological" approach feasible for regional assessment.

Results from use of open-top chambers. A comparison of current field techniques encourages the use of open-top chambers to develop data necessary for regional assessment of crop loss from air pollutants, including SO$_2$. This technique has been used with success for the regional (national) assessment of crop loss by O$_3$ in the National Crop Loss Assessment Network (NCLAN) program (Heck *et al.*, 1982b). The following example demonstrates how open-top field chambers were used to measure the effects of seasonal SO$_2$ and O$_3$ exposures on the yield of soybean cv 'Davis' (Heagle *et al.*, 1983). Four concentrations of SO$_2$ and six of O$_3$ (24 possible combinations) were established for each of two replicates (24 chambers in each replication). The experiment required 48 open-top chambers in addition to dispensing and monitoring equipment (Heagle *et al.*, 1979). The four SO$_2$ treatments were set at approximately 0.00, 0.03, 0.12, and 0.48 ppm (1.25, 5.0, and 20 μmol m^{-3}) added for 4 hours per day (1000 to 1400 hrs) from July 4 through October 12. For each SO$_2$ level, the resulting concentration in the chamber was partly dependent on the wind velocity, which affected the rate of dilution by ingress of ambient air into the open top. Thus, although constant additions of SO$_2$ were made, the resulting concentrations were variable, much as could occur from day-to-day variations in a regional field situation. The four seasonal 4-hour-per-day mean SO$_2$ treatment concentrations were 0.000, 0.026, 0.085, and 0.367 ppm (1.08, 3.54, and 15.3 μmol m^{-3}). The six O$_3$ treatments were charcoal-filtered air and five nonfiltered air treatments, to which approximately 0.00, 0.02, 0.03, 0.05, and 0.07 ppm (0.83, 1.25, 2.08, and 2.91 μmol m^{-3}) of O$_3$ was added for 7 hours per day (1000 to 1700 hours) to the variable O$_3$ concentrations in nonfiltered ambient air. Ozone exposures began on June 24 and continued through October 12. The six seasonal 7-hour-per-day O$_3$ concentrations were 0.025, 0.055, 0.069, 0.086, 0.106, and 0.125 ppm (1.04, 2.23, 2.87, 3.58, 4.41, and 5.20 μmol m^{-3}). These protocols have been used throughout the NCLAN program, but generally without the added SO$_2$.

Regression analyses of yield results and pollutant concentrations showed highly significant relationships between the O$_3$ dose and yield, and the SO$_2$ dose and yield, and a significant linear interaction between O$_3$ and SO$_2$ doses and yield. The nature of the interaction was for higher concentrations of one pollutant to reduce the effect of the other. The O$_3$ dose-yield relationships for each of the four SO$_2$ doses are shown in Fig. 25.4, along with the quadratic regression equations. The two lowest SO$_2$ concentrations (0.00 and 0.026 ppm, or 1.08 μmol m^{-3}) and the two intermediate O$_3$ levels (0.055 and 0.069 ppm, or 2.23 and 2.87 μmol m^{-3}) represent the range of concentrations that occur in ambient

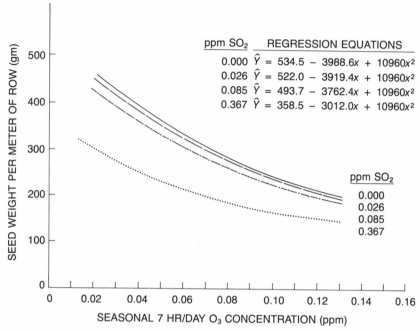

Fig. 25.4. Regression lines from significant regression equations for different O_3 doses at each of four SO_2 doses.

air of soybean production areas. These results show the type of information that is necessary before a regional assessment can be made. Any regional assessment for SO_2 effects on crop yield would be suspect if O_3 effects were not considered.

Regional Assessment of SO_2 Effects on Crops

An area (i.e., region) assessment is the final step in the study of the direct effects of SO_2 on crop yields, and requires as input the type of information contained in previous sections of this paper. It is an iterative process, being continually updated as new information is added that improves assessment capabilities. Assessment, in its current context, is concerned only with the plant parts that are of economic value (e.g., yield of fruit, seed, leaf, some part of biomass). In the air pollution system, this is the step before an economic appraisal and produces data that is essential to the economist. Although of importance, current interest in assessment does not include visual symptoms on ornamental plantings or changes in crop quality. Assessments should be considered on a crop-by-crop basis. If the results are to be used for a national economic study in the U.S., then the major field-production crops should be highlighted (i.e., corn, soybean, wheat, forage). If the southeastern U.S. were of primary interest, crops such as tobacco and peanut should be added. The crops for study must reflect the purpose of the assessment; at some level, all crops are of concern.

Regional assessment of SO$_2$ effects on crop yield requires, at a minimum, three primary inputs. These are (1) SO$_2$-dose/crop-yield functions on a relative-yield basis, (2) analyses of air-quality monitoring data in a form acceptable for the dose-response functions, and (3) agricultural-crop census data for the region. The accuracy of the assessment will depend not only on the accuracy of the three inputs, but also on the amount of detailed data available (i.e., monitoring and crop census data on a county level will give a better assessment than on a state level). The use of these three databases, when available, permits the assessment of relative yield losses, for the crops studied, during past seasons. This empirical approach to crop-loss assessment is the only reasonable way to measure regional effects of SO$_2$ on crops. The accuracy of these empirical assessments will improve with increased accuracy of the SO$_2$-dose/crop-yield functions, with an increase in number and reliability of the air-monitoring data, and with improved accuracy of the census data.

Currently, we do not have sufficient information to make reasonable regional assessments of SO$_2$ effects on crop yields for any crop or region within the U.S. Our primary lack is in SO$_2$-dose/crop-yield functions and in SO$_2$ air-quality data in agricultural areas. It is possible to make some estimates of yield effects around point sources where known fumigations do occur, provided they are severe. Currently, most point-source assessments are made for reimbursing farmers for their losses; visible injury is used to subjectively estimate yield effects.

It is important, when possible, to add some mechanistic concepts to the assessment effort. When a regional assessment is desired, the research on SO$_2$ effects should include studies with the major environmental factors that might affect the response of the crop to SO$_2$. These factors are both biotic and abiotic, and would include soil moisture, temperature, pathogens, and insects, at a minimum. The strictly empirical assessment (discussed earlier) assumes no interactions between SO$_2$ and these environmental factors. Thus, use of a relative yield response and not the actual yield assures a reasonable assessment of loss. If interactions do occur, then the more mechanistic approach is essential to assure a reasonable assessment, but relative yield would still be used for the model.

There is interest in many scientific quarters in developing a detailed mechanistic growth-yield model for use in assessing the regional and national effects of air pollutants on crop production. This approach would be much more sophisticated than that discussed here. The sophistication would come in the SO$_2$-dose/crop-response function, which could include a complete growth-yield model and thus incorporate the effects of all environmental factors on crop production. Although this type of modeling may eventually be possible, there is no evidence that it can be an effective tool in regional assessment efforts. It would use a direct-yield measurement and not a relative-yield value.

Currently, there is some discussion of the value of developing predictive capabilities for crop-yield losses. Some scientists talk about predicting when they mean assessing the effects. Basically, an assessment is after the fact, and predicting is before the fact occurs. The ability to predict effects depends on our ability to predict weather (including pollutant dose) during the growing season. Until this becomes a reality, we can find ways to make only regional assessments, not predictions.

Directions for Future Research

Future research directions for regional assessments of SO_2 effects on crop production follow readily from the previous discussions. The two primary needs include (1) field programs to derive SO_2-dose/crop-loss functions for important crops at several locations around each region, and (2) either developing additional rural monitoring sites or finding out if adequate monitoring data is already available in the private sector. The field studies could use the NCLAN protocols, and incorporate some of the more recent developments in the ZAPS research approach (Unsworth, personal communication). Whatever decision is reached, it should include a network of major research sites that cooperate in the assessment efforts. This field work must be closely coordinated with efforts to improve the air-quality database and to find the best ways to average the air-quality data.

The field research should consider the presence of other gaseous (or aerosol) pollutants (e.g., O_3 and NO_2) that may affect the response of crops to SO_2. This is a critical research area, since no areas in the eastern U.S. are free of O_3 or NO_2, and both affect the response of crops to SO_2. These designs should permit the separation of the SO_2 effects from the effects of the other gases, or incorporate the effects of the three gases into the response function (the latter is more reasonable).

The field research, or associated greenhouse/phytotron research, should incorporate studies of influence of both biotic and abiotic factors on the response of the crop to SO_2 (and to the gas mixtures). There is ample information that at least some biotic and abiotic factors do affect the response of crops to gaseous pollutants.

A commitment to such approaches should permit a first-cut regional assessment of the effects of SO_2 on crop productivity within 2 to 5 years.

26

SO$_2$ Effects on Plant Community Structure

T. T. Kozlowski

Under the stress of SO$_2$ pollution, both the stability and the productivity of plant ecosystems undergo changes that may range from drastic to negligible. The ultimate effects of SO$_2$ on community structure are the result of different responses of competing plants at the cellular, tissue, organ, and species levels (Kozlowski, 1980a). Reduced photosynthesis and visible leaf symptoms such as chlorosis, lesions, necrosis, and abscission are preludes to growth inhibition and to mortality of the more SO$_2$-sensitive plants, leading to alteration of community structure. Biochemical characteristics that are important to photosynthesis and are influenced variously by SO$_2$ include the activity of carbon fixing and chloroplastic enzymes, chlorophyll synthesis and breakdown, phosphorylation rate, pH buffering capacity, potential to oxidize and reduce SO$_2$, and levels of superoxide dismutase (Hällgren, 1978).

Community structure is also influenced by changes in reproductive capacity, because plant vigor and capacity for flowering and fruiting are highly correlated, and SO$_2$ reduces plant vigor (Kramer and Kozlowski, 1979). In the northeastern U.S., production of *Pinus ponderosa* cones by trees was decreased close to an SO$_2$-producing smelter (Scheffer and Hedgcock, 1955). In Ohio cone size and seed weight as well as germination of *P. resinosa* seeds were lowered by proximity to pollution sources (Houston and Dochinger, 1977). Effects of SO$_2$ on reproductive growth often reflect inhibition of pollen germination and elongation of pollen tubes, with the latter more sensitive (Varshney and Varshney, 1981).

Pollutants also influence plant community structure by their interactions with plant diseases and insects. Sulfur dioxide may either decrease or intensify disease responses. The pollutant may act directly on fungi or bacteria, thereby generally inhibiting parasites, or may alter host physiology, to render it either more or less sensitive to a given pathogen. If the host is weakened by SO$_2$, it usually becomes more sensitive to weak pathogens, but less sensitive to obligate parasites. If SO$_2$ physically injures the host, infection usually is facilitated (Treshow, 1980a). Whereas SO$_2$ increased infection by *Armillarea mellea* in *Pinus sylvestris* forests, it decreased infection by *Fomes annosus* (Grzywacz and Wazny, 1973). Interactions of air pollutants and plant disease are discussed in more detail by Treshow (1975).

Trees that are weakened by air pollution often are predisposed to injury or death by certain insects. For example, in the San Bernardino Mountains of California, *Pinus ponderosa* trees with advanced symptoms of oxidant injury were most frequently infected and killed by bark beetles (Cobb and Stark, 1970). Interrelations between air pollutants and insects are discussed further by Sierpinski (1977) and Dahlsten and Rowney (1980).

Effects of SO_2 on Forest Stands

Even without the threat of air pollution, mature forest ecosystems are not completely stable. For example, mature forests are maintained in an oscillating steady state, in which old and suppressed trees are continuously eliminated, and new ones are added. These changes reflect severe competition for light, water, and mineral nutrients in dense forest stands. Additional and severe periodic stresses, such as drought, flooding, fire, and insect or disease attacks, tend to accelerate elimination of the species that are most susceptible to such stresses. Catastrophic disturbances in climax forests cause reversion to pioneer stages of succession, whereas mild disturbances tend to maintain a mature forest in a relatively steady state.

In the course of normal succession in an unpolluted atmosphere, the number of species in a community usually increases. Productivity, biomass, community height, and structural complexity also increase, with the maximum for these often in the climax stage. Superimposed pollution stress usually sets in motion a retrogression characterized by reduction in biomass and productivity, in coverage and structural complexity, in species diversity, in environmental modification, and in nutrient control by the community. Such retrogression involves not a backward retracing of the stages of a normal succession, but rather a reduction in some of the properties that normally increase with time (Whittaker, 1975). Hence the response of a forest ecosystem in polluted air is an integrated response to the pollutant as one of many environmental stresses present. Sudden imposition of a severe air-pollution stress sometimes occurs so rapidly that feedback mechanisms cannot operate to select for resistant species or ecotypes within species. Nevertheless, even under severe plant competition and superimposition of air-pollution stress, some species may increase their growth, if they are given a competitive advantage by the relatively greater impact of the pollutant on other species in the ecosystem, and the growth of other species in the same ecosystem may be greatly reduced because of their lowered competitive potential (Kozlowski, 1980a). For example, following SO_2 pollution in the mixed conifer forest of the Sierra Nevada, basal area of *Pinus ponderosa* was reduced, but that of *Abies concolor* increased (Kercher, Axelrod, and Bingham, 1980).

Particularly relevant is the observation that associated species that are very tolerant to SO_2 may be very intolerant to other stress factors, such as shading, drought, flooding, or low temperature; or the reverse may be true. Whereas

Betula alleghaniensis, *B. papyrifera*, and *B. populifolia* are all sensitive to SO$_2$, their shade tolerance varies from intermediate (*B. alleghaniensis*) to intolerant (*B. papyrifera*) to very intolerant (*B. populifolia*). Among the maples *Acer saccharum* and *A. saccharinum* are tolerant to SO$_2$, but the former species is very tolerant to shading, the latter only intermediately tolerant. *Betula papyrifera*, *Fraxinus pennsylvanica*, and *Salix nigra* are sensitive to SO$_2$. *B. papyrifera* is also very sensitive to flooding (Tang and Kozlowski, 1982), but *F. pennsylvanica* and *S. nigra* are highly tolerant of flooding (Kozlowski, 1982).

SO$_2$ Dosage and Community Structure

The effect of SO$_2$ may range from total destruction of plant communities under high SO$_2$ dosages to no or imperceptible changes under low dosages. The specific changes will depend not only on the amount of SO$_2$, but also on the condition of the receptor ecosystem. Several investigators, including Linzon (1978), Guderian and Kueppers (1980), and Smith (1981), described changes in structure of forest ecosystems that may be expected under various SO$_2$ dosages. These will be discussed separately.

As is summarized in Table 26.1, high dosages break down community structure, with the change in species composition tending toward simplification of the system. Degradation of the ecosystem is characterized by rapid changes in structure, including composition, and is accompanied by secondary succession (Class III responses of Smith, 1974). Direct and chronic injury, especially to leaves, first affects the most sensitive species of the tree stratum, often leading to total destruction of the canopy. With an intermediate pollution dosage, various subtle direct and indirect effects lead to reduced vigor of individual species (Class II responses). Under low pollution dosages, adverse effects on the structure of forest ecosystems may not be obvious, and there may even be a slight beneficial effect of the added sulfur on plants (Class I responses). Some specific examples of each class of response will be given.

High SO$_2$ dosage. Under a very high SO$_2$ load, severe and dramatic degradation of an ecosystem occurs primarily near major point sources of pollution such as energy-production plants, smelters, cement plants, or pulp mills. The effects of the high pollutant dosage usually are limited to a zone a few km around the polluting source and several km downwind. The most sensitive species of the tree layer are first affected, and canopy trees often are killed. Tall shrubs, low shrubs, and herbs are then sequentially destroyed, until a barren zone results. Such general changes are similar to those caused by other severe environmental stresses, such as ionizing radiation or herbicides (Woodwell, 1970). Simplification of the ecosystem in response to high SO$_2$ dosage is accompanied by changes in ecosystem functions, such as a reduced rate of energy fixation, reduced biomass, and increased losses of mineral nutrients (Table 26.1).

TABLE 26.1
Influence of Air Pollution on Forest Ecosystems

Response of vegetation	Impact on ecosystem
Class III: High dosage	
1. Acute morbidity	1. Simplification; increased erodibility, nutrient attrition, altered microclimate and hydrology
2. Mortality	2. Reduced stability
Class II: Intermediate dosage	
1. Reduced growth	1. Reduced productivity, lessened biomass
(a) decreased nutrient availability	
(i) depressed litter decomposition	
(ii) acid-rain leaching	
2. Reduced reproduction	2. Altered species composition
(a) pollinator interference	
(b) abnormal pollen, flower, seed, or seedling development	
3. Increased morbidity	3. Increased insect outbreaks, microbial epidemics; reduced vigor
(a) predisposition to entomological or microbial stress	
(b) direct disease induction	
Class I: Low dosage	
1. Act as a sink for contaminants	1. Pollutants shifted from atmospheric to organic or available nutrient compartment
2. No or minimal physiological alteration	2. Undetectable influence, fertilizing effect

SOURCE: Smith (1974).

There are many specific examples of how high SO_2 concentration can destroy the structure of forest ecosystems, and only a few will be discussed here. For additional examples, see Miller and McBride (1975), who present a comprehensive review of effects of SO_2 on North American and European forest ecosystems at various distances from point sources of pollution.

A classic example of SO_2 effects on plant community structure is shown by studies around an iron-sintering plant near Wawa, Ontario, Canada. Dominant species in the forest around this point source of pollution included *Picea glauca*, *P. mariana*, *Abies balsamea*, *Pinus banksiana*, *Thuja occidentalis*, *Larix laricina*, and *Pinus strobus*. *Acer spicatum* and *Pyrus decora* occurred often as understory species (Gordon and Gorham, 1963).

Severe SO_2 injury to plants was primarily restricted to a narrow strip northeast from the point source, because southwest winds predominated (Fig. 26.1). Gordon and Gorham (1963) found that SO_2 induced successive deterioration of tree, shrub, and microflora layers of the plant community. They reported that the forest was "peeled off in layers" as the smelter was approached from the northeast. There was a striking decline in floristic variety close to the pollution source.

Whereas the number of macrophyte species averaged 43 at distances greater than 16 km from the source, the number declined to two to four species within 5 km from the plant. Also, the number per quadrat declined from 28 species beyond 16 km to 0.2 inside 5 km (Fig. 26.2).

The tree structure was intact at 37 km from the point source, discontinuous within 37 km, and absent within 27 km. The shrub layer dominated from 27 to 19 km, and there was no continuous plant cover within 8 km. *Pinus strobus* was the most sensitive tree species. Seedlings of this species were not observed within 48 km from the sintering plant, and seedlings of *Picea glauca, P. ma-*

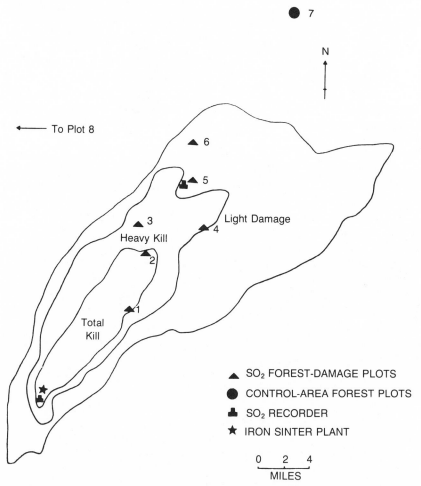

Fig. 26.1. SO₂ injury to a forest ecosystem at various distances from an iron-sintering plant near Wawa, Ontario. Reproduced from Linzon (1978); after Gordon and Gorham (1963).

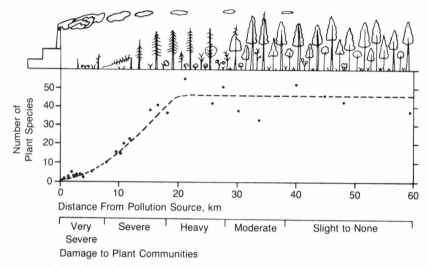

Fig. 26.2. Numbers of plant species at various distances from an SO_2-producing iron-sintering plant near Wawa, Ontario. Reproduced from Whittaker (1975); after Gordon and Gorham (1963).

riana, and *Populus tremuloides* were not recorded within 24 km. Examples of pollution damage at various distances from the polluting source are shown in Figs. 26.3 to 26.7.

A complex pattern of distribution of ground flora with distance from the Wawa, Ontario, iron-sintering plant reflected responses to SO_2, as well as an altered environment caused by changes in composition of overstory trees (Scale, 1980). Fig. 26.8 emphasizes that there were few ground-flora species near the pollution source, and transitory increases in abundance of several species at certain distances from the source. Such increases occurred despite their own greater exposure to SO_2 than at greater distances. For example, *Vaccinium angustifolium* was very abundant 20 to 30 km from the source, yet exhibited appreciable SO_2 injury in that area. The first substantial change in understory abundance occurred at 40 km, where *Corylus cornuta* became very abundant before showing a rapid decline. This apparently contributed to the decreased abundance of *Aster macrophyllus* at that distance.

Excessive production of SO_2 by a smelter in Ducktown, Tennessee, resulted in elimination of all trees and shrubs in a 27 m² area closest to the polluting source. Sheet and gully erosion were excessive (Hedgcock, 1914). Summer air temperatures were 1 to 2°C higher, and winter air temperatures 0.3 to 1°C lower, in the affected zone. Soil temperatures were as much as 11°C higher in the summer, and wind velocity up to 15 times higher (Hursh, 1948). Nickel and copper smelters in Sudbury, Ontario, Canada, released up to several thousand tons of

Figs. 26.3–26.7. Categories of SO_2 injury to a forest ecosystem at various distances from an iron-sintering plant near Wawa, Ontario. Reproduced from Gordon and Gorham (1963).

Fig. 26.3. At 0 to 6 km from the pollution source. Very severe injury. Most vegetation destroyed. Primarily bare rock in exposed situations.

Fig. 26.4. At 6.4 to 19 km from the pollution source. Severe injury. All overstory and most understory gone. Extensive mortality of overstory and shrubs. Ground vegetation dominating, predominantly *Polygonum cilinode*. Erosion evident and increasing toward inner boundary.

Fig. 26.5. At 19 to 26 km from the pollution source. Considerable injury. Overstory almost completely dead. Some *Betula papyrifera* and *Picea glauca* still alive but dying. Up to one-third of the trees remain in some places. Understory is vigorous as a result of release from overstory canopy, but mortality of shoot tips and suckering are extensive. Erosion is not evident.

Fig. 26.6. At 26 to 38 km from the polluting source. Moderate injury. Mortality of shoot tips and crown thinning of the overstory are extensive. Injury to the understory is minimal, and ground flora appears normal.

Fig. 26.7. More than 38 km from the polluting source. No obvious injury. The overstory, understory, shrub layer, and ground flora are normal. Major tree species are *Picea glauca*, *Abies balsamea*, and *Betula papyrifera*.

TABLE 26.2

Maximum Ranges of Annual Average SO_2 Concentrations Tolerable to Epiphytic
Bryophytes on the Noneutrophiated Bark of *Acer, Quercus,
Prunus, Ginkgo*, etc., in and around Tokyo, Japan

SO_2 conc., range, ppm (μmol m^{-3}); and tolerant epiphytic bryophytes	SO_2 conc., range, ppm (μmol m^{-3}); and tolerant epiphytic bryophytes
0.03–0.05 (1.25–2.08) *Hypnum yokohamae* var. *kusatsuense* *Glyphomitrium humillimum* *Entodon compressus* *Cololejeunea japonica* *Venturiella sinesis* 0.02–0.03 (0.83–1.25) *Sematophyllum subhumile* subsp. *japonicum* *Hypnum plumaeforme* var. *minus* *Lejeunea punctiformis* *Bryum argenteum* *Frullania muscicola*	0.01–0.02 (0.416–0.83) *Eurohypnum leptothallum* *Fabronia matsumurae* 0.005–0.01 (0.208–0.416) *Aulacopilum japonicum* *Bryum capillare* *Haplohymenium sieboldii* *Herpetineuron toccoae* *Trocholejeunea sandvicinsis* *Macromitrium japonicum* *Schwetschkea matsumurae* *Lophocolea minor* *Cheilolejeunea* sp.

SOURCE: LeBlanc and Rao (1975); modified from Taoda (1972).

TABLE 26.3

Maximum Ranges of Annual Average SO_2 Concentrations Tolerable to Lichens
and Bryophytes on the Noneutrophiated Bark of *Populus balsamifera*
in the Sudbury Area, Canada

SO_2 conc., range, ppm (μmol m^{-3}); and tolerant epiphytic lichens and bryophytes	SO_2 conc., range, ppm (μmol m^{-3}); and tolerant epiphytic lichens and bryophytes
0.03–0.05 (1.25–2.08) *Bacidia chlorococca* *Lecanora saligna* *Lepraria aeruginosa* *Parmelia sulcata* 0.02–0.03 (0.83–1.25) *Candelariella vitellina* *Cetraria pinastri* *C. saepincola* *Cladonia coniocraea* *Evernia mesomorpha* *Hypogymnia physodes* *Lecanora expallens* *L. subfusca variolosa* *L. subintricata* *Parmelia sulvata* *P. exasperatula* *Parmeliopsis ambigua* 0.01–0.02 (0.416–0.83) *Cladonia cristatella* *Parmelia olivacea* *Pertusaria multipuncta* *Physcia adscendens* *P. grisea* *P. stellaris* *Rinodina hallei*	0.005–0.01 (0.208–0.416) *Caloplaca cerina* *Lecanora symmicta* *Parmelia septentrionalis* *Physcia aipolia* *Alectoria americana* *A. nidulifera* *Usnea* sp. >0.005 (>0.208); relatively "pure" air *Buellia stillingiana* *Candelaria concolor* *Cetraria ciliaris* *Cladonia fimbriata* *Lecidea nylanderi* *Parmelia reducta* *P. trabeculata* *Physcia farrea* *P. orbicularis* *Ramalina fastigiata* *Rinodina papillata* *Xanthoria fallax* *X. polycarpa*

SOURCE: LeBlanc and Rao, (1975); modified from LeBlanc, Rao and Comeau (1972).

De Sloover and LeBlanc (1968) used a vegetation map based on a mathematically derived "index of atmospheric purity" (I.A.P.) to express the long-term effect of air pollution on epiphytic lichens and bryophytes. High values of I.A.P. were associated with a rich epiphytic flora characteristic of unpolluted areas, and low I.A.P. values with poor epiphytic vegetation in polluted environments (Fig. 26.10). Examples of maps using the I.A.P. techniques are those prepared for the Montreal and Sudbury areas in Canada (LeBlanc and Rao, 1975). As emphasized by Hawksworth (1973), the method is more laborious than other species-distribution methods or more directly derived zone maps. Also, the numerical values of the I.A.P. method often do not provide a better index of air pollution than do simpler techniques.

Another way to measure the effect of SO_2 on lower plants involves transplanting lichens and bryophytes, together with their substrates, from unpolluted to polluted sites. For example, corticolous lichens transplanted onto trees near New York City died within 3 to 5 months, whereas those transplanted into SO_2-free areas continued to grow (Brodo, 1961, 1966). SO_2 released in the Sudbury, Ontario, area caused abnormal responses in transplanted lichens and bryophytes,

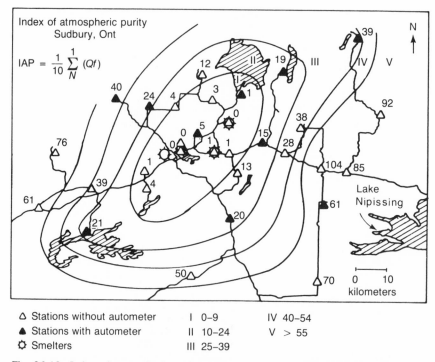

Fig. 26.10. Index of atmospheric purity (IAP) zones in the Sudbury, Ontario, area. Reproduced from LeBlanc and Rao (1975); after LeBlanc, Rao, and Comeau (1972).

including reduction in thallus thickness, as well as plasmolysis and chlorophyll breakdown in algal cells (LeBlanc and Rao, 1966). When the lichen *Hypogymnia physodes* was transplanted into areas of the Ruhr Valley that were polluted with SO_2, it died within a few weeks (Schönbeck, 1969). Corticolous lichens died within 2 years after they were transferred to the industrial area of Port Talbot, England (Pyatt, 1970), and foliose lichens exposed to the SO_2-polluted air of Dovedale, England, died after 19 months (Hawksworth, 1971). Several mosses, including *Comptothecium sericium*, *Grimmia pulvinata*, *Hypnum cupressiforme*, and *Tortula muralis*, deteriorated within a few months after they were transplanted to SO_2-polluted sites in Newcastle-upon-Tyne, England (Gilbert, 1968).

There is considerable variation in the tolerance of lichens and bryophytes to low SO_2 pollution. Lichens that appear to be rather tolerant of SO_2 pollution include *Stereocaulon pileatum*, *Lecanora conizaeoides*, *Buellia punctata*, *Candelariella aurella*, *Lecanora dispersa*, *Bacidia chlorococca*, *Endocarpon pusillum*, and *Micraria trisepta*. Bryophytes tolerant to SO_2 include *Bryum argenteum*, *Lunularia cruciata*, and *Tortula muralis* (LeBlanc and Rao, 1975). Species that occur only in areas with SO_2 in the range of 0.005 to 0.01 ppm (0.203 to 0.406 μmol m^{-3}) or less have essentially no SO_2 tolerance and can be used as indicators of SO_2.

Factors Influencing the Effect of SO_2 on Plant Community Structure

In addition to pollutant damage, the specific responses of ecosystems to SO_2 are modified by several factors. When one is dealing with a heavy SO_2 load from a point source, the SO_2 effect on community structure predominates near the source and often results in near-complete elimination of vegetation. As distance from the pollution source increases and SO_2 load decreases, such factors as life form, species and genetic materials, combinations of pollutants, age of plants, and various site factors and stresses become increasingly important in the nexus controlling plant succession and ecosystem stability.

Some idea of the complexity of plant responses to SO_2 at various distances from a point source may be gained from the work of Winner and Bewley (1978), who studied sensitivity to SO_2 of vascular plants and mosses in the understory of a *Picea glauca* association surrounding a natural-gas refining plant in Alberta, Canada. Mosses were more sensitive than vascular plants in the heavily stressed areas. However, the ecological importance (coverage) of many understory species changed in relation to SO_2 stress. Response to the SO_2 gradient varied in one of several patterns, depending on species (Fig. 26.11). Changes in coverage for *Cornus canadensis*, *Linnaea borealis*, *Mitella nuda*, and *Pyrola asarifolia* declined at a constant rate (pattern A). Coverage of *Gramineae* and *Equisetum arvense* increased with an increase in SO_2 stress (pattern B). Species such as *Adenocaulon bicolor*, *Osmorhiza occidentalis*, *Vaccinium vitis-idaea*, and *Loni-*

Fig. 26.11. Model illustrating species variations in coverage along a gradient of SO_2 stress: A, SO_2-sensitive species; B, species that invade disturbed stands or have a high sulfur requirement; C and D, species that show increases and subsequent decline in coverage with increasing SO_2 stress; e.g., weeds (C) or endogenous community members (D). Reproduced from Winner and Bewley (1978).

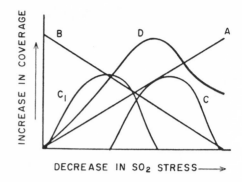

DECREASE IN SO_2 STRESS———➤

cera villosus had high coverage at localized regions of a stress gradient (patterns C and D). The moss species decreased in coverage with increasing SO_2 stress (pattern A) with the exception of *Hylocomium splendens* and *Ptilium crista-castrensis*. Both of these species reflected the D pattern.

Species and genetic materials. Species and genetic materials can differ greatly in susceptibility to SO_2. Gymnosperms are generally more sensitive than angiosperms, but both groups can exhibit large differences in SO_2 tolerance (Table 26.4). Species within the same genus may differ greatly in SO_2 tolerance. For example, *Atriplex triangularis* was more sensitive than *A. sabulosa* to SO_2 (Winner and Mooney, 1980c).

A complicating factor in measuring the effect of SO_2 on different plant species in an ecosystem is that there may be large intraspecific differences in response to SO_2; hence some differences in SO_2 tolerance between seed sources, ecotypes, clones, and cultivars may be expected.

Intraspecific differences in SO_2 tolerance have been reported for *Pinus strobus* (Houston and Stairs, 1973) and *P. ponderosa* (Karpen, 1970). In addition, marked differences in susceptibility to SO_2 of *Populus tremuloides* clones have been demonstrated (Kimmerer and Kozlowski, 1981).

Differences in susceptibility to SO_2 usually result from avoidance of uptake of SO_2 or from biochemical tolerance (resistance to the toxic effect). Additionally, SO_2 tolerance may reflect incorporation of SO_2 into less toxic substances or dilution of SO_2 by its rapid redistribution in the plant (Kozlowski, 1980b).

Differences in avoidance of uptake of SO_2 are most commonly associated with differences in stomatal conductance. For example, leaves of *Fraxinus americana*, with large stomata, absorbed more SO_2 than leaves of *Acer saccharum*, with small stomata (Jensen and Kozlowski, 1975).

The fact that *Atriplex triangularis* is more sensitive to SO_2 than *A. sabulosa* was attributed to the higher stomatal conductance of the former (Winner and Mooney, 1980c). Differences in stomatal conductance were important in determining susceptibility to SO_2 of *Populus tremuloides* clones. Pollution stress avoidance appeared to involve two types of responses: passive avoidance, the

TABLE 26.4
Relative Susceptibility of Trees to SO_2

Sensitive	Intermediate	Tolerant
Acer negundo var. *interius*	*Abies balsamea*	*Abies amabilis*
Amelanchier alnifolia	*Abies grandis*	*Abies concolor*
Betula alleghaniensis	*Acer glabrum*	*Acer platanoides*
Betula papyrifera	*Acer negundo*	*Acer saccharinum*
Betula pendula	*Acer rubrum*	*Acer saccharum*
Betula populifolia	*Alnus tenuifolia*	
		Crataegus douglasii
Fraxinus pennsylvanica	*Betula occidentalis*	*Ginkgo biloba*
Larix occidentalis	*Picea engelmannii*	*Juniperus occidentalis*
Pinus banksiana	*Picea glauca*	*Juniperus osteosperma*
Pinus resinosa	*Pinus contorta*	*Juniperus scopulorum*
Pinus strobus	*Pinus monticola*	
Populus grandidentata	*Pinus nigra*	*Picea pungens*
	Pinus ponderosa	*Pinus edulis*
Populus nigra 'Italica'		*Pinus flexilis*
Populus tremuloides	*Populus angustifolia*	*Platanus* × *acerifolia*
Rhus typhina	*Populus balsamifera*	*Populus* × *canadensis*
Salix nigra	*Populus deltoides*	
Sorbus sitchensis	*Populus trichocarpa*	*Quercus gambelii*
Ulmus parvifolia	*Prunus armeniaca*	*Quercus palustris*
	Prunus virginiana	*Quercus rubra*
	Pseudotsuga menziesii	*Rhus glabra*
		Thuja occidentalis
	Quercus alba	*Thuja plicata*
	Sorbus aucuparia	*Tilia cordata*
	Syringa vulgaris	
	Tilia americana	
	Tsuga heterophylla	
	Ulmus americana	

SOURCE: Davis and Gerhold (1976).

maintenance of low gas-exchange capacity under favorable as well as unfavorable conditions; and active avoidance, the capacity to respond to pollution stress by reducing gas exchange during stress periods (Kimmerer and Kozlowski, 1981). Winner, Koch, and Mooney (1982) found that leaf conductance values of ten species of broad-leaved trees and shrubs in California provided an index of SO_2 uptake and tolerance.

Evidence is also available for nonstomatal components of SO_2 tolerance. Both stomatal and nonstomatal components were important in explaining why *Diplacus aurantiacus* is more sensitive to SO_2 than *Heteromeles arbutifolia* (Winner and Mooney, 1980b). Some pollution-tolerant *Picea abies* clones fixed more sulfur in organic fractions following fumigation with SO_2 than susceptible clones did (Braun, 1977a,b). More SO_2 was absorbed by grafted clones of SO_2-tolerant *Pinus strobus* trees than by grafted clones of susceptible trees, further emphasizing the importance of nonstomatal components of SO_2 tolerance (Roberts, 1976).

Combinations of pollutants.   Only seldom does SO_2 alone influence the structure of plant communities. Rather, SO_2 is usually released to an atmosphere in which other pollutants are also present. When plants are injured by a combination of pollutants, it usually is difficult to identify the pollutants involved. For example, injury from SO_2-NO_2 combinations often is similar to that caused by O_3. The effects of combined pollutants may be synergistic, additive, or antagonistic. Quite often, what would be tolerable concentrations of a single pollutant will injure plants when present with another pollutant at an equally low concentration. Among the important factors influencing plant responses to pollutant combinations are: (1) the concentration of each gas in the combination exposure relative to the injury thresholds of the individual pollutants, (2) the ratio of the concentration of each gas to the other, and (3) whether there is simultaneous or intermittent application of the combined pollutant stress (Reinert, Heagle, and Heck, 1975). A few examples of the effects of SO_2 in combination with another air pollutant will be given.

The widespread occurrence of both SO_2 and O_3 has led to considerable interest in the combined effect of these pollutants. Several investigators have reported synergistic effects of SO_2-O_3 mixtures on *Pinus strobus* seedlings (e.g., Dochinger *et al.*, 1970; Houston, 1974). The response of *Populus tremuloides* to a mixture of SO_2-O_3 also was synergistic (Karnosky, 1976). Although 0.35 ppm (14.6 μmol m^{-3}) of SO_2 alone for 3 h or 0.5 ppm (20.8 μmol m^{-3}) of O_3 for 3 h was necessary to injure foliage, a combination of 0.20 ppm (8.3 μmol m^{-3}) of SO_2 and 0.05 ppm (2.08 μmol m^{-3}) of O_3 for 3 h induced even more injury than the sum of the individual pollutants.

Constantinidou and Kozlowski (1979a,b) demonstrated differences in susceptibility of *Ulmus americana* seedlings to SO_2 and to O_3, and showed that these combined pollutants had more effect on metabolite pools and growth than either pollutant alone. For example, injury to leaves from SO_2-O_3 mixtures was evident within 24 h; from O_3 alone within 36 to 48 h; and from SO_2 alone within 48 h after fumigation ceased. Expansion of young leaves, number of emerging leaves, and dry-weight increment of seedlings were reduced more by combined SO_2-O_3 than by either pollutant alone. Root growth was reduced more than shoot growth.

Synergistic responses of plants to combinations of SO_2 and HF, SO_2 and O_3, and SO_2 and peroxyacetyl nitrate (PAN) were discussed by Reinert, Heagle, and Heck (1975). These investigators emphasized a need for much more research on the long-term importance of pollutant combinations on plant communities. They suggested that (1) air pollutants may cause irreversible changes within natural ecosystems that influence productivity or plant succession, and (2) plant ecosystems at great distances from pollutant sources may be affected.

Age of plants. Young plants are very susceptible to SO_2, with seedlings in the cotyledon stage of development being especially sensitive. Epigeous germinating species that push their cotyledons above the ground by elongation of the hypocotyl (e.g., gymnosperms, *Acer*, *Fagus*) may be more severely injured than hypogeous germinating species (e.g., *Quercus*, *Juglans*, *Aesculus*).

When seedlings of *Pinus resinosa* in the cotyledon stage of development were exposed to one of four concentrations of SO$_2$ (0.5, 1, 3, and 4 ppm, or 20.8, 41.6, 124.7, and 166.3 μmol m^{-3}) for four exposure times (15, 30, 60, or 120 min), subsequent seedling growth was inhibited. The adverse effects were proportional to SO$_2$ concentrations and duration of exposure. The fumigations induced chlorosis, slowed expansion of primary needles, inhibited dry-weight increase of seedlings, and caused death of leaf tips. The great sensitivity of young seedlings was shown by significant growth inhibition following exposure to SO$_2$ for only 15 min (Constantinidou, Kozlowski, and Jensen, 1976). Despite the high susceptibility of young trees to air pollution, as shown in controlled environments, adult trees in forests often are eliminated faster than young trees by pollution because the canopy serves as a filter; hence the young seedlings are exposed to less of the polluting substance.

Environmental factors. The effect of a given dosage of SO$_2$ on structure of a plant community may be influenced in a complex manner by environmental regimes before, during, and after a pollution episode (Kozlowski, 1980a). Prepollution environmental regimes often modify the effect of SO$_2$ by influencing stomatal aperture, plant metabolism, and the rate of plant development; hence the preconditioning effects of light intensity, temperature, drought, and flooding may be expected to influence plant responses to SO$_2$. Norby and Kozlowski (1982a) studied the effects of flooding of soil for 5 weeks and fumigation with 0.35 ppm (14.6 μmol m^{-3}) of SO$_2$ for 30 h, alone and in combination, on *Betula papyrifera* (flood-intolerant) and *B. nigra* (flood-tolerant) seedlings. Both species were adversely affected by flooding, with growth of *B. papyrifera* reduced more. Stomatal conductance and SO$_2$ uptake were reduced more in flooded *B. nigra* seedlings than in unflooded seedlings; hence SO$_2$ caused less injury and less growth inhibition in the flooded seedlings. Stomatal aperture and SO$_2$ uptake were reduced even more in flooded *B. papyrifera* seedlings; yet flooded and unflooded seedlings were similarly affected by SO$_2$. Hence flooding stress apparently influenced mechanisms of pollution avoidance and pollution tolerance differently in the two species.

As was mentioned, environmental regimes during a pollution episode influence plant response to SO$_2$. Conditions that permit greater stomatal opening, such as high temperature, low vapor-pressure deficit, or high light intensity, can increase plant susceptibility to SO$_2$ by increasing leaf uptake of the pollutant (Kozlowski, 1980a). The influences of environmental factors may also be mediated by their effects on plant metabolism and biochemical tolerance of the pollutants.

Absorption of air pollutants is greater when the soil is wet because of high turgor of guard cells and open stomata (Noland and Kozlowski, 1979). A pollution episode in the morning often is more injurious than one in the afternoon, because leaf turgor and stomatal aperture are greater in the morning. Variations in both soil and air temperature also influence uptake of air pollutants by opening or closing stomata (Kozlowski and Pallardy, 1979). Seedlings of *Betula papyrifera*, *Fraxinus pennsylvanica*, and *Pinus resinosa* fumigated with 0.20 ppm (8.3 μmol m^{-3}) of SO$_2$ for 30 h at 30°C had more open stomata and absorbed more sulfur than seedlings fumigated at 12°C (Norby and Kozlowski, 1981b). Stomatal closure at low soil temperature may be the result of decreased absorption of water, leading to leaf water deficits (Kramer and Kozlowski, 1979). Low temperature may also induce changes in carboxylating and other enzymes, and in dark respiration (Bauer, Larcher, and Walker, 1975). After an air pollutant is absorbed by leaves, there may be direct effects of prevailing environmental factors on tissue sensitivity to the pollutant or on the rate of its detoxification.

Responses of some species of plants to SO$_2$ are greatly influenced by the effects of air humidity on stomatal aperture. There apparently are great differences between species in stomatal sensitivity to humidity. Nevertheless, a direct response of stomata to changes in ambient humidity has been demonstrated in a variety of plants (Davies and Kozlowski, 1974). *Betula papyrifera* seedlings were more sensitive to SO$_2$ when fumigated at high humidity than at low humidity, because of greater uptake of SO$_2$ at high humidity. The greater sensitivity at high humidity was manifested in less growth, more injury, and more leaf abscission (Norby and Kozlowski, 1981c).

Postfumigation environmental regimes that favor rapid growth of plants often allow for rapid recovery from a pollution episode. Norby and Kozlowski (1981c) first exposed *Pinus resinosa* seedlings to 0.20 ppm (8.1 μmol m^{-3}) of SO$_2$ for 91 h or to 0.50 ppm (20.3 μmol m^{-3}) of SO$_2$ for 30 h, and to postfumigation temperatures of 12, 22, or 32°C. Seedlings grown at the two higher temperatures recovered partially and initiated secondary needles, thereby providing a new source of photosynthate and further reducing growth inhibition caused by SO$_2$ fumigation. By comparison, at a postfumigation temperature of 12°C, shoot growth was slow and injured tissue was not replaced. There may also be interactions between SO$_2$ dosage and postfumigation temperature regimes (Norby and Kozlowski, 1981a). These observations emphasize that to interpret effects of temperature on SO$_2$ toxicity, one should consider the effects of prefumigation and postfumigation temperatures as well as temperature regimes during the SO$_2$ episode.

Conclusions

There is ample reason for concern with effects of SO$_2$ on plant community structure. Historically, marked changes in community structure were set in motion by high concentrations of SO$_2$ near large point sources of pollution. Despite many efforts to minimize accumulation of air pollutants, the amount of SO$_2$ in the air is expected to increase appreciably because of the growing use of highly polluting fuels. Rubin (1981) predicted that thermal coal demand during the next two decades will increase to between two to four times 1981 levels in industrial countries in North America, Europe, and the Pacific. The U.S. Environmental Protection Agency (1978c) predicted that overall emissions of sulfur oxides would be approximately 10% higher in 1990 than they were in 1975. The amounts of SO_2 may be even higher in some industrial areas. Lincoln and Rubin (1980) estimated effects of air pollution that might result from increased use of coal by industrial facilities in the northeastern United States during 1975 to 1990. Assuming 1978 emission regulations, they found that release of SO_2 in 1990 may increase by 61% over 1975 levels. Even with more stringent emission regulations for new facilities, the increase in SO_2 was projected as near 24%.

Accurately predicting the effects of SO_2 on community structure will be difficult because of differences in pollution stress, in the structure and stability of ecosystems, and in how many environmental stress factors influence the effect of SO_2 pollution (Kozlowski, 1980b).

Ecosystems differ greatly in species composition, age, stability, and capacity to recover from disturbance. The major forests of the world have been broadly classified into tropical hardwood forests, temperate hardwood forests, and coniferous forests. However, there is much diversity within each of these groups. For example, the forests of North America alone (exclusive of Mexico) have been classified into 106 distinct forest-cover types in the east and 50 in the west. And even within each of these types, there are differences in site, climate, age-class, density, etc., that will influence response to superimposed SO_2; hence data obtained on effects of SO_2 on the structure of one ecosystem may not always be accurate for the structure of another ecosystem.

Direct effects of air pollutants can trigger indirect effects throughout an ecosystem. Understanding how air pollutants affect community structure will necessitate identifying these effects, as well as characterizing the various changes in characteristics of individual ecosystems that precede changes in community structure.

As emphasized by Smith (1981), there are two conspicuous gaps in our knowledge of effects of pollution on ecosystems: (1) lack of adequate characterization of air quality over a prolonged period, and (2) lack of information that will enable partitioning of reduced ecosystem productivity to the various class II (intermediate-dosage) interactions that may cause it.

The paucity of ambient measurements of air quality renders it difficult to draw conclusions about effects of threshold doses of SO_2 on growth and, ultimately, on community structure. More continuous measurements are needed of the various air pollutants that impinge on a given ecosystem. Field plots for vegetation analysis should be carefully located in accordance with recommendations for single-event point sources, line sources, continuous or long-term point sources, or area and regional sources of pollution (Skelly, Krupa, and Chevone, 1979).

Reviews of effects of air pollutants on ecosystems generally have emphasized responses of plants exposed to acute high concentrations of pollutants. More attention should now be given to studying the effects of chronic long-term exposures to low concentrations of air pollutants on ecosystems. An example of a useful study is that of Legge (1980), who investigated the effects of low concentrations of SO_2 on a forest ecosystem in west-central Alberta. He found that SO_2 emissions modified the ecosystem in various ways. Although the rate of photosynthesis and ATP production were reduced, mineral nutrient cycling was the major ecological process affected by SO_2. Progressive alteration of mineral nutrient balance of ecosystem components provided an index of environmental deterioration following chronic exposure to SO_2.

A problem in separating effects of SO_2 stress from that of other environmental stresses on community structure has been a lack of suitable unpolluted control plots. There often has been an implicit assumption that experimental plots located at various distances from a point source of pollution are similar except for the amount of pollutant received. However, at a large distance from the polluting source, an ecosystem may be influenced by other pollutants, site changes, and other stresses. SO_2 from a local source often is augmented by pollutants that move for great distances in air masses (Kozlowski, 1980a). Hence it is desirable to obtain baseline descriptions of the preexisting ecosystem. When possible, studies of ecosystem structure should be conducted before a polluting source is constructed. Experience gained from both prepollution and postpollution studies will provide data that will make it possible to minimize or prevent injury at other potential sites of pollution.

If prepollution studies cannot be made, exposure chambers can often be used advantageously to create control plots in the field when one is studying such ecosystems as grasslands, tundra, bog plants, and shrub-dominated communities. In field chambers with closed tops, the temperature will be higher and the light intensity lower than ambient; hence open-top chambers, which do not have these problems, are advised (Heagle and Philbeck, 1979).

Because of the complexity of the effects of SO_2 on community structure, more emphasis on systems analysis is advisable. Prediction of the fate of an ecosystem under SO_2 stress and amelioration of the injury will necessitate such sequential procedures as systems measurement, data analysis, systems modeling, systems simulation, and systems optimization (Miller and McBride, 1975). As Cooper (1980) emphasized, the unaided human mind is not easily able to explore the consequences of all the interactions that occur between air pollutants and other stress factors; hence computer modeling may be very useful. An example of an enlightening study is that of West, McLaughlin, and Shugart (1980), whose model evaluated the time-integrated effects of air pollution on forest stands. They applied various degrees of growth inhibition to trees in different pollution-sensitivity classes to simulate changes in biomass. Their analysis demonstrated quantitatively how the response of individual trees in a forest stand differed from results that were extrapolated from responses measured without plant competition. Another useful study is that by Kercher, Axelrod, and Bingham (1980).

They modeled growth of the mixed conifer forest of the Sierra Nevada, and calculated effects of air pollution superimposed on other stress factors. For a 10% reduction in growth of *Pinus ponderosa* from pollution stress, and with reductions in growth of other species as determined by their relative sensitivities, standing crops of *P. ponderosa* were reduced and those of *Abies concolor* increased. Such studies are useful in gaining insight into changes in community structure.

Finally, more attention should be given to the use of sensitive indicator plants, such as certain lichens and mosses, to obtain early insights into potential threats of SO$_2$ to forested ecosystems. The methods developed by LeBlanc and Rao in Canada, Hawksworth and Pyatt in Great Britain, and Skye in Scandinavia are effective and relatively inexpensive. Much wider applications of their techniques appear to be in order.

27

SO$_2$ Effects on
Plant Community Function

W. K. Lauenroth and D. G. Milchunas

Two problems confront the person interested in the potential effects of SO$_2$ on plant communities. First, there is very little information in the literature about the experimental responses of plant communities to SO$_2$ or to any other air pollutant. Consequently, much of what has been written about how plant communities respond to SO$_2$ was induced from results on individual plant populations and/or individual processes (viz., photosynthesis or translocation). Second, since ecological systems fit a hierarchical model of organization much more closely than a serial model, higher-level responses (plant communities) will not necessarily be simple linear combinations of lower-level responses (photosynthesis, translocation, etc.). We do not claim to have solved the latter problem, but we believe that specific acknowledgment of it as a constraint leads to a more insightful view of the potential responses of plant communities to SO$_2$ air pollution.

Our concept of a plant community is that of an ecological system composed of plant populations and their interactions. Moreover, the interactions themselves are influenced by the activities of herbivores, parasites, and pathogens.

The objective of this chapter is to evaluate and discuss potential effects of SO$_2$ air pollution on plant communities. Although we will address this topic broadly, our major focus will be on grasslands and on a five-year series of field experiments on the effects of SO$_2$.

Most studies of SO$_2$ effects on plant communities have been conducted in the vicinity of an air-pollution source (e.g., Winner and Bewley, 1978; Linzon, 1978; Archibold, 1978; McClenahen, 1978; Freedman and Hutchinson, 1980; Kozlowski, 1980a; Legge, 1980; Treshow, 1980b; and Amiro and Courtin, 1981; to name a few). This may be the only practical way to study structurally complex forest ecosystems. Miller and McBride (1975) commented that the total effect of long-term pollutant exposure on preexisting forest ecosystems will never be known. The relatively simple canopy structure of grasslands and agricultural croplands allows the use of open-air exposure systems, so that one can establish controls and controlled exposure levels, but without disturbing macroenvironmental and microenvironmental influences on dose.

Experimental approaches to the effects of SO$_2$ on plant communities are few in number. The zonal air-pollution system used in our grassland study (Preston and

Lee, 1982) and similar approaches used by Bonté (1977), Bovenkerk *et al.* (1979), and Miller *et al.* (1980) are the only experiments we know of in which controlled, open-air gas-delivery systems were used to expose relatively large areas.

Experimental regimes commonly employed in the study of the botanical effects of SO$_2$ air pollution include controlled chamber experiments, controlled field experiments, and on-site evaluations in the vicinity of a pollution source. Pollutant concentrations in controlled experiments range from a reasonable approximation of those encountered around SO$_2$ sources to relatively high concentrations, with concentrations usually rather high in order to obtain a measurable response in a short period of time. Controlled laboratory experiments are the appropriate means to explain biochemical and organismal responses. They are not appropriate for investigating response at the plant community level. Controlled field exposures and on-site evaluations are the appropriate means to address questions about responses of plant communities to SO$_2$.

Long-Term-Exposure Field Experiment

The potential positive and negative effects of a pollutant such as SO$_2$ necessitate studies of responses to a variety of concentrations, as well as to exposures for long periods of time. Our approach to assessing the community-level effects of SO$_2$ involved long-term exposure (five years) of a native mixed prairie grassland to three controlled levels of SO$_2$.

Description. The study area was a mixed prairie site on the divide between the Powder and Tongue River drainage basins in Custer National Forest, Montana, U.S.A. (45°15′N, 106°E). *Agropyron smithii* Rydb. made up 44% of the total aboveground net primary production; other important plant species included *Koeleria cristata* (L.) Pers., *Poa secunda* Presl., *Stipa comata* Trin and Rupr., *Achillea millefolium* L., *Taraxacum officinale* Weber, and *Tragopogon dubius* Scop. Soils were classified as Farland silty clay loams at Site I. Farland soils are generally deep, well-drained, and medium in texture. Soils at Site II were described as Thurlow clay loams, which are deep, light-colored, and moderately fine in texture.

The climate in this region is characteristic of temperate midcontinental semiarid grasslands. Average annual precipitation is approximately 400 mm, with the greatest quantities received in spring, but there can be large differences from year to year and from month to month. Growing-season precipitation on the study site ranged from a 5-year high of 415 mm in 1978 to a low of 170 mm in 1979. In semiarid regions, most rain events result in rather small accumulations (1 to 5 mm day^{-1}). Over a typical 7-month growing period, air temperatures ranged from near 10°C in April, May, and October to above 20°C in July and August. Relative humidity averaged 50%, and wind speed averaged 2.3 m s^{-1}.

Three concentrations of sulfur dioxide, and air to the controls, were delivered

TABLE 27.1

Seasonal Average SO$_2$ Concentrations on Sites I and II, 1975–79
(low run–high run, $\mu g/m^{-3}$)

Year; and measure[a]	Site I				Site II			
	Control	Low	Medium	High	Control	Low	Medium	High
1975								
G.M.	25[b]	50[b]	90[b]	155[b]				
S.G.D.	1.5	2.4	2.7	2.7				
A.M.	25	75	145	255				
1-h	255	740	1,245	2,055				
3-h	255	460	1,085	1,595				
1976								
G.M.	5–20	40–50	85–90	150–155	5–30	60–65	95–100	150–160
S.G.D.	3.0, 1.5	3.6, 2.5	3.1, 2.7	2.9, 2.5	3.4, 1.2	2.3, 1.8	2.3, 1.9	3.0, 2.3
A.M.	10–25	90	170	265	10–30	85	125	245
1-h	300	1,410–1,455	2,935–3,740	6,100–8,155	140	1,570–1,710	2,540–2,890	4,595–5,195
3-h	175–180	855–925	1,895–2,285	3,950–4,990	90	925–995	1,550–1,665	2,700–3,005
1977								
G.M.	5–35	75–80	120–130	190–220	10–35	65–70	110–115	175–190
S.G.D.	4.5, 1.7	2.2, 2.1	2.7	3.1, 2.5	4.2, 1.2	2.2, 2.1	2.7, 2.4	
A.M.	25–50	105–110	220–235	325–355	25–35	95–100	155–175	275–300
1-h	600–645	1,180–1,295	2,265–2,355	3,165–3,720	760–880	1,525–1,915	2,840–3,695	4,410–5,845
3-h	460–485	880–945	1,480–1,640	2,220–2,680	275–325	880–1,040	1,895–2,285	3,465–4,295
1978								
G.M.	25	65	120	200–205	20–30	60	105	210–215
S.G.D.	1.7, 2.1	2.7	3.1	3.2, 3.4	1.3, 2.3	2.3	2.6	2.8, 2.9
A.M.	30	125	270	445	25–30	90	185	410
1-h	600	2,540	3,950–3,975	10,650–10,740	140	695	1,315	1,890
3-h	255	1,615	1,520	6,790–6,815	90	510	1,085	2,055
1979								
G.M.	20–30	55–60	115	170	20–30	55–60	115–120	220–225
S.G.D.	0.8, 1.6	2.4, 2.1	2.5	2.4	2.9, 1.3	2.6, 2.1	2.7, 2.5	3.0, 2.8
A.M.	35–40	85	195	1,030	30–35	85–90	195–200	420
1-h	440	1,050–1,095	2,020–2,025	3,190–3,200	450	920	1,400	3,530
3-h	305	635–640	1,215–1,220	2,030–2,060	240	595	1,095	2,625

SOURCE: Preston and Lee (1982).

NOTE: Single values are presented when low-run and high-run values are rounded to the same value. To convert to $\mu mol\ m^{-3}$, divide all values by 64.

[a] Abbreviations: G.M. (geometric mean), S.G.D. (standard geometric deviation), A.M. (arithmetic mean), 1-hr (highest 1-hour average concentration observed during the growing season), 3-h (highest 3-hour average concentration observed during the growing season).

[b] High run only for 1975.

to 0.52-ha treatment plots through a network of aluminum pipes placed 0.75 m above ground surface at two site replications (Sites I and II). Each of the four 0.52-ha plots on both Site I and Site II was divided into replicates. Sulfur-dioxide concentrations were monitored with a Meloy flame photometric sulfur-gas analyzer through teflon lines with intakes at canopy height (30 to 35 cm above the ground). By using a time-sharing device, we could analyze the concentration of SO_2 within each line, and record it on a strip chart for a 6-min period every 48 min. Treatment was continuous throughout the April-to-October growing season except for 1975, when treatments were begun in June.

The seasonal averages of SO_2 concentrations for each of the five years of treatment are presented in Table 27.1. The geometric means across all five years were 18, 60, 105, and 175 μg m^{-3} (0.28, 0.94, 1.64, and 2.73 μmol m^{-3}) of SO_2 on Site I and 20, 61, 108, and 193 μg m^{-3} (0.31, 0.95, 1.69, and 3.02 μmol m^{-3}) of SO_2 on Site II, for the control, low, medium, and high treatments, respectively. Since SO_2 was delivered to the plots at a constant rate, diurnal fluctuations and variations within the canopy profile of SO_2 concentration occurred in response to changes in abiotic conditions. Geometric mean concentrations of SO_2 during daylight hours were one-third less than the reported 24-hour-day values. A detailed description of the design and performance of the field-exposure system is given by Preston and Lee (1982). Their evaluation of the exposure system indicated that the geometric means, standard geometric deviations, and peaks of SO_2 concentrations accurately simulated real sources, but that the diurnal concentration pattern from point sources is greater during daylight hours. The temporal pattern of exposures from the zonal air-pollution system was more similar to that of an area source.

Results and discussion. During our investigations of the effect of SO_2 on a native grassland, we monitored many structural and functional attributes of the abiotic, primary producer, and consumer components of the system. We will touch briefly on some of the responses of the plant community and on other responses that directly relate to the plant community. The deposition and dynamics of sulfur in the system is the interface between exposure and response, and will be dealt with first.

The seasonal dynamics of sulfur in the aboveground tillers of *A. smithii* exposed to SO_2 during a typical growth cycle showed increased sulfur content with increased SO_2 concentrations (Fig. 27.1a). The seasonal patterns of SO_4-S and organic S (Fig. 27.1b,c) indicate a subsidy-stress gradient (Odum, Finn, and Franz, 1979) along both seasonal and SO_2-concentration gradients. During the first part of the growing season, organic S was generally higher on all SO_2 treatments than on the control, and SO_4-S was higher only on the high-SO_2 treatment. At this time, SO_2-S was apparently being used for organic-S synthesis on all SO_2 treatments, and SO_4-S was accumulating only on the high-SO_2 treatment. During the later part of the growing season, organic S was higher than in controls on

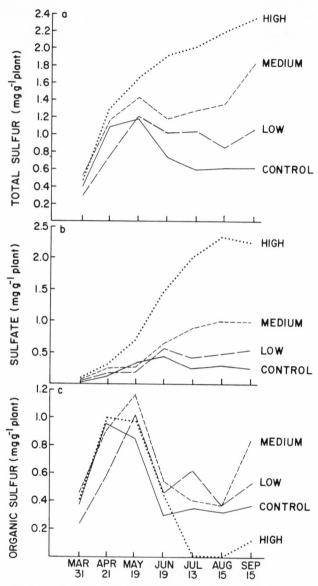

Fig. 27.1. Seasonal total sulfur (*a*), sulfate sulfur (*b*), and organic sulfur (*c*) concentrations (mg g^{-1}) of *Agropyron smithii* tillers on the control, low, medium, and high SO$_2$ treatments for 1979, Site I.

the low- and medium-SO$_2$ treatments, but lower than in controls on the high-SO$_2$ treatment, and SO$_4$-S was higher on the low- and medium-SO$_2$ treatments and proportionately higher than the increase in SO$_2$ concentration on the high-SO$_2$ treatment. At this time the synthesis of organic sulfur compounds was stimulated by the low- and medium-SO$_2$ exposure, even though SO$_4$-S was in excess. On the high-SO$_2$ treatment, however, so much SO$_4$-S accumulated that it impaired the metabolic functioning of the plant, and organic sulfur compounds were either broken down or used up faster than they could be produced.

Although the increments in plant sulfur concentration increased with increasing SO$_2$ concentration, the rate of this increase decreased at high-SO$_2$ concentrations (Fig. 27.2). This would indicate a decreased capacity for sulfur uptake at high SO$_2$ concentrations. We do not know whether this was in response to a feedback inhibition of root sulfate uptake when shoot sulfur concentrations were high (Schiff and Hodson, 1973), a stomatal response, or a decrease in the metabolic activity of the plant, as might be suggested by the organic S, SO$_4$-S concentrations on the high-SO$_2$ treatment.

We did not detect accumulation of sulfur in live plants with successive years of fumigation; either year-to-year productivity-induced fluctuations in sulfur concentrations masked successive year accumulations, or—and this is what the data suggest—year-to-year accumulation did not occur. Since plants in this grassland are annual or perennial, all growth each season occurs from existing belowground stores or from seed. Accumulated sulfur would thus have to come from the buildup of soil sulfur or from root reserves. No trend of increasing sulfur concentration in tillers during May was observed with successive years of exposure. The similar sulfur concentration of newly dead tissue and live tissue indicated negligible redistribution of sulfur with senescence, and belowground organs did not significantly accumulate sulfur with increasing SO$_2$ concentration.

As mentioned above, no significant SO$_2$-treatment effects were observed for total sulfur concentrations in *A. smithii* roots or rhizomes. However, seasonal mean sulfate concentrations of rhizomes were 181, 201, 223, and 256 μg of SO$_4$-S g^{-1}, and seasonal mean root sulfate concentrations were 154, 194, 231, and 273 μg of SO$_4$-S g^{-1} for the control, low-, medium-, and high-SO$_2$ treatments, respectively. These results concur with our observation of similar, and at times less, ^{35}S translocation to belowground organs with SO$_2$ treatment (Coughenour et al., 1979) and an inhibition of root sulfur uptake when shoot sulfur concentrations were high (Milchunas, Lauenroth, and Dodd, 1983).

Sulfur content varied among plant species on both the control and SO$_2$ treatments. Increases in sulfur concentration with the high-SO$_2$ treatment ranged from 101% in *Vicia americana* to 248% in *Astragalus crassicarpus* (Milchunas, Lauenroth, and Dodd, 1983). Plants that accumulated the largest amounts of sulfur under control conditions did not necessarily accumulate greater quantities of sulfur when exposed to SO$_2$.

Total sulfur concentration of litter significantly increased with treatment and

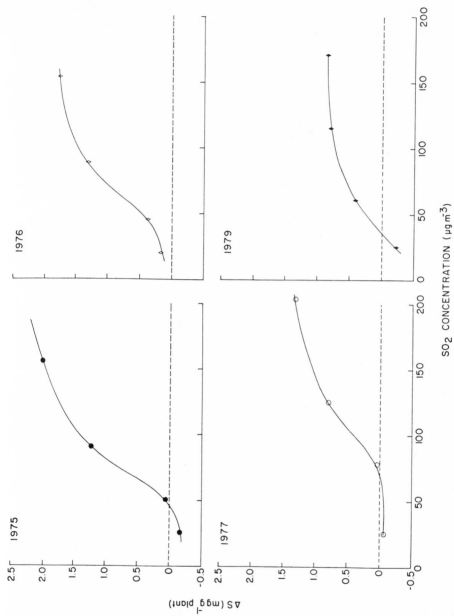

Fig. 27.2. June to August increments (Δs) in *Agropyron smithii* tiller sulfur concentration (mg g^{-1}) in relation to atmospheric SO$_2$ concentration (μg m^{-3}) for 1975–79.

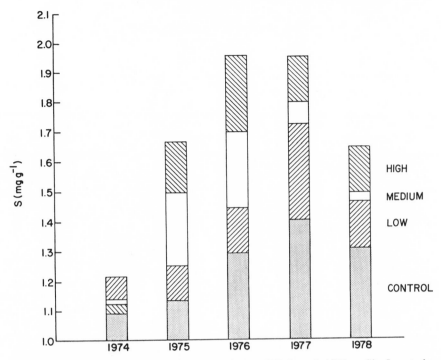

Fig. 27.3. Litter sulfur concentrations (mg g^{-1}) for 1974 through 1978 on Site I control, low, medium, and high SO$_2$ treatments.

with each successive year (except for 1978) of exposure (Fig. 27.3), but not by dates within a year. The decline in litter sulfur in 1978 was attributed to a dry 1977 growing season and a very wet spring in 1978. It is probable that year-to-year litter sulfur concentration would increase on SO$_2$-exposed areas until previously uncontaminated litter has decomposed and been replaced by sulfur-laden litter, after which an input-output equilibrium would be reached, and year-to-year changes would be a function of current and recent past-year abiotic conditions and productivities. On the other hand, we would expect soil sulfur to continually increase as deposition to any or all components of the system continued. Because of the very massive existing soil-sulfur pools, we could not detect sulfur inputs to the soil on a total sulfur basis. We did, however, observe increasing soil sulfate levels (Leininger and Taylor, 1981) and decreasing soil pH (Dodd and Lauenroth, 1981). The pH of high-SO$_2$-treatment soils to a depth of 1 cm declined one pH unit after 5 years of exposure (Table 27.2).

In response to increased litter and soil sulfur and lower soil pH, an inverse relationship was observed between SO$_2$ exposure and rates of litter disappearance (Dodd and Lauenroth, 1981). Again, the abiotic conditions during a partic-

TABLE 27.2
Soil pH after Five Years of Exposure to SO_2

SO_2 concentration	October 1979, Soil depth of:		May 1980, soil depth of 0–1 cm
	0–1 cm	1–5 cm	
Control	5.59[a]	5.35[a]	5.63[a]
Low	5.20[b]	5.23[a]	5.16[b]
Medium	5.25[b]	5.49[a]	5.11[b]
High	4.63[c]	5.35[a]	4.61[c]

NOTE: Values represent a mean of four and six measurements for each SO_2 treatment for 1979 and 1980, respectively, and each measurement is of a pool of four soil samples.
[a-b] Means for each depth not followed by the same letter are significantly different ($p = 0.05$).

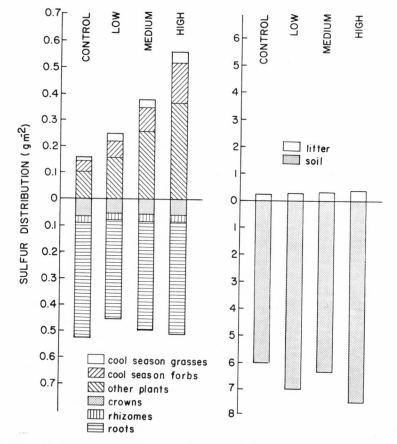

Fig. 27.4. The distribution of sulfur (g m^{-2}) in (1) aboveground organs of plant groups (cool season grasses, cool season forbs, and other plants) and belowground organs (crowns, rhizomes, and roots) at peak standing crop, and (2) litter and soil for the control and SO_2 treatments.

ular growing season exerted an influence on overall decomposition rates, as did the relative differences between treatments.

The pool distribution of sulfur at peak standing crop is pictured in Fig. 27.4. The increases in live plant aboveground sulfur pools were 55, 135, and 294%, and increases in litter sulfur pools were 19, 40, and 62%, for the low-, medium-, and high-SO₂ treatments, respectively. Adjusting peak-standing-crop pools by current standing dead and current litter [calculated by summing the loss in individual leaf areas after their peak (Milchunas *et al.*, 1981)] gave values for total (soil and atmospheric) sulfur uptake by vegetation of 0.18, 0.27, 0.42, and 0.63 g of S m^{-2} yr^{-1} on Site I, and 0.20, 0.30, 0.48, and 0.85 g of S m^{-2} yr^{-1} on Site II for the control, low-, medium-, and high-SO₂ treatments.

The plant-community response to the increased sulfur loading displayed a remarkable degree of compensation. Sulfur dioxide had no measurable effect on annual aboveground net primary production (Table 27.3). Belowground biomass dynamics showed no significant SO₂-treatment effects (Dodd, Lauenroth, and Heitschmidt, 1983). No consistent trends were discerned for the average standing crop of crowns, rhizomes, or roots across SO₂ treatments (Table 27.4), although biomass of rhizomes on Site I appeared to be affected by SO₂ exposure. An exception, however, was observed for the introduced annual grass *Bromus japonicus* (Fig. 27.5), whose biomass appeared to decrease with increasing SO₂ concentration during two later years of treatment.

The canopy cover of vascular plants displayed no detectable response to the SO₂ treatments (Taylor, Leininger, and Hoard, 1980). *Agropyron smithii* tiller density, a measure related to cover, showed no response to SO₂ treatment. However, plasmolysis in two lichen species (*Cladonia* spp.) ranged from 10–14%, 15–18%, to over 30% on the control and low-, medium-, and high-SO₂ treatments, respectively (Eversman, 1978), and lichen cover showed a negative response to SO₂ after just one year of exposure (Taylor, Leininger, and Hoard,

TABLE 27.3
Annual Aboveground Net Primary Production for Sites I and II, 1975–78
(g m^{-2})

SO₂ concentration	1975	1976	1977	1978
Site I				
Control	150 ± 14	199 ± 13	126 ± 12	136 ± 8
Low	149 ± 17	186 ± 9	131 ± 12	156 ± 14
Medium	165 ± 13	195 ± 11	131 ± 7	123 ± 6
High	156 ± 14	199 ± 8	119 ± 5	129 ± 6
Site II				
Control	—	177 ± 8	137 ± 7	268 ± 18
Low	—	218 ± 17	210 ± 13	219 ± 18
Medium	—	205 ± 14	169 ± 9	214 ± 13
High	—	227 ± 15	190 ± 11	253 ± 8

SOURCE: Adapted from Dodd, Lauenroth, and Heitschmidt (1983).
NOTE: All values are $\bar{x}$ ± SE.

TABLE 27.4
Average Standing Crop of Belowground Biomass by Morphological Category
(0–10 cm depth)
$(g\ m^{-2})$

Category	Year			SO$_2$ treatment			
	1975	1976	1978	Control	Low	Medium	High
Site I							
Crown	51[a]	70[b]	97[c]	74[a]	77[a]	74[a]	66[a]
Rhizome	25[a]	32[b]	30[b]	37[b]	26[a]	27[a]	26[a]
Roots	549[a]	528[a]	677[b]	586[a]	573[a]	582[a]	596[a]
Total	624[a]	629[a]	803[a]	697[a]	676[a]	682[a]	687[a]
Site II							
Crown		71[a]	99[b]	79[a]	88[a]	84[a]	88[a]
Rhizome		28[a]	32[b]	26[a]	35[b]	31[a,b]	29[a]
Roots	593[a]	682[b]	593[a]	669[a]	661[a]	626[a]	
Total	692[a]	813[b]	697[a]	791[a]	775[a]	744[a]	

SOURCE: Adapted from Dodd, Lauenroth, and Heitschmidt (1983).

[a,b,c] Means for each site within a row for each year or treatment not followed by the same letter are significantly different ($p = 0.05$).

1980). An increase in the number of leaves on individual tillers in response to the SO$_2$ treatments was reported by both Heitschmidt, Lauenroth, and Dodd (1978) and Milchunas *et al.* (1981). Individual leaves on a tiller responded to the SO$_2$ treatments according to the subsidy-stress gradient associated with the organic S, sulfate-S concentrations described earlier.

Although no effects of the SO$_2$ treatments were observed at the community level, i.e., total biomass, total cover, etc., minor changes were occurring at the population level, and effects on the individual organism and on individual organs became increasingly more apparent.

The conversion of chlorophyll to phaeophytin has been shown to be a frequent consequence of exposure of plants to SO$_2$ (Malhotra, 1977). The response pattern of leaf chlorophyll content of *A. smithii* early in the growing season on our control and low-SO$_2$ treatment was similar to that observed for individual leaf areas late in the growing season—i.e., highest concentrations on the low-SO$_2$ treatment (Fig. 27.6). However, the lowest concentrations were observed on the high-SO$_2$ treatment. Chlorophyll concentrations during midseason were similar for SO$_2$-treated plants, but significantly lower than in controls. No differences were apparent by the end of the growing season. The effect of the SO$_2$ treatments on the chlorophyll content of various species on the sites ranged from positive, to no response, to negative (Lauenroth and Dodd, 1981b). The most sensitive species sampled was *Bromus japonicus*, which had also displayed a decrease in biomass in response to SO$_2$ treatment.

Carbon uptake in *A. smithii*, as measured by $^{14}CO_2$ uptake, was stimulated, not affected, or depressed by SO$_2$ treatment, depending on moisture conditions. Measurements of leaf water potential, stomatal conductance, and carbon uptake spanned a moist growing season, a dry growing season, and a very dry period. Leaf water potentials, unaffected by SO$_2$ treatment, were 0.5 to 1.0 MPa higher during the moist growing season. When abiotic conditions favored relatively high stomatal conductances, SO$_2$ further increased the conductances. The degree to which this occurred regulated SO$_2$ dose and affected carbon uptake. The large SO$_2$ doses received during wet conditions decreased carbon uptake, whereas small SO$_2$ doses increased carbon uptake, and the very small SO$_2$ dose received under dry conditions did not affect carbon uptake.

The distribution of photoassimilated carbon was followed by exposing 0.5 m^2 swards and individual leaf blades of three ages to $^{14}CO_2$ on the control and high-SO$_2$ treatments. Labeling the entire swards resulted in proportionally greater belowground carbon translocation rates on the SO$_2$ treatment during midgrowing

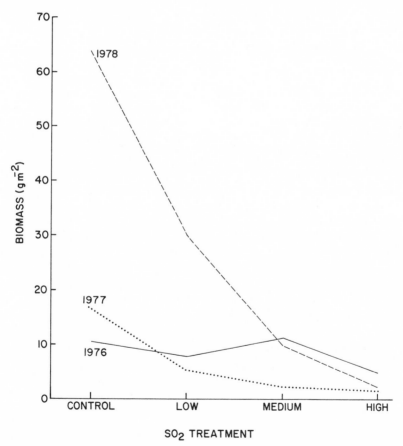

Fig. 27.5. Biomass of *Bromus japonicus* on the control and three SO$_2$ treatments for 1976, 1977, and 1978. Redrawn from Dodd, Lauenroth, and Heitschmidt (1983).

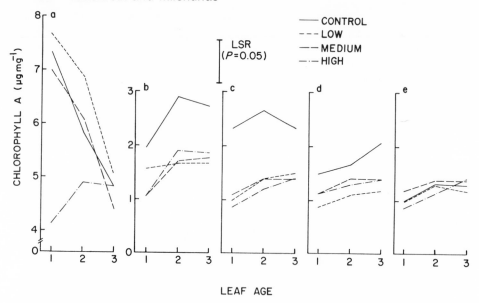

Fig. 27.6. Seasonal chlorophyll *a* concentration (μg mg^{-1}) of *Agropyron smithii* leaves of three different ages (1 = oldest leaf) exposed to the control, low, medium, and high SO$_2$ treatments: (*a*) May 20, (*b*) June 18, (*c*) July 9, (*d*) August 7, (*e*) August 18. LSR = least significant range. A similar response was observed for chlorophyll *b*. Redrawn from Lauenroth and Dodd (1981a).

season (Coughenour *et al.*, 1979). No treatment effects were apparent during spring or fall, when total canopy assimilation was lowest. The labeling of an individual leaf of tillers also indicated the presence of greater sink demands on the SO$_2$ treatment (Milchunas, Lauenroth, and Dodd, 1982). Labeled leaves exposed to SO$_2$ exported an average of 12% more carbon than the leaves of control tillers. The greater carbon export by an individual source leaf under SO$_2$ treatment resulted in greater relative ^{14}C concentrations in roots and rhizomes of SO$_2$-treated plants, but the largest difference between control and SO$_2$-treated plants in tiller carbon partitioning was observed in developing leaves. The concentrations and quantities of ^{14}C in developing leaves were 1.5 and 1.7 times greater, respectively, for the SO$_2$-treated plants than for the controls.

Young developing leaves rely on translocated carbon from other leaves until they develop sufficient photosynthetic capacity to satisfy their own needs (Swanson, Hoddinott, and Sij, 1976). Increased translocation to developing leaves even from leaves that are fixing less carbon is possible because sink demand can override the availability of assimilate in regulating translocation at above-maintenance assimilation rates (Ho, 1979). Our observations of increased translocation rates with high sulfur accumulation, reductions in chlorophyll, increased rates of senescence, greater number of leaves, and no effect on biomass or cover suggests that *A. smithii* can partially compensate for the stress of our high-SO₂ treatment by increasing leaf turnover rate.

Several general conclusions can be drawn from our experimental results. Responses to SO₂ often follow a subsidy-stress gradient along both a time axis and an SO₂-concentration axis. Compensatory mechanisms were in operation; because of this, effects that were manifested at a lower level of organization were dampened in their transmission up the hierarchy. Turnover time, of organs, individuals, or communities, is important in determining effect, because time is one of the important dimensions along which a subsidy-stress gradient operates. These points will be discussed further in the last two sections of this paper.

Simulation-Modeling Experiments

A promising tool for the analysis of complex ecological systems such as plant communities is found in mathematical modeling and computer simulation. We employed simulation modeling in our analysis of the responses of a grassland to SO₂ exposure in order to explore short- and long-term responses to SO₂ (Heasley, Lauenroth, and Dodd, 1981).

Modeling exercises were conducted for two sets of conditions. The first used data from the field experiments for initial conditions and ambient SO₂ concentrations and meteorological data from the years of the field experiments. Simulations were carried out for validation purposes, and also to obtain values for variables that were not measured in the field. The second experiments were conducted to explore system responses during long periods of SO₂ exposure (30 years). SO₂ concentrations were generated by a gaussian plume-dispersion model. A time-series of SO₂ concentrations was generated under the constraint that the federal secondary standard was not violated. This resulted in an annual average concentration of 78 μg (1.22 μmol) of SO₂ m^{-3}.

Neither modeling exercise produced any counter-intuitive results (Table 27.5). Simulations of the field experiment resulted in small decreases in plant carbon and nitrogen, and substantial increases in sulfur. Decomposer carbon, nitrogen, and sulfur were uninfluenced by SO₂ exposure. Soil carbon and nitrogen were unchanged, and inorganic sulfur increased slightly. Both gross primary production and plant respiration were depressed slightly by SO₂ exposure. The result was that net primary production was unchanged by SO₂. Carbon reserves of plants and decomposer respiration and production were very slightly decreased by SO₂ exposure. To a large extent these simulations reinforced the conclusions reached by the field experiments.

We were interested in long-term simulations because we thought many of the effects of SO₂ exposure documented in the field experiment would have substantial influence on system structure and function if allowed to accumulate through time.

Net primary production and plant respiration behaved very predictably during

TABLE 27.5
System-Level Indicators of SO_2 Effect on a Grassland, Simulation Results
(g m^{-2})

Ecosystem variable	Year; and SO_2 treatments							
	1976				1977			
	Control	Low	Medium	High	Control	Low	Medium	High
Producer carbon[a]	286	283	282	277	241	241	241	239
Decomposer carbon[a]	108	108	108	108	111	111	111	111
Soil organic carbon[a]	4,874	4,874	4,874	4,874	4,883	4,883	4,883	4,883
Producer nitrogen[a]	5.9	5.8	5.7	5.6	4.35	4.31	4.36	
Decomposer nitrogen[a]	9.8	9.8	9.8	9.8	10.1	10.1	10.1	10.0
Soil organic nitrogen[a]	554	553	554	554	554	554	554	554
Inorganic nitrogen[a]	4.4	4.4	4.4	4.4	5.2	5.2	5.2	5.2
Producer sulfur[a]	0.42	0.5	0.63	0.67	0.34	0.41	0.49	0.52
Decomposer sulfur[a]	0.87	0.87	0.87	0.87	0.9	0.9	0.9	0.9
Soil organic sulfur[a]	78.18	78.18	78.18	78	78.9	78.9	78.9	78.9
Inorganic sulfur[a]	17.38	17.54	17.57	17.65	17.72	18.08	18.29	18.41
Gross primary production[b]	437	430	426	416	285	285	285	282
Primary producer respiration[b]	309	304	301	291	194	194	194	191
Carbon reserves[b]	65	65	64.9	64.7	45.3	45.3	45.3	45.5
Decomposer respiration[b]	121	121	121	121	125	124	124	124
Decomposer production[b]	171	171	171	171	181	180	179	179
P-R ratio[b]	1.02	1.01	1.01	1.01	0.89	0.90	0.90	0.90
N-cycling ratio[b]	0.99	0.99	0.99	0.99	1.00	1.00	1.00	1.00
S-cycling ratio[b]	1.04	1.06	1.07	1.08	1.03	1.06	1.07	1.09

[a] Value at time of peak standing crop of carbon.
[b] End of calendar year values.

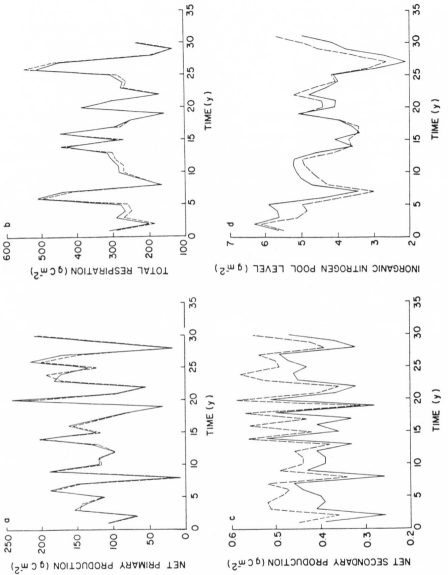

Fig. 27.7. Simulated (a) net primary production, (b) total respiration, (c) net secondary production (cattle), and (d) inorganic nitrogen pool for a grassland exposed to control (solid line) and SO$_2$ (dashed line) concentrations at the Federal Standard level. Redrawn from Heasley, Lauenroth, and Dodd (1981).

the 30 simulation years (Fig. 27.7a,b). Interactions with specific weather years were never large enough to divert the trajectories of the two conditions (with or without SO_2). Net secondary production, which for this model was cattle production, showed substantial differences between the control and SO_2 simulations especially in certain weather years (Fig. 27.7c). We did not have data to validate this response from the field experiment. The explanation is that increased sulfur content of forage improves nitrogen retention by ruminants.

Positive effects were also found for soil inorganic nitrogen. Soil inorganic nitrogen was decreased slightly during the first 10 to 15 years of simulation (Fig. 27.7d). The increase after 30 years probably represents a new steady state resulting from the decrease in aboveground decomposition caused by SO_2.

The most important negative effect of SO_2 was its influence on carbon reserves stored in the crowns of perennial grasses. We measured a decrease in rhizome biomass of *A. smithii* (Table 27.4). This was almost the only response from the field experiment that seemed to accumulate through time, that was not neutralized by the system rapidly seeking a new steady state. After 30 years the difference between the SO_2 simulation and the control was 2 g of C m^{-2} or almost 20% of the control value. The consequences of continuation of that trend would be either extinction of the present perennial grasses or selection for those components of the population most resistant to SO_2 exposure.

Placing the Experiments in Context

We can draw two kinds of conclusions from this work. The first kind consists of inferences from the field experiments. These are relatively straightforward, and concern the short- and long-term consequences of exposing northern mixed prairie ecosystems to sulfur-dioxide concentrations similar to our experimental concentrations. The second set of conclusions concerns a broader topic: What are the likely consequences of exposing northern mixed prairie grasslands to the sulfur-dioxide concentrations that result from the current level of energy development in the region and to concentrations projected for future levels of energy development? Drawing these latter conclusions involves both interpolation and extrapolation of our experimental and simulation-modeling results, and requires us to speculate about the balance between positive and negative effects of SO_2 in plant communities.

One of the most difficult tasks in generalizing our findings to energy-development scenarios is to find a common basis for expressing SO_2 exposure. Our experimental treatments resulted in nearly continuous exposure of the plots to SO_2. Concentrations varied with weather conditions, particularly temperature and wind, but we still had a much lower frequency of zero SO_2 concentrations than one might expect for any location in the northern Great Plains even under maximum development scenarios. The wind data that we used as input to a gaussian diffusion model to generate time-series of SO_2 data for modeling exercises

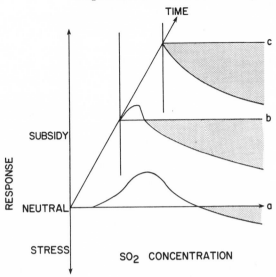

Fig. 27.9. Model of SO$_2$ concentration subsidy-stress response gradients and the influence of length of exposure.

conifer *Larix occidentalis* had greater survival capabilities in an SO$_2$-polluted environment because foliage was not present in the wintertime. In contrast to conifers, the expected functional life of a grass blade is often a fraction of a single growing season. Injury to or even complete loss of a blade as a result of SO$_2$ exposure is likely to have a small negative effect on growing-season carbon balance. A similar loss to a conifer, with its highly deterministic leaf initiation capabilities, will represent a much more significant effect on annual carbon balance.

The relationship between length of exposure and turnover rates of nutrients tied up in dead organic matter may be the opposite of that with nutrients in live components of the community. For example, rapid rate of turnover of below-ground organic matter may lessen the residence time of sulfur in the litter component and result in a more rapid lowering of soil pH. A plant community in which a large aboveground biomass is dependent on the rapid turnover of a limited nutrient pool would be most adversely affected by decreased decomposition rates.

Earlier, we commented that the response of a plant community to SO$_2$ exposure would reflect the responses of the various levels in the community hierarchy, modified by the degree of determinism or the connectedness of the hierarchy. The impact of a specific SO$_2$ concentration will likely influence the various levels in the hierarchy differently. The response of the plant community will be influenced by the way the responses of the lower levels in the hierarchy are transferred upward. These transfers are determined by the way the levels in the hierarchy are connected. The more direct the connections among the lower levels and the properties we attribute to the plant community, the higher the probability we will see an effect, and the looser the connectedness, the lower the probability we will see an effect. For example, in our grassland experiment we measured a decrease in chlorophyll *a* and *b* at our high-concentration SO$_2$ treatment. We were unable to measure a decrease in net primary productivity associated with the decrease in chlorophyll. An obvious conclusion is that the connection between chlorophylls *a* and *b* and net primary productivity is not direct enough to result in a transfer of this effect.

Allen and Starr (1982) consider both underconnectedness and overconnectedness in an analysis of stability and resilience of systems. Our discussion of connectedness in terrestrial plant communities is most appropriately limited to an analysis of relative connectedness. Holling (1978) described an appealing hypothesis about the relationship between the variability of the environment in which a system developed and the ability of the system to adjust to perturbations: systems that develop in fluctuating environments should be better able to adjust to perturbations than those that develop in relatively constant environments. In terms of connectedness, plant communities that develop in relatively stable environments may have a higher connectedness than plant communities in fluctuating environments. We can speculate that the greater the connectedness of the plant-community hierarchy, the more likely it is that effects of SO$_2$ exposure will be found at the plant-community level. Additionally, the more variable the environment of a plant community is, the lower is the probability that community-level effects will be observed. This idea is consistent with the results from our grassland study. Semiarid grasslands survive under widely varying environmental conditions. They should have (relatively) low connectedness and therefore a large ability to adjust to perturbations, such as SO$_2$ air pollution; and that is what we found.

Future Directions for Plant Community Research

Our definition of a plant community as a hierarchical system composed of several plant populations and their interactions suggests an important point for future directions in air-pollution research. Plant communities are not simple linear aggregates of lower levels in the hierarchy. They include not only their constituent plant populations, but also the interactions among plant populations, as well as influences of herbivores, parasites, and pathogens. The existence of interactions as components of plant communities makes it imperative to investigate plant-community responses to air pollution at the plant-community level. We cannot use organism- or population-level responses to answer questions about community responses.

One direction for this research that appears to have a high potential is the open-air exposure experiment, such as our grassland study. Such experiments have the advantage of permitting investigations of plant communities under relatively undisturbed conditions. The important disadvantage is the very small degree of control the investigator has over the system.

Open-air exposure systems provide a means for hypothesis testing in community-level air-pollution studies not possible with experiments conducted in the vicinity of an air-pollution source. The exposure system offers the experimenter the opportunity to manipulate concentrations at a single location. Depending on the degree of sophistication of equipment, virtually any time-series of pollutant exposures could be approached.

28

Modeling SO$_2$ Effects on
Forest Growth and Community Dynamics

H. H. Shugart and S. B. McLaughlin, Jr.

The need to characterize, quantify, and project the effects that SO$_2$ and other air-pollution stress may have on forest productivity has become increasingly important in the past decade, as costs and benefits of various energy technologies have been ecologically and economically evaluated. To identify and to measure the effects that anthropogenic stress may have on growth of forest trees are formidable tasks, because of the complexity of tree-growth processes and their sensitivity to natural environmental stresses, and because of the complexity of the anthropogenic stresses themselves (Kozlowski, 1980a). One important aspect of this complexity is the qualitative and quantitative changes in regional air quality in recent decades, since these have raised new and unanswered questions.

Problems and Questions

Much of our information about the effects of air pollution on forest trees comes from investigations near large smelting operations in the first half of this century. Both mortality and decline of forests were found over thousands of hectares in the immediate vicinity of these unregulated operations (see reviews by Scurfield, 1960; Hepting, 1968; Miller and McBride, 1975; Smith, 1981). However, although those studies provide evidence that relatively high concentrations of pollutants such as SO$_2$, heavy metals, and hydrogen fluoride can cause severe effects, they provide little detailed information that is relevant to today's air-pollution problems. The application of abatement technology, increased urbanization and industrialization, and the shift to large fossil-fueled electric-generating plants with tall stacks have altered both the nature of air-quality problems and the type of research needed to address those problems. Major focus has shifted: from site-specific studies of the effects of gaseous pollutants from individual point sources, to evaluation of effects of chronic exposure, to concern about effects of multiple pollutants from multiple sources within regional airsheds. The recognition that strongly acidic rainfall—an apparent product of long-distance transport and transformation of sulfur and nitrogen oxides from fossil-fuel combustion—is falling over increasingly large areas of the eastern

United States (Cogbill and Likens, 1974; Likens and Butler, 1981) has added significant new possibilities for effects on forest productivity and for appropriate methods to evaluate them.

To deal with the effects of SO$_2$ on forests of the United States, we must address both the local effects of individual point sources and the combined effects of multiple point sources within subregional areas where either fuel supply or demand for electricity is high. Long-distance transport of SO$_2$ is apparently accompanied by oxidation to SO$_4^{2-}$, so that regional-scale effects may be primarily a function of acidification of precipitation rather than effects of SO$_2$ *per se*. Within the U.S., a map of the distribution of coal- and oil-fired power plants (Fig. 28.1) aptly illustrates both the regional and the subregional distribution of these primary sources of SO$_2$. Approximately 80% of these sources are in the eastern U.S., with distinct subregional concentrations being apparent in several areas. These sources are in a region where frequent air stagnation, highly acid precipitation, and high oxidant concentrations combine to increase the potential sensitivity of vegetation to air pollution (McLaughlin, 1981). An overlay of forested regions of the U.S. with isopleths of air stagnation and acid deposition demonstrates the potential for SO$_2$ and other air pollutants to impinge upon these forests (Fig. 28.2). Thus it can be seen that responses will (1) be functions of exposure to both SO$_2$ and other air pollutants, principally O$_3$ and acid rain, and (2) differ greatly. In the immediate vicinity of point sources, fluctuating con-

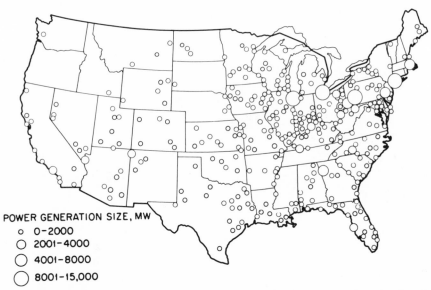

POWER GENERATION SIZE, MW
- 0–2000
- 2001–4000
- 4001–8000
- 8001–15,000

Fig. 28.1. Distribution of coal- and oil-fired power plants in the U.S.; 1981 update based on original figures from U.S. E.P.A. (1971).

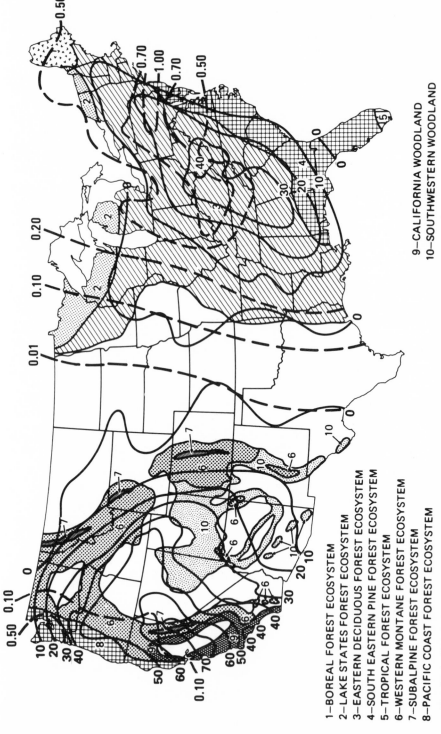

1–BOREAL FOREST ECOSYSTEM
2–LAKE STATES FOREST ECOSYSTEM
3–EASTERN DECIDUOUS FOREST ECOSYSTEM
4–SOUTH EASTERN PINE FOREST ECOSYSTEM
5–TROPICAL FOREST ECOSYSTEM
6–WESTERN MONTANE FOREST ECOSYSTEM
7–SUBALPINE FOREST ECOSYSTEM
8–PACIFIC COAST FOREST ECOSYSTEM

9–CALIFORNIA WOODLAND
10–SOUTHWESTERN WOODLAND

Fig. 28.2. Distribution of frequency isopleths (solid lines) for total numbers of forecast days with high meteorological potential for air pollution over a five-year period. Isopleths are shown in relation to major forest types of the United States (after Miller and McBride, 1975) and in relation to mean annual hydrogen-ion (kg ha^{-1} yr^{-1}; dashed lines) deposition in precipitation (after Henderson *et al.*, 1982).

centrations of SO$_2$ are superimposed on other, background pollutants, but in relatively "remote" areas chronic exposure to low concentrations of multiple pollutants is the norm, and SO$_2$ exerts much less significant influence.

Quantifying the effects of these types of exposures on forest productivity will require several types of data that are either not available or not accurate or detailed enough. The needed data include the following.

1. Better characterization of atmospheric chemistry at and within forest stands in rural areas.

2. Documentation of the effects that chronic exposure to SO$_2$ and associated regional-scale pollutants has on the growth of forest trees in the field under a range of pollutant concentrations.

3. Short- and long-term characterizations of the effects of SO$_2$ and associated pollutants on forest-stand growth, competition, and successional dynamics.

In the U.S., pollutants have not been described carefully in rural areas, because air-quality monitoring is concerned with measuring the exposures that urban populations have been subjected to. The initiative of the National Crop Loss Assessment Program (see Heck and Heagle, this volume) has increased the monitoring and analysis of nonurban exposures of vegetation to SO$_2$ and other pollutants, and should provide useful data for evaluating these exposures. In addition, the recent emphasis on quantifying wet and dry deposition of sulfur in forested ecosystems, which has developed because of interest in the effects of acid rain, should provide much more complete descriptions of inputs of sulfur dioxide to forests than have previously been available.

The effects of sulfur dioxide on forest-tree growth have been addressed most recently in reviews by Miller and McBride (1975), Kozlowski (1980a), and Smith (1981), as well as by Keller and by Kozlowski (this volume). Effects documented primarily from studies in the vicinity of point sources range from total destruction to changes in growth and composition; yet few data exist for subtle growth effects, which may be either positive, neutral, or negative, that will surely be caused by chronic exposure of forests to sulfur oxides and other air pollutants. Although we do not yet have an adequate data base from which to quantify these effects, we can find some tentative answers to some very important questions, by using existing models of forest growth and succession. Among these questions are the following.

1. How much air-pollution stress would be required to significantly alter forest growth and development?

2. How are stress effects integrated over time?

3. How important is competition in moderating or reinforcing induced stresses on individual species?

4. How are species responses integrated into the response of forest systems?

5. How important might site factors be in altering the rate or direction of these changes?

We can begin to address these questions, even at this early stage in our quantification of effects, largely because they concern the dynamics of forest growth

TABLE 28.1
Tests Performed on Various GAP Models

Type of model test	Structural response	Functional response
Verification Model can be made to predict known feature of a forest.	1. Consistent with structure and composition of forests in New Hampshire (JABOWA), Tennessee (FORET), Puerto Rico (FORICO), floodplain of the Mississippi River (FORMIS). 2. Compares to forest of known age in subtropical rainforests (KIAMBRAM). 3. Predict Arkansas upland forests based on 1859 reconnaissance (FORAR).	1. Predict forestry yield tables for Loblolly Pine (*Pinus taeda*) in Arkansas (FORAR). 2. Predict relations of forest types in succession in middle altitudinal zone in Australian Alps (BRIND). 3. Predict clearcut response in Arkansas wetlands (SWAMP). 4. Predict forest types changing as a function of flood frequency in Arkansas wetlands (SWAMP) and the Mississippi floodplain (FORMIS).
Validation Model independently predicts some known feature of a forest.	1. Predict frequency of trees of various diameters in rainforests in Puerto Rico (FORICO) and uplands in Arkansas (FORAR). 2. Predict vegetation change in response to elevation in New Hampshire (JABOWA) and in the Australian Alps (BRIND). 3. Calculate effects of hurricanes on the diversity of Puerto Rican rainforest (FORICO).	1. Predict response of *Eucalyptus* forests to fire (BRIND). 2. Assess effects of the Chestnut Blight on forest dynamics in southern Appalachian forests (FORET). 3. Predict forestry yield tables for Alpine Ash (*Eucalyptus delegatensis*) in New South Wales (BRIND).
Application Model is used to predict a response of a forest to changed conditions.	1. Predict changes in a 16,000-year pollen chronology from East Tennessee in response to climate change (FORET). 2. Assess habitat management schemes for endangered species (FORAR), non-game bird species (FORET), and ducks (SWAMP).	1. Predict response of northern hardwood forest to increased levels of CO_2 in atmosphere (JABOWA). 2. Predict response of southern Appalachian hardwood forest to decreased growth due to air pollutants (FORET). 3. Predict response of northern hardwood forest (JABOWA), southern Appalachian forest (FORET), Arkansas upland forests (FORAR), Arkansas wetlands (SWAMP), and Australian subtropical rainforest (KIAMBRAM) to various timber management schemes.

NOTE: The mnemonic in parentheses indicates the following models: BRIND, a model of Australian *Eucalyptus* Forests (Shugart and Noble, 1981); FORAR, a model of Arkansas Mixed Pine–Oak Forests (Mielke, Shugart, and West, 1978); FORET, a model of Tennessee Appalachian Hardwood Forests (Shugart and West, 1977); FORICO, a model of Puerto Rican Tabonuco Montane Rainforest (Doyle, 1981); FORMIS, a model of Mississippi River Floodplain Deciduous Forest (Tharp, 1978); JABOWA, a model of Northern Hardwood Forest (Botkin, Janak, and Wallis, 1972); KIAMBRAM, a model of Australian Subtropical Rainforest (Shugart et al., 1981); SWAMP, a model of Arkansas Wetlands Forest (Phipps, 1979).

and development, and the nature of basic forest-stand responses to changes in tree growth. These relationships have been the subject of much basic and applied research designed to develop and apply models of forest-stand growth. As a result, a wide variety of such models now exist, and can be used as tools to begin to address these questions.

Development and Applications of Some Forest-Growth Models

For the past two decades, models of forest growth have been developed enough to be useful as tools for predicting the responses of forest ecosystems. These models have been recently reviewed both in general (Munro, 1974; Shugart and West, 1980) and in terms of their usefulness for assessing stress effects on forests (Shugart, McLaughlin, and West, 1980). The principal points of these reviews may be summarized as follows.

1. The general quantitative technique of combining realistic models of individual tree growth with a population-dynamics approach that explicitly takes age (or size) structure into account has been successful. Several derivations of this approach appear to have been made simultaneously and seemingly independently both by foresters and by plant ecologists. The development of these models has been greatly facilitated by the increased availability of large digital computers.

2. These realistic models have been tested in applications and in their ability to reproduce independent data. For the most part, these tests have been successful, even when the models have been used to predict new conditions.

3. There are problems in obtaining parameters for some of these models, particularly for noncommercial tree species. Some of the models are structured to simulate conditions found in plantations, and are difficult to apply to more natural forests.

4. There seems to be a trade-off between extremely realistic model parameterizations that require the sort of highly detailed data usually available only for commercial species and the more general model parameterizations that can be used in mixed-species forests with many less well-studied species.

Our own interests have been in learning how to predict the effects of stress on "natural" mixed-species and mixed-aged forests, and we have used a category of forest models called "gap" models (Shugart and West, 1980) toward this end. Gap models simulate the annual changes in diameter of each tree on a plot of a fixed area (ca. 1/100 to 1/10 ha). The models do not account for the exact position of each tree. From a tree's diameter, its height can be computed, and these height relationships are used to calculate competition due to shading. This one-dimensional approach to the modeling of competition greatly reduces the computer time required to solve the models, and also reduces the amount of detail needed to parameterize the models. These models have been used in a variety of natural forests (Table 28.1), and were the first detailed succession simulators to be applied to air pollution effects research (McLaughlin *et al.*, 1978; West, McLaughlin, and Shugart, 1980).

The Use of Gap Models to Forecast Stress Effects on Forest Ecosystems

Gap models have been used to evaluate stresses of various sorts on ecosystems. Botkin, Janak, and Wallis (1972) drove the JABOWA Model with different climatic conditions to test it against altitudinal gradients. This and other gradient tests of gap models (Table 28.1) provide some support for the idea that the models may be able to predict the response of a forest that has been stressed by novel environmental conditions. Botkin (1973) has investigated the effects of CO_2 increases, and found that an arbitrarily assumed percentage change in the rate of photosynthate production was not manifested directly as a change in forest growth, because competition and shading among trees, for example, lowered the magnitude of the simulated forest response. We (McLaughlin *et al.*, 1978; West, McLaughlin, and Shugart, 1980) have performed gap-model experiments on the effect of chronic air-pollution stress (expressed as a systematic change in the growth rates of pollutant-sensitive trees), as have Kercher, King, and Bingham (1982). The results of these two studies are compatible. Let us consider some specific examples of results from West, McLaughlin, and Shugart (1980), as follows.

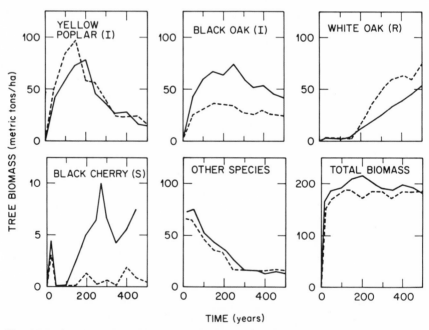

Fig. 28.3. Species and stand dynamics of a forest with (dashed line) and without (solid line) continuous exposure to air pollution stress. Stress was applied as a differential chronic growth inhibition to sensitive ($S = 20\%$), intermediate ($I = 10\%$), and resistant ($R = 0\%$) species within the forest stand.

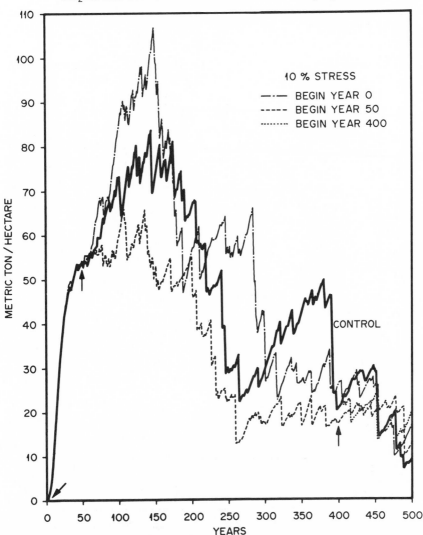

Fig. 28.4. Biomass changes for yellow poplar in a mixed forest stand in which air-pollution stress was initiated at years 0, 50, and 400 during a 500-year simulation. Growth stress imposed on yellow poplar was 10%.

1. Profound effects on the species' population can be caused by even relatively small (5 to 10%) changes in the annual growth rate. For example, yellow poplar actually increases in simulated total-stand biomass even in the face of a 10% growth reduction, but Black Oak is drastically reduced in total biomass given the same 10% reduction (Fig. 28.3).

2. The total ecosystem response to pollutants can be greater than that pre-

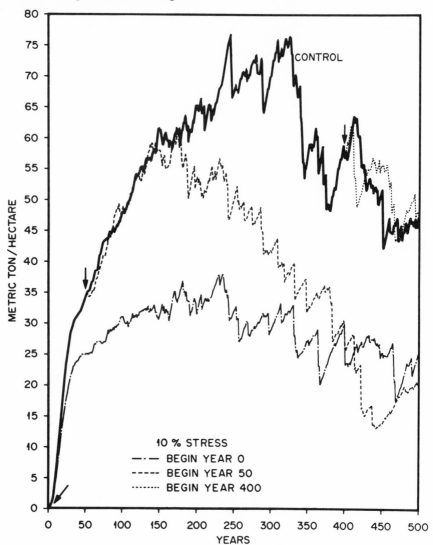

Fig. 28.5. Biomass changes for black oak in a mixed forest stand in which air-pollution stress was initiated at years 0, 50, and 400 during a 500-year simulation. Growth stress imposed on black oak was 10%.

dicted by the tree species responses. In Fig. 28.3, because the reduction in growth of sensitive species is not totally compensated for by resistant species, the maximum response of the forest stand (Total Biomass) is often greater than the additive response of the component species. In a 200-year-old stand, the predicted stand-level response that one might project from the average growth reduction of the seven most important species (which make up 87% of the biomass) would be 7%. The actual reduction in stand biomass at year 200 is 20%, or a threefold increase in the effect at that time (Fig. 28.3, Total Biomass).

3. The age (or stage of development) of a forest stand when the pollution is introduced can radically alter the response of the stand. The nature of this response can be different among different species even when they are equivalently reduced in growth rate. Figs. 28.4 and 28.5 show the response of yellow poplar and black oak with a 10% stress applied to forests of different ages. With yellow poplar (Fig. 28.4), total growth was actually increased when a seedling forest was stressed. When the same level of stress (a 10% growth reduction for pollution-intermediate sensitivity) was applied to a 50-year-old stand and a 400-year-old stand, growth was reduced. Black oak (Fig. 28.5) when similarly stressed shows an immediate negative response in the seedling forest. When the stress is introduced in the 50-year-old forest, there is an approximately 100-year delay in this negative response. Thus black oak and yellow poplar respond to pollutant stress very differently, even though the growth reductions for the two species were identical.

Possible Influence of Site Factors on Forest Response

The work summarized in the preceding indicates that changes (of about 5 to 20%) in the growth rates of individual trees can be manifested as forest-stand effects that differ in both magnitude and sign. It is reasonable to ask whether these synergistic effects are regular within a suite of forest-tree species across natural environmental gradients. To address this problem, an extended version of the FORET model (Shugart and West, 1977) that considers the moisture response of each species was used to inspect the air-pollutant response that is manifested as a zero, 5, and 10% reduction in annual growth for species of apparent pollutant tolerance, relative tolerance, and sensitivity (as tabulated in West, McLaughlin, and Shugart, 1980). The soil-moisture addition to the FORET model has been developed by Solomon, Blasing, and Post (unpublished manuscript), who have used the model to reconstruct palaeolandscapes associated with a 32,000-year-old fossil pollen chronology in east Tennessee. We will outline in general how soil moisture and growth response is simulated in their model.

There are two problems in developing a proper annual model of soil moisture/ tree growth on an annual time-scale. The first problem is to develop a soil-moisture simulator; the second is to calibrate the growth response to moisture stress of the mixture of species in a diverse mixed forest. Because Solomon, Blasing, and Post were interested in palaeoreconstruction, the formulation of models to treat these two problems had to be very general (to accommodate the wide range of simulated conditions) and reasonably simple (to accommodate the great number of tree species considered). They therefore dealt with ecological phenomena to a degree of detail compatible with that used in other parts of the FORET model.

Soil moisture is simulated using a Thornthwaite equation to predict evapotranspiration from the soil rooting zone. Since the Thornthwaite equation requires monthly temperature and precipitation, a stochastic simulator with monthly mean and variance appropriate to the Oak Ridge, TN, meteorological record was added to the model for this application. Soil physical properties and depth were used to compute the number of days during the growing season that the soil-moisture tension was below −15 bars. This drought index was used to adjust the growth rate of trees in a manner proportional to their sensitivity to moisture stress.

The growth of trees was diminished linearly, from optimal growth at no moisture stress (drought index = 0.0) to no growth at the maximum moisture stress that the species appeared to tolerate. This maximum moisture stress was calculated by computing the moisture index for every county in eastern North America, and by then computer-overlaying the range of a given species to find the most extreme value of the moisture index at which the species seems to grow. This rather simple scheme for estimating a parameter has also been used by Weinstein, Shugart, and West (1982) to independently predict the annual increment of trees remeasured annually for 12 years on Walker Branch Watershed (a research site in the Oak Ridge DOE Reservation); they also used this algorithm in the FORET model to predict the gradient response of forested landscapes in the Great Smoky Mountains National Park. Solomon, Blasing, and Post have also successfully applied this function to reconstruct forested landscapes during, e.g., 10,000's of years in response to climatic changes.

To inspect how pollutant-induced growth reductions might affect forests of different ages and on different sites, we devised the following model experiment. A set of 200 simulated plots (each of 1/12 ha) were projected through a successional sequence, starting with a bare soil with ample regeneration for all species and running for 500 years. Of these plots, 50 were grown on a soil with an effective rooting zone of 24 inches; 50 with a rooting zone of 12 inches; 50 with 6 inches; and 50 with a 3-inch rooting zone. The plots were all subjected to the variations in mean monthly temperature and precipitation for Oak Ridge, TN. By rerunning the model several times for these conditions, we were able to introduce the growth-reducing effect of chronic stress at year 0 (young stand), year 100, year 200, and year 300. These runs were compared to a control simulation that included no stress effects. To assess the effects of pollutants, we expressed the integral of total biomass for a species and for the stand as a percentage of the integral of the control for 50 years, 150 years, and 200 years. This comparison reduces the rather large data set to fewer, more easily compared relative percentages.

For example, the results for black oak (*Q. velutina*), a species of intermediate sensitivity to air-pollutant stress (5% growth reduction), showed a tendency for this growth reduction to be amplified in almost all conditions regardless of the site factors or age of the stand (Table 28.2). However, particularly in young stands on dry sites, the effect of growth stage is more intense than in other cases.

TABLE 28.2
Black Oak Intermediate Stress (5% yr^{-1})

Stand age at stress initiation (years)	Biomass change (percent of control)				Duration of stress (years)
0	0.85%	0.87%	0.85%	0.78%	
100	0.99	0.98	0.96	0.96	50
200	1.02	0.93	0.89	1.00	
300	0.98	1.01	0.99	0.95	
0	0.81%	0.81%	0.82%	0.74%	
100	0.99	0.94	0.91	0.92	100
200	0.88	0.89	0.84	0.91	
300	0.94	1.13	1.03	0.99	
0	0.85%	0.83%	0.84%	0.73%	
100	0.95	0.96	0.89	1.04	200
200	0.88	0.83	0.89	0.89	
300	0.88	1.20	0.93	0.95	

SITE MOISTURE STATUS

Wet ————————————————————→ Dry

Fig. 28.6 summarizes the responses of several important species and the total biomass of the stand for young forests subjected to simulated pollutant stress, expressed as the integral of biomass over 50 years. White oak, a pollutant-tolerant species, grows better under pollutant stress. The increase in white oak is particularly pronounced on intermediate sites along the simulated moisture gradient. Yellow poplar is largely unaffected by pollutant stress on mesic sites, but is strongly affected and eventually eliminated as site conditions become dryer. Black oak is strongly affected by pollutant-caused growth reductions. The reduction in biomass for the total forest is about the same as the overall growth reduction. When the same comparison of responses is made for forests stressed at age 100, when competition for canopy position is particularly intense (Fig. 28.7), the effects of pollutant-induced growth reductions are most pronounced in mesic sites. White oak is aided tremendously in its competition for canopy dominance, and yellow poplar is strongly reduced in a similar manner (Fig. 28.7). Total biomass of the stand is strongly reduced relative to the control on mesic sites, because of elevated mortality of already-stressed canopy competitors and because of the composition shift of the forest. Black oak, which was strongly affected in young forests (Fig. 28.6), is not particularly affected in intermediate-aged forests (Fig. 28.7).

Thus there is a strong interaction effect between stand age and site conditions that can amplify the already documented interaction between pollutant response of a forest and stand age (see Fig. 5, West, McLaughlin, and Shugart, 1980). In general, these results follow a pattern in which a strong interaction intensifies the growth reduction attributed to pollutants. A pollutant response interacts with either age or site factors in forests in which one species is engaged in a strong but

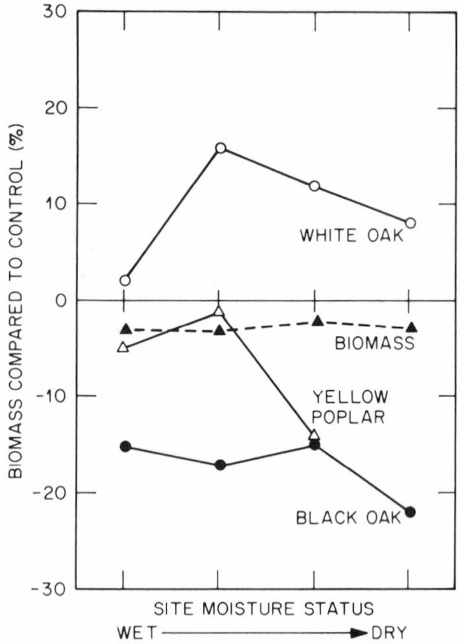

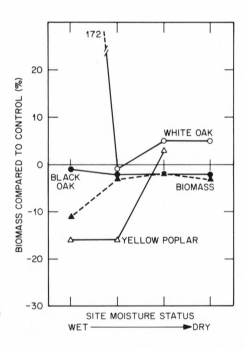

Fig. 28.6. Seedling forest 50 years later. Responses of white oak, black oak, yellow poplar, and total stand biomass (expressed as percent of an unpolluted control case) for 50 years following initiation of the stress for young (year 0) stands.

Fig. 28.7. Hundred-year-old forest 50 years later. Responses of white oak, black oak, yellow poplar, and total stand biomass (expressed as percent of an unpolluted control case) for 50 years following initiation of the stress for intermediate-aged (year 100) stands.

apparently fairly even competition with another species (or suite of species). Such cases, in natural systems, are sensitive to relatively small changes in growth rates, particularly when one of the species is favored by the change. When a species is already disadvantaged in its competitive interactions because of its own ecological attributes, the stand age, the degree of moisture stress, or, more often, a complex combination of species attributes, stand, and site factors, then the effects can virtually eliminate it; such species are, therefore, typically less common.

Conclusions

The influence of site factors (soil-moisture conditions) and stand age on systematic growth reductions in trees can sometimes greatly amplify these growth reductions. The pattern of the changes shows strong interaction effects. We see two important implications of the theoretical model results developed in this chapter. First, if these effects are found in actual forest systems, it may be very difficult to project the consequences of pollutant stress on tree species from laboratory to field conditions. This potential difficulty can compound the already difficult problem of understanding the whole-tree response to pollutant stress. Hence we need a better understanding of forest ecosystems as whole systems before we can evaluate more physiologically oriented studies on pollutant stress on plants.

Second, extreme care is needed in the design of field studies of ecosystem effects of pollutants. The complex web of interactions between site factors, age, and species indicated by this study could be diabolically difficult to unravel in studies that did try hard to control them. In particular, indiscriminate comparisons between forests that differ in stand age, stand structure, and site factors could well produce a tangle of seemingly conflicting results. Our results indicate that forest responses to SO₂ and other stresses will be strongly controlled by, and will control, the successional dynamics of the affected stands. For this reason, efforts to better quantify and describe the dynamics of forest-stand growth and development will continue to be essential to predicting effects of chronic pollution on forest-stand growth and development.

Acknowledgment

The research in this paper has been sponsored jointly by the National Science Foundation and the U.S. Environmental Protection Agency under Interagency Agreements DEB80-21024 and 40-740-78 with the U.S. Department of Energy under contract W-7405-eng-26 with Union Carbide Corporation. Publication No. 2264, Environmental Sciences Division, Oak Ridge National Laboratory.

Part 5
Summary

29

An Assessment of Experimental Methods in Research into Air-Pollutant Effects on Vegetation

R. J. Oshima

This paper assesses experimental methods used in air-pollution research on vegetation. It contains an evaluation of exposure methods, of the strengths and limitations of experimental designs, of air-pollution exposure statistics, and of methods used to characterize plant response, and proposes a standard for defining loss.

Exposure Systems

Research on the effects of air pollution on plants and ecosystems has unique problems because of the physical properties of the phytotoxic gaseous pollutants and the characteristics of their occurrence. Exposure chambers or other apparatus to maintain controlled exposures range from sophisticated, microprocessor-controlled cuvettes (Bingham and Coyne, 1977; Legge, Savage, and Walker, 1979) to a series of tubes with calibrated orifices spatially distributed throughout a field to emit gaseous pollutants (Lee, Preston, and Lewis, 1978). Each exposure system operates most efficiently in the environment for which it was developed. Each has advantages and limitations, and must be evaluated in terms of the objectives it was designed to meet.

Laboratory systems. Laboratory exposure systems range in complexity from simple bubblers that allow gases to permeate solutions (Mudd and McManus, 1969) to gas-exchange chambers designed to provide controlled environmental conditions (Tingey *et al.*, 1979; Winner and Mooney, 1980a). Modifications to environmental chambers may be as simple as inserting a tube to introduce the pollutant into the air-intake system and filtering the exhaust (Fong and Heath, 1981). More sophisticated systems use mass-flow detectors and sophisticated valving to control computer-programmed exposure regimes (McLaughlin *et al.*, 1979). Rogers *et al.* (1977) developed the CSTR (continuous stirred tank reactor), a cylindrical chamber system with mixing properties similar to a sphere, to study pollutant uptake. This system has been widely adapted for general use as an exposure chamber in greenhouse and field exposure systems (Heck, Dunning, and Johnson, 1968; Oshima, 1978).

Laboratory systems typically use artificial lighting and controlled environments. Most were designed to identify and measure effects at anywhere from the subcellular level to that of the whole plant, and are used for basic research.

Mechanistic studies remain the key to defining and understanding the mode of action of pollutant stress. They provide the information describing the perturbations of biological processes that are expressed as injury and reduced growth and yield. Ultimately, understanding interactions with other pollutants and environmental variables, progress in plant breeding for greater resistance, and the formulation of sound ecological strategies will require the knowledge that can be derived only from mechanistic studies. When viewed in perspective, research documenting plant effects defines the magnitude of the problem, but only mechanistic studies define its nature.

Greenhouse exposure systems. Greenhouse exposure systems usually consist of a series of chambers built with various framework materials and covered with a transparent film. The air-exchange systems normally use a negative-pressure, single-direction air flow, and an activated-carbon filtration device on both air entry and air exhaust. Early systems were usually modifications of the system developed by Heck, Dunning, and Johnson (1968), but a variety of designs were constructed. These systems were all designed to have commonly desirable chamber characteristics, and succeeded to various degrees. Exposure chambers were designed to provide uniform pollutant concentrations with minimal environmental alteration. The development of the CSTR design stimulated the development of exposure systems that incorporated its desirable mixing properties and its use of FEP Teflon film as an inert polymer film.

Greenhouse systems are used in studies focusing on the organ and plant level of biological organization. Plants are usually grown in containers in greenhouses, moved in and out of chambers for exposures, and allowed to grow on greenhouse benches during interim periods. Exposures occur under natural or artificial lighting in chambers residing in the greenhouse. Normally, a single plant or a small group of plants is the experimental unit. These systems are used to identify and quantify physiological, growth, and yield responses. Although more closely related to field studies than to laboratory experiments, greenhouse studies differ sufficiently to make direct extrapolations inappropriate.

Field exposure systems. Normally, it is desirable to minimize deviations from the ambient environment, and to simulate as closely as possible the conditions characteristic of agriculture or natural ecosystems. Field exposure systems range from extrapolations of the greenhouse and laboratory chamber designs to the use of chemical protectants. The progress of research on air-pollution effects on vegetation is paralleled by the evolution of innovative designs in field exposure systems.

Field chamber systems. The open-top chamber system (Heagle, Body, and Heck, 1973; Mandl *et al.*, 1976) is the most popular field-exposure design in use. Essentially an upright cylinder with a clear polymer film as a covering, this design has the advantages of portability, moderate cost, and ease of maintenance. The system uses high-volume filtered air flow to control exposures. Although some intrusion of ambient air occurs over the upper lip of the cylinder, innovations in design (Kats, Thompson, and Kuby, 1976; Davis and Rogers, 1980) reduce the intrusions to acceptable levels. The system's portability facilitates storage and maintenance during the winter or in periods of inactivity, and allows standard agricultural practices to be carried out during field preparation, seeding, and early crop growth, before chambers are set in place.

Other field exposure systems have used chambers of various designs, but all have been fully enclosed by film (Thompson and Taylor, 1966; Oshima, 1978). These designs rely on high air-exchange rates in the chamber to minimize temperature alterations. Most of these designs are an extrapolation or alteration of greenhouse exposure systems. Chamber shapes range from a square design, originally developed by Heck, Dunning, and Johnson (1968), to the CSTR cylinder developed by Rogers *et al.* (1977).

Field exposure systems without chambers. The desire to be able to expose large field plots to increase sample size and to remove environmental alterations caused by enclosing plants in chambers led to the development of field-exposure systems without chambers. The advantages of these systems lie in their ability to expose plants to pollutants under ambient-like conditions. This advantage is offset by the disadvantage of losing some control over the concentration of fumigants and the nature of the exposure. These systems are highly influenced by wind speed and direction.

The zonal air pollution system (ZAPS) (Lee, Preston, and Lewis, 1978) consists of an array of tubing or pipes that are distributed throughout the field plot. The tubes contain precalibrated orifices through which the pressurized fumigant is released. Actual air concentrations are a function of wind speed and source strength. Arrays with a uniform distribution of tubes within a plot have less variation in fumigant concentration, but do not provide the exposure control characteristic of chambers. The original ZAPS design was subsequently altered to meet specific exposure requirements. One basic alteration has been the release of the fumigants from uprights, i.e., tubing rising vertically from the ground, with orifices releasing the fumigant along the vertical axis (DeCormis, Borte, and Tisne, 1975). This innovation allows for a more uniform vertical distribution of fumigant, which is especially important when one is exposing trees or other tall perennials. Others have used only a single tube to emit the gases, and have taken advantage of exposures along the decreasing gradient of fumigant concentration downwind of the tube (Miller *et al.*, 1980). A modification of the ZAPS design may be the best system for exposure of perennial or tree crops, because it would cost too much to develop and construct a chamber system large enough to accommodate an acceptable sample size.

A linear gradient systems (LGS) has been used as both an exclusion and a fumigation system in the field (Reich, Amundson, and Lassoie, 1980; Laurence *et al.*, 1982). This system uses a series of inflated polymer-film tubes under positive pressure to distribute either carbon-filtered air or a known mixture of fumigant and filtered air. The inflated tubes are positioned between rows of plants, and have precalibrated holes directed at plant canopies. The filtered or fumigant gas mixtures are emitted from the holes and permeate the canopies, allowing either fumigant exposure or filtered-air exclusion. This system, like the ZAPS, is sensitive to wind speed and direction.

Field-exposure studies most closely approximate ambient conditions and may be representative of small agricultural plots. Plants can be grown to yield under commercial cultural practices without the use of containers. Only relatively minor chamber influences and/or artificial exposure patterns differentiate them from field plots. However, no formal comparison of field exposure plots with commercial agricultural fields has yet been conducted.

Gas-exchange systems for field use. Two sophisticated cuvette and porometer systems have been developed for gas-exchange studies in the field (Bingham and Coyne, 1977; Legge, Savage, and Walker, 1979). Both can control temperature and humidity while measuring leaf gas-exchange parameters. Although these systems are not exposure systems in the truest sense, they represent significant accomplishments in methods development.

Summary. The exposure systems discussed in this section share many common characteristics. Each uses a monitoring system to sample pollutant concentrations continuously, or incorporates a timesharing system that sequentially samples chambers or field-exposure sites. Systems normally use inert Teflon tubing for sampling lines, and continuous air flow to reduce sampling time lags.

Additionally, they are equipped with monitoring instruments that employ EPA-equivalent approved methods of detection. Recently, quality-assurance programs have been developed in order to insure that high-quality, standardized air-monitoring data will be available and readily comparable. Fumigants are generated artificially and dispensed to exposure chambers or field plots, or else proportional activated-carbon filtration is used to provide different ambient concentrations of pollutants.

The systems described in this section represent significant advances in the methods used in air-pollution research on vegetation. As new systems are devised, incorporating the latest technological advances, it is easy to be caught up in the rapid progress of their evolution and lose sight of their limitations. Even the most sophisticated and advanced systems are only as good as the experimenters who use them. The systems do not insure that the research conducted with them will be of superior quality. They only provide the potential for better understanding the impact of air pollutants on vegetation.

Experimental Design

It is important to understand the limitations implicit in an experimental design in order to properly evaluate results. Historically, replicated factorial designs that lend themselves readily to analyses of variance have been the most popular. However, these designs have generally been used only in descriptive experiments using treatment-mean separations with multiple range tests. Designs of this type assign a certain probability to treatment separation, and are usually used to measure plant responses to different amounts of pollutant exposure. Experiments using these designs do not quantify responses in terms of a functional relationship unless they have been specifically designed to incorporate multiple treatment levels and orthogonal polynomials. Only recently have analysis-of-variance designs been used to partition interactions of interest (Oshima, 1978; Reinert, Shriner, and Rawlings, 1982).

Experimental designs amenable to regression analyses are used for a great variety of objectives, such as empirical modeling, convenient data summarization and representation, and hypothesis generation. Despite their utility, they are also the most misused statistical tool. Data must meet several important criteria before regression analyses can be used (Draper and Smith, 1966). These techniques do not establish a causal relationship. Correlation is only a measure of the strength of a relationship in terms of a correlation coefficient (r). The value of establishing the relationship lies in the function produced by the regression. This function provides an empirical model that may be used to predict a plant response from the exposure within the range of the experimental data.

Model validation must be properly defined, since two distinct processes are involved. In one, models must be tested to see if they are representative of the data they were derived from. Statistical tests can be employed to measure goodness of fit. The uncertainty surrounding model estimates can also be calculated. This first process is called model validation. A second type of validation is needed if the model is to be applied to other independent populations. For example, if a crop-loss model is derived from experimental data and is to be used to predict commercial field losses, it should be tested to see if it is representative of commercial fields. This can be done by comparing actual losses from commercial fields with the estimated losses predicted by the model. This second type of validation might be called application validation. The two validation procedures are distinct, and have different objectives.

The current controversy about the merits of replicated versus unreplicated experimental designs has lost proper perspective. Each type of design can be appropriate, depending on the specific experimental objectives. When the development of empirical models for crop-loss estimates is identified as the objective, unreplicated designs have several advantages. They allow more efficient use of exposure systems, since treatment replicates are unnecessary and resources can be used to increase treatments to better define functional forms. Although the uncertainty in treatment responses cannot be estimated, the error associated with the response function and predicted values can be.

Replicated designs are preferable when the experimental objectives require establishing causality or when it is necessary to make specific comparisons between treatments. In these situations, where it is necessary to estimate the error associated with a treatment response, replication is required.

The experimental design focuses an experiment on its specific objective, but, in doing so, limits the application of the results. No design has universal application. The selection of the appropriate design for specific objectives is therefore the most critical step for both the success of the study and the applicability of the results.

Characterizing Pollutant Exposures and Plant Responses

Schwela and Junker (1978) described two procedures to develop air-quality standards: (1) by agreement between different interested groups and the administration or (2) by derivation from suitable relationships between the risks that effects might occur and corresponding dose values.

The federal air-quality standards in the United States have historically evolved by the first procedure. Shoji and Tsukatani (1973) commented on the same procedure in Japan and its shortcomings. It is extremely difficult to justify or attack a standard evolved by agreement, since no documented scientific process is followed. It is, however, easy to defend such a standard in terms of economic feasibility.

The adoption of the first procedure, generation of air-quality standards by agreement, should not exclude progress in deriving standards from a logical scientific process. The sequence of steps outlined by Schwela and Junker (1978) appears to be an excellent starting place. Research with objectives focused on such a sequence would, at the very least, provide the basis for evaluating the current air-quality standards.

Two problem areas of immediate concern are (1) the statistical representation used to summarize plant exposures to air pollutants for meaningful analysis, and (2) the methods by which plant responses are characterized. Both problem areas currently impede progress toward the derivation of secondary air-quality standards.

Characterization of pollutant exposures. Current federal secondary air-quality standards are formatted in time steps of short duration, 1- to 3-hour averaging times. Although convenient for air-quality summarization in terms of percentiles, the short time steps are not relevant to plant growth (which takes place in seasonal time steps), and cannot be aggregated to represent longer time periods. It is this disparity that frustrates both decision makers in regulatory agencies and scientists doing research on effects of air pollution on vegetation. The following discussion comments on theoretical considerations that affect the choice of statistics to summarize plant exposures, evaluates the current methods being used to represent seasonal air-pollution exposures, and identifies new approaches.

Plant growth is a progression of identifiable stages, from seed germination to seed production. This process occurs during a seasonal time step, usually of long duration (30 to 180 days). Fig. 29.1 represents a seasonal exposure for a growing season. Daily doses (concentration × time) are plotted over time for convenience, but the following points are equally valid for hourly averages or instantaneous readings. Air-pollution exposures normally occur as a time-series of randomly distributed events within the season. Elevated concentrations of air pollutants or a series of elevated daily exposures may coincide with one or more stages of plant development, and thus create a unique exposure pattern. It is apparent, however, that many other time-series with an equivalent seasonal exposure can occur during a season. Conceptually, this can be illustrated by rearrangements of the daily exposures in Fig. 29.1. Every stage of plant growth

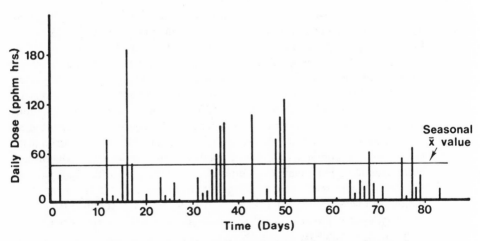

Fig. 29.1. Seasonal distribution of daily pollutant doses and the seasonal mean dose. Daily dose is calculated by summing the hourly averages for a 24-hour period.

could theoretically be affected by the higher pollution concentrations characteristic of one of the many possible time-series. Which stage of plant growth will be most affected by a specific year's pollution time-series depends on chance. Research must be initiated to see if stages of plant growth are equally or differently sensitive to pollutant exposures in terms of ultimate yield. This information is not now known.

What summary statistic should be chosen to represent plant exposure depends on whether one must include plant growth-stage sensitivity. A simple summary statistic, such as a seasonal mean or a cumulative dose, cannot represent the temporal dynamics of the seasonal exposure. This type of summary statistic does not differentiate between specific pollutant time-series of equivalent exposure.

Current methods of representing seasonal pollution exposure. The selection of the most appropriate summary statistic to represent air-pollution exposure should be made after considering the potential differences in sensitivity of various stages of plant growth. This information was not available when exposure summary statistics were selected. The statistics selected for use by investigators were therefore discretionary.

Mean concentrations for various averaging times have been the most popular exposure statistic used. A mean can be represented as a horizontal line parallel with the *x*-axis (Fig. 29.1). The use of this average value implies that each increment of time in the seasonal exposure is equally weighted in producing the plant response, and all stages of plant growth are equally sensitive. Temporal variations within the season, for example, the relative effect of high daily exposures occurring at specific stages of plant growth, are eliminated by the use of this statistic.

The use of a seasonal dose (concentration × time) also ignores the temporal dynamics of pollutants within the season. Seasonal dose can be represented in Fig. 29.1 as a single vertical line whose amplitude is equal to the sum of all the daily doses. Again, the implicit assumption is that exposures within the season occurring at specific and perhaps critical periods of plant growth are unimportant.

Specific distribution functions have also been used to characterize seasonal exposures (Georgopoulos and Seinfeld, 1982). Log normal and Weibull functions, for example, are very useful for summarizing the concentration distributions within a specific period, but do not describe the temporal dynamics. Each distribution function comprises an infinite number of time-series with different temporal dynamics. For example, a series of rearrangements of the daily doses in Fig. 29.1 would produce different pollutant time-series with the same distribution parameters as the original. One could theoretically continue rearranging the daily doses to produce an almost infinite number of time-series with the same distribution parameters.

When a large time period, such as a growing season, is considered, we can take one of two approaches.

30

Toward a Mechanistic Understanding of Plant Responses to SO$_2$

Richard L. Olson, Jr., and Peter J. H. Sharpe

In all science there is a dichotomy between pure and applied research. This dichotomy can be particularly troublesome when the line between the two types of research is blurred, as it is in work on the effects of SO$_2$ pollution on plants. In a sense, all pollution research is applied: the aim is to control adverse effects. A more meaningful distinction for SO$_2$ research might be whole-plant dose response versus metabolic mechanism. The former is usually investigated by fumigating plants (or populations) and measuring the pollutant effect on an organism-level variable, such as growth. The metabolic-mechanism approach is often undertaken at the biochemical level *in vitro*. This distinction is readily apparent in the emphases of different sections of these symposium proceedings. At one extreme, research by Bell and others is conducted at the plant-response level and above. At the other, papers by Wellburn and others discussed primarily *in vitro* experimentation at the biochemical level.

In the literature on sulfur-dioxide effects on plants, there has been little effort to integrate our knowledge of whole-plant responses with what we know about biochemical reactions, and that little seems inconclusive at best. Even more disturbing is a tendency to compartmentalize rather than synthesize research knowledge.

Several factors have contributed to the gap between the two approaches. One is the shrinking research dollar. In a time of reduced funding levels, granting agencies are reluctant to fund long-term biochemical studies that cannot yet be integrated with higher-level work. They tend to fund studies that have a probability of rapid applicability, in short, whole-plant dose-response work. Another factor is the political nature of pollution problems. Regulatory agencies are charged with arriving at standards for air and water quality in a (relatively) short time. The public safety, not to mention that of natural and cultivated plant resources, depends on a rapid setting of these standards. One might add that political careers can also depend on this. Naturally, money is channeled into research that appears able to satisfy these demands. Chapters in this volume by Heck and Heagle and by Jones present schemes for the systematic quantification of whole-plant dose responses. However, these authors seem to advocate essentially the abandonment of basic metabolic studies. Obviously, one cannot advocate the reverse, that is, the concentration of all effort into biochemical research at the expense of whole-plant dose-response studies. However, knowledge of basic biochemical and cellular mechanisms is essential if we are to understand the various responses of plants to pollutants over a wide range of environmental growth conditions and plant genotypes.

Kercher and King (this volume) describe the modeling process as it relates to pollution effects on the productivity and growth of individual plants. They describe two broad categories of models: descriptive and process. The latter category is subdivided into mechanistic, physiological, and phenomenological classes. For the purposes of our discussion, we would like to divide models into two categories: empirical and mechanistic. As always, there are authors who champion each method. Thornley (1976), an advocate of the mechanistic model, described it as "couched in terms of mechanisms or how the parts of the system work together, as in a machine." The empirical model, however, "simply redescribes the data and does not give rise to any information that is not contained in the data." Hunt (1978) argues for the use of empirical models, in particular, fitted growth equations. He claims that mechanistic models rarely approach reality, and that "whole plants rarely behave as machines." He also points out that for mechanistic models to approach reality, "great elaboration" (beyond that of a few simple equations) is necessary.

Although Hunt is correct that plants do not behave like machines, a mechanistic model can have advantages that far outweigh its disadvantages. It can, if carefully constructed, provide enough realism to enable one to make useful predictions. It can also enable one to manipulate biotic and abiotic component interactions symbolically, rather than by means of (or, better, in addition to) long and costly experiments. The empirical approach can certainly be useful for some needs of the modeler. It enables one to get a rapid description of data in a manner that can point out important correlations. However, this approach is limited, in that biological information beyond correlation (e.g., causality) cannot be represented in an empirical model.

Mathematical models can provide the necessary links between basic and applied research. We will call these research models. In formulating such models, the objective is to represent basic knowledge of the dynamics at the biochemical and physiological levels. Such models are usually complex, but considerably less so than the underlying metabolism itself. Some degree of abstraction is mandatory; otherwise biochemical detail will overwhelm the modeler. The attempt to construct basic research models plays a number of roles. These include integrating metabolic components, and stating knowledge in a form in which it is immediately available for use in solving more applied research problems.

For some applications, a research model may be too cumbersome, especially for routine forecasting, optimization, or economic analysis. Under these circumstances, the research model can be further abstracted into what might be termed an "applications model." To be useful, the applications model must include the essential predictions of the basic research studies. The parameters should be interpretable in terms of the underlying processes and have clearly defined units.

Only rarely do pollution models satisfy such criteria. Most mathematical models ignore basic biochemical processes and physiological interactions. They generally represent a mathematical summary or regression of experimental results (i.e., empirical). Although this type of model is a useful summary of given experiments or data, it does not include any knowledge not contained in the data itself. It is invalid to extrapolate a model of this type beyond its database.

Synthesis of dose-response and metabolic mechanism research must begin with concepts defined from basic investigations. These can be linked via mechanistic models to studies under less-controlled conditions. In forging this linkage, at least two types of experimental data are required. In the basic-research phase, experimental data are collected under rigidly controlled environments. Data may be collected on the dynamics of all or part of the system. The scientific method then dictates that, from these observations, one or more hypotheses be formulated. It is from these hypotheses that the basic research model is constructed. A schematic for integrating whole-plant dose-response and metabolic-mechanism research results is shown in Fig. 30.1. Arrows provide feedback for applied researchers and regulators to guide the direction of basic research when hypotheses are shown to be inadequate.

This chapter is not meant to be a comprehensive review of the other chapters of this volume. In the following discussion, we attempt to identify several areas of SO_2 research where mechanistic modeling can be especially useful. We start with effects within the leaf, touch briefly on the organismal level, and then leap to the community level.

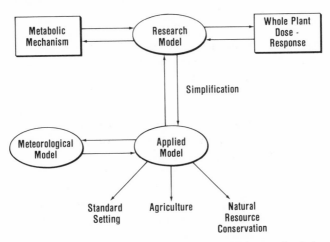

Fig. 30.1. Postulated linkage between metabolic mechanism and whole-plant dose-response research by a mechanistic research model. The research model can be simplified into an applied one, which can interact with, for example, a meteorological model.

Modeling Pollutant Effects

A useful conceptual model of SO_2 interaction with plants has been proposed by Tingey and Olszyk, in Chapter 11. They have divided pollutant and plant interactions into four compartments: leaf conductance, perturbation, homeostasis, and injury. Rather than review the state of knowledge about each compartment (as has been done extensively elsewhere in this book), we will use the first two compartments as a framework for our discussion of plant-pollution response.

First, however, we must consider exactly what it is we are going to examine. There is a tendency among some researchers to equate plant responses with injury. It should be remembered that there is a continuum of responses to SO_2, from normal use as a nutrient to irreversible injury. For example, reduced photosynthetic capacity, leading to reduction in growth and yield, can be caused by SO_2 action within the normal context of mesophyll biochemistry. It could also be caused by a decrease in stomatal aperture in response to the pollutant, although Winner *et al.* (this volume) have stated that SO_2 exposure may uncouple conductance and photosynthesis. Sulfur is a normal constituent of plant biochemistry, and many effects of small sulfur dosages may be due to shifting internal metabolic balances within the plant. Biologists should be concerned not only with injury, but also with the spectrum of responses exhibited. The eventual, admittedly ambitious goal of a pollution-modeling effort should be to describe in some manner all significant responses of the plant to the pollutant. In this way, research at many different levels of organization can be integrated into a meaningful whole.

Fig. 30.2 summarizes our view of some of the processes that should be included in a research model of SO_2 effects on leaf function. There is a resistance pathway to the mesophyll, with internal pollutant concentration controlled dynamically by a continuum of biochemical responses. At one end is normal use of sulfur as a nutrient. The continuum leads eventually to irreversible inhibition as a possible, but not necessary, end to the process. Along the way are homeostasis and reversible inhibition, connected by k values (k_i; $i = 1, 2, \ldots, 5$) analogous to rate constants in chemical expression. The transition rates between the four response types are controlled by the leaf biochemistry. All the responses along the continuum feed back to mesophyll pollutant concentration. We sometimes tend to view concentration as a fixed property of a particular dosage of SO_2. It is not: internal concentration is a dynamic thing, constantly being altered by continued pollutant uptake and biochemical processes within the cell.

Leaf conductance. As Tingey and Olszyk point out, the biochemical sites of the leaf interior are where many plant responses to SO_2 occur. Therefore the conductivity of the leaf is very important to any overall examination of pollutant effects. Two pathways exist for sulfur admittance into the leaf: through the cuticle and through the stomata. We will concentrate on the latter pathway.

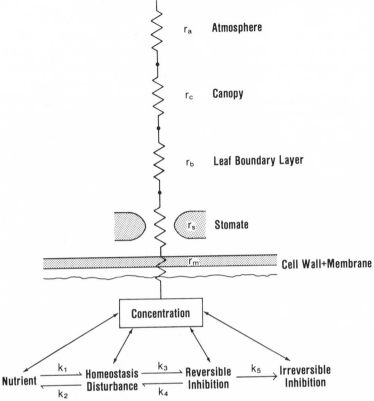

Fig. 30.2. The path of SO_2 into the leaf through the stomata, and its subsequent action within; r_a, r_c, r_b, r_s, and r_m are atmospheric, canopy, leaf boundary layer, stomatal, and cell wall and membrane resistances, respectively; and k_i ($i = 1, 2, \ldots, 5$) are "rate constants" between components of the response continuum.

The pathways are usually thought of as resistance analogs (Black and Unsworth, 1979b; Unsworth, 1981), and Fig. 30.2 contains such a representation. In calculating resistances, one assumes the final concentration in the cell to be zero. Although this is a convenient assumption for calculations, it may be a conceptual block to understanding biochemically complex system dynamics. In modeling such a system, the dynamics of SO_2 influx become extremely important. As exposure to the gas is prolonged, internal concentrations may become substantially greater than zero, altering the quantities of pollutant that reach the mesophyll cytosol. It has been shown by Sharpe (1983) that if the zero-concentration assumption is made for CO_2 uptake, mesophyll resistance becomes a complex function of light, CO_2, oxygen, leaf nitrogen, and temperature. There is no physical justification for this complexity, because mesophyll resistance is caused by the physical resistance to diffusion across the cell wall and plasmalemma.

Therefore, in modeling the uptake of SO_2 into the mesophyll, one should use an internal pollutant concentration calculated from the normal rate of uptake into the mesophyll. Any abnormal response of the cell to the pollutant must be considered to be overlaid on this normal response. The mathematical and biochemical complexity of such relations will depend on the modeler's purposes in constructing the model. For micrometeorological studies, a set of equations that simply embodies the functional uptake response may be all that is required. If the modeler were examining photosynthetic-capacity alterations by the pollutant, a more complex set of equations describing the biochemistry in more detail might be needed.

More than anything else, stomatal conductance controls the uptake of gas into the leaf. As Black has described in her chapter, SO_2 effects on stomatal conductance are not well-understood. At certain concentrations of pollutants, some plants show increased, others decreased, conductance, and still others exhibit little sensitivity to SO_2 at all. What we can facetiously call "Murphy's Law of Stomatal Response" might be appropriate here: "Every stomatal response possible will be observed if we look over the spectrum of species and environmental conditions."

How does one begin to make sense out of the experimental results described by Black? It is our contention that mechanistic models of possible hypotheses could be of use. Such models, if built around an adequate description of normal stomatal action, can examine the various ideas discussed by Black (e.g., pollutant interactions with ABA concentration, ion movement/production interference). Mechanistic models cannot test these hypotheses, because they are, after all, mathematical approximations of reality. They can, however, facilitate judgment on the validity of postulated mechanisms, suggest new areas of experimental research, and offer alternative hypotheses about the mechanisms. Models allow us to follow complex system-component interactions (for example, multiple feedback loops), and establish their proper roles and relative importance. Complex systems are very difficult to conceptualize, and often exhibit counterintuitive behavior. At Texas A&M, work has been done on biophysical models of stomatal mechanics (e.g., Wu and Sharpe, 1979), and a model of sulfur biochemistry coupled to this model might enable us to identify interactions that control SO_2 uptake by the leaf.

Perturbation. Perturbation occurs within the leaf and other living portions of the plant. It encompasses the range of effects on photosynthesis, respiration, membrane structure, carbon allocation, etc. The biochemical effects of SO_2 have been well documented by Wellburn, Peiser and Yang, and Garsed in this volume. Here we will concentrate on several aspects and possible approaches for synthesis.

As Wellburn points out in Chapter 8, most work on SO_2 biochemical effects has been done on photosynthesis. The picture that emerges lacks coherence, probably because there is no standardization of experimental techniques. Different researchers use different *in vitro* methods of measuring phenomena. For example, Wellburn describes the disparity that results from two different methods of measuring photosynthetic oxygen evolution. Because of assay optimization and lack of sufficient treatment replication, results of polarographic measurements of SO_2 effects on O_2 evolution are contradictory. When its effects on 3-PGA-dependent evolution are measured, however, the results are consistent: SO_2 inhibits O_2 evolution by competing with orthophosphate (references given in Chapter 8).

Many biochemical assays are conducted under very artificial conditions. Polarographic O_2 evolution measurements, for example, are performed on isolated thylakoid residues. The difficulty in reconciling these types of experiments with whole-plant dose-response work is therefore not surprising. A sufficiently detailed and tested biophysical model can become, in effect, an experimental plant. For instance, a model of chloroplast function can be used in exactly the same manner as an isolated plastid. Working from the bottom up, pollutant effects can be tested on individual components of the chloroplast system. These components can then be coupled at successively higher levels of organization, and pollutant effects tested at each step. In this manner, SO_2 effects on chloroplast behavior can be investigated on a "surrogate plastid" in far more detail than is possible in *in vitro* experiments on real plants.

Perhaps the most important component of a chloroplast model is photosynthesis. Models of this process in the past have tended to be too simplistic for detailed exploration of biochemical pollutant effects. Most, for example, assume that the light and dark reactions occur at one site, which is patently false. Many leave out either NADPH or ATP, assuming that one or the other is rate-limiting. Some leave out vital components, such as inorganic phosphate, altogether; such a model certainly could not deal with the orthophosphate interactions described by Wellburn in Chapter 8. For reviews of current photosynthesis models, see Farquar and von Caemmerer (1982) and Sharpe (1983). To be useful on the scale demanded here, a model should describe the biochemical mechanisms of photosynthesis and other systems as closely as possible. We are developing a photosynthesis model that describes the interactions of important constituents of photosynthesis (e.g., ATP, NADPH, phosphate, oxygen) in sufficient detail to explore the effects of SO_2 pollution on the photosynthetic process.

Organism effects. Winner *et al.* (Chapter 7) point out that SO_2 pollution undoubtedly alters carbon allocation in plants. We feel that this is an important area of SO_2 research. It is, however, one that is difficult to quantify. Mechanisms of action remain unknown, although the observed changes in growth and allocation are similar to the effects of water stress or insect pests. Water stress, insect pests, and pollution (Bennett, Resh, and Runeckles, 1974) can cause increases in plant production at low intensities. Winner *et al.* state in Chapter 7 that the increase could be due to shifting allocation patterns that favor the leaves.

To understand the mechanisms by which SO_2 changes carbon allocation, we must know how stress influences plant growth and allocation in general; but here our knowledge of mechanisms is extremely limited. We can obtain this critical information by using ^{11}C techniques (Fares *et al.*, 1978), the advantage of which is that allocation patterns can be studied by means of nondestructive, real-time measurements. The plant can be observed for both steady-state and transient phenomena, and mechanisms underlying translocation processes can be unravelled.

The mechanisms of SO_2 action on carbon allocation could be investigated easily by using ^{11}C techniques. Labeled or unlabeled SO_2 could be introduced together with ^{11}C through an appropriate leaf chamber. If labeled SO_2 were used, one could then investigate the allocation pattern of both carbon and sulfur. Mechanistic models of phloem transport for steady-state (Goeschl *et al.*, 1976) and time-dependent conditions (Smith *et al.*, 1980) are available for investigation of SO_2 transport dynamics. Coupled with basic research via the new ^{11}C technology, this model could provide a good tool for the examination of SO_2 effects on carbon allocation and transport.

Long-Term Ecological Effects

Sulfur dioxide as a major atmospheric component of the environment has in the geological past been limited to regions surrounding active volcanoes. Natural plant populations or communities in these areas evolved to adjust to these high SO_2 concentrations during long time periods. Elsewhere, most plant communities have not been forced to adjust to high SO_2 concentrations, now a reality in most of the worldwide industrialized regions.

Predicting long-term effects of SO_2 on natural ecosystems is much more complex than assessing short-term effects on annual crop yields. The immediate and obvious difference is that no data are available to validate predictions. The impact of a stressor on a system can be very subtle, and may require hundreds of years to become clearly visible.

If models of long-term processes cannot be checked, at least within the modeler's lifetime, then how can their predictions be justified? Given that we cannot check the long-term performance of an ecological model, three criteria can be identified by which predictions can be evaluated. Perhaps the most important aspect is the completeness of the model. It must include all the important factors likely to affect the system. This requirement is not easily met, because missing factors are often not apparent until the model has been thoroughly stressed and checked against field conditions. Only after sufficient testing can a model reveal hidden deficiencies in the range of included factors. Simple models that neglect significant biological and biophysical processes can therefore be both misleading and counterproductive.

The second criterion in judging the usefulness of a long-term model is "robustness." A model is considered robust when it has the appropriate functional relationships to account for a wide range of conditions. It has to be able to predict consequences involving interactions between factors, as well as those arising from extreme events. Responses to extreme conditions can be evaluated in terms of their rational expectation, internal consistency, and functional explanation. It is this capability that gives a researcher confidence where corroborating observations are lacking.

The third criterion is the need to incorporate changes in the local or global environment other than the ones the model is expressly designed to address. Fluctuations in SO_2 concentrations are not the only changes anticipated during the next century. Temperature and concentrations of CO_2, NO_2, and other gaseous pollutants are expected to change significantly. These need to be considered, because, as has been shown elsewhere in this volume, these factors often show strong interactions; the overall response is not equal to the sum of responses to individual factors. Uncertainty about expected changes in any subsidiary factor makes long-term predictions difficult.

Given such uncertainty, the most useful approach is to apply models to "what if" scenarios. In the following sections, we would like to investigate several such scenarios, and point out where modeling could be of use in the process.

Environment-pollutant interactions. One commonly held concept of community structure can be called the *continuum* view (Fig. 30.3). In it, the floristic composition of a community changes gradually along environmental gradients. A question addressed by several chapters in this volume (those of Shugart and McLaughlin, Westman *et al.*, Kozlowski, and Lauenroth and Milchunas, for example) is what effects environmental fluctuations have on plant susceptibility to SO_2. Westman *et al.* point out that natural communities are subject to levels of environmental

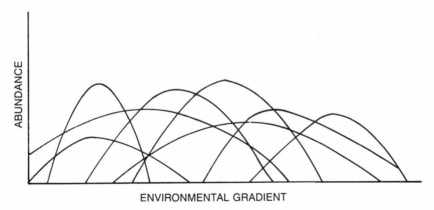

Fig. 30.3. The continuum view of community structure. The *x*-axis represents an environmental gradient, and the *y*-axis represents abundance. Each curve represents a population of a different species.

stress not found in cultivated monocultures. They note that temperature and moisture affect plant SO_2 susceptibility in ways that differ from species to species. Thus, in a polluted community, there is a third layer of complexity overlying what is found in a "pristine," or unpolluted, one. In a pristine community, interactions within and between species are complicated by environmental interactions and the differences in species response to environmental stress. The effects of environmental stress on pollutant susceptibility can alter the course of interactions within and between species. For example, species A and B might be of equal abundance in an area of abundant soil moisture. Under conditions of limiting moisture, A might have a competitive advantage. If B is more pollution-resistant, or if water stress increases A's pollution susceptibility, the results of the environmental stress under pollution might be an increase in B. Increasing the number of species in question, as in a rich community, increases greatly the number and complexity of effects.

It would be useful to investigate SO_2 pollution effects on the distributions of plants along environmental gradients. If other environmental factors are held constant, and pollutant dosage is the same over the gradient, one could examine the effects of the factor in question on community susceptibility to SO_2. Let us assume the gradient is one of temperature. One might expect, under a moderate pollutant load, to see some plants reduced in abundance and some disappear altogether. These effects may be due to any number of factors, such as reduced growth potential, reproductive capacity, or competitive ability. If the effects of SO_2 along the gradient were approximately the same everywhere along its length, species diversity would be decreased by a similar amount all along the gradient. One could conclude, therefore, that there was no overall interaction between temperature and pollutant affecting the structure of this community.

The effects of SO_2 pollution along the gradient might, alternatively, be heavier in one portion of the gradient than in another. This would lead one to suspect a synergistic effect of pollutant and temperature on the community structure. The kind of experiment described in the preceding would be extremely difficult, if not impossible, to carry out in the field. Community structure is not two-, but three-dimensional. Matters are thus very much more complex than shown in Fig. 30.3. In addition, a single-variable gradient with other environmental factors held constant might be difficult to find. Finally, the problems of controlling SO_2 dosage throughout a large area are tremendous. Here is where computer modeling can be useful, and Shugart and McLaughlin (this volume) have shown how it might be done.

Shugart and McLaughlin, using the gap model FORET, examined the effect of soil moisture on pollution-depressed stand growth in eastern deciduous forest. They found that there seemed to be a synergistic effect between moisture and pollutant on the growth rates of certain species (e.g., Black Oak, *Quercus velutina*). Furthermore, this effect varied when the stand age at the onset of pollution stress was altered. Despite the fact that the results were given only for several species, and not the entire community, Shugart and McLaughlin's work shows the power of such a modeling approach. As they point out in their conclusions, the complex interactions between various community factors and pollution are extremely difficult to work out. They also see the need for extreme care in construction and control of experiments relating to pollutant effects on community structure.

As Kozlowski (this volume) points out, "the unaided human mind is not easily able to explore the consequences of all the interactions that occur between air pollutants and other stress factors." He goes on to offer computer modeling as a partial solution to the problem. We agree, and feel there is a need for more studies of the kind described by Shugart and McLaughlin. Because their model (FORET) is a gap-succession model, it is limited in the kinds of problems it can address. Repeated many times, it can simulate large-area pollution effects in spite of its small plot size (1/100 to 1/10 ha). Because of this limitation, however, and because it has no way to represent a community structure that changes along a gradient, it is not well-suited to the kind of analysis represented in Fig. 30.3. To produce an analysis of effects that are functions of environmental gradients, a simulation needs to be able to represent communities spatially throughout a large area. Such a simulation would be costly to build and run, but not as much as equivalent carefully designed field experiments. Once a simulation like this has been built, it can be modified many times to serve many different needs. FORET is based on JABOWA (Botkin, Janak, and Wallis, 1972), a model that has been modified many times and used successfully for various applications. Models of this type are thus very cost-effective tools for research.

Evolution of tolerance to SO_2. It was noted by Bell (this volume) that there is little evidence for significant damage to grasslands by SO_2 pollution. The contributions of Kozlowski and Keller, on the other hand, show that forests can be very sensitive to the pollutant. Keller gives several reasons for this phenomenon. The large crown volumes of forest trees mean that they absorb much more pollutant than grasses or herbs. Tree crowns are generally exposed to higher winds than those of low-growing life forms. Wind reduces leaf boundary layer resistance and increases filtered air volume. The longevity of forest trees allows for repeated pollution exposures. Finally, Keller suggests that because many forests are restricted to areas with limited water and nutrients, they could be harder hit by pollution. We would like to discuss another reason why forests might be more sensitive to SO_2 than grasslands: forests probably cannot evolve SO_2 resistance as quickly as grasslands.

In his review of the literature about SO_2 effects on grassland productivity, Bell (this volume) describes research showing selection for SO_2 tolerance in grass populations. Westman *et al.* (this volume) report evolved SO_2 resistance in brome grass (*Bromus rubens*). As Bell notes, a grassland community exhibits intense competition within and between species, which facilitates changes in community structure in response to environmental change. In addition, grasslands have a relatively high biomass turnover rate compared to forest communities. Evolution can proceed, therefore, at higher rates. Bell saw significant adaptation to pollution in as little as four years; Westman *et al.* (this volume) saw it in 25, a short span of time in the life of a mature forest. Many plant ecologists (e.g., Barbour, Burk, and Pitts, 1980) define a climax forest as one in which significant changes in species composition do not occur within 500 years. Competition within or between species in such a forest can be very weak. If significant changes in species composition occur only over the span of half a millennium, then we can expect significant adaptation to a stressor to take about as long. It is therefore reasonable to expect that grasses, subject to higher turnover rates and competition levels, are much more adaptable to long-term SO_2-imposed stress than are forests.

Because most crop species are annuals, and do not experience the levels of competition discussed by Bell, they cannot respond to rising atmospheric SO_2 levels by evolving tolerance. However, as Karnosky (this volume) pointed out, there is much evidence that different species respond differently to SO_2. This genetic difference can be exploited, using artificial selection techniques, to produce pollution-resistant crop strains. Preceding this, as Karnosky has said, should come more research to identify and describe pollution-resistant strains in crops as well as in wild plants. In an age of pollution, efforts should be made to not only curtail pollution, but also to develop means of ameliorating its effects on exposed crops.

Pollution and stability. The concept of stability is very important in systems ecology, and there are almost as many definitions of it as there are ecologists. As Ricklefs (1973) points out, actually two basic concepts are involved. The first, constancy, embodies the idea normally associated with stability: lack of significant change. This alone, however, is not a sufficient definition; the idea of recoverability is also needed. Recoverability can be thought of as the ability of a system (population, community, or ecosystem) to return to a similar equilibrium following perturbation. A system may be said to be stable if it returns to an equilibrium quickly after a point perturbation, such as an acute SO_2 dosage (Fig. 30.4a). If we envision a pollutant regime of occasional dosage spikes embedded in a constant, low background pollution, it may be that some systems (even the "stable" ones) might never return to an equilibrium state (Fig. 30.4b).

Acute dosages, however, are not the only phenomena that must be dealt with. Pollution often remains low, but slowly fluctuates with time. May (1974) describes the concept of "structural stability" in dealing with cases like this (Fig. 30.4c). A population, community, or ecosystem is structurally stable if, in response to slow, continuous environmental change, its response is also continuous. If system response is discontinuous, it is structurally unstable. It may be that grasslands, able to respond more quickly to stress, are more structurally

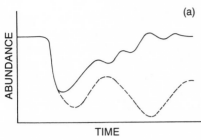

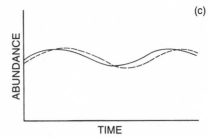

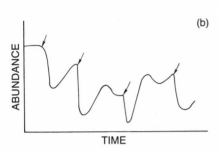

Fig. 30.4. (*a*) Stable versus nonstable populations. Solid and dotted lines are stable and unstable populations, respectively. (*b*) Population rendered unstable by occasional acute pollution dosages. Arrows represent points of perturbance onset. (*c*) A structurally stable population. Solid line represents changing environmental variable; dotted line is population numbers (not on same scale).

stable against stress from long-term, low SO_2 concentrations than forests are. They may be able to continuously "track" gradual pollutant changes, in terms of favorable response, more efficiently than mature forests. In his contribution, Bell points out that evidence of grassland injury from air pollution is rare. Keller and Kozlowski, in their chapters, cite many examples of damage to forests. As Horn (1974) emphasizes, by the criterion of recoverability, mature forests are extremely unstable and fragile. Because of this fragility, it is imperative that SO_2 control measures take forests into account as much as agricultural concerns.

Conclusions

We have attempted to provide a rationale for the use of mechanistic mathematical modeling as a tool for SO_2 research. Such an approach can provide vital links between long-term ecological, whole-plant dose-response, and biochemical mechanism research (Fig. 30.1). The question is, how far are we from being able to formulate such a model? We are a long way from a *totally* mechanistic model of plant function, but there is much that can be done now. Where biological knowledge permits, we can model the biochemistry mechanistically. Elsewhere, empirical functions can sometimes suffice. Certainly, attempting a semimechanistic model of plant-SO_2 interactions is no small task, and it is a good way to discover areas where basic research is lacking. Some of these problems have been discussed elsewhere in this book. They include the question of concentration *versus* dosage. What external dosage results in what internal concentrations? What internal concentrations can we reasonably expect in nature? At what concentrations do pollution effects become important? Also, we need to know more about the pollution responses of cellular processes other than photosynthesis. Mitochondrial function, for example, could be very important to the overall plant response. Until we have answers to these and other questions, a truly mechanistic model of plant response to pollution at the biochemical and organismal levels is infeasible.

Lauenroth and Milchunas, in their contribution to this volume, speak of the "nested hierarchy" model of nature. One can see this hierarchy as starting at the biochemical level and proceeding up through the cellular, organismal, population, community, and ecosystems levels. For our purposes, the biochemical level is the lowest practical level, although to be complete one would have to include even lower (e.g., atomic and subatomic) levels. A plant community, then, contains all the complexity brought about by its component populations, the individuals within those populations, the cellular interactions within those individuals, and the biochemical processes underlying it all. A truly integrated view of SO_2 effects on plant communities requires knowledge of its effects on all these levels.

Whatever the modeling approach, and whatever the level, the benefits of modeling should be carefully weighed against its costs. Questions should be asked about the kind of model needed, and whether one is needed at all. JABOWA (Botkin, Janak, and Wallis, 1972) has been modified many times and used for many applications, and is thus quite cost-effective. Other models in the literature have not been so useful. However, with careful evaluation, and a sure grasp of their uses and limitations, models can still be useful aids for understanding how and why plants respond to SO_2.

Bibliography

Bibliography

Abel, E. 1951. Zur theorie der Oxydation von Sulfit zu Sulfat durch Sauerstoff. *Monatshefte Chemie* 82: 815–34.

Adams, D. F., *et al.* 1980. Estimates of natural sulfur source strengths. *In* Shriner, Richmond & Lindberg (1980), pp. 35–45. Coauthors: S. O. Farwell, M. R. Pack & E. Robinson.

Adams, J. A. S., M. S. M. Montovan & L. L. Sundell. 1977. Wood versus fossil fuel as a source of excess carbon dioxide in the atmosphere: A preliminary report. *Science* 196: 45–46.

Adams, R. M., and T. D. Crocker. 1980. Analytical issues in economic assessments of vegetation damages. *In* Teng and Krupa (1980), pp. 198–207.

———. 1981. Dose response information and environmental damage assessments: An economic perspective. *Journal of the Air Pollution Control Association* 32: 1062–67.

Adams, R. M., T. D. Crocker & N. Thanavibulchai. 1982. An economic assessment of air pollution damages to selected crops in southern California. *Journal of Environmental Economics and Management* 9: 42–58.

Adedipe, N. O., R. E. Barrett & D. P. Ormrod. 1972. Phytotoxicity and growth responses of ornamental bedding plants to ozone and sulfur dioxide. *Journal of the American Society for Horticultural Science* 97: 341–45.

Ahn, Y.-S., and D. Y. Yeam. 1977. Studies on the reduction of SO_2 injury to *Hibiscus syriacus* L. by use of growth regulant, CCC. *Journal of the Korean Society for Horticultural Science* 18: 203–4.

Aketagawa, J., and G. Tamura. 1980. Ferrodoxin-sulfite reductase from spinach. *Agricultural Biological Chemistry* 44: 2371–78.

Akita, S., and D. N. Moss. 1972. Differential stomatal response between C_3 and C_4 species to atmospheric CO_2 concentration and light. *Crop Science* 12: 789–93.

———. 1973. Photosynthetic responses to CO_2 and light by maize and wheat leaves adjusted for constant stomatal apertures. *Crop Science* 13: 234–37.

Akita, S., and I. Tanaka. 1973. Studies on the mechanisms of differences in photosynthesis among species, IV: The differential response in dry matter production between C_3 and C_4 species to atmospheric carbon dioxide. *Proceedings of the Crop Science Society of Japan* 43: 288–95.

Akoyunoglou, G., ed. 1981. *Proceedings of the Fifth International Congress on Photosynthesis*. Philadelphia, Pa.: Balaban.

Allen, T. F. H., and T. B. Starr. 1982. *Hierarchy—Perspectives for Ecological Complexity*. Chicago, Ill.: Univ. Chicago Press.

Alscher-Herman, R. 1982. The effect of sulfite on light activation of chloroplast fructose 1,6-bisphosphatase in two cultivars of soybean. *Environmental Pollution* 27: 83–96.

Altwicker, E. R., and A. J. Johannes. 1983. Wet and dry deposition into Adirondack watersheds. *In* E.P.R.I. (1983), pp. 2: 1–31.

Amiro, B. D., and G. M. Courtin. 1981. Patterns of vegetation in the vicinity of an industrially disturbed ecostem, Sudbury, Ontario. *Canadian Journal of Botany* 59: 1623–39.

Amundsen, R. G., and L. H. Weinstein. 1981. Joint action of sulfur dioxide and nitrogen dioxide on foliar injury and stomatal behaviour in soybean. *Journal of Environmental Quality* 10: 204–6.

Anderson, L. E., and J. X. Duggan. 1977. Inhibition of light modulation of chloroplast enzyme activity by sulfite, one of the lethal effects of SO_2. *Oecologia* (Berl.), 28: 147–51.

Anderson, L. S., and T. A. Mansfield. 1979. The effects of nitric oxide pollution on the growth of tomato. *Environmental Pollution* 20: 113–21.

Andrew, S. P. S. 1955. A simple method of measuring gaseous diffusion coefficients. *Chemical Engineering Science* 4: 269–72.

Andrussow, L. 1969. Diffusion. *In* Borchers *et al.* (1969), II (Part 5a), 513–701.

Antonovics, J., A. D. Bradshaw & R. G. Turner. 1971. Heavy metal tolerance in plants. *Advances in Ecological Research* 7: 1–85.

Archibold, O. W. 1978. Vegetation recovery following pollution control at Trail British Columbia. *Canadian Journal of Botany* 56: 1625–37.

Armentano, T. V. 1980. Calculation of yield-loss coefficients for major crops in the Ohio River basin. *In* Institute of Ecology (1980), pp. 77–120.

Arndt, U. 1974. The Kautsky effect: A method for the investigation of the actions of air pollutants in chloroplasts. *Environmental Pollution* 6: 181–94.

Arnold, G. W., and C. T. de Wit, eds. 1976. *Critical Evaluation of Systems Analysis in Ecosystem Research and Management.* Wageningen, Netherlands: Center for Agricultural Publishing and Documentation.

Asada, K. 1980. Formation and scavenging of superoxide in chloroplasts, with relation to injury by sulfur dioxide. *In* N.I.E.S. (1980), pp. 165–79.

Asada, K., R. Deura & Z. Kasai. 1968. Effect of sulfate ions on photophosphorylation by spinach chloroplasts. *Plant and Cell Physiology* 9: 143–46.

Asada, K., and K. Kiso. 1973. Initiation of aerobic oxidation of sulfite by illuminated spinach chloroplasts. *European Journal of Biochemistry* 33: 253–57.

Ashenden, T. W. 1978. Growth reductions in cocksfoot (*Dactylis glomerata* L.) as a result of SO_2 pollution. *Environmental Pollution* 15: 161–66.

———. 1979a. Effects of SO_2 and NO_2 pollution on transpiration in *Phaseolus vulgaris* L. *Environmental Pollution* 18: 45–50.

———. 1979b. The effects of long-term exposures to SO_2 and NO_2 pollution on the growth of *Dactylis glomerata* L. and *Poa pratensis* L. *Environmental Pollution* 18: 249–58.

Ashenden, T. W., and T. A. Mansfield. 1977. Influence of wind speed on the sensitivity of ryegrass to SO_2. *Journal of Experimental Botany* 28: 729–35.

———. 1978. Extreme pollution sensitivity of grasses when SO_2 and NO_2 are present in the atmosphere together. *Nature* 273: 142–43.

Ashenden, T. W., T. A. Mansfield & A. R. Wellburn. 1978. Influence of wind on the sensitivity of plants to SO_2. *Verein Deutscher Ingenieure Berichte* 314: 231–35.

Ashenden, T. W., and I. A. D. Williams. 1980. Growth reductions in *Lolium multiflorum* Lam. and *Phleum pratense* L. as a result of SO_2 and NO_2 pollution. *Environmental Pollution* (series A) 21: 131–39.

Ashenden, T. W., *et al.* 1982. A large-scale system for fumigating plants with SO_2 and NO_2. *Environmental Pollution* (series B) 3: 21–26. Coauthors: P. W. Tabner, P. Williams, M. E. Whitmore & T. A. Mansfield.

Ashmore, M. R., J. N. B. Bell & C. L. Reily. 1978. A survey of ozone levels in the British Isles using indicator plants. *Nature* 276: 813–15.

———. 1980. The distribution of phytotoxic ozone in the British Isles. *Environmental Pollution* (series B) 1: 195–216.

Ashmore, M. R., *et al*. 1980. Visible injury to crop species by ozone in the United Kingdom. *Environmental Pollution* (series A) 21: 209–15. Coauthors: J. N. B. Bell, C. Dalpra & V. C. Runeckles.

Awang, M. B. 1979. The effects of sulfur dioxide pollution on plant growth with special reference to *Trifolium repens*. Ph.D. thesis, Univ. of Sheffield.

Ayazloo, M. 1979. Tolerance to sulfur dioxide in British grass species. Ph.D. thesis, Imperial College, London.

Ayazloo, M., and J. N. B. Bell. 1981. Studies on the tolerance to sulfur dioxide of grass populations in polluted areas, I: Identification of tolerant populations. *New Phytologist* 88: 203–22.

Ayazloo, M., J. N. B. Bell & S. G. Garsed. 1980. Modification of chronic sulfur dioxide injury to *Lolium perenne* L. by different sulfur and nitrogen nutrient treatments. *Environmental Pollution* (series A) 22: 295–307.

Ayazloo, M., S. G. Garsed & J. N. B. Bell. 1982. Studies on the tolerance to sulfur dioxide of grass populations in polluted areas, II: Morphological and physiological investigations. *New Phytologist* 90: 109–26.

Baba, I., and S. Sakai. 1976. Physiological studies on the mechanism of the occurrence of air pollution damage in crop plants, III: Ethylene production by wheat and barley leaves injured by fumigation with sulfur dioxide. *Nogaku Kenyu* 55: 199–203.

Bacastow, R., and C. D. Keeling. 1973. Atmospheric carbon dioxide and radiocarbon in the natural carbon cycle, II: Changes from A.D. 1700 to 2070 as deduced from a geochemical model. *In* Woodwell and Pecan (1973), pp. 86–134.

Bache, B. W., and N. M. Scott. 1979. Sulfur emissions in relation to sulfur in soils and crops. *In* Ross (1979), pp. 242–54.

Backstrom, H. L. J. 1927. The chain-reaction theory of negative catalysis. *Journal of the American Chemical Society* 49: 1460–72.

Baenziger, P. S., R. A. Kilpatrick & J. G. Moseman. 1979. Reduced root and shoot growth caused by *Erysiphe graminis tritici* in related wheats grown in nutrient solution culture. *Canadian Journal of Botany* 57: 1345–48.

Baes, C. F., *et al*. 1977. Carbon dioxide and climate: The uncontrolled experiment. *American Scientist* 65: 310–20. Coauthors: H. E. Goeller, J. S. Olson & R. M. Rotty.

Baker, C. K., M. H. Unsworth & P. Greenwood. 1982. Leaf injury on wheat plants exposed in the field in winter to SO_2. *Nature* 299: 149–51.

Baker, D. N. 1965. Effects of certain environmental factors on net assimilation in cotton. *Crop Science* 5: 53–56.

Baldry, C. W., W. Cockburn & D. A. Walker. 1968. Inhibition, by sulfate, of the oxygen evolution associated with photosynthetic carbon assimilation. *Biochimica et Biophysica Acta* 153: 476–83.

Ballantyne, D. J. 1972. Fluoride inhibition of the Hill reaction in bean chloroplasts. *Atmospheric Environment* 6: 267–73.

———. 1977. Sulfite oxidation by mitochondria from green and etiolated peas. *Phytochemistry* 16: 49–50.

Bandy, A. R., and P. J. Marculis. 1980. Impact of recent measurements of OCS, CS_2 and SO_2 in background air on the Global Sulfur Cycle. *In* Shriner, Richmond & Lindberg (1980), pp. 55–65.

Barber, J., ed. 1977. *Primary Processes of Photosynthesis*. Amsterdam, the Netherlands: Elsevier.

Barbour, M. G., J. G. Burk & W. D. Pitts. 1980. *Terrestrial Plant Ecology*. Menlo Park, Calif.: Benjamin-Cummings.

Bardsley, C. E., and J. D. Lancaster. 1960. Determination of reserve sulfur and soluble sulfates in soils. *Proceedings of the Soil Science Society of America* 24: 265–68.

————. 1965. Sulfur. *In* Black (1965), pp. 1102–16.

Barkman, J. 1969. The influence of air pollution on bryophytes and lichens. *Proceedings of the First European Congress on the Influence of Air Pollution on Plants and Animals, 1968*, pp. 197–209.

Barrett, C. F., *et al.* 1982. *Acidity of Rainfall in the United Kingdom: A Preliminary Report*. Warren Spring Laboratory, Stevenage: U.K Review Group on Acid Rain. Coauthors: D. Fowler, J. G. Irwin, A. S. Kallend, A. Martin, R. A. Scriven & A. F. Tuck.

Barrett, L. B., and T. E. Waddell. 1973. *Cost of Air Pollution Damage: A Status Report*. U.S. Environmental Protection Agency Report AP-85. Washington, D.C.

Barton, J. R., S. B. McLaughlin & R. K. McConathy. 1980. The effects of SO_2 on components of leaf resistance to gas exchange. *Environmental Pollution* (series A) 21: 255–65.

Bauer, H., W. Larcher & R. B. Walker. 1975. Influence of temperature stress on CO_2-gas exchange. *In* Cooper (1975), pp. 557–86.

Bayes, T. 1764. An essay towards solving a problem in the doctrine of chances. *Philosophical Transactions of the Royal Society* 45: 293–315. (Reprinted in *Biometrika*, 1958.)

Beckerson, D. W., and G. Hofstra. 1979a. Response of leaf diffusion resistance of radish, cucumber, and soybean to ozone and sulfur dioxide. *Atmospheric Environment* 13: 1263–68.

————. 1979b. Stomatal responses of white bean to O_3 and SO_2 singly and in combination. *Atmospheric Environment* 13: 533–35.

Beckerson, D. W., G. Hofstra & R. Wukasch. 1979. The relative sensitivities of 33 bean cultivars to ozone and sulfur dioxide singly or in combination in controlled exposures and to oxidants in the field. *Plant Disease Reporter* 63: 478–82.

Beekley, P. K., and G. R. Hoffman. 1981. Effects of sulfur dioxide fumigation on photosynthesis, respiration, and chlorophyll content of selected lichens. *Bryologist* 84: 379–90.

Beevers, L., and R. H. Hageman. 1969. Nitrate reduction in higher plants. *Annual Review of Plant Physiology* 20: 495–522.

Belford, R. K. 1979. Collection and evaluation of large soil monoliths for soil and crop studies. *Journal of Soil Science* 30: 363–73.

Bell, J. N. B. 1982. Sulfur dioxide and the growth of grasses. *In* Unsworth and Ormrod (1982), pp. 225–46.

————. 1983. Air pollution problems in Western Europe. *In* Koziol and Whatley (1983), in press.

Bell, J. N. B., M. Ayazloo & G. B. Wilson. 1982. Selection for sulfur dioxide tolerance in grass populations in polluted areas. *In* Bornkamm, Lee & Seaward (1982), pp. 171–80.

Bell, J. N. B., and W. S. Clough. 1973. Depression of yield in ryegrass exposed to sulfur dioxide. *Nature* 241: 47–49.

Bell, J. N. B., and C. H. Mudd. 1976. Sulfur dioxide resistance in plants: A case study of *Lolium perenne*. *In* Mansfield (1976), pp. 86–103.

Bell, J. N. B., A. J. Rutter & J. Relton. 1979. Studies on the effects of low levels of sulfur dioxide on the growth of *Lolium perenne* L. *New Phytologist* 83: 627–43.

Benedict, H. M., C. J. Miller & R. E. Olson. 1971. *Economic Impact of Air Pollution on Plants in the United States*. Menlo Park, Calif.: Stanford Research Institute.

Bennett, J. H., and A. C. Hill. 1973a. Absorption of gaseous air pollutants by a standardized plant canopy. *Journal of the Air Pollution Control Association* 23: 203–6.

————. 1973b. Inhibition of apparent photosynthesis by air pollutants. *Journal of Environmental Quality* 2: 526–30.

————. 1974. Acute inhibition of apparent photosynthesis by phytotoxic air pollutants. *American Chemical Society Symposium Series* no. 3, pp. 115–27.

————. 1975a. Acute effects of combinations of sulfur dioxide and nitrogen dioxide on plants. *Environmental Pollution* 9: 127–32.

————. 1975b. Interactions of air pollutants with canopies of vegetation. *In* Mudd and Kozlowski (1975), pp. 273–306.

Bennett, J. H., A. C. Hill & D. M. Gates. 1973. A model for gaseous pollutant sorption by leaves. *Journal of the Air Pollution Control Association* 23: 957–62.

Bennett, J. H. *et al.* 1975. Acute effects of combinations of sulfur dioxide and nitrogen dioxide on plants. *Environmental Pollution* 9: 127–32. Coauthors: A. C. Hill, A. Soleimani & W. H. Edwards.

Bennett, J. P. 1981. Rare plants and the Clean Air Act. *Proceedings of the Rocky Mountain Regional Rare Plant Conference*, pp. 37–39.

————. 1982. A computerized flora of the national parks. *George Wright Society Forum* 2: 14–16.

Bennett, J. P., and R. J. Oshima. 1976. Carrot injury and yield response to ozone. *Journal of the American Society of Horticultural Science* 101: 638–39.

Bennett, J. P., J. Resh & V. Runeckles. 1974. Apparent stimulation of plant growth by air pollutants. *Canadian Journal of Botany* 52: 35–41.

Benson, F. J., *et al.* 1982. *Economic Assessment of Air Pollution Damage to Agricultural and Silvicultural Crops in Minnesota.* St. Paul: Minnesota Pollution Control Agency. Coauthors: S. V. Krupa, P. S. Teng & D. E. Welsh.

Berry, C. R. 1973. The differential sensitivity of eastern white pine to three types of air pollution. *Canadian Journal of Forest Research* 3: 543–47.

Berry, C. R., and L. A. Ripperton. 1963. Ozone a possible cause of white pine evergreen tipburn. *Phytopathology* 53: 552–57.

Bialobok, S. 1979. Identification of resistance of tolerant strains and artificial selection or production of such strains in order to protect vegetation from air pollution. *In* United Nations Economic Commission for Europe (1979).

————. 1980. Forest genetics and air pollution stress. *In* Miller (1980), pp. 100–102.

Bialobok, S., P. Karolewski & J. Oleksyn. 1980. Sensitivity of Scots pine needles from mother trees and their progenies to the action of SO_2, O_3, a mixture of these gasses, NO_2 and HF. *Arboretum Kornickie* 25: 289–303.

Biggs, A. R., and D. D. Davis. 1980. Stomatal response of three birch species exposed to varying acute doses of SO_2. *Journal of the American Society for Horticultural Science* 105: 514–16.

————. 1981a. Effect of SO_2 on growth and sulfur content of hybrid poplar. *Canadian Journal of Forest Research* 11: 830–33.

————. 1981b. Foliar response of ten tree species exposed to SO_2 pollution. The Pennsylvania State Univ. College of Agriculture Progress Report no. 375.

————. 1981c. Sulfur-dioxide injury, sulfur content, and stomatal conductance of birch foliage. *Canadian Journal of Forest Research* 11: 69–72.

Bingham, G. E., and P. I. Coyne. 1977. A portable, temperature-controlled, steady-state porometer for field measurements of transpiration and photosynthesis. *Photosynthetica* 11: 148–60.

Biscoe, P. V., R. K. Scott & J. L. Monteith. 1975. Barley and its environment, III: Carbon budget of the stand. *Journal of Applied Ecology* 12: 269–93.

Biscoe, P. V., M. H. Unsworth & H. R. Pinckney. 1973. The effects of low concentrations of sulfur dioxide on stomatal behaviour in *Vicia faba*. *New Phytologist* 72: 1297–1306.

Bishop, P. M., and C. P. Whittingham. 1968. The photosynthesis of tomato plants in a carbon dioxide enriched atmosphere. *Photosynthetica* 2: 31–38.

Bjorkman, E. 1970. The effect of fertilization on sulfur dioxide damage to conifers in industrial and built-up areas. *Studia Forestalia Suecica* 78: 1–48.

Black, C. A., ed. 1965. *Methods of Soil Analysis, Part 2*. Madison, Wis.: American Society of Agronomy.

Black, C. R., and V. J. Black. 1979a. The effects of low concentrations of sulfur dioxide on stomatal conductance and epidermal cell survival in field beans (*Vicia faba* L). *Journal of Experimental Botany* 30: 291–98.

———. 1979b. Light and scanning electron microscopy of SO₂-induced injury to leaf surfaces of field bean (*Vicia faba* L). *Plant, Cell and Environment* 2: 329–33.

Black, V. J. 1982. Effects of sulfur dioxide on physiological processes in plants. *In* Unsworth and Ormrod (1982), pp. 67–91.

Black, V. J., D. P. Ormrod & M. H. Unsworth. 1982. Effects of low concentration of ozone, singly, and in combination with sulphur dioxide on net photosynthesis rates of *Vicia faba* L. *Journal of Experimental Botany* 137: 1302–11.

Black, V. J., and M. H. Unsworth. 1979a. Effects of low concentrations of SO₂ on net photosynthesis and dark respiration of *Vicia faba* L. *Journal of Experimental Botany* 30: 473–83.

———. 1979b. Resistance analysis of SO₂ fluxes to *Vicia faba*. *Nature* 282: 68–69.

———. 1979c. A system for measuring effects of sulfur dioxide on the gas exchange of plants. *Journal of Experimental Botany* 30: 81–88.

———. 1980. Stomatal responses to sulfur dioxide and vapor pressure deficit. *Journal of Experimental Botany* 31: 667–77.

Bleasdale, J. K. A. 1952a. Air pollution and plant growth. *Nature* 169: 376–77.

———. 1952b. *Atmospheric Pollution and Plant Growth*. Ph.D. diss., Univ. of Manchester, England.

Bolin, B. 1977. Changes of land biota and their importance for the carbon cycle. *Science* 196: 613–15.

Bonas, U., *et al.* 1982. Phloem transport of sulfur in *Ricinus*. *Planta* 155: 82–88. Coauthors: K. Schmitz, H. Rennenberg & L. Bergmann.

Bonté, J. 1975. Interrelation entre la pollution par le dioxyde de soufre et le mouvement des stomates chez le *Pelargonium*. Ph.D. thesis, Univ. of Pierre and Marie Curie, Paris.

———. 1977. *Effets du SO₂ sur les végétaux en plein champ, à faible concentration et appliqué d'une façon permanente*. Morlass, France: Ministère de l'Agriculture de France, INRA.

———. 1982. Effects of air pollutants upon flowering and fruiting. *In* Unsworth and Ormrod (1982), pp. 207–24.

Bonté, J., L. DeCormis & P. Louguet. 1975. Influence d'une pollution par le dioxyde de soufre sur le degré d'overture des stomates du *Pelargonium X hortorum*. *Comptes Rendus de L'Academie des Sciences, Paris* (series D) 280: 2377–80.

———. 1977. Inhibition, en anaérobiose, de la réaction de fermeture des stomates du *Pelargonium* en présence de dioxyde de soufre. *Environmental Pollution* 12: 125–33.

Bonté, J., *et al.* 1977. Contribution à l'étude des caractères de résistance de Pelargonium à un pollutant atmosphérique le dioxyde de soufre. *Physiologie Végetale* 15: 15–27. Coauthors: C. Bonté, L. DeCormis & P. Houguet.

Boone, G. C. 1978. Responses of white pine and ponderosa pine stomata to ozone and sulfur dioxide. Ph.D. dissertation, W. Virginia Univ.

Borchers, H. *et al.*, eds. 1969. *Landolt-Bornstein Zahlenwerte und Funktionen aus Physik, Chemie, Astronomie, Geophysik, und Technik*. 6th ed. Berlin: Springer-Verlag. Coeditors: H. Hausen, K.-H. Hellwege, K. Schafer & E. Schmidt.

Borka, G. 1980. The effect of cement dust pollution on growth and metabolism of *Helianthus annuus*. *Environmental Pollution* 22: 75–79.

Bornkamm, R., J. A. Lee & M. R. D. Seaward, eds. 1982. *Urban Ecology: Proceedings of the Second European Ecological Symposium.* Oxford: Blackwell Scientific Publications.

Börtitz, S., and M. Vogl. 1965. Physiologische und biochemische Beiträge zur Rauchschandenforschung, 5: Versuche zur Erarbielung eines schnelltestes für die zuchterische Vorselektion auf Raucharte bei Larchen. *Züchter* 35: 307–11.

Botkin, D. B. 1973. Estimating the effects of carbon fertilization on forest composition by ecosystem simulation. *In* Woodwell and Pecan (1973), pp. 328–44.

Botkin, D. B., J. F. Janak & J. R. Wallis. 1972. Some ecological consequences of a computer model of forest growth. *Journal of Ecology* 60: 849–72.

Bovenkerk, M. L., *et al.* 1979. *De involoed van luchtverontreining op vegetaties.* Wageningen, Netherlands: Instituut voor Plantenziektenkundig Onderzoek. Coauthors: J. M. van der Erden, E. H. T. Kruit & H. Uppelschoten.

Boyer, J. S. 1982. Plant productivity and environment. *Science* 218: 443–47.

Bradshaw, A. D. 1976. Pollution and evolution. *In* Mansfield (1976), pp. 135–59.

Braun, G. 1977a. Causes and criteria of resistance to air pollution in Norway spruce, II: Tolerance of toxic material ("internal" resistance). *European Journal of Forest Pathology* 7: 236–49.

———. 1977b. Causes of resistance against air pollution in Norway spruce and conclusions in respect of resistance breeding. *Forstwissenschaft Zentralblatt* 96: 62–66.

———. 1977c. Über die Ursachen und Kriterien der Immissions-resistenz bei Fichte, *Picea abies* (L.) Karst, I: Morphologisch-anatomische Immissionresistenze. *European Journal of Forest Pathology* 7: 23–43.

———. 1977d. Über die Ursachen und Kriterien der Immissionsresistenz bei Fichte, *Picea abies* (L.) Karst, II: Reflecktorische Immissionsresistenz. *European Journal of Forest Pathology* 7: 129–52.

———. 1977e. Über die Ursachen und Kriterien der Immissionsresistenz bei Fichte, *Picea abies* (L.) Karst, III: Schadstoffverträglichkeit ("innere" Resistenz). *European Journal of Forest Pathology* 7: 303–19.

———. 1978. Über die Ursachen und Kriterien der Immissionresistenz bei Fichte, *Picea abies* (L.) Karst, IV: Erholungsfähigkeit der Nadeln, Wechselwirkungen zwischen der Resistenzformen und zusammenfassendes Gesamtergebnis. *European Journal of Forest Pathology* 8: 83–96.

Brennan, E., and P. M. Halisky. 1970. Response of turf grass cultivars to ozone and sulfur dioxide in the atmosphere. *Phytopathology* 60: 1544–46.

Brennan, E., and I. A. Leone. 1968. The response of plants to SO_2 and O_3 polluted air supplied at varying flow rates. *Phytopathology* 58: 1661–64.

———. 1970. The response of English Holly selections to ozone and sulfur dioxide. *Holly Letter* 37: 6–7.

———. 1972. Chrysanthemum response to sulfur dioxide and ozone. *Plant Disease Reporter* 56: 85–87.

Bressan, R., L. Wilson & P. Filner. 1978. Mechanisms of resistance to sulfur dioxide in the Cucurbitaceae. *Plant Physiology* 61: 761–67.

Bressan, R., *et al.* 1979. Emission of ethylene and ethane by leaf tissue exposed to injurious concentrations of sulfur dioxide or bisulfite ion. *Plant Physiology* 63: 924–30. Coauthors: L. LeCureux, L. G. Wilson & P. Filner.

Bressan, R. A., *et al.* 1981. Inheritance of resistance to sulfur dioxide in cucumber. *Hortscience* 16: 332–33. Coauthors: L. LeCureux, L. G. Wilson, P. Filner & L. R. Baker.

Brimblecombe, P. 1978. Dew as a sink for SO_2. *Tellus* 30: 151–57.

Brodo, I. M. 1961. Transplant experiments with corticolous lichens using a new technique. *Ecology* 42: 838–41.

————. 1966. Lichen growth and cities: A study on Long Island, New York. *Bryologist* 69: 427–49.

Brogan, J. C., ed. 1978. *Sulfur in Forages*. Dublin: An Foras Taluntais.

Brough, A., M. A. Parry & A. C. Kendall. 1975. Effects of aerial pollutants on cereal growth. *Rothamsted Experiment Station Report* 1: 41–42.

Brown, K. W. 1976. Sugar beet and potatoes. *In* Monteith (1976), 2: 65–86.

Brymer, P. 1982. Effects of ozone and sulfur dioxide singly or in combination on net photosynthesis, transpiration, and growth of *Glycine max* L. Merr. M.Sc. thesis, Univ. of Guelph, Canada.

Bull, J. N., and T. A. Mansfield. 1974. Photosynthesis in leaves exposed to SO_2 and NO_2. *Nature* 250: 443–44.

Butler, L. K., and T. W. Tibbitts. 1979. Stomatal mechanisms determining genetic resistance to ozone in *Phaseolus vulgaris* L. *Journal of the American Society for Horticultural Science* 104: 213–16.

Butler, L. K., T. W. Tibbitts & F. A. Bliss. 1979. Inheritance of resistance to ozone in *Phaseolus vulgaris* L. *Journal of the American Society for Horticultural Science* 104: 211–13.

Bystrom, B. G., et al. 1968. Leaf surface of *Beta vulgaris*: Electron microscope study. *Botanical Gazette* 129: 133–38. Coauthors: R. B. Glater, F. M. Scott & E. S. C. Bowler.

Calvert, J. G., et al. 1978. Mechanisms of the homogeneous oxidation of sulfur dioxide in the troposphere. *Atmospheric Environment* 12: 197–226. Coauthors: F. Su, J. W. Bottenheim & O. P. Strausz.

Campbell, D. E. 1972. A comparison of SO_2 tolerance between two populations of *Muhlenbergia asperifolia*. M.S. thesis, Univ. of Utah.

Capannelli, G., et al. 1977. Nitrogen oxides: Analysis of urban pollution in the city of Genoa. *Atmospheric Environment* 11: 719–27. Coauthors: E. Gollo, S. Munari & G. Ratto.

Capron, T. M., and T. A. Mansfield. 1976. Inhibition of net photosynthesis in tomato in air polluted with NO and NO_2. *Journal of Experimental Botany* 27: 1181–86.

————. 1977. Inhibition of growth in tomato by air polluted with nitrogen oxides. *Journal of Experimental Botany* 28: 112–16.

Caput, C., et al. 1978. Absorption of sulfur dioxide by pine seedlings leading to acute injury. *Environmental Pollution* 16: 3–15. Coauthors: Y. Belot, D. Auclair & N. Decourt.

Carlson, R. W. 1979. Reduction in the photosynthetic rate of *Acer*, *Quercus*, and *Fraxinus* species caused by sulfur dioxide and ozone. *Environmental Pollution* 18: 159–70.

————. 1983a. The effect of SO_2 on photosynthesis and leaf resistance at varying concentrations of CO_2. *Environmental Pollution* (series A) 30: 309–21.

————. 1983b. Interaction between SO_2 and NO_2 and their effects on photosynthetic properties of soybean (*Glycine max*). *Environmental Pollution* (series A) 32: 11–38.

Carlson, R. W., and F. A. Bazzaz. 1980. Elevated CO_2 concentrations differentially increase growth, photosynthesis, and water use efficiency of plants—ecological implications. *In* Singh and Deepak (1980), pp. 609–22.

————. 1982. Photosynthetic and growth response to fumigation with SO_2 at elevated CO_2 for C_3 and C_4 plants. *Oecologia* (Berl.) 54: 50–54.

Carlson, R. W., F. A. Bazzaz & J. J. Stukel. 1976. Physiological effects, wind reentrainment, and rainwash of Pb aerosol particulate deposited on plant leaves. *Environmental Science Technology* 10: 1139–42.

Case, J. W., and H. R. Krouse. 1980. Variations in sulfur content and stable sulfur isotope composition of vegetation near a SO_2 source at Fox Creek, Alberta, Canada. *Oecologia* (Berl.) 44: 248–57.

Heagle, A. S., W. W. Heck & D. Body. 1971. Ozone injury to plants as influenced by air velocity during exposure. *Phytopathology* 61: 1209–12.

Heagle, A. S., and J. W. Johnston. 1979. Variable responses of soybeans to mixtures of ozone and sulfur dioxide. *Journal of the Air Pollution Control Association* 29: 729–32.

Heagle, A. S., and R. B. Philbeck. 1979. Exposure techniques. *In* Heck, Krupa & Linzon (1979), Chap. 14.

Heagle, A. S., *et al.* 1979. Dispensing and monitoring ozone in open-top field chambers for plant effect studies. *Phytopathology* 69: 15–20. Coauthors: R. B. Philbeck, H. H. Rogers & M. B. Letchworth.

Heagle, A. S., *et al.* 1982. Response of soybean to ozone–sulfur dioxide mixtures. *In* U.S. Environmental Protection Agency (1982b). Coauthors: R. B. Philbeck, M. J. Lee, V. M. Lesser, W. W. Heck & J. O. Rawlings.

Heagle, A. S., *et al.* 1983. Effects of chronic doses of ozone and sulfur dioxide on injury and yield of soybean in open-top field chambers. *Crop Science* 23: 1184–91. Coauthors: W. W. Heck, J. O. Rawlings & R. B. Philbeck.

Health and Safety Executive. 1982. *118th Annual Report on Alkali, etc., Works.* Her Majesty's Stationery Office.

Healy, C., and D. H. F. Atkins. 1975. *The Determination of Atmospheric Sulfur Dioxide after Collection on Impregnated Filter Paper.* Report A.E.R.E.-R7956. Harwell, Berks.: U.K. Atomic Energy Authority.

Heasley, J. E., W. K. Lauenroth & J. L. Dodd. 1981. Systems analysis of potential air pollution impacts on grassland ecosystems. *In* Mitsch, Bosserman & Klopatek (1981), pp. 347–60.

Heath, R. L. 1975. Ozone. *In* Mudd and Kozlowski (1975), pp. 23–55.

———. 1980. Initial events in injury to plants by air pollutants. *Annual Review of Plant Physiology* 31: 395–431.

Heck, W. W. 1968. Factors influencing expression of oxidant damage to plants. *Annual Review of Phytopathology* 6: 165–88.

———. 1982. Future directions in air pollution research. *In* Unsworth and Ormrod (1982), pp. 411–35.

———. 1983. Defining gaseous pollution problems: North America. *In* Koziol and Whatley (1983), in press.

Heck, W. W., and C. S. Brandt. 1977. Effects on vegetation: Native, crops, forests. *In* Stern (1977), II, 157–229.

Heck, W. W., and J. A. Dunning. 1978. Response of oats to SO_2: Interactions of growth exposure with exposure temperature and humidity. *Journal of the Air Pollution Control Association* 28: 241–46.

Heck, W. W., J. A. Dunning & I. J. Hindawi. 1966. Ozone: Nonlinear relation of dose and injury in plants. *Science* 151: 577–78.

Heck, W. W., J. A. Dunning & H. Johnson. 1968. *Design of a Simple Plant Exposure Chamber.* Washington, D.C.: U.S. Dept. Health, Education and Welfare, National Center for Air Pollution Control.

Heck, W. W., S. V. Krupa & S. N. Linzon, eds. 1979. *Handbook of Methodology for the Assessment of Air Pollution Effects on Vegetation.* St. Paul, Minn.: Air Pollution Control Association.

Heck, W. W., R. B. Philbeck & J. A. Dunning. 1978. *A Continuous Stirred Tank Reactor (CSTR) System for Exposing Plants to Gaseous Air Pollutants: Principles, Specifications, Construction, and Operation.* Washington, D.C.: U.S.D.A. Agricultural Research Service, Publication no. ARS-S-181.

Heck, W. W., *et al.* 1981. National Crop Loss Assessment Network (NCLAN) 1980 Annual Report. Corvallis, Oregon: Corvallis Environmental Research Laboratory,

Environmental Protection Agency. Coauthors: O. C. Taylor, R. M. Adams, G. E. Bingham, J. E. Miller, E. M. Preston & L. H. Weinstein.

Heck. W. W., *et al.* 1982a. Ozone impacts on the productivity of selected crops. *In* Jacobson and Millen (1982), appendix pp. 147–76. Coauthors: W. W. Cure, D. S. Shriner, R. J. Olsen & A. S. Heagle.

Heck, W. W., *et al.* 1982b. Assessment of crop loss from ozone. *Journal of the Air Pollution Control Association* 32: 353–61. Coauthors: O. C. Taylor, Richard Adams, Gail Bingham, Joseph Miller, Eric Preston & Lenard Weinstein.

Hedgcock, G. C. 1912. Winter-killing and smelter-injury in the forests of Montana. *Torreya* 2: 25–30.

———. 1914. Injuries by smelter smoke in southeastern Tennessee. *Journal of the Washington Academy of Science* 4: 70–71.

Heggestad, H. E., and J. H. Bennett. 1981. Photochemical oxidants potentiate yield losses in snap beans attributable to sulfur dioxide. *Science* 213: 1008–10.

Heggestad, H. E., J. J. Bennett & L. H. Douglass. 1982. Response of soybean to ozone–sulfur dioxide mixtures. *In* U.S. Environmental Protection Agency (1982b).

Heggestad, H. E., J. J. Bennett & E. H. Lee. 1981. Effects of mixtures of sulfur dioxide and ozone on tomato. *In* U.S. Environmental Protection Agency (1981).

Heggestad, H. E., and J. T. Middleton. 1959. Ozone in high concentrations as cause of tobacco leaf injury. *Science* 129: 208–10.

Heggestad, H. E., K. L. Tuthill & R. N. Stewart. 1973. Differences among poinsettias in tolerance to sulfur dioxide. *HortScience* 8: 337–38.

Heitschmidt, R. K., W. K. Lauenroth & J. L. Dodd. 1978. Effects of controlled levels of sulfur dioxide on western wheatgrass in a southeastern Montana grassland. *Journal of Applied Ecology* 14: 693–702.

Heldt, H. W., *et al.* 1973. Alkalization of the chloroplast stroma caused by light-dependent proton flux into the thylakoid space. *Biochimica et Biophysica Acta* 314: 224–41. Coauthors: K. Werden, M. Milovancev & G. Geller.

Henderson, R., *et al.* 1982. Map cited in *Perspectives on the Issue of Acid Precipitation.* Washington, D.C.: U.S. Dept. of Energy.

Hennies, H. H. 1975. Die Sulfitreduktase aus *Spinacea oleracea*: Ein Ferredoxin-abhängiges Enzym. *Zeitschrift für Naturforschung* (C)30: 359–62.

Hepting, G. H. 1968. Diseases of forest and tree crops caused by air pollutants. *Phytopathology* 58: 1098–1101.

Hess, B., and H. J. Staudinger, eds. 1968. *Biochemie des Sauerstoffs.* Berlin: Springer-Verlag.

Hicks, B. B., M. L. Wesely & J. L. Durham. 1982. Critique of methods to measure dry deposition. *In* Keith (1982), II, 205–23.

Hicks, B. B., *et al.* 1982. Some direct measurements of atmospheric sulfur fluxes over a pine plantation. *Atmospheric Environment* 16: 2899–2903. Coauthors: M. L. Wesely, J. L. Durham & M. A. Brown.

Hill, A. C., and J. H. Bennett. 1970. Inhibition of apparent photosynthesis by nitrogen oxides. *Atmospheric Environment* 4: 341–48.

Hill, A. C., *et al.* 1974. Sensitivity of native desert vegetation to SO_2 and to SO_2 and NO_2 combined. *Journal of the Air Pollution Control Association* 24: 153–57. Coauthors: S. Hill, C. Lamb & T. W. Barrett.

Hill, G. R., and M. D. Thomas. 1933. Influence of leaf destruction by sulfur dioxide and by clipping on yield of alfalfa. *Plant Physiology* 8: 223–45.

Ho, L. C. 1977. Effects of CO_2 enrichment on the rates of photosynthesis and translocation of tomato leaves. *Annals of Applied Biology* 87: 191–200.

———. 1979. Regulation of assimilate translocation between leaves and fruits in the tomato. *Annals of Botany* 43: 437–48.

Hocking, D., and M. B. Hocking. 1977. Equilibrium solubility of trace atmospheric sulfur dioxide in water and its bearing on air pollution injury to plants. *Environmental Pollution* 9: 57–64.

Hodges, G. H., H. A. Menser & W. B. Ogden. 1971. Susceptibility of Wisconsin Havana tobacco cultivars to air pollutants. *Agronomy Journal* 63: 107–11.

Hodgkin, S. E., and D. Briggs. 1981. The effect of simulated acid rain on two populations of *Senecio vulgaris* L. *New Phytologist* 90: 687–91.

Hoffmann, J., E. Pahlich & L. Steubing. 1976. Enzymatischanalytische Untersuchungen zum Adenosinphosphatgehalt SO_2-begaster Erbsen. *International Journal of Environmental Analytical Chemistry* 4: 183–96.

Hofstra, G., and D. W. Beckerson. 1981. Foliar responses of five plant species to ozone and a sulphur dioxide/ozone mixture after a sulphur dioxide pre-exposure. *Atmospheric Environment* 15: 383–89.

Hofstra, G., and D. P. Ormrod. 1977. Ozone and sulphur dioxide interaction in white bean and soybean. *Canadian Journal of Plant Science* 57: 1193–98.

Holling, C. S., ed. 1978. *Adaptive Environmental Assessment and Management.* New York: Wiley.

Holmes, J. A., E. C. Franklin & R. A. Gould. 1915. *Report of the Selby Smelter Commission.* U.S. Bureau of Mines, Bulletin 98.

Horn, H. S. 1974. The ecology of secondary succession. *Annual Review of Ecology and Systematics* 5: 25–37.

Horng, A. J., and S. F. Yang. 1973. Peroxidase-catalyzed oxidation of indole-3-acetaldehyde to 4-hydroxyquinoline in the presence of bisulfite ion: Elimination of pyrrole ring C_2 as formic acid. *Biochimica et Biophysica Acta* 321: 456–60.

———. 1975. Aerobic oxidation of indole-3-acetic acid with bisulfite. *Phytochemistry* 14: 1425–28.

Horsfall, J. G., and E. B. Cowlings, eds. 1980. *Plant Disease: An Advanced Treatise.* New York: Academic Press.

Horsman, D. C., T. M. Roberts & A. D. Bradshaw. 1978. Evolution of sulfur dioxide tolerance in perennial ryegrass. *Nature* 276: 493–94.

———. 1979. Studies on the effects of sulphur dioxide on perennial ryegrass (*Lolium perenne* L.), II: Evolution of sulphur dioxide tolerance. *Journal of Experimental Botany* 30: 495–501.

Horsman, D. C., et al. 1979. Studies on the effect of sulphur dioxide on perennial ryegrass (*Lolium perenne* L.), I: Characteristics of fumigation system and preliminary experiments. *Journal of Experimental Botany* 30: 485–93. Coauthors: T. M. Roberts, M. Lambert & A. D. Bradshaw.

Horsman, D. C., and A. R. Wellburn. 1975. Synergistic effect of sulfur dioxide and nitrogen dioxide polluted air upon enzyme activity in pea seedlings. *Environmental Pollution* 8: 123–33.

———. 1977. Effect of SO_2-polluted air upon enzyme activity in plants originating from areas with different mean atmospheric SO_2 concentrations. *Environmental Pollution* 13: 33–39.

Hosker, R. P., and S. E. Lindberg. 1982. Review: Atmospheric deposition and plant assimilation of gases and particles. *Atmospheric Environment* 16: 889–910.

Hou, L.-Y., A. C. Hill & A. Soleimani. 1977. Influence of CO_2 on the effects of SO_2 and NO_2 on alfalfa. *Environmental Pollution* 12: 7–16.

Houston, D. B. 1970. Physiological and genetic response of *Pinus strobus* L. clones to sulfur dioxide and ozone exposures. Ph.D. thesis, Univ. of Wisconsin.

———. 1974. Response of selected *Pinus strobus* L. clones to fumigations with sulfur dioxide and ozone. *Canadian Journal of Forest Research* 4: 65–68.

Houston, D. B., and L. S. Dochinger. 1977. Effects of ambient air pollution on cone, seed, and pollen characteristics in eastern white and red pines. *Environmental Pollution* 12: 1–5.

Houston, D. B., and G. R. Stairs. 1972. Physiological and genetic response of *Pinus strobus* L. clones to sulfur dioxide and ozone exposures. *Mitteilungen aus den Forstlichen Bundesversuchsanstalt Wien* 97: 387–98.

————. 1973. Genetic control of sulfur dioxide and ozone tolerance in eastern white pine. *Forest Science* 19: 267–71.

Howe, T. K., and S. S. Woltz. 1981. Resistance of tomato cultivars to sulfur dioxide and accumulation of foliar sulfite related to sulfur dioxide susceptibility. *HortScience* 16: 413.

————. 1982a. Sensitivity of tomato cultivars to sulfur dioxide. *HortScience* 17: 249–50.

————. 1982b. Symptomology and relative sensitivity of marigold cultivars exposed to acute sulfur dioxide. *HortScience* 17: 596–98.

Hsieh, S. T. 1973. Response of plants to air pollutants: Effects of sulfur dioxide and its synergism with ozone on *Nicotiana tabacum*. *Journal of the Chinese Agricultural Chemistry Society* 11: 74–81.

H.S.E. *See* Health and Safety Executive.

Hultberg, H., P. Grennfelt & B. Olsson. 1982. Sulphur and chloride deposition and ecosystem transport in a strongly acidified lake watershed. *In Proceedings of the International Conference on Coal-Fired Power Plants and the Aquatic Environment.* Hoersholm, Denmark: VKI, Water Quality Institute.

Hunt, R. 1978. The fitted curve in plant growth studies. *In* D. A. Rose and D. A. Charles-Edwards, eds., *Mathematics and Plant Physiology*, pp. 283–98. London: Academic Press.

Hursh, C. R. 1948. Local climate in the copper basin of Tennessee as modified by removal of vegetation. *U.S. Department of Agriculture Circular*, no. 744.

Hutchinson, T. C., and M. Havas, eds. 1980. *Effects of Acid Precipitation on Terrestrial Ecosystems*. New York: Plenum Press.

Huttenen, S. 1978. The effect of air pollution on provenances of Scots pine and Norway spruce in northern Finland. *Silva Fennica* 12: 1–16.

Huttenen, S., P. Havas & K. Laine. 1981. Effects of air pollutants on the wintertime water economy of the Scots pine *Pinus sylvestris*. *Holarctic Ecology* 4: 94–101.

Huttunen, S., and H. Törmälehto. 1982. Air pollution resistance of some Finnish *Pinus sylvestris* L. provenances. *Aquilo Ser Botanica* 18: 1–9.

I.E.R.E. (International Electric Research Exchange). 1981. *Effects of SO₂ and Its Derivatives on Health and Ecology, II: Natural Ecosystems, Agriculture, Forestry, and Fisheries*. London: Central Electricity Generating Board. Available from E.P.R.I., Palo Alto, Calif.

Inglis, F., and D. Hill. 1974. The effect of sulfite and fluoride on carbon dioxide uptake by mosses in the light. *New Phytologist* 73: 1207–13.

Inoue, M., and H. Hayatsu. 1971. The interactions between bisulfite and amino acids: The formation of methionine sulfoxide from methionine in the presence of oxygen. *Chemical Pharmaceutical Bulletin* 19: 1286–88.

Inoue, M., H. Hayatsu & H. Tanooka. 1972. Concentration effect of bisulfite on the inactivation of transforming activity of DNA. *Chemical-Biological Interactions* 5: 85–95.

Institute of Ecology. 1980. *Crop and Forest Losses Due to Current and Projected Emissions from Coal-fired Power Plants in the Ohio River Basin*. Indianapolis, Ind.: The Institute of Ecology, Butler University.

International Atomic Energy Agency. 1972. *Symposium on the Use of Isotopes and Radiation in Research on Soil-Plant Relationships, Including Applications in Forestry*. Vienna: International Atomic Energy Agency.

International Electric Research Exchange. *See* I.E.R.E.

Jacobson, J. S. 1982. Economics of biological assessment. *Journal of the Air Pollution Control Association* 32: 145–46.

Jacobson, J. S., and L. J. Colavito. 1976. The combined effect of sulfur dioxide and ozone on bean and tobacco plants. *Environmental and Experimental Botany* 16: 277–85.

Jacobson, J. S., and A. A. Millen, eds. 1982. *Effects of Air Pollution on Farm Commodities: Proceedings of the Symposium, Washington, D.C., February 18, 1982.* Arlington, Va.: Izaak Walton League of America.

Jäger, H.-J., and H. Klein. 1977. Biochemical and physiological detection of sulfur-dioxide injury to pea plants (*Pisum sativum*). *Journal of the Air Pollution Control Association* 27: 464–66.

————. 1980. Biochemical and physiological effects of SO_2 on plants. *Angewandte Botanik* 54: 337–48.

Jäger, H.-J., and L. Steubing. 1970. Fractionierte Schwefelbestimmung in Pflanzenmaterial zur Beurteilung einer SO_2-Einwirkung. *Angewandte Botanik* 44: 209–21.

James, W. C., et al. 1972. The quantitative relationship between late blight of potato and loss in tuber yield. *Phytopathology* 62: 92–96. Coauthors: C. S. Shih, W. A. Hodgson & L. C. Callbeck.

Jarman, P. D. 1974. The diffusion of carbon dioxide and water vapor through stomata. *Journal of Experimental Botany* 25: 927–36.

Jarvis, P. G. 1971. The estimation of resistances to carbon dioxide transfer. *In* Sesták, Cătský & Jarvis (1971), pp. 566–631.

————. 1981. Stomatal conductance, gaseous exchange, and transpiration. *In* Grace, Ford & Jarvis (1981), pp. 175–204.

Jarvis, P. G., G. B. James & J. J. Landsberg. 1976. Coniferous forest. *In* Monteith (1976), I, 171–240.

Jarvis, P. G., and T. A. Mansfield, eds. 1981. *Stomatal Physiology.* London: Cambridge Univ. Press.

Jarvis, P. G., et al. 1971. General principles of isometric methods and the main aspects of installation design. *In* Sesták, Cătský & Jarvis (1971), pp. 49–110. Coauthors: J. Cătský, F. E. Eckart, W. Koch & D. Koller.

Jensen, K. F. 1981a. Air pollutants affect the relative growth rate of hardwood seedlings. *U.S.D.A. Forest Service Research Paper* NE-470.

————. 1981b. Growth analysis of hybrid poplar cuttings fumigated with ozone and sulfur dioxide. *Environmental Pollution* 26: 243–50.

Jensen, K. F., and L. S. Dochinger. 1979. Growth responses of woody species to long- and short-term fumigation with sulfur dioxide. *U.S.D.A. Forest Service Research Paper* NE-422.

Jensen, K. F., and T. T. Kozlowski. 1975. Absorption and translocation of sulfur dioxide by seedlings of four forest-tree species. *Journal of Environmental Quality* 4: 379–82.

Jensen, K. F., et al. 1976. Pollution responses. *In* Miksche (1976), pp. 189–216. Coauthors: L. S. Dochinger, B. R. Roberts & A. M. Townsend.

Johnson, D. W. 1980. Site susceptibility to leaching by H_2SO_4 in acid rainfall. *In* Hutchinson and Havas (1980), pp. 525–35.

Johnson, D. W., J. Turner & J. M. Kelly. 1982. The effects of acid rain on forest nutrient status. *Water Resources Research* 18: 449–61.

Johnson, D. W., et al. 1982. Cycling of organic and inorganic sulfur in a chestnut oak forest. *Oecologia* (Berl.) 54: 141–48. Coauthors: G. S. Henderson, D. D. Huff, S. E. Lindberg, D. D. Richter, D. S. Shriner, D. E. Todd & J. Turner.

Johnstone, H., and P. Leppla. 1934. The solubility of sulfur dioxide at low partial pressures: The ionization constant and heat of ionization of sulfurous acid. *Journal of the American Chemical Society* 56: 2233–38.

Jones, H. C., and J. C. Noggle. 1980. Ecological effects of atmospheric emissions from coal-fired power plants. *Proceedings of the Symposium on Air-Quality Management in the Electric Power Industry, Austin, Texas, January 1980.*

Jones, H. C., *et al.* 1979. Power-plant siting: Assessing risks of sulfur dioxide effects on agriculture. Presented at the 72d Annual Meeting of the Air Pollution Control Association, Cincinnati, Ohio, June 24–29, 1979. Coauthors: F. P. Weatherford, J. C. Noggle, N. T. Lee & J. R. Cunningham.

Jones, T., and T. A. Mansfield. 1982a. The effect of SO_2 on growth and development of seedlings of *Phleum pratense* under different light and temperature environments. *Environmental Pollution* (series A) 27: 57–71.

————. 1982b. Studies on dry-matter partitioning and distribution of ^{14}C-labelled assimilates in plants of *Phleum pratense* exposed to SO_2 pollution. *Environmental Pollution* (series A) 28: 199–207.

Just, R. E., D. L. Hueth & A. Schmitz. 1982. *Applied Welfare Economics and Public Policy.* Englewood Cliffs, N.J.: Prentice-Hall.

Kaji, M., *et al.* 1980. Absorption of atmospheric NO_2 by plants and soils, VI: Transformation of NO_2 absorbed in the leaves and transfer of the nitrogen in the plants. *In* N.I.E.S. (1980), pp. 51–58. Coauthors: T. Yoneyama, T. Tosuka & H. Iwaki.

Kalbfleisch, J. D., and R. L. Prentice. 1980. *The Statistical Analysis of Failure Time Data.* New York: Wiley.

Kanna, R., *et al.* 1981. NMR and ESR studies of thylakoid membranes. *In* Akoyunoglou (1981), I, 147–53. Coauthors: S. Rajan, Govindjee & H. S. Gutowsky.

Kaplan, D., C. McJilton & D. Luchtel. 1975. Bisulfite-induced lipid oxidation. *Archives of Environmental Health* 30: 507–9.

Kärenlampi, L., ed. 1976. *Proceedings of the Kuopio Meeting on Plant Damages Caused by Air Pollution.* Kuopio, Finland: Univ. of Kuopio.

Kärenlampi, L., and S. Soikkeli. 1980. Morphological and fine structural effects of different pollutants on plants. *In* U.N.E.C.E. (1980), pp. 92–102.

Karnosky, D. F. 1976. Threshold levels for foliar injury to *Populus tremuloides* by sulfur dioxide and ozone. *Canadian Journal of Forest Research* 6: 166–69.

————. 1977. Evidence for genetic control of response to sulfur dioxide and ozone in *Populus tremuloides*. *Canadian Journal of Forest Research* 7: 437–40.

————. 1981a. Chamber and field evaluations of air-pollution tolerances of urban trees. *Journal of Arboriculture* 7: 99–105.

————. 1981b. Changes in eastern white-pine stands related to air-pollution stress. *Mitteilungen der Forstlichen Bundesversuchsanstalt Wien* 137: 41–45.

Karnosky, D. F., and G. R. Stairs. 1974. The effects of SO_2 on *in vitro* forest-tree pollen germination and tube elongation. *Journal of Environmental Quality* 3: 406–9.

Karnosky, D. F., and K. C. Steiner. 1981. Provenance and family variation in response of *Fraxinus americana* and *F. pennsylvanica* to ozone and sulfur dioxide. *Phytopathology* 71: 804–7.

Karpen, D. N. 1970. Ozone and sulfur dioxide synergism: Foliar injury to a ponderosa pine geographic race plantation in the Puget Sound region. *Plant Disease Reporter* 44: 945–48.

Kats, G., C. R. Thompson & W. C. Kuby. 1976. Improved ventilation of open-top greenhouses. *Journal of the Air Pollution Control Association* 26: 1089–90.

Katz, M. 1939. Sulfur dioxide in the atmosphere of industrial areas. *In* Katz and Lathe (1939), pp. 14–50.

Katz, M., and F. E. Lathe, eds. 1939. *Effect of Sulfur Dioxide on Vegetation.* Ottawa: National Research Council of Canada.

Katz, M., and G. E. Ledingham. 1939. Effect of environmental factors on the susceptibility of barley and alfalfa to sulphur dioxide. *In* Katz and Lathe (1939), pp. 262–97.

Katz, R. W., A. H. Murphy & R. L. Winkler. 1982. Assessing the value of frost forecasts to orchardists: A dynamic decision-making approach. *Journal of Applied Meteorology* 21: 72–85.

Kays, W. M. 1966. *Convective Heat and Mass Transfer.* New York: McGraw-Hill.

Keith, L. W., ed. 1982. *Energy and Environmental Chemistry.* Ann Arbor, Mich.: Ann Arbor Science.

Keller, T. 1978a. Wintertime atmospheric pollutants—do they affect the performance of deciduous trees in the ensuing growing season? *Environmental Pollution* 16: 243–47.

———. 1978b. Der Einfluss einer SO$_2$-Belastung zu verschiedenen Jahreszeiten auf CO$_2$-Aufnahme und Jahrringbau der Fichte. *Schweizerische Zeitschrift für Forstwesen* 129: 381–93.

———. 1979. Der Einfluss langdauernder SO$_2$-Begasungen auf das Wurzelwachstum der Fichte. *Schweizerische Zeitschrift für Forstwesen* 130: 429–35.

———. 1980. The effect of a continuous springtime fumigation with SO$_2$ on CO$_2$ uptake and structure of the annual ring in spruce. *Canadian Journal of Forest Research* 10: 1–6.

———. 1981a. Folgen einer winterlichen SO$_2$—Belastung für die Fichte. *Gartenbauwissenschaft* 46: 170–81.

———. 1981b. Winter uptake of airborne SO$_2$ by shoots of deciduous species. *Environmental Pollution* (series A) 26: 313–17.

———. 1982. Physiological bioindications of an effect of air pollution on plants. *In* Steubing and Jäger (1982), pp. 85–95.

———. 1983. Physiological bioindications of a wintertime SO$_2$ fumigation of spruce. Sixth World Congress on Air Quality, Paris, *Proceedings* 2: 569–76.

Keller, T., and H. Beda-Puta. 1981. Luftverunreinigungen und Schneedruck: Eine Beobachtung an Buche. *Forstarchiv* 52: 13–16.

Kercher, J. R. 1977. GROW1: A crop-growth model for assessing impacts of gaseous pollutants from geothermal technologies. UCRL-52247. Livermore, Calif.: Lawrence Livermore National Laboratory.

———. 1978. A model of leaf photosynthesis and the effects of simple gaseous sulfur compounds. UCRL-52579. Livermore, Calif.: Lawrence Livermore National Laboratory.

———. 1980. Developing realistic crop loss models for air-pollutant stress. *In* Teng and Krupa (1980), pp. 90–97.

———. 1982. An assessment of the impact on crops of effluent gases from geothermal energy development in the Imperial Valley, California. *Journal of Environmental Management* 15: 213–28.

Kercher, J. R., and M. C. Axelrod. 1981. A model for forecasting the effects of SO$_2$ pollution on growth and succession in a western coniferous forest. UCRL-53109. Livermore, Calif.: Lawrence Livermore National Laboratory.

Kercher, J. R., M. C. Axelrod & G. E. Bingham. 1980. Forecasting effects of SO$_2$ pollution on growth and succession in a western conifer forest. *In* Miller (1980), pp. 200–202.

Kercher, J. R., D. A. King & G. E. Bingham. 1982. Approaches for modeling crop-pollutant interactions in the NCLAN program. *Proceedings of the 75th Annual Meeting of the Air Pollution Control Association*, paper 82-69.3. Pittsburgh, Pa.: Air Pollution Control Association.

Kharasch, M., E. May & F. Mayo. 1938. The peroxide effect in the addition of reagents to unsaturated compounds, XVIII: The addition and substitution of bisulfite. *Journal of Organic Chemistry* 3: 175–92.

Kickert, R. N., and F. Benenati. 1983. Models of chemical stress in agriculture. *In Encyclopedia of Systems and Control.* New York: Pergamon Press.

Kimmerer, T. W., and T. T. Kozlowski. 1981. Stomatal conductance and sulfur uptake of five clones of *Populus tremuloides* exposed to sulfur dioxide. *Plant Physiology* 67: 990–95.

Kimpton, D. D., and F. T. Wall. 1952. Determination of diffusion coefficients from rates of evaporation. *Journal of Physical Chemistry* 56: 715–17.

King, D. A., J. R. Kercher & G. E. Bingham. 1982. A model assessment of ozone-caused soybean losses for Illinois and Indiana. UCID-19584. Livermore, Calif.: Lawrence Livermore National Laboratory.

———. 1983. Modeling the effects of air pollutants on soybean yield. *In* Lauenroth, Skogerboe & Flug (1983), in press.

Kirchner, T. B., J. E. Heasley & W. K. Lauenroth. 1981. A model of carbon flow in lichens. *In* Mitsch, Bosserman & Klopatek (1981), pp. 347–60.

Kitamura, N., and H. Hayatsu. 1974. Cleavage of the glycosidic linkage of pyrimidine ribonucleosides by the bisulfite-oxygen system. *Nucleic Acids Research* 1: 75–86.

Klebanoff, S. J. 1961. The sulfite-activated oxidation of reduced pyridine nucleotides by peroxidase. *Biochimica et Biophysica Acta* 48: 93–103.

Klein, H., and H.-J. Jäger. 1976. Einfluss der Nährstoff versorgung auf die SO_2-Empfindlichkeit von Erbsenpflanzen. *Zeitschrift für Pflanzenkrankheit und Pflanzenschutz* 83: 555–68.

Klein, H., *et al.* 1978. Mechanisms contributing to differential sensitivities of plants to SO_2. *Oecologia* (Berl.) 33: 203–8. Coauthors: H.-J. Jäger, W. Domes & C. H. Wong.

Knabe, W. 1968. Experimentelle Prüfung der Fluoranreicherung in Nadeln und Blättern von Pflanzen in Abhängigkeit von deren Expositionshöhe über Grund. *Materialy VI Miedzyn. Konf. Katowice*, pp. 101–16.

———. 1970. Distribution of Scots pine forest and sulfur-dioxide emissions in the Ruhr area. *Staub-Reinhaltung der Luft* 30: 43–47.

Körner, C., J. A. Scheel & H. Bauer. 1979. Maximum leaf diffusive conductance in vascular plants. *Photosynthetica* 13: 45–142.

Kommedahl, T. H. J., and P. H. Williams, eds. 1983. *Challenging Problems in Plant Health*. St. Paul, Minn.: American Phytopathological Soc.

Kondo, N., I. Maruta & K. Sugahara. 1980. Abscisic acid–dependent changes in transpiration rate with SO_2 fumigation and the effects of sulfite and pH and stomatal aperture. *Research Report from the National Institute for Environmental Studies, Japan* 11: 127–36.

Kondo, N., and K. Sugahara. 1978. Changes in transpiration rate of SO_2-resistant and -sensitive plants with SO_2 fumigation and the participation of abscisic acid. *Plant and Cell Physiology* 19: 36–73.

Kondo, N., *et al.* 1980. Sulfite-oxidizing activities in plants. *In* N.I.E.S. (1980), pp. 137–50. Coauthors: Y. Akiyama, M. Fujiwara & K. Sugahara.

Koritz, H. G., and F. W. Went. 1953. The physiological action of smog on plants. 1: Initial growth and transpiration studies. *Plant Physiology* 28: 50–62.

Koziol, M. J. 1980. Monitoring gas concentrations in pollutant exposure systems: Defining exposure concentrations. *Journal of Experimental Botany* 31: 1413–23.

Koziol, M. J., and F. R. Whatley, eds. 1984. *Gaseous Air Pollutants and Plant Metabolism*. London: Butterworth.

Kozlowski, T. T. 1980a. Impacts of air pollution on forest ecosystems. *BioScience* 30: 88–93.

———. 1980b. Responses of shade trees to pollution. *Journal of Arboriculture* 6: 29–41.

———. 1982. Water supply and tree growth, II: Flooding. *Forestry Abstracts* 43: 145–61.

Kozlowski, T. T., and S. G. Pallardy. 1979. Effects of low temperature on leaf-diffusion resistance of *Ulmus americana* and *Fraxinus pennsylvanica* seedlings. *Canadian Journal of Botany* 57: 2466–70.

Kramer, P. J., and T. T. Kozlowski. 1979. *Physiology of Woody Plants*. New York: Academic Press.

Krause, G. H. M., and H. Kaiser. 1977. Plant response to heavy metals and sulfur dioxide. *Environmental Pollution* 12: 63–71.

Kress, L. W. 1978a. Growth impact of O_3, NO_2, and SO_2 singly and in combination on two hybrid lines of loblolly pine. *Proceedings of the American Phytopathological Society* 4: 120. (Abstract.)

———. 1978b. Growth impact of O_3, NO_2, and SO_2 singly and in combination on two maternal lines of American sycamore. *Proceedings of the American Phytopathological Society* 4: 120. (Abstract.)

Kress, L. W., J. M. Skelly & K. H. Hinkelmann. 1982a. Growth impact of O_3, NO_2, and/or SO_2 on *Pinus taeda*. *Environmental Monitoring and Assessment* 1: 229–39.

———. 1982b. Growth impact of O_3, NO_2, and/or SO_2 on *Platanus occidentalis*. *Agriculture and Environment* 7: 265–74.

Krizek, D. T., and P. Semeniuk. 1981. Role of anthocyanin and flavonol pigments in differential sensitivity of poinsettia in SO_2. *HortScience* 16: 413.

Krizek, D. T., W. P. Wergin & P. Seminiuk. 1982. Physiological and morphological properties of leaves and bracts of poinsettia in relation to sulfur-dioxide tolerance. *HortScience* 17: 519.

Krouse, H. R. 1977. Sulfur isotope abundances elucidate uptake of atmospheric sulfur emissions by vegetation. *Nature* 265: 45.

Kruis, A. 1976. Gleichgewicht der Absorption von Gasen in Flussigkeiten. *In* Borchers *et al.* (1976), Vol. 4, Part 4C.

Krupa, S. V., and A. H. Legge, eds. 1980. *Proceedings of the International Conference on Air Pollutants and Their Effects on the Terrestrial Ecosystem, May 1980*. New York: Wiley-Interscience.

Kudo, I., A. Miura & H. Hayatsu. 1977. Effects on bacteriophage λ of free radicals generated by autooxidation of bisulfite. *Ecotoxicology Environmental Safety* 1: 271–74.

Kunishige, M., and Y. Hirata. 1975. Collection and propagation of indicator plants for gaseous sulfurous acid, II: Interspecies and intervarietal differences of susceptibility to sulfur dioxide in Ericacea, and genera of acacia and osmanthus. *Vegetable and Flowers Conference Proceedings, Kurame City, Japan*, p. 2485.

Kylin, A. 1960. The incorporation of radio-sulfur from external sulfate into different sulfur fractions of isolated leaves. *Physiologia Plantarum* 13: 366–79.

Lampadius, F., E. Pelz & E. Pohl. 1970. Beitrag zum Problem der Beurteilung und des Nachweises der Resistenz von Waldbaümen gegenüber Immissionen. *Biologisches Zentralblatt* 89: 301–26.

Lang, K. J., P. Neumann & P. Schütt. 1971. The effect of seed-source and fertilizing on SO_2-sensitivity of *Pinus contorta* seedlings. *Flora* 160: 1–9.

Lange, O. L., *et al.* 1971. Responses of stomata to changes in humidity. *Planta* 100: 76–86. Coauthors: R. Lösch, E. D. Schulze & L. Kappen.

Lange, O. L., *et al.*, eds. 1982. *Physiological Plant Ecology, II: Water Relations and Carbon Assimilation*. Encyclopedia of Plant Physiology Series, vol. 12D. Berlin: Springer-Verlag. Coeditors: P. S. Nobel, C. B. Osmond & H. Ziegler.

Langer, R. H. M. 1972. *How Grasses Grow*. London: Edward Arnold.

Larcher, W. 1980. *Physiological Plant Ecology*. 2d ed. Berlin: Springer.

Larson, R. I., and W. W. Heck. 1976. An air-quality data-analysis system for interrelating effects, standards, and needed source reductions, III: Vegetation injury. *Journal of the Air Pollution Control Association* 26: 325–33.

Last, F. T. 1982. Analysis of the effects of *Erysiphe graminis* D.C. on the growth of barley. *Annals of Botany* 26: 279–89.

Lathe, F. E., and A. W. McCallum. 1939. The effect of sulfur dioxide on the diameter increment of conifers. *In* Katz and Lathe (1939), pp. 174–206.

Lauenroth, W. K., and J. L. Dodd. 1981a. Chlorophyll reduction in western wheatgrass (*Agropyron smithii* Rydb.) exposed to sulfur dioxide. *Water, Air, and Soil Pollution* 15: 309–15.

———. 1981b. The impact of sulfur dioxide on the chlorophyll content of grassland plants. *In* Preston, O'Guinn & Wilson (1981), pp. 66–73.

Lauenroth, W. K., G. V. Skogerboe & M. Flug, eds. 1983. *Analysis of Ecological Systems: State-of-the-Art in Ecological Modeling*. Louisville, Ky.: ISEM Press.

Laurence, J. A. 1979. Response of maize and wheat to sulfur dioxide. *Plant Disease Reporter* 63: 468–71.

Laurence, J. A., *et al.* 1982. Field tests of a linear gradient system for exposures of row crops to SO$_2$ and HF. *Water, Air and Soil Pollution* 17: 399–407. Coauthors: D. C. MacLean, R. E. Schneider & K. S. Hansen.

LeBlanc, F. 1969. Epiphytes and air pollution. *Proceedings of the First European Congress on the Influence of Air Pollution on Plants and Animals*, pp. 211–21. Wageningen, the Netherlands: Center for Agriculture Publishing and Documentation.

LeBlanc, F., and D. N. Rao. 1966. Réaction de quelques lichens et mousses épiphytiques à l'anhydride sulfureux dans la région de Sudbury, Ontario. *Bryologist* 69: 338–46.

———. 1973. Effects of sulfur dioxide on lichen and moss transplants. *Ecology* 54: 612–17.

———. 1975. Effects of air pollutants on lichens and bryophytes. *In* Mudd and Kozlowski (1975), pp. 237–72.

LeBlanc, F., D. N. Rao & G. Comeau. 1972. The epiphytic vegetation of *Populus balsamifera* and its significance as an air-pollution indicator in Sudbury, Ontario. *Canadian Journal of Botany* 50: 519–28.

Lee, E. H., and J. H. Bennett. 1982. Superoxide dismutase: A possible protective enzyme against ozone injury in snap beans (*Phaseolus vulgaris* L.). *Plant Physiology* 69: 1444–49.

Lee, J. J., and R. A. Lewis. 1978. Zonal air pollution system, design and performance. *In* Preston and Lewis (1978), pp. 332–44.

Lee, J. J., E. M. Preston & R. A. Lewis. 1978. A system for the experimental evaluation of the ecological effects of sulfur dioxide. *Proceedings of the American Chemical Society, 4th Joint Conference on Sensing of Environmental Pollutants*, pp. 49–53.

Lee, S. D., ed. 1977. *Biochemical Effects of Environmental Pollutants*. Ann Arbor, Mich.: Ann Arbor Science.

Lefohn, A. S., and H. M. Benedict. 1982. Development of a mathematical index that describes ozone concentration, frequency, and duration. *Atmospheric Environment* 16: 2529.

Legg, B. J. 1981. Aerial environment and crop growth. *In* Rose and Charles-Edward (1981), pp. 129–49.

Legge, A. H. 1980. Primary productivity, sulfur dioxide, and the forest ecosystem: An overview of a case study. *In* Miller (1980), pp. 51–62.

Legge, A. H., D. J. Savage & R. B. Walker. 1979. Special techniques: A portable gas-exchange leaf chamber. *In* Heck, Krupa & Linzon (1979), pp. 16/12–16/24.

Legge, A. H., *et al.* 1977. Field studies of pine, spruce, and aspen periodically subjected to sulfur-gas emissions. *Water, Air, and Soil Pollution* 8: 105–29. Coauthors: D. R. Jaques, R. G. Amundson & R. B. Walker.

Leininger, W. C., and J. E. Taylor. 1981. Germination and seedling establishment as affected by sulfur dioxide. *In* Preston, O'Guinn & Wilson (1981), pp. 115–30.

Lemon, E., D. W. Stewart & R. W. Shawcroft. 1971. The sun's work in a cornfield. *Science* 174: 371–78.

Lendzian, K. J. 1983. Permeability of plant cuticles to gaseous air pollutants. *In* Koziol and Whatley (1983), in press.

Leone, I. A., and E. Brennan. 1972a. Sensitivity of begonias to air pollution. *Horticultural Research* 9: 112–16.

———. 1972b. Sulfur nutrition as it contributes to the susceptibility of tobacco and tomato to SO_2 injury. *Atmospheric Environment* 6: 259–66.

Lepp, N. W., and N. M. Dickinson. 1976. The *p*H of leaf surfaces and its modification by atmospheric pollution. In Kärenlampi (1976), pp. 138–39.

Lerman, S. L., and E. F. Darley. 1975. Particulates. *In* Mudd and Kozlowski (1975), pp. 141–58.

Levitt, J. 1972. *Responses of Plants to Environmental Stresses.* New York: Academic Press.

Lewis, E., and E. Brennan. 1978. Ozone and sulfur dioxide mixtures cause PAN-type injury to petunia. *Phytopathology* 68: 1011–14.

Libera, W., I. Ziegler & H. Ziegler. 1975. The action of sulfite on the HCO_3^- fixation and the fixation pattern of isolated chloroplasts and leaf tissue slices. *Zeitschrift für Pflanzenphysiologie* 74: 420–33.

Likens, G. E., and T. J. Butler. 1981. Recent acidification of precipitation in North America. *Atmospheric Environment* 15(7), 1103–9.

Likens, G. E., *et al.* 1977. *Biogeochemistry of a Forested Ecosystem.* New York: Springer-Verlag. Coauthors: F. H. Bormann, R. S. Pierce, J. S. Eaton & N. M. Johnson.

Lincoln, D. R., and E. S. Rubin. 1980. Air pollution emissions from increased industrial coal use in the northeastern United States. *Journal of the Air Pollution Control Association* 30: 1310–15.

Lindberg, S. E., *et al.* 1979. *Mechanisms and Rates of Atmospheric Deposition of Selected Trace Elements and Sulfate to a Deciduous Forest Watershed.* Oak Ridge, Tenn.: Oak Ridge National Laboratory. Coauthors: R. C. Harriss, R. R. Turner, D. S. Shriner & D. D. Huff.

Linthurst, R. A., and A. P. Altshuller, eds. 1983. *Critical Assessment Document: The Acidic Deposition Phenomena and Its Effects.* Washington, D.C.: U.S. Environmental Protection Agency.

Linzon, S. N. 1971. Economic effects of sulfur dioxide on forest growth. *Journal of the Air Pollution Control Association* 21: 81–86.

———. 1978. Effects of airborne sulfur pollutants on plants. *In* Nriagu (1978), pp. 109–62.

Lizada, M. C. C., and S. F. Yang. 1981. Sulfite-induced lipid peroxidation. *Lipids* 15: 189–94.

Lockyer, D. R., D. W. Cowling & L. H. P. Jones. 1976. A system for exposing plants to atmospheres containing low concentrations of sulphur dioxide. *Journal of Experimental Botany* 27: 397–409.

Lorenc-Plucinska, G. 1982. Effects of sulfur dioxide on CO_2 exchange in SO_2-tolerant and SO_2-susceptible Scots pine seedlings. *Photosynthetica* 16: 140–44.

Lorimer, G. H., M. R. Badger & T. J. Andrews. 1976. The activation of ribulose-1,5-bisphosphate carboxylase by carbon dioxide and magnesium ions: Equilibria, kinetics, a suggested mechanism, and physiological implications. *Biochemistry* 15: 529–36.

Louwerse, W. 1980. Effects of CO_2 concentration and irradiance on the stomatal behavior of maize, barley, and sunflower plants in the field. *Plant Cell Environment* 3: 391–98.

Lowe, L. E. 1969. Distribution and properties of organic fractions in selected Alberta soils. *Canadian Journal of Soil Science* 49: 129–41.

Lüttge, U., *et al.* 1972. Bisulfite compounds as metabolic inhibitors: Nonspecific effects on membranes. *Plant and Cell Physiology* 13: 505–14. Coauthors: C. B. Osmond, E. Ball, E. Brinckmann & G. Kinze.

Luxmoore, R. J. 1980. Modeling pollutant uptake and effects on the soil-plant-litter system. *In* Miller (1980), pp. 174–80.

Lyons, D., and G. Nickless. 1968. The lower oxy-acids of sulfur. *In* G. Nickless, ed., *Inorganic Sulfur Chemistry*, pp. 509–33. New York: Elsevier.

Lyric, R. M., and I. Suzuki. 1970. Enzymes involved in the metabolism of thiosulfate by *Thiobacillus thioparus*. I: Survey of enzymes and properties of sulfite: cytochrome *c* oxidoreductase. *Canadian Journal of Biochemistry* 48: 334–43.

Ma, T.-H., and S. H. Khan. 1972. Pollen-tube mitosis and pollen-tube growth inhibited by SO_2 in cultured pollen tubes of *Tradescantia*. *Journal of Cell Biology* 55: 160a.

———. 1976. Pollen mitosis and pollen-tube growth inhibition by SO_2 in cultured pollen tubes of *Tradescantia*. *Environmental Research* 12: 144–49.

Ma, T.-H., *et al.* 1973. Low levels of SO_2-enhanced chromatid aberrations in *Tradescantia* pollen tubes and seasonal variations of the aberration rates. *Mutation Research* 21: 93–100. Coauthors: D. Isbandi, S. H. Khan & X.-S. Tseng.

Macdowall, F. D. H., and A. F. W. Cole. 1971. Threshold and synergistic damage to tobacco by ozone and sulfur dioxide. *Atmospheric Environment* 5: 553–59.

Machta, L. 1973. Prediction of CO_2 in the atmosphere. *In* Woodwell and Pecan (1973), pp. 21–31.

———. 1979. Atmospheric measurements of carbon dioxide. *Workshop on the Global Effects of Carbon Dioxide from Fossil Fuels, March 1977*, pp. 44–50. Washington, D.C.: U.S. Department of Energy.

MacLeod, R. M., *et al.* 1961. Purification and properties of hepatic sulfite oxidase. *Journal of Biological Chemistry* 236: 1841–46. Coauthors: W. Farkas, I. Fridovich & P. Handler.

Majernik, O., and T. A. Mansfield. 1970. Direct effect of SO_2 pollution on the degree of opening of stomata. *Nature* 227: 377–78.

———. 1972. Stomatal responses to raised atmospheric CO_2 concentrations during exposure of plants to SO_2 pollution. *Environmental Pollution* 3: 1–7.

Malanson, G. P., and W. E. Westman. Unpublished manuscript. Post-fire succession in Californian coastal sage scrub: The role of continual basal sprouting. *Madroño*.

Malcolm, D. C., and M. F. Garforth. 1977. The sulfur:nitrogen ratio of conifer foliage in relation to atmospheric pollution with sulfur dioxide. *Plant and Soil* 47: 89–102.

Male, L. M. 1982. An experimental method for predicting plant yield response to pollution time series. *Atmospheric Environment* 16: 2247–52.

Male, L. M., J. Van Sickle & R. Wilhour. 1978. *Time-Series Experiments for Studying Plant-Growth Response to Pollution*. Corvallis, Ore.: U.S. Environmental Protection Agency.

Malek, I., ed. 1970. *Prediction and Measurement of Photosynthetic Productivity*. Wageningen, the Netherlands: Center for Agricultural Publishing and Documentation.

Malhotra, S. S. 1977. Effects of aqueous sulfur dioxide on chlorophyll destruction in *Pinus contorta*. *New Phytologist* 78: 101–9.

Mamejev, S. A., and O. D. Shkarlet. 1970. Effects of air and soil pollution by industrial waste on the fructification of scotch pine in the Urals. *Proceedings of the International Conference on Air Pollution, Essen, West Germany* 7: 443–50.

Mandl, R. L., *et al.* 1976. A cylindrical open-top chamber for the exposure of plants to air pollutants in the field. *Journal of Environmental Quality* 2: 132–35. Coauthors: L. H. Weinstein, D. C. McCune & M. Keveny.

Mandl, R. H., *et al.* 1980. The response of sweet corn to HF and SO_2 under field conditions. *Environmental and Experimental Botany* 20: 359–65. Coauthors: L. H. Weinstein, M. Dean & M. Wheeler.

Mansfield, T. A. 1973. The role of stomata in determining the responses of plants to air pollutants. *Current Advances in Plant Science* 2: 11–20.

———. 1975. The role of stomata in determining the responses of plants to air pollutants. *In* Smith (1975), I, 13–22.

———, ed. 1976. *Effects of Air Pollution on Plants.* New York: Cambridge Univ. Press.

Mansfield, T. A., and P. H. Freer-Smith. 1981. Effects of urban air pollution on plant growth. *Biological Reviews* 56: 343–68.

———. 1984. The role of stomata in resistance mechanisms. *In* Koziol and Whatley (1984), in press.

Mansfield, T. A., and O. Majernik. 1970. Can stomata play a part in protecting plants against air pollutants? *Environmental Pollution* 1: 149–54.

Margaris, N. S., and H. A. Mooney, eds. 1981. *Components of Productivity of Mediter- ranean-Climate Regions: Basic and Applied Aspects.* The Hague: Junk.

Markowski, A., S. Grzesiak & M. Schramel. 1975. Indexes of susceptibility of various species of cultivated plants to sulfur-dioxide action. *Bulletin de L'Académie Polanaise des Sciences* 23: 637–46.

———. 1975a. Susceptibility of 6 species of cultivated plants to sulfur-dioxide action under optimum soil-moisture and drought conditions. *Bulletin de l'Académie Polonaise des Sciences* 22: 889–98.

Marshall, P. E., and G. R. Furnier. 1981. Growth responses of *Ailanthus altissima* seed- lings to SO_2. *Environmental Pollution* (series A) 25: 149–53.

Martin, A., and F. R. Barber. 1981. Sulfur oxides, oxides of nitrogen, and ozone mea- sured continuously for two years at a rural site. *Atmospheric Environment* 15: 567–78.

Martin, J. T., and B. E. Juniper. 1970. *The Cuticles of Plants.* New York: St. Martin's Press.

Masaru, N., F. Syozo & K. Saburo. 1976. Effects of exposure to various injurious gases on germination of lily pollen. *Environmental Pollution* 11: 181–87.

Materna, J., and R. Kohout. 1967. Stickstoff-Düngung und Schwefeldioxid-Aufnahme durch Fichtennadeln. *Naturwissenschaften* 54: 351–52.

———. 1969. Uptake of sulfur dioxide into the leaves of some tree species. *Communica- tiones Instituti Forestalis Cechosloveniae* 6: 38–47.

Matsushima, J. 1971. On composite harm to plants by sulfurous-acid gas and oxidant. *Sangyō kōgai* (Industrial report) 7: 218–24.

Maugh, T. H. 1979. SO_2 pollution may be good for plants. *Science* 205: 383.

May, R. M. 1974. *Stability and Complexity in Model Ecosystems.* Princeton, N.J.: Prince- ton Univ. Press.

Mayer, R., and B. Ulrich. 1980. Input to soil, especially the influence of vegetation in intercepting and modifying inputs: A review. *In* Hutchinson and Havas (1980), pp. 173–82.

McClenahan, J. R. 1978. Community changes in a deciduous forest exposed to air pollu- tion. *Canadian Journal of Forest Research* 8: 432–38.

McCool, P. M., and J. A. Menge. 1978. Interaction of air pollutants and mycorrhizae. *California Air Environment* 7: 1 et seq.

McCord, J. M., and I. Fridovich. 1968. The reduction of cytochrome *c* by milk xanthine oxidase. *Journal of Biological Chemistry* 243: 5753–60.

———. 1969a. Superoxide dismutase: An enzymic function for erythrocuprein (hemo- cuprein). *Journal of Biological Chemistry* 244: 6049–55.

———. 1969b. The utility of superoxide dismutase in studying free-radical reactions, I: Radicals generated by the interaction of sulfite, dimethyl sulfoxide, and oxygen. *Jour- nal of Biological Chemistry* 244: 6056–63.

McCune, D. C. 1973. Summary and synthesis of plant toxicology. *In* Naegele (1973), pp. 48–62.

———. 1980. Terrestrial vegetation air pollution interactions: Gaseous pollutants hydrogen fluoride and sulfur dioxide. *In* Krupa and Legge (1980).

———. 1983. Terrestrial vegetation air pollutant interactions: Gaseous pollutant–hydrogen fluoride and sulfur dioxide. *In Proceedings of the International Conference on Air Pollutants and Their Effects on Terrestrial Ecosystems: Current Status and Future Needs of Research on Effects and Technology, May 10–17, 1980.* Banff, Alberta.

McCune, D. C., and T. C. Weidensaul. 1978. Effects of atmospheric sulfur oxides and related compounds on vegetation. *In* National Academy of Sciences (1978), pp. 80–129.

McLaren, R. G., and R. S. Swift. 1977. Changes in soil organic sulfur fractions due to long-term cultivation of soils. *Journal of Soil Science* 28: 445–53.

McLaughlin, S. B. 1981. SO_2, vegetation effects, and the air-quality standard: Limits of interpretation and application. *Proceedings of the Air Pollution Control Association Specialty Conference on Air Quality Standards for Particulate Matter and Sulfur Dioxide, Atlanta, Georgia, September 16–18, 1980*, pp. 62–83.

McLaughlin, S. B., H. C. Jones & V. J. Schorne. 1976. A programmable exposure chamber for kinetic dose-response studies with air pollutants. *Journal of the Air Pollution Control Association* 26: 132–35.

McLaughlin, S. B., and N. T. Lee. 1974. Botanical studies in the vicinity of Widows Creek Steam Plant: Review of air-pollution effects studies, 1952–1972, and results of 1973 surveys. Tennessee Valley Authority Internal Report I-EB-74-1, Muscle Shoals, Alabama.

McLaughlin, S. B., and R. K. McConathy. 1979. Temporal and spatial patterns of carbon allocation in the canopy of white oak. *Canadian Journal of Botany* 57: 1407–13.

———. 1984. Effects of SO_2 and O_3 on allocation of ^{14}C-labeled photosynthate in *Phaseolus vulgaris*. *Plant Physiology*, in press.

McLaughlin, S. B., and D. S. Shriner. 1980. Allocation of resources to defense and repair. *In* Horsfall and Cowling (1980), Chap. 7.

McLaughlin, S. B., and G. E. Taylor. 1981. Relative humidity: Important modifier of pollutant uptake by plants. *Science* 211: 167–69.

McLaughlin, S. B., *et al.* 1978. Air-pollution effects on forest growth and succession: Application of a mathematical model. *Proceedings of the 71st Annual Meeting of the Air Pollution Control Association* 78-24.5. Coauthors: D. C. West, H. H. Shugart & D. Shriner.

McLaughlin, S. B., *et al.* 1979. The effects of SO_2 dosage kinetics and exposure frequency on photosynthesis and transpiration of kidney beans (*Phaseolus vulgaris* L.). *Environmental and Experimental Botany* 19: 179–91. Coauthors: D. S. Shriner, R. K. McConathy & L. K. Mann.

McLaughlin, S. B., *et al.* 1982. Effects of chronic air-pollution stress on photosynthesis, carbon allocation, and growth of white pine trees. *Forest Science* 28: 60–70. Coauthors: R. K. McConathy, D. Duvick & L. K. Mann.

Meagher, J. F., *et al.* 1981. Cross-sectional studies of plume from a partially SO_2-scrubbed power plant. Presented at the Symposium on Plumes and Visibility: Measurements and Model Components, Grand Canyon, Arizona, November 10–14, 1980. Coauthors: L. Stockburger III, R. J. Bonanno & M. Luria.

Meidner, H. 1975. Water supply, evaporation, and vapor diffusion in leaves. *Journal of Experimental Botany* 26: 666–73.

———. 1981. Measurements of stomatal aperture and responses to stimuli. *In* Jarvis and Mansfield (1981), pp. 25–49.

Meiwes, K. J., and P. K. Khanna. 1981. Distribution and cycling of sulfur in the vegetation of two forest ecosystems in an acid rain environment. *Plant and Soil* 60: 369–75.

Mejnartowicz, L., S. Bialobok & P. Karolewski. 1978. Genetic characteristics of Scots pine specimens resistant and susceptible to SO_2 actions. *Arboretum Kórnickie* 13: 233–38.

Menser, H. A., and H. E. Heggestad. 1966. Ozone and SO_2 synergism: injury to tobacco plants. *Science* 153: 424–25.

Menser, H. A., and G. H. Hodges. 1970. Effects of air pollutants on burley tobacco cultivars. *Agronomy Journal* 62: 265–69.

Menser, H. A., G. H. Hodges & C. G. McKee. 1973. Effects of air pollution on Maryland (Type 32) tobacco. *Journal of Environmental Quality* 2: 253–58.

Metson, A. J. 1973. Sulfur in forage crops. *Sulfur Institute Technical Bulletin* 20.

Mielke, D. M., H. H. Shugart & D. C. West. 1978. *A Stand Model for Upland Forests of Southern Arkansas.* Oak Ridge, Tenn.: Oak Ridge National Laboratory. (Report no. ORNL/TM 6225.)

Mikkelsen, E. P., *et al.* 1981. Response of petunia plants to SO_2 and of detached leaves, leaf discs, and callus to sodium sulfite. *Journal of the American Society for Horticultural Science* 106: 708–12. Coauthors: L. S. Schnabelrauch, L. W. LeCureux & K. C. Sink.

Miksche, J. P., ed. 1976. *Modern Methods in Forest Genetics.* Berlin: Springer-Verlag.

Milchunas, D. G., W. K. Lauenroth & J. L. Dodd. 1982. The effect of SO_2 on ^{14}C translocation in *Agropyron smithii* Rydb. *Environmental and Experimental Botany* 22: 81–91.

————. 1983. The interaction of atmospheric and soil sulfur on the sulfur and selenium concentration of range plants. *Plant and Soil* 72: 117–25.

Milchunas, D. G., *et al.* 1981. Effects of SO_2 exposure with nitrogen and sulfur fertilization on the growth of *Agropyron smithii*. *Journal of Applied Ecology* 18: 291–302. Coauthors: W. K. Lauenroth, J. L. Dodd & T. J. McNary.

Miller, C. A., and D. D. Davis. 1981. Response of pinto bean plants exposed to O_3, SO_2, or mixtures at varying temperatures. *HortScience* 16: 548–50.

Miller, H. G., and J. D. Miller. 1980. Collection and retention of atmospheric pollutants by vegetation. *In* Drabløs and Tollan (1980), pp. 33–40.

Miller, J. E., and P. Xerikos. 1979. Residence time of sulfite in SO_2 "sensitive" and "tolerant" soybean cultivars. *Environmental Pollution* 18: 259–64.

Miller, J. E., *et al.* 1978. Yield Response of Field-grown Soybeans to an Acute SO_2 Exposure. Argonne National Laboratory, ANL-78-65. Coauthors: H. Y. Smith, D. G. Sprugel & P. B. Xerikos.

Miller, J. E., *et al.* 1980. Open-air fumigation system for investigating sulfur-dioxide effects on crops. *Phytopathology* 70: 1124–28. Coauthors: D. G. Sprugel, R. N. Muller, H. J. Smith & P. B. Xerikos.

Miller, P. C., and W. A. Stoner. 1979. Canopy structure and environmental interaction. *In* Solbrig *et al.* (1979).

Miller, P. R., ed. 1980. *Effects of Air Pollutants on Mediterranean and Temperate Forest Ecosystems.* Berkeley, Calif.: U.S. Forest Service. (General Technical Report PSW-43.)

Miller, P. R., and J. R. McBride. 1975. Effects of air pollutants on forests. *In* Mudd and Kozlowski (1975), pp. 196–235.

Miller, P. R., *et al.* 1963. Ozone injury to the foliage of *Pinus ponderosa*. *Phytopathology* 53: 1072–76. Coauthors: J. R. Parmeter, Jr., O. C. Taylor & E. A. Cardiff.

Miller, V. L., R. K. Howell & B. E. Caldwell. 1974. Relative sensitivity of soybean genotypes to ozone and sulfur dioxide. *Journal of Environmental Quality* 3: 35–37.

Mills, J. D., R. E. Slovacek & G. Hind. 1978. Cyclic electron transport in isolated intact chloroplasts. *Biochimica et Biophysica Acta* 504: 298–309.

Misra, R., and B. Gopal, eds. 1968. *Proceedings of the Symposium on Recent Advances in Tropical Ecology*. Varanasi, India.

Miszalski, Z., and H. Ziegler. 1980. "Available SO_2": A parameter for SO_2 toxicity. *Phytopathologische Zeitschrift* 97: 144–47.

Mitchell, P. 1966. Chemiosmotic coupling in oxidative and photosynthetic phosphorylation. *Biological Reviews* 41: 445–502.

Mitsch, W. J., R. W. Bosserman & J. M. Klopatek, eds. 1981. *Energy and Ecological Modeling*. New York: Elsevier.

Mittelheuser, C. J., and R. F. M. van Steveninck. 1971. Rapid action of abscisic acid on photosynthesis and stomatal resistance. *Planta* 97: 83–86.

Miyake, Y., and Y. Uno. 1973. Studies on physiological tobacco leaf spot, VII: Effects of sulfur dioxide on the growth and chemical components of tobacco plant. *Bulletin of the Utsunomiya Tobacco Experiment Station* 2: 67–76.

Moller, D. 1980: Kinetic model of atmospheric SO_2 oxidation based on published data. *Atmospheric Environment* 14: 1067–76.

Monsi, M. 1968. Mathematical models of plant communities. *In* Eckardt (1968), pp. 131–50.

Monteith, J. L. 1973. *Principles of Environmental Physics*. London: Edward Arnold.

———, ed. 1976. *Vegetation and the Atmosphere*. 2 vols. London: Academic Press.

Mooi, J. 1981. Influence of ozone and sulfur dioxide on defoliation and growth of poplars. *Mitteilungen der Forstlichen Bundesversuchsanstalt Wien* 137: 47–51.

Moon, R. B., and J. H. Richards. 1973. Determination of intracellular pH by ^{31}P magnetic resonance. *Journal of Biological Chemistry* 248: 7276–78.

Mooney, H. A., and S. L. Gulmon. 1982. Constraints on leaf structure and function in reference to herbivory. *BioScience* 32: 198–206.

Morgan, M. 1975. Sulfur-dioxide damage to alfalfa: A case study of the limits to single-valued damage functions. *ERDA/National Laboratories Workshop in Environmental Effects of Energy*, pp. 102–15. Upton, N.Y.: Brookhaven National Laboratory. (Report no. BNL-20701.)

Moss, D. N., R. B. Musgrave & E. R. Lemon. 1961. Photosynthesis under field conditions, III: Some effects of light, carbon dioxide, temperature, and soil moisture on photosynthesis, respiration, and transpiration of corn. *Crop Science* 1: 83–87.

Mourioux, G., and R. Douce. 1978. Transport spécifique du sulfate à travers l'enveloppe des chloroplastes d'Épinard. *Compte rendu de l'Académie des Sciences* 286: 277–80.

Mudd, J. B. 1973. Biochemical effects of some air pollutants on plants. *In* Naegele (1973), pp. 31–47.

———. 1975a. Sulfur dioxide. *In* Mudd and Kozlowski (1975), pp. 9–22.

———. 1975b. Peroxyacetyl nitrates. *In* Mudd and Kozlowski (1975), pp. 97–119.

———. 1982. Effects of oxidants on metabolic functions. *In* Unsworth and Ormrod (1982), pp. 189–203.

Mudd, J. B., and W. M. Dugger. 1963. The oxidation of reduced pyridine nucleotides by peroxyacetyl nitrates. *Archives of Biochemistry and Biophysics* 102: 52–58.

Mudd, J. B., and T. T. Kozlowski, eds. 1975. *Responses of Plants to Air Pollution*. New York: Academic Press.

Mudd, J. B., and T. T. McManus. 1969. Products of the reaction of peroxyacetyl nitrate with sulfhydryl compounds. *Archives of Biochemistry and Biophysics* 132: 237–41.

Mueller, P. K., et al. 1980. The occurrence of atmospheric aerosols in the Northeastern United States. *Annals of the New York Academy of Science* 338: 463–82. Coauthors: G. M. Hidy, K. Warren, T. F. Lavery & R. L. Baskett.

Muller, F., and V. Massey. 1969. Flavin-sulfite complexes and their structures. *Journal of Biological Chemistry* 244: 4007–16.

Muller, R. N., J. E. Miller & D. G. Sprugel. 1979. Photosynthetic response to field-grown soybeans to fumigations with sulfur dioxide. *Journal of Applied Ecology* 16: 567–76.

Munro, D. D. 1974. Forest growth models: A prognosis. *In* Fries (1974), pp. 7–21.

Murdy, W. H. 1979. Effect of SO_2 on sexual reproduction in *Lepidium virginicum* L. originating from regions with different SO_2 concentrations. *Botanical Gazette* 140: 299–303.

Murphy, C. E., T. R. Sinclair & K. R. Knoerr. 1974. Modeling the photosynthesis of plant stands. *In* Strain and Billings (1974), pp. 125–47.

Murphy, M. D. 1978. Responses to sulfur in Irish grassland. *In* Brogan (1978), pp. 95–109.

―――. 1979. Much Irish grassland is deficient in sulfur. *Farm and Food Research* 10: 190–92.

Murray, D. R., and J. W. Bradbeer. 1971. Inhibition of photosynthetic CO_2 fixation in spinach and chloroplasts by α-hydroxypyridine methanesulfonate. *Phytochemistry* 10: 1999–2003.

Murray, J. J., R. K. Howell & A. C. Wilton. 1975. Differential response of seventeen *Poa pratensis* cultivars to ozone and sulfur dioxide. *Plant Disease Reporter* 59: 852–54.

Naegele, J. A. 1973. *Air Pollution Damage to Vegetation*. Washington, D.C.: American Chemical Society.

Naik, B. I., and S. K. Srivastava. 1978. Effect of polyamines on tissue permeability. *Phytochemistry* 17: 1885–87.

Nakamura, S. 1970. Initiation of sulfite oxidation by spinach ferredoxin NADP reductase and ferredoxin system: A model experiment on the superoxide anion radical production by metalloflavoproteins. *Biochemical Biophysical Research Communications* 41: 177–83.

Nash, T. 1975. Absorption of sulfur dioxide by aqueous solutions. *Atmospheric Environment* 9: 1129–30.

National Academy of Sciences. 1978. *Sulfur Oxides*. Washington, D.C.: N.A.S.

National Institute for Environmental Studies, Japan. 1980. *Studies on the Effects of Air Pollutants on Plants and Mechanisms of Phytotoxicity*. Environment Agency, Japan.

National Research Council of Canada. 1977. *Sulfur and Its Inorganic Derivatives in the Canadian Environment*. Ottawa: N.R.C.C.

Natural Environment Research Council. 1976. *Research on Pollution of the Natural Environment*. London: N.E.R.C.

Navon, G., *et al.* 1972. Phosphorus-31 nuclear magnetic resonance studies of wild-type and glycolytic pathway mutants of Saccharomyces cerevisiae. *Biochemistry* 18: 4487–99. Coauthors: R. G. Shulman, T. Yamane, T. R. Eccleshall, K.-B. Lam, J. J. Baronofsky & J. Mamur.

Neely, G. E., D. T. Tingey & R. G. Wilhour. 1977. Effects of ozone and sulfur dioxide singly and in combination on yield, quality, and N-fixation of alfalfa. *Proceedings of the International Conference on Photochemical Oxidant Pollution and Its Control, 1977*, pp. 663–73. (Report no. ETA-600/3-77-001B.)

N.E.R.C. *See* Natural Environment Research Council.

Neurath, H. 1963. *The Proteins*. 2d ed. New York: Academic Press.

Nicholson, I. A., I. S. Patterson & F. T. Last, eds. 1977. *Proceedings of the Workshop on Methods for Studying Acid Precipitation in Forest Ecosystems*. Cambridge, Eng.: Institute of Terrestrial Ecology.

Nicholson, I. A., *et al.* 1980a. pH and sulfate content of precipitation over Northern Britain. *In* Drabløs and Tollan (1980), pp. 84–92. Coauthors: J. N. Cape, D. Fowler, J. W. Kinnaird & I. S. Paterson.

Nicholson, I. A., *et al.* 1980b. Continuous monitoring of airborne pollutants. *In* Drabløs and Tollan (1980), pp. 144–45. Coauthors: D. Fowler, I. S. Paterson, J. N. Cape & J. W. Kinnaird.

Nieboer, E., J. D. MacFarlane & D. H. S. Richardson. 1983. Modification of plant cell buffering capacities by gaseous air pollutants. *In* Koziol and Whatley (1983), in press.

Nielsen, J. P. 1938. A study of the action of sulfur dioxide on growing plants under various conditions. Ph.D. thesis, Stanford Univ.

N.I.E.S. (National Institute for Environmental Studies, Japan). 1980. *Studies on the Effects of Air Pollutants on Plants and Mechanisms of Phytotoxicity.* Environment Agency, Japan.

Noack, K. 1929. Untersuchungen über die Rauchgasschaden der Vegetation. *Zeitschrift für Angewandte Chemie* 42: 123–29.

Noble, R. D., and K. F. Jensen. 1980a. Effects of sulfur dioxide and ozone on growth of hybrid poplar leaves. *American Journal of Botany* 67: 1005–9.

———. 1980b. Effects of SO_2 and ozone on photosynthesis and leaf growth in hybrid poplar. *In* Miller (1980), p. 246.

Noggle, J. C., and H. C. Jones. 1981. Regional effects of multiple air pollutants on plants. 74th Annual Meeting of the Air Pollution Control Association, Philadelphia. Paper 81-42.5.

Noland, T. L., and T. T. Kozlowski. 1979. Effect of SO_2 on stomatal aperture and sulfur uptake of woody angiosperm seedlings. *Canadian Journal of Forest Research* 9: 57–62.

Norby, R. J., and T. T. Kozlowski. 1981a. Interactions of SO_2 concentrations and post-fumigation temperature on growth of five species of woody plants. *Environmental Pollution* (series A) 25: 27–39.

———. 1981b. Relative sensitivity of three species of woody plants to SO_2 at high or low exposure temperature. *Oecologia* (Berl.) 51: 33–36.

———. 1981c. Response of SO_2-fumigated *Pinus resinosa* seedlings to post-fumigation temperature. *Canadian Journal of Botany* 59: 470–75.

———. 1982a. Flooding and SO_2 stress interaction in *Betula papyrifera* and *B. nigra* seedlings. *Forest Science*, in press.

———. 1982b. The role of stomata in sensitivity of *Betula papyrifera* Marsh. seedlings to SO_2 at different humidities. *Oecologia* (Berl.) 53: 34–39.

Noyes, R. D. 1980. The comparative effects of sulfur dioxide on photosynthesis and translocation in bean. *Physiological Plant Pathology* 16: 73–79.

Nriagu, J. O., ed. 1978. *Sulfur in the Environment, II: Ecological Impacts.* New York: Wiley.

O'Connor, J. A., D. G. Parbery & W. Strauss. 1974. The effects of phytotoxic gases on native Australian plant species, I: Acute effects of sulfur dioxide. *Environmental Pollution* 7: 7–23.

O'Dell, R. A., M. Taheri & R. L. Kabel. 1977. A model for uptake of pollutants by vegetation. *Journal of the Air Pollutant Control Association* 27: 1104–9.

Odum, E. P., J. T. Finn & E. H. Franz. 1979. Perturbation theory and the subsidy-stress gradient. *BioScience* 29: 349–52.

O.E.C.D. *See* Organization for Economic Cooperation and Development.

O'Gara, P. J. 1922. Sulfur dioxide and fume problems and their solutions. *Journal of Industrial Engineering Chemistry* 14: 744. (Abstract.)

Okoloko, G. E., and J. D. Brewley. 1982. Potentiation of sulfur-dioxide-induced inhibition of protein synthesis by dessication. *New Phytologist* 91: 169–75.

Olson, J. S., H. Pfuderer & Y. H. Chan. 1978. *Changes in the Global Carbon Cycle and the Biosphere.* Oak Ridge, Tenn.: Oak Ridge National Laboratory.

Olson, R. J., C. J. Emerson & M. K. Nungesser. 1980. *Geoecology: A County-Level Environmental Data Base for the Coterminous United States.* Oak Ridge, Tenn.: Oak Ridge National Laboratory, Environmental Science Division.

Olszowski, J. 1980. Variability of selected morphological features of pine under the influence of fertilization in a region affected by air pollution. *In* U.N.E.C.E. (1980), p. 109.

Olszyk, D. M., and T. W. Tibbitts. 1981a. Stomatal response and leaf injury of *Pisum sativum* L. with SO₂ and O₃ exposures, I: Influence of pollutant level and leaf maturity. *Plant Physiology* 67: 539–44.

————. 1981b. Stomatal response and leaf injury of *Pisum sativum* L. with SO₂ and O₃ exposures, II: Influence of moisture stress and time of exposure. *Plant Physiology* 67: 545–49.

————. 1982. Evaluation of injury to expanded and expanding leaves of pea exposed to sulfur dioxide and ozone. *Journal of the American Society of Horticultural Science* 107: 266–71.

Omasa, K., *et al.* 1980a. Analysis of air-pollutant sorption by plants, II: A method for simultaneous measurement of NO₂ and O₃ sorptions by plants in environmental control chambers. In N.I.E.S. (1980), pp. 195–211. Coauthors: F. Abo, S. Funada & I. Aiga.

Omasa, K., *et al.* 1980b. Analysis of air pollutant sorption by plants, III: Sorption under fumigation with NO₂, O₃ or NO₂ and O₃. In N.I.E.S. (1980), pp. 213–24. Coauthors: F. Abo, T. Natori & T. Totsuka.

Omura, T., H. Sato & K. Sugahara. 1978. Differences among local rice varieties in resistance to sulfur dioxide. *Kokuritsu Kōgai Kenkyōsho Tokubetsu Kenkyū Seika Hōkoku* 2: 135–44.

Omura, T., *et al.* 1980. Inheritance of sensitivity to sulfur dioxide in rice, *Oryza sativa* L. *In* N.I.E.S. (1980), pp. 263–65. Coauthors: H. Sato, Y. Fujinuma & I. Aiga.

Ordin, L., *et al.* 1962. Use of antioxidants to protect plants from oxidant-type air pollutants. *International Journal of Air and Water Pollution* 6: 223–37. Coauthors: O. C. Taylor, B. E. Propst & E. A. Cardiff.

Organization for Economic Cooperation and Development. 1979. *O.E.C.D. Programme on the Long-Range Transport of Air Pollutants.* 2d ed. Paris: O.E.C.D.

————. 1981. *The Costs and Benefits of Sulfur Oxide Control: A Methodological Study.* Paris: O.E.C.D.

Ormrod, D. P. 1982. Air pollutant interactions in mixtures. *In* Unsworth and Ormrod (1982), pp. 307–31.

————. 1983. Effects of pollutants on horticultural plants. *Proceedings of the 21st International Horticultural Congress, Hamburg*, pp. 646–56.

Ormrod, D. P., N. O. Adedipe & G. Hofstra. 1973. Ozone effects on growth of radish plants as influenced by nitrogen and phosphorus nutrition and by temperature. *Plant and Soil* 39: 437–39.

Ormrod, D. P., V. J. Black & M. H. Unsworth. 1981. Depression of net photosynthesis in *Vicia faba* L. exposed to sulfur dioxide and ozone. *Nature* 291: 585–86.

Ormrod, D. P., D. T. Tingey & M. Gumpertz. 1983. Covariate measurements for increasing the precision of plant response to O₃ and SO₂. *HortScience.*

Ormrod, D. P., *et al.* 1983. Utilization of a response surface technique in the study of physiological responses to ozone and sulfur dioxide mixtures. *Plant Physiology.* Coauthors: D. T. Tingey, M. L. Gumpertz & D. M. Olszyk.

Oshima, R. J. 1975. Development of a system for evaluating and reporting economic crop losses caused by air pollution in California, II: Ozone dosage–crop loss conversion function for alfalfa and sweet corn. California Air Resources Board. (Report no. ARB-3-690.)

562   Bibliography

———. 1978. The impact of sulfur dioxide on vegetation: A sulfur-dioxide/ozone response model. California Air Resources Board. (Report no. A6-162-30.)

Oshima, R. J., J. P. Bennett & P. K. Braegelmann. 1978. Effect of ozone on growth and assimilate partitioning in parsley. *Journal of the American Society for Horticultural Science* 103: 348–50.

Oshima, R. J., *et al.* 1976. Ozone dosage–crop loss function for alfalfa: A standardized method for assessing crop losses from air pollutants. *Journal of the Air Pollution Control Association* 26: 861–65. Coauthors: M. P. Poe, P. K. Braegelmann, D. W. Baldwin & V. Van Way.

Oshima, R. J., *et al.* 1979. The effects of ozone on the growth, yield, and partitioning of dry matter in cotton. *Journal of Environmental Quality* 8: 474–79. Coauthors: P. K. Braegelmann, R. Flagler & R. R. Teso.

Oshino, N., and B. Chance. 1975. The properties of sulfite oxidation in perfused rat liver: Interaction of sulfite oxidase with the mitochondrial respiratory chain. *Archives of Biochemistry and Biophysics* 170: 514–28.

Padgett, J., and H. Richmond. 1983. The process of establishing and revising national ambient air-quality standards. *Journal of the Air Pollution Control Association* 33: 13–16.

Pahlich, E. 1975. Effect of SO_2-pollution on cellular regulation: A general concept of the mode of action of gaseous air contamination. *Atmospheric Environment* 9: 261–63.

Patton, R. L. 1981. Effects of ozone and sulfur dioxide on height and stem specific gravity of populus hybrids. *U.S.D.A. Forest Service Research Paper* 471.

Peet, R. K. 1974. The measurement of species diversity. *Annual Review of Ecology and Systematics* 5: 285–307.

Peiser, G., C. Lizada & S. F. Yang. 1982. Sulfite-induced lipid peroxidation in chloroplasts as determined by ethane production. *Plant Physiology* 70: 994–98.

Peiser, G., and S. F. Yang. 1977. Chlorophyll destruction by the bisulfite-oxygen system. *Plant Physiology* 60: 277–81.

———. 1978. Chlorophyll destruction in the presence of bisulfite and linoleic acid hydroperoxide. *Phytochemistry* 17: 79–84.

———. 1979a. Ethylene and ethane production from sulfur-dioxide-injured plants. *Plant Physiology* 63: 142–45.

———. 1979b. Sulfite-mediated destruction of β-carotene. *Journal of Agricultural Food Chemistry* 27: 446–49.

Pelz, E. 1962. Untersuchungen über die individuelle Rauchärte von Fichten. *Wissenschaftliche Zeitschrift Technischen Universität Dresden* 11: 595–600.

Pena, J. A., *et al.* 1982. SO_2 content in precipitation and its relationship with surface concentrations of SO_2 in the air. *Atmospheric Environment* 16: 1711–15. Coauthors: R. G. De Pena, V. C. Bowersox & J. F. Takacs.

Perkins, D. F., R. O. Millar & P. E. Neep. 1980. Accumulation of airborne fluoride by lichens in the vicinity of an aluminium reduction plant. *Environmental Pollution* (series A) 21: 155–68.

Petering, D. H. 1977. Sulfur dioxide: A view of its reactions with biomolecules. *In* Lee (1977), pp. 293–306.

Petit, C., M. LeDoux & M. Trinite. 1977. Transfer resistances to SO_2 capture by some plane surfaces, water, and leaves. *Atmospheric Environment* 11: 1123–26.

Phipps, R. L. 1979. Simulation of wetlands forest vegetation dynamics. *Ecological Modeling* 7: 257–88.

Plesničar, M., and R. Kalezić. 1980. Sulfite inhibition of oxygen evolution associated with photosynthetic carbon assimilation. *Periodicum Biologorum* 82: 297–301.

Posthumus, A. C. 1978. New results from SO_2 fumigations of plants. *VDI-Berichte* 314: 225–30.

————. 1980. Monitoring of levels and effects of airborne pollutants on vegetation: Use of biological indicators and other methods: National and international programmes. *In* U.N.E.C.E. (1980), pp. 296–311.

————. 1982. Biological indicators of air pollution. *In* Unsworth and Ormrod (1982), pp. 27–42.

Preston, E. M., and D. W. J. Guinn, eds. 1980. *The Bioenvironmental Impact of a Coal-Fired Power Plant*. Corvallis, Ore.: U.S. Environmental Protection Agency.

Preston, E. M., and T. L. Gullett, eds. 1979. *The Bioenvironmental Impact of a Coal-fired Power Plant: Fourth Interim Report, Colstrip, Montana*. Corvallis, Ore.: U.S. Environmental Protection Agency.

Preston, E. M., and J. J. Lee. 1982. Design and performance of a field exposure system for evaluation of the ecological effects of SO_2 on native grassland. *Environmental Monitoring Assessment* 1: 213–28.

Preston, E. M., and R. A. Lewis, eds. 1978. *The Bioenvironmental Impact of a Coal-fired Power Plant*. Corvallis, Ore.: U.S. Environmental Protection Agency.

Preston, E. M., D. W. O'Guinn & R. A. Wilson, eds. 1981. *The Bioenvironmental Impact of a Coal-fired Power Plant*. Corvallis, Ore.: U.S. Environmental Protection Agency.

Preston, K. P. 1980. Effects of sulfur dioxide pollution on Californian coastal sage scrub. M.A. thesis, Univ. of California, Los Angeles.

————. 1983. Effects of sulfur-dioxide pollution on a Californian coastal sage scrub community. *Oecologia* (Berl.), in press.

Preston, R. D., and H. W. Woolhouse, eds. 1977. *Advances in Botanical Research*. London: Academic Press.

Priebe, A., H. Klein & H.-J. Jäger. 1978. Role of polyamines in SO_2-polluted pea plants. *Journal of Experimental Botany* 29: 1045–50.

Prince, R., and F. F. Ross. 1972. Sulfur in air and soil. *Water, Air and Soil Pollution* 1: 286–302.

Puckett, K., *et al.* 1973. Sulfur dioxide: Its effect on photosynthetic ^{14}C fixation in lichens and suggested mechanisms of phytotoxicity. *New Phytologist* 72: 141–54. Coauthors: E. Nieboer, W. P. Flora & D. H. S. Richardson.

Puckett, K., *et al.* 1974. Photosynthetic ^{14}C fixation by the lichen *Umbilicaria muhlenbergii* (Ach.) Tuck following short exposures to aqueous sulfur dioxide. *New Phytologist* 73: 1183–92. Coauthors: D. H. S. Richardson, W. P. Flora & E. Nieboer.

Puckett, K., *et al.* 1977. Potassium efflux by lichen thalli following exposure to aqueous sulfur dioxide. *New Phytologist* 79: 135–45. Coauthors: F. Tomassini, E. Nieboer & D. Richardson.

Pyatt, F. B. 1970. Lichens as indicators of air pollution in a steel-producing town in South Wales. *Environmental Pollution* 1: 45–56.

Radmer, R., and G. Cheniae. 1977. Mechanisms of oxygen evolution. *In* Barber (1977), pp. 303–51.

Rajput, C. B. S., D. P. Ormrod & W. D. Evans. 1977. The resistance of strawberry to ozone and sulfur dioxide. *Plant Disease Reporter* 61: 222–25.

Rand, R. H. 1977. Gaseous diffusion in the leaf interior. *Transactions of the American Society of Agricultural Engineering* 20: 701–4.

————. 1978. A theoretical analysis of CO_2 absorption in sun versus shade leaves. *Journal of Biomechanical Engineering* 100: 20–24.

Rao, D. N., and F. Le Blanc. 1967. Influence of an iron-sintering plant and corticolous epiphytes in Wawa, Ontario. *Bryologist* 70: 141–57.

Raschke, K. 1960. Heat transfer between the plant and the environment. *Annual Review of Plant Physiology* 11: 111–26.

————. 1975a. Stomatal action. *Annual Review of Plant Physiology* 26: 309–40.

————. 1975b. Simultaneous requirement of carbon dioxide and abscisic acid for stomatal closing in *Xanthium strumarium* L. *Planta* 125: 243–59.

Raybould, C. C., M. H. Unsworth & P. J. Gregory. 1977. Sources of sulfur in rain collected below a wheat canopy. *Nature* 267: 146–47.

Regehr, D. L., F. A. Bazzaz & W. R. Boggess. 1975. Photosynthesis, transpiration, and leaf conductance of *Populus deltoides* in relation to flooding and drought. *Photosynthetica* 9: 52–61.

Reich, P. B., R. G. Amundson & J. P. Lassoie. 1980. Multiple pollutant fumigations under near-ambient environmental conditions using a linear gradient technique. *In* Miller (1980), p. 247.

————. 1982. Reduction in soybean yield after exposure to ozone and sulfur dioxide using a linear gradient exposure technique. *Water, Air, and Soil Pollution* 17: 29–36.

Reinert, R. A., and T. N. Gray. 1981. The response of radish to nitrogen dioxide, sulfur dioxide, and ozone, alone and in combination. *Journal of Environmental Quality* 10: 240–43.

Reinert, R. A., A. S. Heagle & W. W. Heck. 1975. Plant responses to pollutant combinations. *In* Mudd and Kozlowski (1975), pp. 159–77.

Reinert, R. A., H. E. Heggestad & W. W. Heck. 1982. Response and genetic modification of plants for tolerance to air pollutants. *In* Christiansen and Lewis (1982), pp. 259–92.

Reinert, R. A., and P. V. Nelson. 1980. Sensitivity and growth of five *Elatior begonia* cultivars to SO_2 and O_3, alone and in combination. *Journal of the American Society for Horticultural Science* 105: 721–23.

Reinert, R. A., and J. S. Sanders. 1982. Growth of radish and marigold following repeated exposure to nitrogen dioxide, sulfur dioxide, and ozone. *Plant Disease* 66: 122–24.

Reinert, R. A., D. S. Shriner & J. O. Rawlings. 1982. Responses of radish to all combinations of three concentrations of nitrogen dioxide, sulfur dioxide, and ozone. *Journal of Environmental Quality* 11: 52–57.

Reinert, R. A., and D. E. Weber. 1980. Ozone and sulfur dioxide induced changes in soybean growth. *Phytopathology* 70: 914–16.

Rennenberg, H., K. Schmitz & L. Bergmann. 1979. Long-distance transport of sulfur in *Nicotiana tabacum*. *Planta* 147: 57–62.

Rennenberg, H., *et al.* 1982. Evidence for an intracellular sulfur cycle in cucumber leaves. *Planta* 154: 516–24. Coauthors: J. Sekija, L. G. Wilson & P. Filner.

Revelle, R. 1977. *Energy and Climate: Report of the Panel*. Washington, D.C.: National Academy of Sciences.

Revelle, R., and W. Munk. 1977. The carbon dioxide cycle and the biosphere. *In* Revelle (1977), pp. 243–81.

Reynolds, J. F., G. C. Cunningham & J. P. Syvertsen. 1979. A net CO_2 exchange model for *Larrea tridentata*. *Photosynthetica* 13: 279–86.

Rice, P. M., *et al.* 1979. The effects of "low level SO_2" exposure on sulfur accumulation and various plant life responses of some major grassland species on the ZAPS sites. *In* Preston and Gullett (1979), pp. 494–591. Coauthors: L. H. Pye, R. Boldi, J. O'Loughlin, P. C. Tourangeau & C. C. Gordon.

Rich, S., and N. C. Turner. 1972. Importance of moisture on stomatal behavior of plants subjected to ozone. *Journal of the Air Pollution Control Association* 22: 718–21.

Richards, B. L., J. H. Middleton & W. B. Hewitt. 1958. Air pollution with relation to agronomic crops. V: Oxidant stipple of grape. *Agronomy Journal* 50: 559–61.

Richardson, D. H. S., and K. J. Puckett. 1973. Sulfur dioxide and photosynthesis in lichens. *In* Ferry *et al.* (1973), pp. 283–98.

Ricks, G. R., and J. H. Williams. 1974. Effects of atmospheric pollution on deciduous woodland, II: Effects of particulate matter upon stomatal diffusion resistance in leaves of *Quercus petraea* (Mattuschka) Leibl. *Environmental Pollution* 6: 87–109.

Riely, C. A., G. Cohen & M. Lieberman. 1974. Ethane evolution: A new index of lipid peroxidation. *Science* 183: 208–10.

Ripley, E. A., and R. E. Redman. 1976. Grassland. *In* Monteith (1976), II, 351–98.

Rist, D. L., and D. D. Davis. 1979. The influence of exposure temperature and relative humidity on the response of pinto-bean foliage to sulfur dioxide. *Phytopathology* 69: 231–35.

Roberts, B. R. 1975. The influence of sulfur dioxide concentration on growth of potted white birch and pin oak seedlings in the field. *Journal of the American Society for Horticultural Science* 100: 640–42.

———. 1976. The response of field-grown white pine seedlings to different sulfur-dioxide environments. *Environmental Pollution* 11: 175–80.

Roberts, J. K. M., N. Wade-Jardetzky & O. Jardetzky. 1981. Intracellular *p*H measurements by ^{31}P nuclear magnetic resonance: Influence of factors other than *p*H on ^{31}P chemical shifts. *Biochemistry* 20: 5389–94.

Roberts, J. K. M., *et al.* 1980. Estimation of cytoplasmic and vacuolar *p*H in higher plant cells by ^{31}P NMR. *Nature* 283: 870–72. Coauthors: P. M. Ray, N. Wade-Jardetzky & O. Jardetzky.

Roberts, T. M., *et al.* 1983. The use of open-top chambers to study the effects of air pollutants, in particular sulfur dioxide, on the growth of ryegrass, *Lolium perenne* L., I: Characteristics of modified open-top chambers used for both air-filtration and SO$_2$-fumigation studies. *Environmental Pollution* (series A), 31: 9–33. Coauthors: R. M. Bell, D. C. Horsman & K. E. Colvill.

Robinson, D. C., and A. R. Wellburn. 1983. Light-induced changes in the quenching of 9-amino-acridine fluorescence by photosynthetic membranes due to atmospheric pollutants and their products. *Environmental Pollution* (series A) 32: 109–20.

Robinson, E., and R. C. Robbins. 1968. Gaseous sulfur pollutants from urban and natural sources. *Journal of the Air Pollution Control Association* 20: 233–35.

Rodes, C. E., and D. M. Holland. 1981. Variations of NO, NO$_2$, and O$_3$ concentrations downwind of a Los Angeles freeway. *Atmospheric Environment* 15: 243–50.

Rogers, H. H., J. C. Campbell & R. J. Volk. 1979. Nitrogen-15 dioxide uptake and incorporation by *Phaseolus vulgaris* (L.). *Science* 206: 333–35.

Rogers, H. H., *et al.* 1977. Measuring air-pollutant uptake by plants: A direct kinetic technique. *Journal of the Air Pollution Control Association* 27: 1192–97. Coauthors: H. E. Jeffries, E. P. Stahel, W. W. Heck, L. A. Ripperton & A. M. Witherspoon.

Rohmeder, E., W. Merz & A. von Schönborn. 1962. Breeding spruce and pine varieties relatively resistant against industrial exhaust fumes. *Forstwissenschaftliches Centralblatt* 81: 321–32.

Rohmeder, E., and A. von Schönborn. 1965. Der Einfluss von Unwelt und Erbgut auf die Widerstandsfähigkeit der Waldbäume gegenüber Luftverunreinigung durch Industrieabgase ein Beitrag zur Züchtung einer relative Rauchresistenten Fichtensorte. *Forstwissenschaftliches Centralblatt* 84: 1–13.

———. 1967. The breeding of spruce with increased resistance to exhaust gas. *International Union of Forest Research Organizations Proceedings* 5: 556–66.

Roose, M. L., A. D. Bradshaw & T. M. Roberts. 1982. Evolution of resistance to gaseous air pollutants. *In* Unsworth and Ormrod (1982), pp. 379–409.

Rose, D. A., and D. A. Charles-Edwards, eds. 1981. *Mathematics and Plant Physiology*. London: Academic Press.

Rose, F. 1973. Detailed mapping in southeast England. *In* Ferry *et al.* (1973), pp. 77–88.

Rosen, P. M., R. C. Musselman & W. J. Kender. 1978. Relationship of stomatal resistance to sulfur dioxide and ozone injury in grapevines. *Scientia Horticulturae* 8: 137–42.

Rosenberg, C. R., R. J. Hutnik & D. D. Davis. 1979. Forest composition at varying distances from a coal-burning power plant. *Environmental Pollution* 19: 307–17.

Ross, F. F., ed. 1979. *Symposium on Sulfur Emissions and the Environment.* London: Society for Chemical Industry.

Rowlands, J. R., C. J. Allen-Rowlands & E. M. Gause. 1977. Effects of environmental agents on membrane dynamics. *In* Lee (1977), pp. 203–46.

Rowlatt, S., D. V. Crawford & M. H. Unsworth. 1978. Sulfur cycle in wheat and other farm crops. *In* Brogan (1978), pp. 1–14.

Rubin, S. 1981. Air pollution constraints on increased coal use by industry. *Journal of the Air Pollution Control Association* 31: 349–60.

Runeckles, V. C. 1974. Dosage of air pollutants and damage to vegetation. *Environmental Conservation* 1: 305–8.

Runeckles, V. C., and P. M. Rosen. 1974. Effects of ambient ozone pretreatment on transpiration and susceptibility to ozone injury. *Canadian Journal of Botany* 55: 193–97.

Russell, R. S. 1977. *Plant Root Systems.* London: McGraw-Hill.

Ryder, E. H. 1973. Selecting and breeding plants for increased resistance to air pollutants. *In* Naegele (1973), pp. 75–84.

Ryrie, I. J., and A. T. Jagendorf. 1971. Inhibition of photophosphorylation in spinach chloroplasts by inorganic sulfate. *Journal of Biological Chemistry* 246: 582–88.

Sanders, J. S., and R. A. Reinert. 1982. Screening azalea cultivars for sensitivity to nitrogen dioxide, sulfur dioxide, and ozone alone and in mixtures. *Journal of the American Society for Horticultural Science* 107: 87–90.

Santamour, F. S., Jr. 1969. Air pollution studies of *Platanus* and American elm seedlings. *Plant Disease Reporter* 53: 482–84.

Scale, P. R. 1980. Changes in plant communities with distance from an SO_2 source. *In* Miller (1980), p. 248.

Scheffer, T. C., and G. G. Hedgcock. 1955. Injury to northwestern forest trees by sulfur dioxide from smelters. *U.S. Forest Service Technical Bulletin* 1117.

Schertz, R. D., W. J. Kender & R. C. Musselman. 1980a. Effects of ozone and sulfur dioxide on grapevines. *Scientia Horticulturae* 13: 37–45.

———. 1980b. Foliar response and growth of apple trees following exposure to ozone and sulfur dioxide. *Journal of the American Society for Horticultural Science* 105: 594–98.

Schiff, J. A., and R. C. Hodson. 1973. The metabolism of sulfate. *Annual Review of Plant Physiology* 24: 381–414.

Schindlbeck, W. E. 1977. Biochemische Beiträge zur Immissionsforschung. *Forstwissenschaftliches Centralblatt* 96: 67–71.

Schmutz, D., and C. Brunold. 1982. Regulation of sulfate assimilation in plants, XIII: Assimilatory sulfate reduction during ontogenesis of primary leaves of *Phaseolus vulgaris* L. *Plant Physiology* 70: 524–27.

Scholz, F. 1981. Genecological aspects of air-pollution effects on northern forests. *Silva Fennica* 4: 384–91.

Scholz, F., and S. Reck. 1976. Effects of acids on forest trees as measured by titration in vitro inheritance of buffer capacity in *Picea abies. Proceedings of the First International Symposium on Acid Precipitation and the Forest Ecosystems (U.S.D.A. Forest Service General Technical Report* NE-23), pp. 971–76.

Schönbeck, H. 1969. Eine Methode zur Erfassung der biologischen Wirkung von Luftverunreinigungen durch transplantierte Flechten: Staub-Reinhalt. *Luft* 29: 14–18.

Schönback, H., *et al.* 1964. The differential effect of sulfur dioxide upon needles of a variety of two-year larch cross-breedings. *Züchter* 34: 312–16. Coauthors: H. G. Dassler, H. Enderlein, E. Belmann & W. Kastner.

Schönherr, J. 1982. Resistance of plant surfaces to water loss: Transport properties of cutin, suberin and associated lipids. *In* Lange *et al.* (1982), vol. 12B, 153–80.

Schramel, M. 1975. Influence of sulfur dioxide on stomatal apertures and diffusive resistance of leaves in various species of cultivated plants under optimum soil moisture and drought conditions. *Bulletin de l'Académie Polonaise des Sciences. Série des Sciences de la Terre* 23: 57–63.

Schreiber, U., *et al.* 1978. Chlorophyll fluorescence assay for ozone injury in intact plants. *Plant Physiology* 61: 80–84. Coauthors: W. Vidaver, V. C. Runeckles & P. Rosen.

Schroeder, J. 1873. Die Einwirkung der schwefligen Säure auf die Pflanzen. *Tharander forstliches Jahrbuch* 23: 217–67.

Schroeter, L. C. 1966. *Sulfur Dioxide*. Oxford: Pergamon Press.

Schuldiner, S., H. Rottenberg & M. Avron. 1972. Determination of ΔpH in chloroplasts. *European Journal of Biochemistry* 25: 64–70.

Schwela, D. H. 1979. An estimate of deposition velocities of several air pollutants on grass. *Ecotoxicology and Environmental Safety* 3: 174–89.

Schwela, D., and A. Junker. 1978. Derivation of air-quality standards on the basis of risk considerations. *Water, Air, and Soil Pollution* 10: 255–68.

Scott, N. M., and G. Anderson. 1976. Organic sulfur fractions in Scottish soils. *Journal of the Science of Food and Agriculture* 17: 358–66.

Scurfield, G. 1960. Air pollution and tree growth. *Forestry Abstracts* 21: 339–49.

Seiler, W., and P. J. Crutzen. 1980. Estimates of gross and net fluxes of carbon between the biosphere and the atmosphere from biomass burning. *Climate Change* 2: 207.

Sekiya, J., L. Wilson & P. Filner. 1982. Resistance to injury by sulfur dioxide: Correlation with its reduction to, and emission of, hydrogen sulfide in curcurbitaceae. *Plant Physiology* 70: 437–41.

Sekiya, J., *et al.* 1982. Emission of hydrogen sulfide by leaf tissue in response to L-cysteine. *Plant Physiology* 70: 430–36. Coauthors: A. Schmidt, L. G. Wilson & P. Filner.

Semeniuk, P., and H. E. Heggestad. 1981. Differences among *Coleus* cultivars in tolerance to ozone and sulfur dioxide. *Journal of Heredity* 72: 459–60.

Sesták, Z., J. Cátský & P. G. Jarvis. 1971. *Plant Photosynthetic Production: Manual of Methods*. The Hague: Junk.

Shapiro, R. 1977. Genetic effects of bisulfite (sulfur dioxide). *Mutation Research* 39: 149–76.

Shapiro, R., R. E. Servis & M. Welcher. 1970. Reactions of uracil and cytosine derivatives with sodium bisulfite: A specific deamination method. *Journal of the American Chemical Society* 92: 422–24.

Sharkey, T. D., *et al.* 1982. A direct confirmation of the standard method of estimating intercellular partial pressure of CO_2. *Plant Physiology* 69: 657–59. Coauthors: K. Imai, G. D. Farquhar & I. R. Cowan.

Sharma, G. K. 1975. Leaf surface effects of environmental pollution on sugar maple (*Acer saccharum*) in Montreal. *Canadian Journal of Botany* 53: 2312–14.

Sharma, G. K., and J. Butler. 1975. Environmental pollution: Leaf cuticular patterns in *Trifolium pratense* L. *Annals of Botany* 39: 1087–90.

Sharpe, P. J. H. 1983. Responses of photosynthesis and dark respiration to temperature. *Annals of Botany* 52: 325–43.

Sheffer, T. C., and G. G. Hedgcock. 1955. Injury to northwestern forest trees by sulfur dioxide from smelters. *U.S.D.A. Forest Service Technical Bulletin* 1117.

Shibuya, A., and S. Horie. 1980. Studies on the composition of the mitochondrial sulfite oxidase system. *Journal of Biochemistry* 87: 1773–84.

Shimazaki, K., and K. Sugahara. 1979. Specific inhibition of photosystem II activity in chloroplasts by fumigation of spinach leaves with SO_2. *Plant Cell Physiology* 20: 947–55.

Shimazaki, K., *et al.* 1980. Active oxygen participation in chlorophyll destruction and lipid peroxidation in SO_2-fumigated leaves of spinach. *Plant Cell Physiology* 21: 1193–204. Coauthors: T. Sakaki, N. Kondo & K. Sugahara.

Shimizu, H., A. Furukawa & T. Totsuka. 1980. Effects of low concentrations of SO_2 on the growth of sunflower. In N.I.E.S. (1980), pp. 9–17.

Shinn, J. H., B. R. Clegg & M. L. Stuart. 1977. A linear gradient chamber for exposing field plants to constant levels of air pollutants. Dept. of Energy preprint UCRL-80411.

Shoji, H., and T. Tsukatani. 1973. Statistical model of air-pollutant concentration and its application to air-quality standards. *Atmospheric Environment* 7: 485–501.

Showman, R. E. 1975. Lichens as indicators of air quality and a coal-fired power-generating plant. *Bryologist* 78: 1–6.

Shriner, D. S., and G. S. Henderson. 1978. Sulfur distribution and cycling in a deciduous forest watershed. *Journal of Environmental Quality* 7: 392–97.

Shriner, D. S., C. R. Richmond & S. E. Lindberg, eds. 1980. *Atmospheric Sulfur Deposition.* Ann Arbor, Mich.: Ann Arbor Science.

Shriner, D. S., *et al.* 1984. *An Analysis of Potential Agriculture and Forest Impacts of Long Range Transport of Air Pollutants.* Report to the Office of Technology Assessment. Washington, D.C., in press. Coauthors: W. W. Cure, W. W. Heck, D. W. Johnson, R. J. Olson & J. N. Skelly.

Shugart, H. H., S. B. McLaughlin & D. C. West. 1980. Forest models: Their development and potential applications for air pollution effects research. *In* Miller (1980), pp. 203–14.

Shugart, H. H., and I. R. Noble. 1981. A computer model of succession and fire response of the high-altitude Eucalyptus forest of the Brindabella Range, Australian Capital Territory. *Australian Journal of Ecology* 6: 149–64.

Shugart, H. H., and D. C. West. 1977. Development of an Appalachian deciduous forest succession model and its application to assessment of the impact of the chestnut blight. *Journal of Environmental Management* 5: 161–79.

———. 1980. Forest succession models. *BioScience* 30: 309–13.

———. 1981. Long-term dynamics of forest ecosystems. *American Scientist* 69: 647–52.

Shugart, H. H., *et al.* 1974. TEEM: A terrestrial ecosystem energy model for forests. *Oecologia Plantarum* 9: 231–64. Coauthors: R. A. Goldstein, R. V. O'Neill & J. B. Mankin.

Shugart, H. H., *et al.* 1981. The development of a succession model for subtropical rain forest and its application to assess the effects of timber harvest at Wiangarree State Forest, New South Wales. *Journal of Environmental Management* 11: 243–65. Coauthors: M. S. Hopkins, I. P. Burgess & A. T. Mortlock.

Shuwen, Y., *et al.* 1981. Relationship between the resistance of plants to SO_2 and pH of leaf tissue. *Kexue Tongbao* 26: 185–87. Coauthors: L. Yu, L. Zhenguo, W. Youmei & Y. Heidong.

Shuwen, Y., *et al.* 1982. Studies on mechanism of SO_2 injury to plants. *In* Unsworth and Ormrod (1982), pp. 507–8. Coauthors: L. Yu, L. Zhenguo, T. Chang, Y. Ziwen, Y. Huidong & W. Youmei.

Siegenthaler, U., and H. Oeschger. 1978. Predicting future atmospheric carbon-dioxide levels. *Science* 199: 388.

Sierpinski, Z. 1977. Economic significance of noxious insects in pine stands under the permanent impact of the industrial air pollution. *Sylwan* 64: 59–71.

Sij, J. W., and C. A. Swanson. 1974. Short-term kinetics on the inhibition of photosynthesis by sulfur dioxide. *Journal of Environmental Quality* 3: 103–7.

Silvius, J. E., M. Ingle & C. H. Baer. 1975. Sulfur-dioxide inhibition of photosynthesis in isolated spinach chloroplasts. *Plant Physiology* 56: 434–37.

Silvius, J. E., *et al*. 1976. Photoreduction of sulfur dioxide by spinach leaves and isolated spinach chloroplasts. *Plant Physiology* 57: 799–801. Coauthors: C. H. Baer, S. Dodrill & H. Patrick.

Simson, J. R., and R. J. Freney, eds. 1982. *Gaseous Loss of Nitrogen from Plant Soil Systems.* The Hague: Martinus Nijhuj.

Sinclair, W. A. 1969. Polluted air: Potent new selective force in forests. *Journal of Forestry* 67: 305–9.

Singh, J. J., and A. Deepak, eds. 1980. *Symposium on Environmental and Climatic Impact of Coal Utilization.* New York: Academic Press.

Sionit, N., H. Helmers & B. R. Strain. 1980. Growth and yield of wheat under CO_2 enrichment and water-stress conditions. *Crop Science* 20: 687–90.

Skelly, J. M., S. V. Krupa & B. I. Chevone. 1979. Field surveys. *In* Heck, Krupa & Linzon (1979), Chapter 12.

Skelly, J. M., L. D. Moore & L. L. Stone. 1972. Symptom expression of eastern white pine located near a source of oxides of nitrogen and sulfur dioxide. *Plant Disease Reporter* 56: 3–6.

Skye, E. 1968. Lichens and air pollution: A study of cryptogamic epiphytes and environment in the Stockholm region. *Acta Phytogeographica Suecica* 53: 1–123.

Slinn, W. G. N., R. G. Semonin & H. R. Pruppacher, eds. 1983. *Proceedings of the Fourth International Conference on Precipitation, Scavenging, Dry Deposition, and Resuspension, Nov. 29–Dec. 3, 1982, Santa Monica, California.* Elsevier.

Slovacek, R. E., and G. Hind. 1981. Correlation between photosynthesis and the trans-thylakoid proton gradient. *Biochimica et Biophysica Acta* 635: 393–404.

Smith, H., ed. 1975. *Commentaries in Plant Science.* Oxford: Pergamon Press.

Smith, H. J., and D. D. Davis. 1978. Susceptibility of conifer cotyledons and primary needles to acute doses of SO_2. *HortScience* 13: 703–4.

Smith, K. C., *et al*. 1980. A time-dependent mathematical expression of the Munch hypothesis of phloem transport. *Journal of Theoretical Biology* 86: 493–505. Coauthors: C. E. Magnuson, J. D. Goeschl & D. W. DeMichele.

Smith, W. H. 1974. Air pollution: Effects on the structure and function of the temperate forest ecosystem. *Environmental Pollution* 6: 111–29.

———. 1981. *Air Pollution and Forests: Interactions between Air Contaminants and Forest Ecosystems.* New York: Springer-Verlag.

Sobotka, A. 1964. Vliv prumyslovych exhalatuna pudni Zivenu Smrkovjch porostu Krusnych hor. *Lesnicky Casopis* 10: 987–1002.

Solbrig, O. T., *et al*., eds. 1979. *Topics in Plant Population Biology.* New York: Columbia Univ. Press.

Spedding, D. J. 1969. Uptake of sulfur dioxide by barley leaves at low sulfur-dioxide concentrations. *Nature* 224: 1229–30.

———. 1974. *Air Pollution.* London: Oxford Univ. Press.

Spedding, D. J., *et al*. 1980. Effect of pH on the uptake of ^{31}S-sulfur from sulfate, sulfite and sulfide by isolated spinach chloroplasts. *Zeitschrift für Pflanzenphysiologie* 96: 351–64. Coauthors: I. Ziegler, R. Hampp & H. Ziegler.

Spierings, F. H. F. G. 1971. Influence of fumigations with NO_2 on growth and yield of tomato plants. *Netherlands Journal of Plant Pathology* 77: 194–200.

Sprugel, D. G., *et al*. 1978. *Effects of Chronic Sulfur Dioxide Fumigation on Development, Yield, and Seed Quality of Field-grown Soybeans: Summary of 1977 and 1978 Experiments.* Argonne National Laboratory. Coauthors: J. E. Miller, P. B. Xerikos & H. J. Smith.

Sprugel, D. G., *et al*. 1980. Sulfur dioxide effects on yield and seed quality in field-grown soybeans. *Phytopathology* 70: 1129–33. Coauthors: J. E. Miller, R. N. Muller, H. J. Smith & P. B. Xerikos.

Squire, G. R., and T. A. Mansfield. 1977. A simple method of isolating stomata on detached epidermis by low *p*H treatment: Observations of the importance of subsidiary cells. *New Phytologist* 71: 1033–43.

Srivastava, H. S., P. A. Jolliffe & V. C. Runeckles. 1975a. The effect of environmental conditions on the inhibition of leaf gas exchange by NO_2. *Canadian Journal of Botany* 53: 475–82.

———. 1975b. Inhibition of gas exchange in bean leaves by NO_2. *Canadian Journal of Botany* 53: 466–74.

Stanhill, G. 1976. Cotton. *In* Monteith (1976), pp. 121–48.

Stern, A. C. 1977. *Air Pollution.* 3d ed. New York: Academic Press.

Steubing, L., and H.-J. Jäger, eds. 1982. *Monitoring of Air Pollutants by Plants: Methods and Problems.* The Hague: Junk.

Stevens, T. H., and T. W. Hazelton. 1976. Sulfur-dioxide pollution and crop damage in the Four Corners region: A simulation analysis. Las Cruces: Agricultural Experiment Station, New Mexico State Univ.

Stewart, D. W. 1970. A simulation of net photosynthesis of field corn. Ph.D. diss., Cornell Univ., Ithaca, New York.

Stoklasa, J. 1923. *Die Beschädigungen der Vegetation durch Rauchgase und Babriksexhalationen.* Berlin: Urban and Schwarzenburg.

Stone, L. L., and J. M. Skelly. 1974. The growth of two forest-tree species adjacent to a periodic source of air pollution. *Phytopathology* 64: 773–78.

Strain, B. R., ed. 1978. *Report of the Workshop on Anticipated Plant Responses to Global Carbon-Dioxide Enrichment, August 1977.* Durham, N.C.: Duke Univ.

Strain, B. R., and T. V. Armentano. 1980. *Environmental and Societal Consequences of CO_2-Induced Climate Change: Response of "Unmanaged" Ecosystems.* AAAS.

Strain, B. R., and W. D. Billings, eds. 1974. *Handbook of Vegetation Science, VI: Vegetation and Environment.* The Hague: Junk.

Stratmann, H. 1963. Field experiments to determine the effects of SO_2 on vegetation, II: Measurements and evaluation of SO_2 ground-level concentration. *Forschungsberichte des Landes Nordrhein-Westfalen* No. 1184.

Stuiver, M. 1978. Atmospheric carbon dioxide and carbon reservoir changes. *Science* 199: 253–58.

Stumm, W., ed. 1977. *Global Chemical Cycles and Their Alterations by Man.* Berlin: Dahlem Institute.

Stumm, W., and J. J. Morgan. 1981. *Aquatic Chemistry: An Introduction Emphasizing Chemical Equilibria in Natural Waters.* 2d ed. New York: Wiley-Interscience.

Sutton, R., and I. P. Ting. 1977. Evidence for the repair of ozone-induced membrane injury. *American Journal of Botany* 64: 404–11.

Suwannapinunt, W., and T. T. Kozlowski. 1980. Effect of SO_2 on transpiration, chlorophyll content, growth, and injury in young seedlings of woody angiosperms. *Canadian Journal of Forest Research* 10: 78–81.

Svensson, B. H., and R. Söderlund, eds. 1976. *Nitrogen, Phosphorus, and Sulfur: Global Cycles.* Stockholm: SCOPE Report no. 7.

Swain, R. E. 1949. Smoke and fume investigations: A historical review. *Industrial and Engineering Chemistry* 41: 2384–88.

Swain, R. E., and A. B. Johnson. 1936. Effect of sulfur dioxide on wheat development: Action at low concentrations. *Industrial and Engineering Chemistry* 28: 42–47.

Swanson, C. A., J. Hoddinott & J. W. Sij. 1976. The effect of selected sink leaf parameters on translocation rates. *In* Wardlaw and Passioura (1976), pp. 347–56.

Swift, R. S., and A. M. Posner. 1972. Nitrogen, phosphorus, and sulfur contents of humic acids fractionated with respect to molecular weight. *Journal of Soil Science* 23: 50–57.

Swoboda, B., and V. Massey. 1966. On the reaction of the glucose oxidase from *Aspergillus niger* with bisulfite. *Journal of Biological Chemistry* 241: 3409–16.

Tabatabai, M. A., and A. A. Al-Khafaji. 1980. Comparison of nitrogen and sulfur mineralization in soils. *Journal of the Soil Science Society of America* 44: 1000–1006.

Tager, J. M., and N. Rautanen. 1956. Sulfite oxidation by plant mitochondrial system: Enzymic and nonenzymic oxidation. *Physiologia Plantarum* 9: 665–73.

Takemoto, B. K., and R. D. Noble. 1982. The effects of short-term SO_2 fumigation on photosynthesis and respiration in soybean *Glycine max*. *Environmental Pollution* (series A) 28: 67–74.

Tamm, C. O., and A. Aronsson. 1972. Plant growth as affected by sulfur compounds in a polluted atmosphere: Literature survey No. 12. Stockholm: Department of Forest Ecology and Forest Soils, Royal College of Forestry.

Tamura, G., T. Hosoi & J. Aketagawa. 1978. Ferredoxin-dependent sulfite reductase from spinach leaves. *Agricultural Biological Chemistry* 42: 2165.

Tanaka, K., and K. Sugahara. 1980. Role of superoxide dismutase in defense against SO_2 toxicity and an increase in superoxide dismutase activity with SO_2 fumigation. *Plant Cell Physiology* 21: 601–11.

Tang, Z. C., and T. T. Kozlowski. 1982. Some physiological and growth responses of *Betula papyrifera* seedlings to flooding. *Physiologia Plantarum* 55: 415–20.

Taniyama, T. 1972. Studies on the development of symptoms and the mechanism of injury caused by sulfur dioxide in crop plants. *Bulletin of the Faculty of Agriculture, Mie University, Tsu, Japan* 44: 11–130.

Taoda, H. 1972. Mapping of atmospheric pollution in Tokyo based upon epiphytic bryophytes. *Japanese Journal of Ecology* 22: 125–33.

Taylor, G. E., Jr. 1976. The evolutionary process of population differentiation of an annual plant species, *Geranium carolinianum* L. in response to sulfur dioxide. Ph.D. thesis, Emory Univ.

———. 1978a. Genetic analysis of ecotypic differentiation within an annual plant species, *Geranium carolinianum* L. in response to sulfur dioxide. *Botanical Gazette* 139: 362–68.

———. 1978b. Plant and leaf resistance to gaseous air pollution stress. *New Phytologist* 80: 523–34.

Taylor, G. E., S. B. McLaughlin & D. S. Shriner. 1982. Effective pollutant dose. *In* Unsworth and Ormrod (1982), pp. 458–60.

Taylor, G. E., Jr., and W. H. Murdy. 1975. Population differentiation of an annual plant species, *Geranium carolinianum*, in response to sulfur dioxide. *Botanical Gazette* 136: 212–15.

Taylor, G. E., Jr., and D. T. Tingey. 1979. A gas-exchange system for assessing plant performance in response to environment stress. United States Environmental Protection Agency, report no. 600/3-79-108.

———. 1981. Physiology of ecotypic plant response to sulfur dioxide in *Geranium carolinianum* L. *Oecologia* (Berl.) 49: 76–82.

———. 1982. Flux of ozone to *Glycine max*: sites of regulation and relationship to leaf injury. *Oecologia* (Berl.) 53: 179–86.

———. 1983. Sulfur dioxide flux into leaves of *Geranium carolinianum* L.: Evidence for a nonstomatal or residual resistance. *Plant Physiology* 72: 237–44.

Taylor, G. S., and S. Rich. 1962. Antiozonant-treated cloth protects tobacco from fleck. *Science* 135: 928.

Taylor, J. E., W. C. Leininger & M. W. Hoard. 1980. Plant community structure on ZAPS. *In* Preston and Guinn (1980), pp. 216–34.

Taylor, J. S., D. M. Reid & R. P. Pharis. 1981. Mutual antagonism of sulfur dioxide and abscisic acid in their effect on stomatal aperture in broad bean (*Vicia faba* L.) epidermal strips. *Plant Physiology* 68: 1504–7.

Taylor, O. C., E. A. Cardiff & J. D. Mersereau. 1965. Apparent photosynthesis as a measure of air pollution damage. *Journal of the Air Pollution Control Association* 151: 71–76.

Taylor, O. C., and F. M. Eaton. 1966. Suppression of plant growth by nitrogen dioxide. *Plant Physiology* 41: 132–35.

Taylor, O. C., *et al.* 1960. Effects of airborne oxidants on leaves of pinto bean and petunia. *Proceedings of the American Society for Horticultural Science* 75: 435–44. Coauthors: E. R. Stephens, E. F. Darley & E. A. Cardiff.

Taylor, O. C., *et al.* 1961. Interaction of light and atmospheric photochemical products (SMOG) within plants. *Nature* 192: 814–16. Coauthors: W. M. Dugger, E. A. Cardiff & E. F. Darley.

Taylor, O. C., *et al.* 1975. Oxides of nitrogen. *In* Mudd and Kozlowski (1975), pp. 121–39. Coauthors: C. R. Thompson, D. T. Tingey & R. A. Reinert.

Teh, K. H., and C. A. Swanson. 1977. Sulfur dioxide inhibition of translocation and photosynthesis in *Phaseolus vulgaris* L. *Plant Physiology* 59 (suppl.): 123.

———. 1982. Sulfur dioxide inhibition of translocation in bean plants. *Plant Physiology* 69: 88–92.

Temple, P. J. 1972. Dose-response of urban trees to SO_2. *Journal of the Air Pollution Control Association* 22: 271–74.

Teng, P. S., and S. V. Krupa, eds. 1980. *Crop Loss Assessment: Proceedings of the E. C. Stakman Commemorative Symposium.* St. Paul: Univ. of Minnesota Agricultural Experiment Station, misc. publication no. 7-1980.

Teng, P. S., and R. J. Oshima. 1983. Identification and assessment of losses. *In* Kommedahl and Williams (1983), pp. 69–81.

Terman, G. L. 1978. Atmospheric sulfur: The agronomic aspects. *Sulphur Institute Technical Bulletin* 23.

Terraglio, F. P., and R. M. Manganelli. 1967. The absorption of atmospheric sulfur dioxide by water solutions. *Journal of the Air Pollution Control Association* 17: 403–6.

Tharp, M. L. 1978. Modeling major perturbations on a forest ecosystem. M.S. thesis, Univ. of Tennessee, Knoxville.

Thom, A. S. 1976. Momentum mass and heat exchange. *In* Monteith (1976), I, 57–109.

Thomas, M. D. 1932. Automatic apparatus for the determination of small concentrations of sulfur dioxide in air. Part III. *Industrial and Engineering Chemistry, Analytical Edition* 4: 253–56.

———. 1933. Precise automatic apparatus for continuous determination of carbon dioxide in air. *Industrial and Engineering Chemistry, Analytical Edition* 5: 193–98.

———. 1948. Agricultural research with radioactive sulfur and arsenic. *Proceedings of the Auburn Conference on the Use of Radioactive Isotopes in Agricultural Research*, pp. 103–17.

———. 1951. Gas damage to plants. *Annual Review of Plant Physiology* 2: 293–322.

———. 1956. The invisible injury theory of plant damage. *Journal of the Air Pollution Control Association* 5: 205–8.

———. 1961. Effects of air pollution on plants. *In* World Health Organization, *Air Pollution*, pp. 233–78. New York: Columbia University Press.

———. 1964. Auswirkungen der Luftverunreinigung auf Pflanzen. *In* World Health Organization, *Die Verunreinigung der Luft*, pp. 229–77. Weinheim, West Germany: Whemie.

Thomas, M. D., and J. M. Abersold. 1929. Automatic apparatus for determination of small concentrations of sulfur dioxide in air, II. *Industrial and Engineering Chemistry, Analytical Edition* 1: 14–15.

Thomas, M. D., and R. J. Cross. 1928. Automatic apparatus for the determination of small concentrations of sulfur dioxide in air, I. *Industrial and Engineering Chemistry* 20: 645–47.

Thomas, M. D., and R. H. Hendricks. 1944. The hydrolysis of cystine and the fractionation of sulfur in plant tissues. *Journal of Biological Chemistry* 153: 313–25.

Thomas, M. D., R. H. Hendricks & G. R. Hill. 1944. Some chemical reactions of sulfur dioxide after absorption by alfalfa and sugar beets. *Plant Physiology* 9: 212–26.

———. 1950. Sulfur metabolism of plants: Effects of sulfur dioxide on vegetation. *Industrial Engineering Chemistry* 42: 2231–35.

Thomas, M. D., and G. R. Hill. 1935. Absorption of sulfur dioxide by alfalfa and its relation to leaf injury. *Plant Physiology* 10: 291–307.

———. 1937a. The continuous measurement of photosynthesis, respiration, and transpiration of alfalfa and wheat growing under field conditions. *Plant Physiology* 12: 285–307.

———. 1937b. Relation of SO$_2$ in the atmosphere to photosynthesis and respiration of alfalfa. *Plant Physiology* 12: 309–83.

Thomas, M. D., *et al.* 1943a. The utilization of sulphate and sulfur dioxide for the sulfur nutrition of alfalfa. *Plant Physiology* 18: 345–71. Coauthors: R. H. Hendricks, T. R. Collier & G. R. Hill.

Thomas, M. D., *et al.* 1943b. An installation of large sand-culture beds surmounted by individual air conditioned greenhouses. *Plant Physiology* 18: 334–44. Coauthors: R. H. Hendricks, J. O. Ivie & G. R. Hill.

Thomas, M. D., *et al.* 1944. A study of the sulfur metabolism of wheat, barley, and corn using radioactive sulfur. *Plant Physiology* 19: 227–44. Coauthors: R. H. Hendricks, L. C. Bryner & G. R. Hill.

Thompson, C. R., and G. Kats. 1978. Effects of continuous H$_2$S fumigation on crop and forest plants. *Environmental Science Technology* 12: 550–53.

Thompson, C. R., G. Kats, and R. W. Lennox. 1980. Effects of SO$_2$ and/or NO$_2$ on native plants of the Mojave Desert and Eastern Mojave-Colorado Desert. *Journal of the Air Pollution Control Association* 30: 1304–9.

Thompson, C. R., and O. C. Taylor. 1966. Plastic-covered greenhouses supply controlled atmospheres to citrus trees. *Transactions of the American Society of Agricultural Engineering* 9: 338–42.

Thor, E., and W. R. Gall. 1978. Variation in air-pollution tolerance and growth rate among progenies of southern Appalachian white pine. *Proceedings of the First Conference of the Metropolitan Tree Improvement Alliance (METRIA), Lanham, Md.*, pp. 80–86.

Thornley, J. H. M. 1976. *Mathematical Models in Plant Physiology*. London: Academic Press.

Tibbitts, T. W., *et al.* 1982. *Impact of Air Pollution on Vegetation Near the Columbia Generating Station–Wisconsin Power Plant Study*. Duluth, Minn.: U.S. Environmental Protection Agency Environmental Research Laboratory. (Report no. EPA-600/3-82-068.) Coauthors: S. Will-Wolf, D. F. Karnosky & D. M. Olszyk.

Tingey, D. T., W. W. Heck & R. A. Reinert. 1971. Effect of low concentrations of ozone and sulfur dioxide on foliage, growth, and yield of radish. *Journal of the American Society for Horticultural Science* 96: 369–71.

Tingey, D. T., and R. A. Reinert. 1975. Effect of ozone and sulfur dioxide singly or in combination on plant growth. *Environmental Pollution* 9: 117–25.

Tingey, D. T., and G. E. Taylor. 1982. Variation in plant response to ozone: A conceptual model of physiological events. *In* Unsworth and Ormrod (1982), pp. 113–38.

Tingey, D. T., *et al.* 1971. Vegetation injury from the interaction of NO_2 and SO_2. *Phytopathology* 61: 1506–10. Coauthors: R. A. Reinert, J. A. Dunning & W. W. Heck.

Tingey, D. T., *et al.* 1973a. Foliar injury responses of eleven plant species to ozone/sulfur dioxide mixtures. *Atmospheric Environment* 7: 201–8. Coauthors: R. A. Reinert, J. A. Dunning & W. W. Heck.

Tingey, D. T., *et al.* 1973b. Chronic ozone or sulfur dioxide exposures, or both, affect the early vegetative growth of soybean. *Canadian Journal of Plant Science* 53: 875–79. Coauthors: R. A. Reinert, C. Wickliff & W. W. Heck.

Tingey, D. T., *et al.* 1979. The influence of light and temperature on isoprene emission rates from live oak. *Physiologia Plantarum* 47: 112–18. Coauthors: M. Manning, L. C. Grothaus & W. F. Burns.

Tinus, R. W. 1974. Impact of the CO_2 requirement on plant water use. *Agricultural Meteorology* 14: 99–112.

Todd, G. W. 1958. Effect of ozone and ozonated 1-Hexene on respiration and photosynthesis of leaves. *Plant Physiology* 33: 416–20.

Tomassini, F. D., *et al.* 1977. The effect of time of exposure to sulfur dioxide on potassium loss from and photosynthesis in the lichen, *Caldina rangiferina* (L.) Harm. *New Phytologist* 79: 147–55. Coauthors: P. Lavoie, K. J. Packett, E. Nieboer & D. H. S. Richardson.

Treshow, M. 1975. Interactions of air pollutants and plant disease. *In* Mudd and Kozlowski (1975), pp. 307–34.

———. 1980a. Interactions of air pollutants and plant disease. *In* Miller (1980), pp. 103–9.

———. 1980b. Pollution effects on plant distribution. *Environmental Conservation* 7: 279–86.

Treybal, R. E. 1980. *Mass Transfer Operations*. 3d ed. New York: McGraw-Hill.

Tuazon, R. T., and S. L. Johnson. 1977. Free-radical and ionic reaction of bisulfite with reduced nicotinamide adenine dinucleotide and its analogues. *Biochemistry* 16: 1183–88.

Tukey, H. B., Jr. 1970. The leaching of substances from plants. *Annual Review of Plant Physiology* 21: 305–24.

Tyree, M. T., and P. Yianopolis. 1980. The site of water evaporation from substomatal cavities, liquid path resistances, and hydroactive stomatal closure. *Annals of Botany* 46: 175–93.

Tzschachsch, O. 1972. Die Häufigkeitsverteilung der individuellen SO_2-Resistenz in Populationen und ihre Bedeutung für die Forstpflanzenzüchtung. *Beiträge für die Forstwirtschaft* 6: 17–20.

Tzschachsch, O., M. Vogl & K. Thummler. 1969. Vorselektion geeigneter Provenienzen von *Pinus contorta* Douglas (*Pinus murrayana* Balf.) für den Anbau in den Rauchschadgebieten des oberen Erzgebirges. *Archiv für Forstwesen* 18: 979–82.

Uchijima, Z. 1976. Maize and rice. *In* Monteith (1976), I, 33–62.

Ulrich, B. 1980. Production and consumption of hydrogen ions in the ecosphere. *In* Hutchinson and Haves (1980), pp. 255–82.

U.N.E.C.E. *See* United Nations Economic Commission for Europe.

United Nations Economic Commission for Europe. 1980. *Symposium on the Effects of Airborne Pollution, Warsaw, Poland*, August 20–24, 1979.

United States Council on Environmental Quality. 1981. *Global Energy Futures and the Carbon Dioxide Problem*. Washington, D.C.: U.S. Government Printing Office.

United States Department of Agriculture Forest Service. 1976. *Proceedings of the First International Symposium on Acid Precipitation and the Forest Ecosystem*. Springfield, Va.: National Technical Information Service.

United States Department of Agriculture. 1981. *Agricultural Statistics, 1980.* Washington, D.C.: U.S. Government Printing Office.

United States Environmental Protection Agency. 1976a. *Progress in the Prevention and Control of Air Pollution in 1975: Annual Report of the Administrator of the E.P.A. to the Congress of the United States.* Washington, D.C.: U.S. Government Printing Office.

————. 1976b. *Sulfate Briefing for Regional Administrators.* New York: U.S.E.P.A.

————. 1977. *National Air Quality, Monitoring, and Emissions Trends Report.* Research Triangle Park, N.C.: U.S.E.P.A. Office of Air Quality Planning and Standards. (Report no. EPA-450/2-78-052.)

————. 1978a. *Environmental Effects of Increased Coal Utilization: Ecological Effects of Gaseous Emissions from Coal Combustion.* Washington, D.C.: U.S.E.P.A. (Report no. EPA-600/7-78-108.)

————. 1978b. *National Air Pollution Emissions Estimates, 1940–1976.* Research Triangle Park, N.C.: U.S.E.P.A. (Report no. EPA-450/1-78-003.)

————. 1978c. *Research Outlook, 1978.* Washington, D.C.: U.S.E.P.A. (Report no. EPA-600/9-78-001.)

————. 1980a. *National Air Pollution Emissions Estimates, 1970–1980.* Research Triangle Park, N.C.: U.S.E.P.A. (Report no. EPA-450/4-80-002.)

————. 1980b. *1977 National Emissions Report.* Research Triangle Park, N.C.: U.S.E.P.A. (Report no. EPA-450/40-80-005.)

————. 1980c. *Air Quality: Volume III of Air Quality Criteria for Particulate Matter and Sulfur Oxides.* Research Triangle Park, N.C.: U.S.E.P.A., Office of Environmental Criteria and Assessments (Draft Report).

————. Research Management Committee. 1981. *The National Crop Loss Assessment Network (NCLAN): 1980 Annual Report.* Corvallis, Ore.: Corvallis Environmental Research Laboratory, U.S.E.P.A.

————. 1982a. *Air Quality Criteria for Particulate Matter and Sulfur Oxides.* Research Triangle Park, N.C.: U.S.E.P.A., Office of Research and Development.

————. Research Management Committee. 1982b. *The National Crop Loss Assessment Network (NCLAN): 1981 Annual Report.* Corvallis, Ore.: Environmental Research Laboratory, U.S.E.P.A.

————. 1982c. *Review of the National Ambient Air Quality Standards for Sulfur Oxides: Assessment of Scientific and Technical Information.* Research Triangle Park, N.C.: U.S.E.P.A., Office of Air-Quality Planning and Standards. (Report no. EPA-450/5-82-007.)

————. 1983. *Volume II of Air Quality Criteria for Particulate Matter and Sulfur Oxides.* Research Triangle Park, N.C.: U.S.E.P.A, Office of Environmental Criteria and Assessments. (Report no. EPA-800/8-82-029B.)

Unsworth, M. H. 1977. Evaluation of atmospheric inputs, I: Dry deposition of gases and particulates onto vegetation: A review. *In* Nicholson, Patterson & Last (1977), pp. 9–15.

————. 1981. The exchange of air pollutants and carbon dioxide between vegetation and the atmosphere. *In* Grace, Ford & Jarvis (1981), pp. 111–38.

————. 1982. Exposure to gaseous pollutants and uptake by plants. *In* Unsworth and Ormrod (1982), pp. 42–63.

Unsworth, M. H., P. V. Biscoe & V. J. Black. 1976. Analysis of gas exchange between plants and polluted atmospheres. *In* Mansfield (1976), pp. 5–16.

Unsworth, M. H., P. V. Biscoe & H. R. Pinckney. 1972. Stomatal responses to sulfur dioxide. *Nature* 239: 458–59.

Unsworth, M. H., and V. J. Black. 1981. Stomatal responses to gaseous air pollutants. *In* Jarvis and Mansfield (1981), pp. 187–203.

Unsworth, M. H., and T. A. Mansfield. 1980. Critical aspects of chamber design for fumigation experiments on grasses. *Environmental Pollution* (series A) 23: 115–20.

Unsworth, M. H., and D. P. Ormrod. 1982. *Effects of Gaseous Air Pollution in Agriculture and Horticulture*. London: Butterworth Scientific.

Valonzona, F. F., F. A. Saladagu & N. L. Silva. 1978. Threshold limits and effects of sulfur dioxide on 22 rice varieties. *NSDB Technology Journal* 3: 35–42.

Van Amerongen, G. J. 1946. The permeability of different rubbers to gases and its relation to diffusivity and solubility. *Journal of Applied Physics* 17: 972–85.

Van Bavel, C. H. M. 1974. Antitranspirant action of carbon dioxide on intact sorghum plants. *Crop Science* 14: 208–12.

Van Breemen, N., *et al*. 1982. Soil acidification from atmospheric ammonium sulfate in forest canopy throughfall. *Nature* 299: 548–50. Coauthors: P. A. Burrough, E. J. Velthorst, H. F. van Dobben, T. de Wit, T. B. Ridder & H. F. R. Reijnders.

Van Haut, H., and H. Stratmann. 1970. *Farbtafelatlas über Schwefeldioxid-Wirkungen an Pflanzen*. Essen: Girardet.

Varshney, S. R. K., and C. K. Varshney. 1981. Effect of sulphur dioxide on pollen germination and pollen tube growth. *Environmental Pollution* (series A) 24: 87–92.

Vogl, M. 1970. Studies on differences in the relative SO_2 resistance in pine progeny. *Archiv für Forstwesen* 19: 3–12.

Von Caemmerer, S., and G. D. Farquhar. 1981. Some relationships between the biochemistry of photosynthesis and the gas exchange of leaves. *Planta* 153: 376–87.

Vonderschmitt, D. J., *et al*. 1967. Addition of bisulfite to folate and dehydrofolate. *Archives of Biochemistry Biophysics* 122: 488–93. Coauthors: K. S. Vitols, F. M. Huennekens & K. G. Scringeow.

Wardlaw, I. F., and J. B. Passioura, eds. 1976. *Transport and Transfer Processes in Plants*. New York: Academic Press.

Weast, R. C., ed. 1972. *Handbook of Chemistry and Physics*. Cleveland, Ohio: Chemical Rubber Co.

Webb, E. K. 1965. Aerial microclimate. *Meteorological Monographs* 6: 27–53.

Weigl, J., and H. Ziegler. 1962. Die räumliche Verteilung von ^{35}S und die Art der markieten verbindungen in Spinatblattern nach Begasung mit $^{35}SO_2$. *Planta* 58: 435–47.

Weinstein, D. A., H. H. Shugart & D. C. West. 1982. *The Long-term Nutrient Retention Properties of Forest Ecosystems: A Simulation Investigation*. Oak Ridge, Tenn.: Oak Ridge National Laboratory.

Weinstein, L. H., and R. Alscher-Herman. 1982. Physiological responses of plants to fluorine. *In* Unsworth and Ormrod (1982), pp. 139–67.

Wellburn, A. R. 1982. Effects of SO_2 and NO_2 on metabolic function. *In* Unsworth and Ormrod (1982), pp. 169–87.

———. 1983. The influence of atmospheric pollutants and their cellular products upon photophosphorylation and related events. *In* Koziol & Whatley (1983), pp. 203–21.

Wellburn, A. R., O. Majernik & F. Wellburn. 1972. Effects of SO_2 and NO_2 polluted air upon the ultrastructure of chloroplasts. *Environmental Pollution* 3: 37–49.

Wellburn, A. R., J. Wilson & P. H. Aldridge. 1980. Biochemical responses of plants to nitric oxide polluted atmospheres. *Environmental Pollution* (series A) 22: 219–28.

Wellburn, A. R., *et al*. 1976. Biochemical effects of atmospheric pollutants. *In* Mansfield (1976), pp. 105–14. Coauthors: T. M. Capron, H. S. Chan & D. C. Horsman.

Wellburn, A. R., *et al*. 1981. Biochemical explanations of more than additive inhibitory effects of low atmospheric levels of sulfur dioxide plus nitrogen dioxide upon plants. *New Phytologist* 88: 223–37. Coauthors: C. Higginson, D. Robinson & C. Walmsley.

Wells, A. E. 1917. Results of recent investigations of the smelter smoke problem. *Industrial and Engineering Chemistry* 9: 640–46.

Wentzel, K. F. 1982. Ursachen des Waldsterbens in Mitteleuropa. *Allgemeine Forstzeitschrift* (München) 37: 1365–68.

Wesley, M. L., *et al.* 1978. Daytime variations of ozone eddy fluxes to maize. *Boundary-Layer Meteorology* 15: 361–73. Coauthors: J. A. Eastman, D. R. Cook & B. B. Hicks.

West, D. C., S. B. McLaughlin & H. H. Shugart. 1980. Simulated forest response to chronic air pollution stress. *Journal of Environmental Quality* 9: 43–49.

West, D. C., H. H. Shugart & D. B. Botkin, eds. 1981. *Forest Succession: Concepts and Applications.* New York: Springer-Verlag.

Westman, W. E. 1979. Oxidant effects on Californian coastal sage scrub. *Science* 205: 1001–3.

———. 1981. Seasonal dimorphism of foliage in Californian coastal sage scrub. *Oecologia* (Berl.) 51: 385–88.

———. 1982. Coastal sage scrub succession. *In Dynamics and Management of Mediterranean-type Ecosystems,* pp. 91–99. U.S.D.A. Forest Service, Pacific Southwest Forest and Range Experiment Station, General Technical Report PSW-43, Berkeley, Calif.

White, K. L., A. C. Hill & J. H. Bennett. 1974. Synergistic inhibition of apparent photosynthesis rate of alfalfa by combinations of sulfur dioxide and nitrogen dioxide. *Environmental Science and Technology* 8: 574–76.

Whitmore, M. E. 1982. A study of the effects of SO_2 and NO_2 pollution on grasses, with special reference to *Poa pratensis* L. Ph.D. thesis, Univ. of Lancaster.

Whitmore, M. E., and P. H. Freer-Smith. 1982. Growth effects of SO_2 and/or NO_2 on woody plants and grasses observed during the spring and summer. *Nature* 300: 55–57.

Whitmore, M. E., and T. A. Mansfield. 1983. Effects of long-term exposure to SO_2 and NO_2 on *Poa pratensis* and other grasses. *Environmental Pollution* (series A) 31: 217–35.

Whittaker, R. H. 1965. Dominance and diversity in land plant communities. *Science* 147: 250–60.

———, ed. 1973. *Handbook of Vegetation Science, Part V: Ordination and Classification.* The Hague: Junk.

———. 1975. *Communities and Ecosystems.* New York: MacMillan.

Whittaker, R. H., *et al.* 1976. The Hubbard Brook ecosystem study: Forest biomass and production. *Ecological Monographs* 44: 233–52. Coauthors: F. H. Bormann, G. E. Likens & T. G. Siccama.

Whittaker, R. H., *et al.* 1979. The Hubbard Brook ecosystem study: Forest nutrient cycling and element behavior. *Ecology* 60: 203–20. Coauthors: G. E. Likens, F. H. Bormann, J. S. Eaton & T. G. Siccama.

Wilkerson, G. G., *et al.* 1983. Modeling soybean growth for crop management. *Transactions of the American Society of Agricultural Engineers* 26: 63–73. Coauthors: J. W. Jones, K. J. Boote, K. T. Ingram & J. W. Mishoe.

Will-Wolf, S. 1980. Effects of a "clean" coal-fired power generating station on four common Wisconsin lichen species. *Bryologist* 83: 296–300.

Wilson, L. G., R. A. Bressan & P. Filner. 1978. Light-dependent emission of hydrogen sulfide from plants. *Plant Physiology* 61: 184–89.

Winkler, R. L. 1972. *Introduction to Bayesian Inference and Decision.* New York: Holt, Rinehart & Winston.

Winner, W. E. 1981. The effect of SO_2 on photosynthesis and stomatal behavior of mediterranean-climate shrubs. *In* Margaris and Mooney (1981), pp. 91–103.

Winner, W. E., and J. D. Bewley. 1978. Contrasts between bryophyte and vascular plant synecological responses in a SO_2-stressed white spruce association in central Alberta. *Oecologia* (Berl.) 33: 311–25.

Winner, W. E., G. W. Koch & H. A. Mooney. 1981. Ecology of SO_2 resistance, IV: Predicting metabolic responses of fumigated shrubs and trees. *Oecologia* (Berl.) 52: 16–22.

Winner, W. E., and H. A. Mooney. 1980a. Ecology of SO_2 resistance, I: Effects of fumigations on gas exchange of deciduous and evergreen shrubs. *Oecologia* (Berl.) 44: 290–95.

———. 1980b. Ecology of SO_2 resistance, II: Photosynthetic changes of shrubs in relation to SO_2 absorption and stomatal behavior. *Oecologia* (Berl.) 44: 296–302.

———. 1980c. Ecology of SO_2 resistance, III: Metabolic changes of C_3 and C_4 *Atriplex* species due to SO_2 fumigations. *Oecologia* (Berl.) 46: 48–54.

———. 1980d. Responses of Hawaiian plants to volcanic sulfur dioxide: Stomatal behavior and foliar injury. *Science* 210: 789–91.

Winner, W. E. *et al.* 1981. Rates of emission of H_2S from plants and patterns of stable isotope fractionation. *Nature* 289: 672–73. Coauthors: C. L. Smith, G. W. Koch, H. A. Mooney, J. D. Bewley & H. R. Krouse.

Wislicenus, H. 1914. *Sammlung von Abhandlungen über Abgase und Rauschschaden.* Berlin: P. Parey.

Wit, C. T. de, *et al.* 1978. *Simulation of Assimilation, Respiration, and Transpiration of Crops.* Wageningen, the Netherlands: Center for Agricultural Publishing and Documentation. Coauthors: J. Goudriaan, H. H. van Laar, F. W. T. Penning de Vries, R. Rabbinge, H. van Keulen, W. Louwerse, L. Sibma & C. de Joge.

Witt, H. T., B. Rumberg & W. Junge. 1968. Electron transfer, field changes, proton translocation and phosphorylation in photosynthesis: Coupling in the thylakoid membrane. *In* Hess and Staudinger (1968), pp. 262–306.

Wittwer, S. H. 1980. Carbon dioxide and climatic change: An agricultural perspective. *Journal of Soil and Water Conservation* 35: 116.

Wong, C. H., H. Klein & H.-J. Jäger. 1977. The effect of SO_2 on the ultrastructure of *Pisum* and *Zea* chloroplasts. *Angewandte Botanik* 51: 311–19.

Wong, C. S. 1978. Atmospheric input of carbon dioxide from burning wood. *Science* 200: 197–99.

———. 1979. Carbon input to the atmosphere from forest fires. *Science* 204: 210.

Wong, S. C. 1979. Stomatal behavior in relation to photosynthesis. Ph.D. thesis, Australian National Univ.

Wong, S. C., I. R. Cowan & G. D. Farquhar. 1979. Stomatal conductance correlates with photosynthetic capacity. *Nature* 282: 424–26.

Wood, M. J., ed. 1979. *Ecological Effects of Acid Precipitation.* Palo Alto, Calif.: Electric Power Research Institute.

Woodwell, G. M. 1970. Effects of pollution on the structure and physiology of ecosystems. *Science* 168: 429–33.

———. 1978. The carbon-dioxide question. *Scientific American* 238: 34–43.

Woodwell, G. M., and R. A. Houghton. 1977. Biotic influences on the world carbon budget. *In* Stumm (1977), pp. 61–72.

Woodwell, G. M., and E. V. Pecan, eds. 1973. *Carbon and the Biosphere.* Springfield, Va.: National Technical Information Service.

Woodwell, G. M., *et al.* 1978. The biota and the world carbon budget. *Science* 199: 141–46. Coauthors: R. H. Whittaker, W. A. Reiners, G. E. Likens, C. C. Delwiche & D. B. Botkin.

Wu, H., and P. J. H. Sharpe. 1979. Stomatal mechanics II: Material properties of guard cell walls. *Plant, Cell and Environment* 2: 235–44.

Wu, L., A. D. Bradshaw & D. A. Thurman. 1975. The potential for evolution of heavy-metal tolerance in plants, III: Rapid evolution of copper tolerance in *Agrostis stolonifera. Heredity* 34: 165–87.

Yang, S. F. 1967. Biosynthesis of ethylene: Ethylene formation from methional by horseradish peroxidase. *Archives of Biochemistry and Biophysics* 122: 481–87.

———. 1970. Sulfoxide formation from methionine or its sulfide analogs during aerobic oxidation of sulfite. *Biochemistry* 9: 5008–14.

———. 1973. Destruction of tryptophan during the aerobic oxidation of sulfite ions. *Environmental Research* 6: 395–402.

Yang, S. F., and M. A. Saleh. 1973. Destruction of indole-3-acetic acid during the aerobic oxidation of sulfite. *Phytochemistry* 12: 1463–66.

Yokoyama, E., R. E. Yoder & N. R. Frank. 1971. Distribution of ^{35}S in the blood and its excretion in urine of dogs exposed to $^{35}SO_2$. *Archives of Environmental Health* 22: 389–95.

Yoneyama, T., and H. Sasakawa. 1979. Transformation of atmospheric NO_2 absorbed in spinach leaves. *Plant and Cell Physiology* 20: 263–66.

Zadoks, J. C. 1980. Yields, losses and costs of crop protection: Three views, with special reference to wheat growing in the Netherlands. *In* Teng and Krupa (1980), pp. 17–23.

Zahn, R. 1961. Effects of sulfur dioxide on vegetation. Results from gas exposure experiments. *Staub* 21: 56–60.

———. 1970. The effects on plants of a combination of subacute and toxic sulfur-dioxide doses. *Staub* 30: 20–23.

Zeevaart, A. J. 1974. Induction of nitrate reductase by NO_2. *Acta Botanica Neerlandica* 23: 345–46.

Zeigler, I. 1973. *In* Coulson and F. Korte (1973), pp. 182–208.

Zellner, A. 1962. An efficient method of estimating seemingly unrelated regressions and tests of aggregation bias. *Journal of the American Statistical Association* 57: 348–68.

———. 1971. *An Introduction to Bayesian Inference in Econometrics.* New York: Wiley.

Ziegler, I. 1972. The effect of SO_3^{2-} on the activity of ribulose-1,5-diphosphate carboxylase in isolated spinach chloroplasts. *Planta* 103: 155–63.

———. 1973. Effect of sulfite on phosphoenolpyruvate carboxylase and malate formation in extracts of *Zea mays*. *Phytochemistry* 12: 1026–30.

———. 1975. The effect of SO_2 pollution on plant metabolism. *Residue Review* 56: 79–104.

———. 1977. Subcellular distribution of ^{35}S-sulfur in spinach leaves after application of $^{35}SO_4^{-2}$, $^{35}SO_3^{-2}$, and $^{35}SO_2$. *Planta* 135: 25–32.

Ziegler, I., and R. Hampp. 1977. Control of $^{35}SO_4^{2-}$ and $^{35}SO_3^{2-}$ incorporation into spinach chloroplasts during photosynthetic CO_2 fixation. *Planta* 137: 303–7.

Zimmerman, P. W., and W. Crocker. 1934. Toxicity of air containing sulfur-dioxide gas. *Contributions of the Boyce Thompson Institute* 6: 455–70.

Index